Applied Building
Performance Simulation

Other related titles:

You may also like

- PBBE008 | Shakir | AI in the Built Environment | under development
- PBBE004 | Swan | Concepts, Methodologies and Technologies for Domestic Retrofit | under development
- PBBE003 | Al-Kodmany | Sustainable High-Rise Buildings: Design, technology, and innovation | 2022
- PBBE001 | Cao | Handbook of Ventilation Design for the Built Environment | 2021
- PBPO155 | Ting | Energy Generation and Efficiency Technologies for Green Residential Buildings | 2019

We also publish a wide range of books on the following topics:
Computing and Networks
Control, Robotics and Sensors
Electrical Regulations
Electromagnetics and Radar
Energy Engineering
Healthcare Technologies
History and Management of Technology
IET Codes and Guidance
Materials, Circuits and Devices
Model Forms
Nanomaterials and Nanotechnologies
Optics, Photonics and Lasers
Production, Design and Manufacturing
Security
Telecommunications
Transportation

All books are available in print via https://shop.theiet.org or as eBooks via our Digital Library https://digital-library.theiet.org.

IET BUILT ENVIRONMENT SERIES 02

Applied Building Performance Simulation

Joe Clarke, Jeremy Cockroft, Jon Hand and Raheal McGhee

The Institution of Engineering and Technology

About the IET

This book is published by the Institution of Engineering and Technology (The IET).

We inspire, inform and influence the global engineering community to engineer a better world. As a diverse home across engineering and technology, we share knowledge that helps make better sense of the world, to accelerate innovation and solve the global challenges that matter.

The IET is a not-for-profit organisation. The surplus we make from our books is used to support activities and products for the engineering community and promote the positive role of science, engineering and technology in the world. This includes education resources and outreach, scholarships and awards, events and courses, publications, professional development and mentoring, and advocacy to governments.

To discover more about the IET please visit https://www.theiet.org/

About IET Books

The IET publishes books across many engineering and technology disciplines. Our authors and editors offer fresh perspectives from universities and industry. Within our subject areas, we have several book series steered by editorial boards made up of leading subject experts.

We peer review each book at the proposal stage to ensure the quality and relevance of our publications.

Get involved

If you are interested in becoming an author, editor, series advisor, or peer reviewer please visit https://www.theiet.org/publishing/publishing-with-iet-books/ or contact author_support@theiet.org.

Discovering our electronic content

All of our books are available online via the IET's Digital Library. Our Digital Library is the home of technical documents, eBooks, conference publications, real-life case studies and journal articles. To find out more, please visit https://digital-library.theiet.org.

In collaboration with the United Nations and the International Publishers Association, the IET is a Signatory member of the SDG Publishers Compact. The Compact aims to accelerate progress to achieve the Sustainable Development Goals (SDGs) by 2030. Signatories aspire to develop sustainable practices and act as champions of the SDGs during the Decade of Action (2020–30), publishing books and journals that will help inform, develop, and inspire action in that direction.

In line with our sustainable goals, our UK printing partner has FSC accreditation, which is reducing our environmental impact on the planet. We use a print-on-demand model to further reduce our carbon footprint.

British Library Cataloguing in Publication Data

A catalogue record for this product is available from the British Library

ISBN 978-1-83953-165-1 (hardback)
ISBN 978-1-83953-166-8 (PDF)

Typeset in India by MPS Limited
Printed in the UK by CPI Group (UK) Ltd, Eastbourne

Cover image: Westend61/Westend61 via Getty Images

Contents

Foreword

Buildings are central to our lives, but they come with a hefty environmental price tag. From the resources used in construction to the emissions generated during operation and demolition, buildings have a significant impact on our planet. Ensuring their sustainability is crucial not only for the environment but also for our wellbeing and economic prosperity.

Unlike other designed objects, buildings present unique challenges. They vary widely in size, involve numerous stakeholders, and are usually custom-built. Each building is a balancing act between available resources and stakeholder desires, with construction methods ranging from DIY to industrial. Furthermore, buildings have long lifespans and must withstand changing usage and climate conditions. As they become more complex and interconnected, they are even transitioning from energy consumers to producers.

To address these challenges, innovative solutions are needed. From building-integrated electricity production and storage, adaptive building skins, and demand response to local energy communities, these innovative solutions must be optimised for existing or new buildings and thoroughly tested to ensure their effectiveness and resilience. This requires collaboration across technical and non-technical disciplines.

Building performance simulation (BPS) has proved to be a valuable tool in this endeavour. By using computer models, BPS indicates how buildings will perform in terms of energy usage, occupant comfort, and environmental impact. This predictive capability empowers architects, engineers, and policymakers to optimise buildings – and the built environment – for sustainability and resilience, paving the way for a more harmonious coexistence with our planet.

The focus of this book is on the effective application of BPS. The authors demonstrate its potential for studies at different spatial and temporal scales using a range of modelling complexities. By this, they show the intricacies of BPS and explore its myriad applications and potential.

From energy efficiency to lifecycle cost analysis, the book offers a glimpse into the transformative power of computational modelling in the built environment.

Through this combination of theoretical insights and real-world case studies, this book serves as a testament to the boundless potential of BPS to shape the future of our cities and communities. It is a call to action for researchers, practitioners, and policymakers alike to embrace this powerful tool and chart a course towards a more sustainable and resilient built environment.

As we embark on this journey together, let us remember that the future of our planet lies in our hands. By harnessing the power of innovation and collaboration, we can build a world where buildings not only stand as testaments to our achievements but also as beacons of sustainability and stewardship.

Jan Hensen
Professor Emeritus
Building Physics & Services Unit
Eindhoven University of Technology
The Netherlands
April 2024

Preface

Acceptable performance at least cost is the basic tenet of the construction industry and building design practitioners have traditionally employed a range of assessment tools to achieve this goal. Separate assessments are applied to minimise heat loss, utilise solar energy, size plant components, and avoid overheating. A meme slowly emerged that it was futile to attempt to optimise, in a piecemeal manner, a system that is dynamic (state variables change at different rates), non-linear (describing parameters depend on the system state), systemic (the different parts interact), and stochastic (events and influences occur randomly). In short, good overall performance is contingent on complex, interacting factors. To address this problem, and thereby provide better design support, computational tools for the simulation of building behaviour emerged in the 1970s and have been under development ever since. These tools[1] allow practitioners to adopt a virtual prototype and test approach, with information on likely system performance supporting the synthesis of appropriate design solutions from myriad competing options.

Whilst significant progress has been made in improving predictive accuracy vis-à-vis reality and aligning modelling and simulation functionality with user needs, there remain barriers to the routine application of performance simulation in practice. These barriers stem from weak agreement on what constitutes a performance assessment, a lack of tool application protocols and standards, semantically incompatible user interfaces, fragmented user training, and an industry reluctance to dictate required developments. This situation, whilst improving, is to be expected given the disruptive nature of a technology that portends a radical change to the traditional design process and demands that users acquire new knowledge and skills.

Over several decades, the simulation approach has been the object of sustained development by the worldwide academic community as evidenced by the growth and impact of the IBPSA[2] organisation since its inception in 1985. With 31 regional affiliates spanning five continents (as of 2024), a high-impact international journal[3], a biennial international conference, and regional conferences of high standing, the organisation has formed strategic partnerships with professional organisations and nurtured a growing student and early career base.

---

[1] Referred throughout as BPS+ signifying 'Building Performance Simulation plus additional capability'.
[2] International Building Performance Simulation Association: https://www.ibpsa.org.
[3] *Journal of Building Performance Simulation*: https://www.tandfonline.com/journals/tbps20.

The next evolutionary stage of BPS+ will benefit immensely from industry involvement in all facets of program development and application – requirements capture, algorithm validation, applicability testing, input model definition and quality assurance, application standardisation, user training, and tool/user certification. Only then will development be fully aligned with business needs. The prospect that a design proposal can be virtually prototyped and extensively tested in a cost-effective manner has the potential to bring together all stakeholders (including building users) in a way that radically improves the process and outcome.

The aim of this book is to illuminate the wide-ranging potential of a computational approach to design. It describes applications that address single- and multi-issue performance assessments as tackled within variously resourced projects undertaken by ESRU[4] staff over several decades. The selected projects exemplify the potential of building modelling and simulation to be applied at different scales to address the issues confronting the construction industry's aspiration to reduce energy use and environmental impact.

Of course, the focus on tool application comes at the expense of attention to the theories underpinning building performance simulation: enquiring readers are directed elsewhere to satisfy their mathematical curiosity (Clarke 2001, De Wilde 2018, Hensen and Lamberts 2019, Beausoleil-Morrison 2021, IBPSA 2023).

The topics covered offer examples of the translation of new design and refurbishment intentions to BPS+ input models that support conceptual or detailed assessments irrespective of the resources available at the time. The aim is to demonstrate how the technology can be used at any stage in the design process and post-occupancy to narrow the gap between design intent and the multivariate performance observed in reality. The targeted readers are those members of the student body wishing to apply or evolve BPS+ functionality, and those practitioners wishing to apply or manage the approach in practice. Whilst the application software used relates mostly to open-source programs developed by ESRU academic and research staff over several decades, and the ESP-r[5] system in particular, there is no substantial reason why the assessment approaches outlined, where valued, cannot be re-enacted within other simulation tools.

Chapters 1–4 summarise the contextual challenges posed by the clean energy systems transition, the issues confronting building performance simulation, the required functionality, and aspects relating to use in practice. Chapters 5–13 address specific topics: high-resolution modelling and simulation, application examples, urban energy schemes, energy action at the regional/national scale, smart grids, urban land use, eliminating the performance gap, linking the virtual and real, and complementing strategic renewable energy schemes. Finally, Chapter 14 indicates possible future developments and changes to the software engineering process intended to overcome barriers caused by some unhelpful aspects of competition.

---

[4]Energy Systems Research Unit at the University of Strathclyde: https://www.esru.strath.ac.uk.
[5]https://www.esru.strath.ac.uk/applications

Whilst we have strived to keep errors and omissions to a minimum, especially when reporting the work of others, these will undoubtedly be spotted by astute readers. We apologise for these errors in advance and will endeavour to make corrections in a future edition of this book where issues are brought to our attention[6].

It is our hope that the reader will not regard this book as a Panglossian view of the role of computers in design. We offer it as a modest contribution to tackling the issues underpinning a problem domain that is complex because it comprises a unique blend of soft and hard science underpinned by pervasive uncertainty. Where this offering falls short, perhaps the reader will be motivated to act to improve the situation. We will be rooting for you in our retirement years.

Joe Clarke, Jeremy Cockroft, Jon Hand, and Raheal McGhee
Glasgow
February 2024

References and further reading

Beausoleil-Morrison I (2021) *Fundamentals of Building Performance Simulation*, Routledge, Milton Park, ISBN 9 780367518059.

Clarke J A (2001) *Energy Simulation in Building Design* (2nd ed.), Butterworth-Heinemann, Oxford, ISBN 0 7506 5082 6.

de Wilde P (2018) *Buildings Performance Analysis*, Wiley, New York, ISBN 9781119341925.

Hensen J L M and Lamberts R (2019) *Building Performance Simulation for Design and Operation* (2nd ed.), Routledge, Milton Park, ISBN 9781138392199.

IBPSA (2023) Proceedings of IBPSA's Building Simulation biennial conference series, *publications.ibpsa.org*.

---

[6] via joe@esru.strath.ac.uk

Acknowledgements

The authors are indebted to many organisations and individuals, who, over five decades, saw merit in building performance simulation and supported our research, consultancy, and training activities; and to our consultancy clients, who asked the unexpected of BPS+ and thereby challenged us to evolve our thinking. We especially thank our research colleagues, past and present[7], who contributed much to these activities, and the ongoing development of the ESP-r system in particular. Our thanks also to the many students and researchers worldwide, who have applied our programs in innovative ways and exposed more than a few deficiencies along the way: you were no less a part of the development effort. Also to the many businesses, small and large, who were early adopters of our software products and provided platforms to test applicability in practice. We are also pleased to register our gratitude to the funding agencies, who resourced our research and in many cases sustained this support over extended periods – the essential prerequisite of attaining useful, practical outcomes from the research process.

Finally, and at the risk of unintended omissions, we acknowledge and thank the following individuals for their distinctive, essential, and unwaivering support.

To Emeritus Professor Tom Maver, who was instrumental in ESRU's inception and focused our efforts on participatory computer-aided design, the raison d'etre of performance simulation in the long term.

To our ESRU colleagues, for their contributions to the development and application of the software tools featured throughout the book and their inputs to research, consultancy, and teaching from which the book draws.

To our families for their dogged support and cheerful tolerance of our spells of absence during prolonged writing sessions and coffee-fuelled debate.

We dedicate this book to our co-author, Jeremy Cockroft, who sadly passed away on Sunday 12 November 2023 after a prolonged illness. He will be remembered by those who knew him as a perfect gentleman and notable scholar.

---

[7]See https://www.esru.strath.ac.uk/Downloads/ESP-r/Contributors/list.html for contributors over a 45-year period.

Acronyms and abbreviations

AC	Alternating Current
ACH	Air Changes per Hour
AI	Artificial Intelligence
AIC	Acceptable Indoor Concentration
ANM	Active Network Manager
API	Application Programming Interface
ASHP	Air Source Heat Pump
ASHRAE	American Society of Heating, Refrigerating, and Air-Conditioning Engineers
BEMS	Building and Energy Management System
BESTEST	Building Energy Simulation Test
BIM	Building Information Model
BIPV	Building Integrated Photovoltaic
Blender	An open source 3D creation program
BPS	Building Performance Simulation
BPS+	BPS plus additional capability
BRE	Building Research Establishment
BREEAM	BRE Environmental Assessment Methodology
CAD	Computer-Aided Design
CAV	Constant Air Volume
CEA	Carbon Emissions Abatement
CEDA	Centre of Environmental Data Analysis
CFC	Chlorofluorocarbons
CFD	Computational Fluid Dynamics
CHP	Combined Heat and Power
CIBSE	Chartered Institution of Building Services Engineers
CO, CO$_2$	Carbon Monoxide, Carbon Dioxide
COP	Coefficient of Performance
CoRMat	Contra-Rotating Marine Turbine
CPD	Continuing Professional Development
CPU	Central Processing Unit

CRC	Carbon Reduction Commitment
DC	Direct Current
DGR	Discomfort Glare Ratio
DH	District Heating
DHW	Domestic Hot Water
DLL	Dynamic Link Library
DSA	Differential Sensitivity Analysis
DST	Direct Solar Transmittance
DWT	Ducted Wind Turbine
EIA	Environmental Impact Assessment
EPBD	Energy Performance of Buildings Directive
EPC	Energy Performance Certificate
EPSRC	Engineering and Physical Sciences Research Council
ERIC	Emissions Reduction Investment Curve
ESP-r	Environmental Systems Performance - Research
ESRU	Energy Systems Research Unit
EU	European Union
EV	Electric Vehicle
EXPRESS	A data modelling language
FC	Fuel Cell
FMU	Functional Mock-up Unit
FORTRAN	FORmula TRANslation
gbXML	Open schema supporting transfer of BIM data to engineering analysis tools
GIS	Geographic Information System
GOMap	Geospatial Opportunity Mapping
GSHP	Ground-source Heat Pump
GT	Gas Turbine
HDM	High Definition Model
HOE	Holographic Optical Elements
HP	Heat Pump
HTC	Heat Transfer Coefficient
HVAC	Heating, Ventilation, and Air Conditioning
IBPSA	International Building Performance Simulation Association
IECC	International Energy Conservation Code
IFC	Industry Foundation Classes
IIBDS	Intelligent, Integrated Building Design System
IDF	Input Data File

IPV	Integrated Performance View
IRR	Internal Rate of Return
IT	Information Technology
JPPD	J index of Previsible Percentage Dissatisfied
LCA	Life Cycle Analysis
LCD	Liquid Crystal Display
LEED	Leadership in Energy and Environmental Design
LED	Light Emitting Diode
LV	Low Voltage
MAC	Marginal Abatement Cost; Maximum Allowable Concentration
MACC	Marginal Abatement Cost Curve
MCA	Monte Carlo Analysis
MVHR	Mechanical Ventilation with Heat Recovery
NCM	National Calculation Methodology
NHER	National Home Energy Rating
NIAM	Nijssen's Information Analysis Method
NMVOC	Non-methane Volatile Organic Compound
NO$_x$	Nitric Oxide (NO) and nitrogen dioxide (NO$_2$)
NPI	Normalised Performance Indicator
NPV	Net Present Value
NRE	New and Renewable Energy
obFMU	Occupant Behavior Functional Mockup Unit
PAM	Performance Assessment Method
PHPP	Passivhaus Planning Package
P, PI, PID	Proportional, Proportional-Integral, Proportional-Integral-Differential control
PIR	Passive Infra-Red
PM	Project Manager, Percentage Match, Particulate Matter
PMV	Predicted Mean Vote
POE	Post Occupancy Evaluation
PP	Primitive Part
PPD	Predicted Percentage of Dissatisfied
PV	Photovoltaic
PVPS	Photovoltaic Power Station
RH	Relative Humidity
RTE	Resilience Testing Environment
SAP	Standard Assessment Procedure (for UK domestic buildings)
SESG	Scottish Energy Systems Group

SET	Standard Effective Temperature
SHOCC	Sub-Hourly Occupancy Control
SME	Small and Medium-sized Enterprises
SMIP	Smart Meter Implementation Programme
SO$_2$	Sulphur Dioxide
TES	Thermal Energy Storage
TIM	Transparent Insulation Material
TIMES	The Integrated MARKAL-EFOM System
TMY	Typical Meteorological Year
TPV	Thermal Preference Vote
TPM	Time Proportional, Modulating
TRV	Thermostatic Radiator Valve
TVT	Total Visible Transmittance
U	Thermal Transmittance
UCTPE	Union for the Coordination of the Transmission of Electricity
UGR	Unified Glare Ratio
VAV	Variable Air Volume
VCP	Visual Comfort Probability
VDL	Vacant and Derelict Land
XML	Extensible Markup Language

About the authors

Joe Clarke's career has focused on the role that computational methods can play in reducing energy demand, accelerating the take-up of sustainable energy technologies, mitigating environmental impacts, and improving human wellbeing. A major aspect of his work has involved the development and dissemination of software tools for energy systems simulation and support for the application of these tools in design, research, teaching, and policy-making contexts. He is the progenitor and code controller of the ESP-r building simulation program, which has been used internationally in research, teaching, and design contexts, and a founder member and past President of the International Building Performance Simulation Association. At the University of Strathclyde, he established and directed the Energy Systems Research Unit (1987–2006 and 2010–2019), and the Energy Systems and the Environment MSc programme (1990–2005). He also directed the Environmental Engineering (EE) and Building Design Engineering (EE component) undergraduate programmes. He is a Fellow of the UK Energy Institute, the UK Society for the Environment, and the International Building Performance Simulation Association. He has been a member of several editorial boards, including most recently the Honorary Editorial Board of the Journal of Building Performance Simulation. He graduated with a BSc degree (1st Class Hons) in Environmental Engineering in 1974, obtained a PhD degree in Energy Systems Simulation in 1977, and was awarded a DSc in 2005, all from the University of Strathclyde, and an Honoris Causa Doctorate from the University of Bratislava in 2008.

Jeremy Cockroft had specialist knowledge in all aspects of building physics and, in particular, in energy systems control. He joined Honeywell in 1980 and held senior positions concerned with the development of smart controls for the domestic market. He joined ESRU in 2010 as Knowledge Exchange Director and helped to place the group's portfolio of CPD courses and industrial consultancy offerings on a fully commercial footing. He served as the Technical Meetings Convener for the Chartered Institution of Building Services Engineers (Scotland) from 2012 to 2023. He held a BSc degree (Hons) in Mechanical Engineering and a PhD degree in Heat Transfer and Air Flow in Buildings, both from the University of Glasgow. Jeremy sadly passed away on 12 November 2023.

Jon Hand was employed as a senior research fellow in the Energy Systems Research Unit at the University of Strathclyde from 1989 to 2023, where he served as an engineering consultant, software developer, and PassivHaus trainer. His research focused on tools and methods to assess the performance of the built environment. He is a major contributor to the ESP-r building simulation suite of programs. His publications have more than 3200 citations (Research Gate), and he is a prolific reviewer. He has been on the scientific committees of scores of conferences and a project committee member of the International Building Performance Simulation Association (IBPSA). He is a Fellow of IBPSA. He holds BSc and MArch degrees in Architecture and a PhD degree in Building Energy Simulation Support.

Raheal McGhee has been a research fellow in the Energy Systems Research Unit since 2013, where his research involves collaborations with Local Authorities and Utilities to assess policy and technical barriers in the deployment of renewable energy schemes within cities. He is the progenitor of the GOMap feasibility assessment and urban opportunity-mapping tool. His areas of expertise relate to modelling and simulation support for refurbishment projects addressing net-zero goals, and the application of open-source programs such as the Quantum Geographic Information System for geospatial analysis and Blender for 3D modelling, animation, and simulation. He holds a BSc (Hons) degree in Physics, an MSc degree in Renewable Energy Systems and the Environment, and a PhD degree in Spatial Planning for Urban Renewable Energy Systems Deployment, all from the University of Strathclyde.

Chapter 1

Sustainable energy systems challenge

This chapter outlines the contextual challenges that give rise to the need for high-resolution performance modelling and simulation. These challenges correspond to the various aspects underpinning government policy and regulations, the competing technical solutions that might be brought to bear, and the often conflicting requirements of stakeholders.

Energy and its myriad related issues have risen to the top of the political agenda, with a clean energy transition underway worldwide that is likely to continue apace in the coming decades. Whilst target setting is a relatively easy task, realising such targets is a seemingly intractable problem. This is due, in part, to the conceptual chasm between non-technical policymakers and technology specialists. The consultation questions that normally accompany a proposed energy strategy (e.g., SG 2023) illustrate the complex nature of the domain, whilst the responses to such consultations (e.g., SG 2017) serve to exemplify the underlying uncertainties and conflicting viewpoints.

From a technical standpoint, a building is a microcosm of complexity due to the many physical processes that interact and evolve under stochastic influences. Determining the ideal combination of materials, mechanical equipment, and control system strategies to maintain satisfactory performance across multiple variables over time is a complex undertaking. This task demands that professionals concerned with the built environment utilise performance assessment tools during the design phase. These tools have traditionally been based on simplified calculation methods intended to 'size' construction and plant components under extreme conditions. This approach, simplification for restricted application, is arguably the principal reason that actual building performance is often far removed from the design intent – the so-called performance gap.

Building performance simulation (BPS), a technological response to this issue, has been under development since the 1973 OPEC oil embargo, which gave impetus to a heightened focus on energy efficiency (Spitler 2006). BPS provides practitioners with a cost-effective means to examine a building's dynamic performance at the design stage or pre-upgrade. It does this by emulating operation under realistic weather and usage conditions, not by predicting any future state. By considering all relevant design and operational parameters and their evolution over time, the approach is generally applicable, allowing practitioners to appraise design options in terms of a range of cost and performance metrics covering occupant wellbeing, energy use, and environmental impact (inside and out). The utility of the approach is expected to improve markedly when and if performance assessment methods are harmonised and barriers to tool interoperability are removed. The explicit design-stage simulation will then become ubiquitous and routine, with results for different design configurations (for the same or different buildings) being directly intercomparable because they arise from similar analyses and generate the same range of output metrics.

The development of the technology will not stop there because the sustainability challenge is an expanding vista that is not confined to a single building with its bespoke environmental control systems. Instead, the target of BPS is increasing in scope and scale, encompassing whole communities, hybrid and renewable energy supplies, and smarter control of energy use. The final destination may well be the routine subjugation of proposals (at whatever scale) to realistic, lifetime simulations to ensure operational resilience under uncertain influences.

It is clear from the current level of research activity that the development of BPS remains a work in progress in relation to matching evolving user requirements. The expectation of Hong *et al.* (2000) remains as prescient today as it was when first stated:

> *"... with the growing trend towards environmental protection and achieving sustainable development, the design of 'green' buildings will surely gain attention. Building simulation serves not only to reveal the interactions between the building and its occupants, HVAC systems, and the outdoor climate, but also to make possible environmentally-friendly design options".*

A significant challenge going forward is to add supply-side considerations to BPS, hereinafter referred to as BPS+ to mark this extension. Fossil-based energy systems will continue to underpin the global economy over the coming decades and will need to be nurtured to buy the time needed to develop, deploy, and optimise replacement energy sources, most notably derived from renewables, nuclear, and synthetic fuels. This presents a challenge due to the magnitude of the task, the incompatible infrastructure required, and the complex interactions between technical, policy, and economic aspects. Blending supply- and demand-side options requires holistic appraisal approaches to identify project-specific solutions. Attaining aggressive reductions in carbon emissions in the short to medium term (e.g., by 2050 or sooner) can only be realised from renewable sources alone at phenomenal cost, monetary and otherwise.

Whatever the approach to energy supply, aggressive energy efficiency measures will be necessary to offset the year-on-year increases in demand associated with population growth and power-hungry communication technology. Employing 'carbon' as the lone driver of change is inappropriate because it detracts from significant issues such as fuel poverty, local environmental impact, indoor environmental quality, and operational robustness. There is a need for option appraisals that support rational planning by matching supply and demand at all scales. To emphasise this point, consider the following deployments that are each equal in capacity terms[1]: Thermal

---

[1]The capacity of the renewable systems would need to be 3–5 times greater if the requirement was to match energy production.

Power Station (1 × 2000 MW); Wind Farm (100 turbines × 20 MW); Marine Power (4000 devices × 0.5 MW); Urban CHP (40,000 units × 0.05 MW); Building-integrated PV (200,000 units × 0.01 MW). How can an appropriate mix be selected to ensure that supply meets demand after the introduction of load reduction and profile reshaping measures – all whilst meeting expectations for human wellbeing, stimulating economic growth, avoiding grid instabilities, and satisfying development guidelines?

Whilst the power of modelling and simulation is widely recognised, it is not generally appreciated that the approach does not generate design solutions per se, optimum or otherwise. Instead, its practicality is twofold. First, it supports user understanding of complex systems by providing (relatively) rapid feedback on the performance implications of proffered designs. This essential attribute of simulation, learning support, is well summarised by Bellinger (2019):

> *After having been involved in numerous modelling and simulation efforts, which produced far less than the desired results, the nagging question becomes; Why? The answer lies in two areas. First, we must admit that we simply don't understand. And, second, we must pursue understanding. Not answers but understanding.*

Second, it supports the resilience testing of proposals through the tracking of performance under realistic operating conditions. This must not be confused with an attempt to predict future performance, which is an entirely futile endeavour. Instead, it is here asserted that the ultimate aim of BPS+ is to support design innovation by providing a high-integrity representation of the dynamic, connected, and non-linear physical processes governing the disparate performance aspects that dictate the overall acceptability of buildings and their related energy supply systems. To support high-integrity modelling, BPS+ programs need to operate with holistic representations that encapsulate the problem parts that impact all performance aspects of importance.

The purpose of this book is to illuminate the BPS+ approach when applied to issues of contemporary relevance. After a summary of the state-of-the-art and the necessary refinements that may be expected in the near term, the book presents typical applications of high-resolution modelling and simulation that might underpin energy-related design and policy action. Most examples are based on the application of the ESP-r system (Clarke 2001) within research and consultancy projects undertaken by the Energy Systems Research Unit (ESRU) at the University of Strathclyde in Glasgow over several decades. They are intended to be generic and therefore reproducible using other tools, of which there are many (BEST 2023). Specific project results, where included, are intended to indicate helpful output types. The intention is to impart insight into what and how to model, not to deliver design paradigms or promote any particular design solution. The power of BPS+ is its ability to deliver bespoke solutions to unique problems.

Chapters 1 through 4 cover the aspects underlying BPS+ when applied to support the clean energy transition: the sustainable energy systems challenge, the required modelling functionality, performance assessment metrics, and best practice tool application.

Chapters 5 through 12 present example applications of BPS+ tools, including high-level descriptions of models and the performance assessments delivered. These examples derive from the application of BPS+ tools within research and consulting projects. They are intended to highlight the different possible approaches to performance assessment depending on the design stage reached, the resources available, and the issue(s) being addressed.

1.1   Energy policies, directives, and action plans

As highlighted by many authors (e.g., Wilkinson *et al.* 2007, Glicksman 2008, Clarke *et al.* 2009), energy efficiency within the built environment makes a significant contribution to the design and operation of sustainable energy systems. In the short term, realising energy efficiency will require greater public awareness of cost-effective options, improvements to building regulations, and the development of low-energy appliances and systems that offer value propositions to consumers. In the longer term, other options can be pursued, such as embedded and distributed renewables, retrofitted heat pumps in the domestic market, and internet-based energy services that bring energy management and control to the mass market. Whilst there are many technology options for energy performance improvement, it is not yet clear which combinations will prove to be cost-effective. Whatever approach is taken, it is important to keep human health and wellbeing at the centre. What is the point of a widely touted low-energy/carbon building that does not satisfy the basic requirements of its occupants?

The challenge is significant, and as of yet only partially met. Taking the UK as an example, the built environment accounts for around 40% of total energy consumption (RAE 2021) and around 40% of CO_2 emissions, including residential and business emissions and power station emissions attributable to the built environment's electricity consumption (SG 2022). The potential contribution of the built environment to a sustainable energy economy has been recognised in an Energy White Paper (BEIS 2020) and the UK National Energy Efficiency Action Plan (ESNZ 2023). The former outlines measures designed to reduce emissions from electricity generation, industry, and buildings by up to 230 MtC/year by 2032. The measures include radical year-on-year improvements to building standards, particularly in the domestic sector, where emissions have reduced by 18 $MtCO_2e$ (17%) over the last 30 years due to higher standards for energy-consuming products, resulting in a 13% decrease in overall energy consumption (BEIS 2022). This improving situation has been helped by the number of homes with cavity wall insulation increasing by 11% between December 2015 and December 2021, such that 14.8 million of the 21.1 million homes with cavities are now insulated (although sometimes at the cost of introducing moisture-related problems).

On the other hand, increasing energy consumption has been driven by a multitude of factors, including:

- a 6.1% increase in the number of households since 2012, now estimated to be 28.2 million households (ONS 2023);

- the deployment of central heating in around 95% of UK households (ONS 2019); and
- a 63% rise in the power consumed by appliances between 1970 and 2005 before gradually falling to 43% in 2021 (BEIS 2022a).

At the same time, the UK has a legacy of poorly performing buildings, with 83% of the housing stock being more than 30 years old (Piddington *et al.* 2017). A consequence of this is that, whilst central heating has increased average indoor air temperatures by 6 °C since 1970 (Shorrock and Utley 2003), it has been at the expense of energy efficiency because of inadequate thermal insulation levels. The poor quality of the building stock has also contributed to a situation where an estimated 4 million households are classified as 'fuel poor' (EAS 2023).

In the non-domestic sector, a similar picture emerges, with energy consumption increasing by 23% between 1990 and 2005 before falling below 1990s values by 2% in 2021 (BEIS 2022). This initial increase was driven by trends in information technology, poorly designed and regulated speculative developments, and the often unnecessary installation of air conditioning, with energy consumption projected to rise by 25% by the early 2020s (AEA Technology 2007). In addition, it has been estimated that 40% of the energy used in non-domestic buildings over 10 years old is wasted due to poorly maintained mechanical plants and building fabrics (Johnson 1993).

The continuing low level of energy efficiency in the built environment offers vast scope for improvement, which could be achieved through the deployment of measures ranging from simple plant and insulation upgrades to advanced energy monitoring and control. Such measures would benefit from the ubiquitous application of BPS+ in support of apt selections.

There are many options on the supply side, and some of these, especially renewable technologies, only become viable after the application of energy efficiency measures. Fossil fuels are presently abundant and, excessive taxation aside, relatively inexpensive. The principal objection to their continued use is the impact on climate change through the related emissions of greenhouse gases. Irrespective of the eventual outcome of the climate change debate (whether the cause is mostly anthropogenic or mostly natural, the efficacy of proposed actions, etc.), one thing is clear: the transition to a non-fossil fuel economy is underway and will accelerate over the coming decades. The pertinent question is how to manage this transition, mitigate impacts, and deploy blends of the various technology options over time: fossil fuel de-carbonisation and CO_2 sequestration, energy efficiency and load management to reduce/reshape demand, new and renewable sources of energy to replace fossil fuels, and the removal of barriers confronting new nuclear plants. The second and third of these options are in the domain of BPS+.

First, a greater level of energy efficiency is required to extend the life of fossil fuels and reshape demand profiles to accommodate the outputs available from renewable energy technologies. Many sector-specific steps may be taken to attain the various targets. Savings of 25%–60% could be realised in the transport sector through journey curbing, congestion alleviation, higher efficiency engines, alternative fuels, electric vehicles, hybrid engines, and fuel cell (hydrogen) technology. Savings of 30%–85% could be realised in the building sector through smart materials, smart control, heat recovery, passive solar devices, and embedded renewable energy components. Whilst savings of 15%–75% could be realised in the industrial sector through heat recovery, more efficient plants, load scheduling, waste reduction, and materials recycling. Such technologies give rise to further complexity in relation to their competitive selection, installation, and operation. For example, a cost-benefit comparison between passive solar and embedded renewable energy options for a building is entirely non-trivial. Where the former option is preferred, how can punitive indoor conditions be avoided in the event of a solar shading device failure? Where the latter is favoured, there exists a dichotomy between power exporting and local utilisation, in addition to the problems related to non-standard installation and operation.

Second, the widespread adoption of new and renewable energy systems is required at both the distributed (network-connected) and embedded (load-connected) scales. Most renewables derive from solar energy: the total energy content of the annual solar radiation incident on the earth has been estimated at 2,895,000 EJ, or approximately 7200 times the present annual global energy consumption. In contrast to its vastness, the resource is difficult to harness because of its intermittent nature and low power density. It is for these reasons that the indirect, higher power density sources of energy dominate. The principal UK energy sources include gas (37.4%), wind (30.4%), nuclear (15%), biomass (4.9%), solar (4.6%), hydro (1.2%), coal (1.1%), and tidal stream (0.4%); here the parenthesised data indicate the generation capacity relative to the UK total electricity demand in 2022 (National Grid Live 2023[2]). Although the least developed, tidal stream offers the major advantage that, like fossil-derived power, it has the potential to meet a portion of a base load because of its predictable nature. Such renewable technologies, although at present more expensive overall than fossil-derived energy, are potentially cheaper, easier to process, and less polluting, with low or no carbon emissions. The present level of renewable-derived energy in many countries is expected to rise rapidly (from a present low of less than 0.1% of consumed energy in some cases) as the technologies become more competitive. Even countries with a high number of deployed wind turbines will face future challenges. For example, the UK electricity demand averages around 50 GW against a peak demand of around 60 GW. In 2023, the UK had 28 GW of installed wind capacity, split evenly between on- and off-shore wind farms. Any further expansion will need to identify locations that are both technically feasible and policy-unconstrained whilst resourcing the substantial capital expenditure associated with grid extension and the replacement of first-generation turbines.

Renewable energy systems have typically been pursued at the strategic level, with distributed hydroelectric stations, biogas plants, and wind farms being connected to the electricity networks at the transmission or distribution levels. To avoid problems with fault clearance, network balancing, and power quality, it has been estimated that the connection of systems with limited control possibilities should be restricted to around 25% of the total installed capacity (EA Workshop 1999). This limitation is due to the intermittent nature of these sources, which require controllable, fast-responding reserve capacity to compensate for

[2]https://grid.iamkate.com/

fluctuations in output, and energy storage to compensate for non-availability. To achieve a greater penetration level, new and renewable energy systems can be embedded within the built environment, where they serve as a demand reduction device (as seen from the supply network).

Embedded generation requires the matching of local supply potentials to optimised demands arrived at by the application of demand reduction and control measures. For example, passive solar components, smart materials, heat recovery, and/or smart control may be used to reduce energy requirements, and combined heat and power, heat pumps, and/or renewable energy components may be used to meet a significant portion of the residual demand. Any energy deficit is met by the public electricity supply operating in cooperative mode. The important point is that, for the embedded approach to be successful, demand reduction and reshaping measures must be deployed alongside the new and renewable energy systems. This requires procedures whereby appropriate technology matches can be selected and assessed in terms of a range of relevant criteria, such as indoor/outdoor air quality, occupant comfort, energy use, environmental impact, and capital/running cost.

There is also a need to involve society in the decision-making process in order to ensure that adopted schemes represent a value proposition for citizens. How acceptable will local combined heat and power plants be if local air quality worsens and maintenance costs rise? How acceptable will heat pumps be if indoor temperature control deteriorates and capital/maintenance costs rise? Productive involvement can best be achieved by consulting with stakeholders at key stages in a modelling study – a form of participatory democracy in the design process.

Whilst governments regularly issue aspirational targets for energy efficiency and clean supply technologies in pursuit of sustainable development, they do little to resolve the formidable challenges that will confront those who must bring forward cost-effective schemes. How can the technical barriers associated with stochastic, distributed renewable energy resources be overcome, especially where these barriers are cross-discipline and highly non-trivial? How can society be helped to appreciate the major infrastructure changes that lie ahead? Should the introduction of new technology be market driven or derived from a step-change investment in R&D? Further, how can the wider environmental costs be incorporated into market prices without retarding economic progress? These are among the pertinent questions of our time.

The energy and environment domains are inherently complex, and consequently, conflicting viewpoints abound, proffered solutions are polarised and consensus is difficult to attain. Indeed, vested interests can render the relationship between sustainability and energy action void. This unacceptable situation gives rise to three engineering challenges:

• how to consider energy systems in a holistic manner in order to address the inherent interactions;
• how to include environmental and social considerations in the assessment of cost-performance in order to satisfy all stakeholders; and
• how to embrace interdisciplinary working in order to derive benefit from the innovative approaches to be found at the interface between the disciplines.

It would be helpful if the efficacy of different approaches to target attainment could be estimated in advance or if the means to do this on a case-by-case basis were made available. BPS+ can contribute in both regards by representing explicitly the underlying engineering uncertainties. Support for policy formulation can be realised by 'soft linking' BPS+ tools with general equilibrium theory modelling tools such as TIMES (The Integrated MARKAL-EFOM System; Loulou *et al.* 2004, 2005) as developed within an International Energy Agency collaboration programme (IEA 2023). The former would be applied to reduce the uncertainties associated with inputs to the latter, which, in turn, would facilitate the exploration of future energy scenarios at the regional or national scale. In this way, robust action plans can be formulated, and the underlying virtual models can evolve in response to observed deficiencies. The process of target setting would thereby improve naturally over time, whilst the options for attaining the targets would be clearer. The conjoining of bottom-up and top-down approaches is discussed elsewhere (Crawley 2019).

As an example, BPS+ can be used to assess the energy reduction potential of applying efficiency measures to national housing stocks (see Sections 7.7 and 8.1). This can be done in a manner that prioritises necessity (poorly performing dwellings first, greatest replication potential, tackling fuel poverty, etc.) and economic constraints (identifying combinations that can be deployed over time as budgets permit). On the supply side, BPS+ can be used to assess the energy supply potential of building-integrated, urban-scale, and national renewable energy schemes (see Chapters 7 and 13), whilst taking into account the barriers relating to land availability and local electricity grid resilience (see Chapter 9). These outputs, in turn, would inform the TIMES input model. The opportunities for utilising BPS+ to support policy development are vast, as discussed by Crawley (2008).

The use of BPS+ in this way would embed international best practices within the policy-making process. A recent global collaboration, for example, involved researchers contributing occupant behaviour data for their building types and climates to ASHRAE's Global Occupant Behaviour Database (Liu *et al.* 2023). This open-source database allows users to query and analyse datasets for use in BPS+ applications. Some of these datasets are applicable to temperate climates where the requirement is mostly for heating; others relate to hot climates with the focus mostly on cooling, or to warm climates that require both heating and cooling. One study focused on Kuwait, concluded that increasing the cooling set point temperature by 1 °C would result in a 10% reduction in energy use and an insignificant impact on heating (Azar *et al.* 2021). The question then is whether occupants will accept this relaxation in the regulation of indoor temperatures.

1.2   Technical solutions and concomitant issues

Before the early 1980s, the energy performance of buildings was essentially unregulated in many countries. Since then, building regulations have steadily evolved and their scope expanded so that the CO_2 emissions associated with heating, hot water, ventilation, and lighting energy use are now included in design compliance regulations (Acts of the Northern Ireland Assembly 2009, Wales Statutory Instruments 2022, Scottish Statutory Instruments 2023, UK Statutory Instruments 2023).

Dwellings contribute about 40% of the UK's carbon emissions, equivalent to around 3.3 tonnes CO_2/year per average dwelling in 2020 (Energy Guide 2022). If such a property is upgraded to the 2022 fabric standards and a modern gas heating system is installed, the emissions would be reduced by around 30%. Previous legislation proposed a target of an 80% reduction from 1990 levels, which has since been updated with new legislation to bring all greenhouse gas emissions to net zero by 2050 (UK Statutory Instruments 2019).

The majority of the existing UK housing stock was constructed prior to the development of building energy standards and, to date, the mechanisms for improvement have been government-sponsored, voluntary initiatives promoting efficiency upgrades or low-carbon supply technologies. The result has been the sporadic application of upgrades. For example, 28% of cavity-wall homes in Scotland have no cavity insulation (SG 2023). A detailed study of a local authority housing stock of around 8,000 dwellings concluded that a 50% reduction in carbon emissions could be achieved by conventional fabric and system upgrades, whilst a further 50% CO_2 reduction was achievable through the deployment of renewable energy systems, heat pumps, and combined heat and power plants (Tuohy *et al.* 2006). This same study illustrated the potential to attain net carbon neutrality through local, commercial-scale wind farms. Another study considered the space heating energy demand of the entire Scottish housing stock and concluded that a 50% CO_2 reduction could be achieved through conventional fabric and system upgrades alone (Clarke *et al.* 2007).

In the non-domestic sector, energy savings are highly dependent on the building type. Consider a large, naturally ventilated office building. Typically, this would have a carbon emission of around 704 tonnes CO_2/year (SG 2023). If this were refurbished to 2020 standards, emissions would fall to 514 tonnes CO_2/year. Further significant improvements in emissions and energy consumption are possible through the implementation of non-standard technologies. For example, controllers linked to motion and daylight sensors to maximise natural lighting, increased passive measures for natural ventilation, and zone-based control, could together reduce carbon emissions by 30% (Electricity North West 2020).

There is significant scope for improvement in the energy used by appliances. For example, high potential reductions have been associated with data centre energy management (40%), server operation (25%), and storage performance (53%) (Zhu *et al.* 2023). Such savings can be realised through new ventilation approaches (natural or mechanical) and by replacing older equipment with energy-efficient models. The positive impact of reduced energy for IT will provide secondary benefits in terms of reduced space heat gain, more comfortable indoor environments, and lower cooling energy use.

The improvements identified above are potentials, with their translation into real energy savings and emissions reductions dependent on uptake and performance in use. Two research projects, the 40% House (Environmental Change Institute 2005) and Building Market Transformation (Boardman 2007), identified the actions required to meet the (then) Government aspiration for a 60% reduction in carbon emissions by 2050, as outlined in the 2003 Energy White Paper (DTI 2003). These projects used a housing stock model that incorporated future population and building occupancy trends to analyse scenarios covering different rates of uptake of low-carbon technologies. The conclusion was that in order to meet the 60% target, there needed to be a profound change within the built environment. Moreover, realising the carbon potential of improvements is also dependent on the quality of the implementation of energy-saving measures. There are many examples where low-energy buildings have been monitored and their energy consumption has been shown to be substantially greater than trumpeted at the design stage (Bordass *et al.* 2001). This discrepancy is often found to be due to poor quality construction (e.g., missing insulation and significant thermal bridging) and ineffective control of plant and systems. Such problems highlight the need for better design stage scrutiny of options, more attention to commissioning, skills improvement through training, and the post-construction monitoring of energy performance to support the enforcement of standards. The incorporation of airtightness sampling in the 2006 and 2007 building regulations served as one means of verifying the implementation of building energy efficiency measures, whilst the introduction of thermographic imaging and the adoption of operational energy certificates in England and Wales served as additional methods.

The energy savings potentials discussed above can only be realised where mechanisms are in place to ensure the upward trend in building energy performance and, where necessary, enforce minimum standards. In new buildings, the main driver towards increased energy efficiency is legislation. The EU Energy Performance of Buildings Directive (EPBD; EC 2002)[3] led to the development of the UK National Calculation Methodology (NCM) for non-domestic buildings, which is embedded in a Simplified Building Energy Model (SBEM) or accredited building energy simulation tools. The NCM is used to provide energy certificates and prove building compliance for non-domestic buildings, whilst the UK Government's Standard Assessment Procedure (SAP[4]; Building Research Establishment 2017) is used for domestic buildings. Kokogiannakis (2008) examined the use of simulation in a regulatory context and concluded that the benefits are significant.

Both the Scottish (Scottish Statutory Instruments 2023) and UK Governments (UK Statutory Instruments 2023) have issued strategy documents outlining proposed net zero targets by 2050. Targets are supported by tax policies and subsidies, the latter aiming to help low-income and vulnerable households by providing insulation grants (Institute for Fiscal Studies 2021). Improvements in regulations were expected to bring the UK into alignment with European benchmarks (Passive House Plus 2017).

---

[3]The revised EPBD (2022), issued as part of the European Green Deal (EGD 2023), requires that all new buildings have zero emissions by 2030 and that all existing buildings be transformed into zero-emission buildings by 2050.

[4]As of 2023, SAP 11 is under development. The new version aims to better reflect the current political and technological situation in the UK by incorporating new technologies such as heat pumps, solar panels, and energy storage.

There are a number of initiatives to deliver energy and CO_2 savings in existing buildings, each reflecting the UK Government's historical preference for employing fiscal incentives rather than legislation for retrospective energy performance improvements. In 2003/2004, the total budget for these initiatives was nearly £270 million (Kelly 2006), with delivery generally devolved to bodies such as the Carbon Trust[5] and the Energy Saving Trust[6]. The former assists large industry consumers in adopting energy-efficiency practices and technologies. The latter offers Energy Efficiency Advice Centres that advise domestic and small commercial consumers. Additionally, power utilities have been obliged to provide help and assistance (usually in the form of grants) to consumers to assist them in reducing their energy bills.

There are increasing indications that the future driver of improvements in the existing building stock will be legislation and enforcement. For example, Energy Performance Certificates are a mechanism for setting mandatory minimum standards for existing dwellings. Other components of the EPBD are the requirements for regular inspection of boilers and air-conditioning equipment and checks on building airtightness. By focusing on actual systems and their operational energy ratings, rather than expected energy use, such measures are intended to ensure that theoretical energy savings are achieved in practice.

Whilst institutions, legislation, and improved practice in the construction sector can go some way towards reducing energy consumption and emissions, a step-change in performance will require the integration of new technologies addressing energy supply and demand. Because of a sustained research effort over several decades, many technology options are available that can help bring about this step-change. Whilst some of these options are yet unable to compete economically, this situation may be expected to change as carbon-related policies develop and energy costs rise. Some examples of 'new' technologies for buildings are mentioned below.

Smart facades

Advanced glazing provides high insulation and/or solar/daylight capture/exclusion (Field and Ghosh 2023). Transparent insulation material reduces construction heat loss and captures solar energy (Peuportier and Michel 1995). Breathable walls passively pre-heat ventilation air (Kalantar and Borhani 2017). Novel shading and light-redirecting devices can alleviate overheating and/or enhance solar energy and daylight capture. Finally, daylight utilisation systems linked to the control of artificial lighting are attractive because they directly displace electricity use whilst improving the indoor visual environment (Mohammad and Ghosh 2023).

Passive solar architecture

Solar energy may be captured passively to contribute to heating, lighting, and ventilation (Pugsley *et al.* 2022, Girard *et al.* 2023). Examples include Trombe-Michel walls (Wang *et al.* 2012), attached sunspaces, mass walls with transparent

[5]https://carbontrust.com
[6]https://energysavingtrust.org.uk

insulation, glazed atria, switchable glazing, light shelves, solar ventilation pre-heat, and thermo-siphon air panels.

Building-integrated renewables

Solar thermal collectors mounted on rooftops are utilised to contribute to hot water, whilst photovoltaic elements are employed to convert solar energy into both electricity and heat, with the former output used locally and/or exported to the electric grid. Such systems minimise the demand placed on the public supply infrastructure. For example, photovoltaic components and encased (ducted) wind turbines, when deployed in cooperative mode in niche cases, have been shown to reduce the consumption of conventionally generated electricity by up to 90% (Born *et al.* 2001).

Heat and power systems

Air- and ground-source heat pumps are seen as an appropriate replacement for gas-fired boilers at both the individual building and district heating scales (Berntsson 2002, Harvey 2006). Ground-source heat pumps, for example, can operate with a coefficient of performance of around 4 (i.e., one unit of electricity is consumed to supply four units of heat). Further, as the primary fuel is electricity, heat pumps are well aligned with a future where natural gas supplies are replaced by electricity derived from renewable energy sources and nuclear power. In the near term, Combined Heat and Power (CHP) systems utilising oil, gas, or biofuels could underpin decentralised energy solutions: in 2022, CHP capacity in the UK reached 27.4 MW (DUKES 2023). Fuel cells for application within buildings, whilst potentially carbon-neutral, are still at the research and development stage, with high capital costs and uncertain performance in the field.

Appliance efficiency

Despite the existence of energy labelling schemes, financial incentives, and increased public awareness of energy, the number and variety of electrical appliances continue to rise. To compound the problem, the power consumption of new devices is often greater than that of the obsolete device they seek to replace (e.g., previous plasma televisions). The challenge is to reverse this upward trend through radical improvements in appliance energy efficiency and the elimination of standby losses.

Internet-enabled energy services

The internet is effectively juxtaposed with the energy network, and this offers the prospect of establishing a range of energy services based on energy use monitoring, cost-effective control, and enhanced user participation in energy-efficiency initia-tives (Clarke *et al.* 2002). By embedding low-cost sensors within buildings, it is possible to transmit high-frequency data on environmental conditions and appli-ance power consumption to service providers, who then deliver appropriate ser-vices: energy use statistics to regulatory bodies, the remote control of appliances to households, the management of micro-grids to utilities, and so on.

Smart metering

There is likely to be a rapid increase in smart metering of energy usage in the future, giving consumers up-to-date information on their consumption. There is some evidence that providing such information can lead to significant reductions in consumption (Darby 2000). That said, it is also possible that the technology will enable services that attract consumer resistance, such as the nudging of energy use through dynamic tariffs. The UK Smart Meter Implementation Programme was introduced in 2013 with the goal of installing 53 million smart electricity and gas meters in UK households by 2020 (Smart Energy Code 2023). However, this target was missed, with only 17.3 million meters (33%) operating by the end of the first quarter of 2020, with the shortfall attributed to consumer reticence (Gosnell and McCoy 2023).

Off-grid supply

In the UK, approximately 15% of households are disconnected from the electricity network, with consumers encouraged to adopt various renewable energy technologies (FREE 2018) in place of oil – such as PV, battery storage, and/or heating via wood pellet boilers. Combining such technologies with an energy-efficient building can allow occupants to maintain a comfortable lifestyle.

Electric vehicles

There were over 26 million electric cars worldwide on the road in 2022, up 60% relative to the previous year and some five times the stock in 2018 (IEA 2023a). It may not be long before a majority of buildings have connected EVs, increasing electricity demand and altering the demand profile. It has also been mooted that the distributed storage potential of EV batteries could be utilised (through favourable tariffs) for grid balancing purposes, given the stochastic nature of renewable sources.

The ability to model a building's energy performance has been a reality since the early 1980s, and the benefits have been extensively documented (McElroy *et al.* 2001). However, the use of modelling has not been widespread in the UK. The adoption of the EPBD in the UK through building regulations was a first step to ensuring that energy modelling and simulation become an integral part of the design process. Explicit modelling allows practitioners to match technologies to building types and contexts. In this way, it is possible to ensure supply approximately matches demand over time, a particular challenge with designs incorporating contingent renewable systems. Further, strategic energy modelling can provide the data needed to develop robust energy policies at the regional and national levels (Clarke 2003, Clarke *et al.* 2004).

1.3   Stakeholder participation

The computational approach to design equates to a value proposition to businesses by enabling operational emulation for resilience at the design stage and better integration of design teams. It provides a means to communicate design intent to clients through experiential outputs that can be more readily understood by all stakeholders. It helps to close the gap between design intent and performance in use. It supports design and policy considerations from the same input model. Moreover, it introduces the concept of a 'digital twin' as a means to connect the design and post-occupancy stages.

In addition, explicit performance simulation foretells an exciting prospect: the participation of building occupants in design decision-making. This is achieved by arranging for building user participation in input model creation and the evaluation of performance outcomes when expressed in a relevant, non-jargon manner. Maver (1985) conducted research in this area against the premise that value judgements ought to be made by those who will be affected by outcomes. One project involved experiments to assess the feasibility of head teacher use of multi-variate performance appraisal tools (addressing daylight level, energy use, circulation efficiency, and capital/running cost) to design 80-place nursery schools.

The principal conclusions from the project were as follows:

- Headteachers are capable of formulating design objectives and producing designs that incorporate these objectives.
- Such designs were considered by the participants themselves to be more acceptable than comparable architect-produced designs.
- Whilst participants evaluated their own designs more highly than other participants' individual designs, they were capable of cooperating to produce a collective design that delivered an improvement in building performance.
- In blind tests, design professionals evaluated the participants' designs as highly as those that were architect-generated and considered the group solutions to be an improvement over the individual solutions upon which they were based.

The overall conclusion from the project was that there were no barriers to effective computer-aided user participation in design. Computer-based models promote effective user participation in the same way as they promote effective participation within professional design teams. Participatory democracy in design, which leads to shared insight and results in convergence and consensus, is viewed by many as the raison d'etre of computer-aided design.

1.4   Chapter summary

This chapter has considered the energy-related impact of the built environment, highlighting the poor performance of buildings and the upward trend in consumption, which has been exacerbated by historically weak performance standards, demographic trends, and the increase in electricity use. Despite the worrying trend, the potential for building energy efficiency is substantial, with even relatively basic measures delivering substantial improvements. The deployment of existing and emerging technologies can bring about a radical step-change in the energy performance of the built environment.

Selecting appropriate technologies will require access to performance assessment tools that respect the dynamic, connected nature of the problem domain. At the same time, the target for such tools is expanding to include, in addition to building design, community energy grids and the energy systems that service them. This situation is driving an increase in application scope and fidelity vis-à-vis the wider built environment. The formulation of energy policy and resilient action plans can be effectively supported by the explicit performance appraisal of options when conducted in a manner that includes the uncertainty associated with principal determining factors. Thereafter, practitioners can employ the same appraisal tools to ensure cost-effective and compliant developments that attract consumer support.

References and further reading

Acts of the Northern Ireland Assembly (2009) *Building Regulations (Amendment) Act (Northern Ireland)* 2009, https://legislation.gov.uk/nia/2009/4/contents/.

AEA Technology (2007) *BNAC18: modelling the energy consumption of air conditioning*, Atomic Energy Agency, Market Transformation Programme Policy Briefing.

Azar E, Alaifan B, Lin M, Trepci E and Asmar M E (2021) 'Drivers of energy consumption in Kuwaiti buildings: Insights from a hybrid statistical and building performance simulation approach', *Energy Policy*, 150, 112154.

BEIS (2020) *Energy White Paper: Powering our Net Zero Future*, CP337, Department for Business, Energy & Industrial Strategy, London, ISBN 978-1-5286-2219-6.

BEIS (2022) *Energy consumption in the UK 2022*, Department for Business, Energy & Industrial Strategy, Stationary Office, London.

BEIS (2022a) *UK Energy in Brief*, Department for Business, Energy and Industrial Strategy, Stationary Office, London.

Bellinger G (2019) 'Simulation is not the answer', *https://systems-thinking.org/simulation/simnotta.htm*.

Berntsson T (2002) 'Heat sources – technology economy and environment', *Refrigeration*, 25, pp. 428–438.

BEST (2023) Building Energy Software Tools directory, *https://buildingenergy-softwaretools.com/*.

Boardman B (2007) *Reducing the environmental impact of housing*, Environmental Change Institute, *https://eci.ox.ac.uk/research/energy/*.

Bordass W, Leaman A and Ruyssevelt P (2001) 'Assessing building performance in use', *Building Research and Information*, 29(2), pp. 144–157.

Born F J, Clarke J A, Johnstone C M, *et al.* (2001) 'On the integration of renewable energy systems within the built environment', *Building Services Engineering Research and Technology*, 22(1), pp. 3–13.

Building Research Establishment (2017) *Changes to the UK Government's Standard Assessment Procedure (SAP) Government Response*, assets.publishing. https://service.gov.uk/media/5a81be69ed915d74e33ffd27/Government_Response_-_Changes_to_SAP_FINAL-v2.pdf.

Clarke J A (2001) *Energy Simulation in Building Design* (2nd Ed.), Butterworth-Heinemann, Oxford, ISBN 0 7506 5082 6.

Clarke J A (2003) 'IT systems for energy and environment monitoring, planning, design and control', *Solar Thermal Technologies for Buildings*, James & James, London.

Clarke J A, Johnstone C M, Kelly N J, Strachan P A and Tuohy P G (2009) 'The role of built environment energy efficiency in a sustainable UK energy economy', *Energy Policy*, 36(12), pp. 4005–4609.

Clarke J A, Johnstone C M, Kim J and Strachan P A (2002) 'On-line energy services for smart homes', *Proc. EPIC'02*, Lyon.

Clarke J A, Johnstone C M, Kondratenko I, *et al.* (2004) 'Using simulation to formulate domestic sector upgrading strategies for Scotland', *Energy and Buildings*, 36, pp. 759–770.

Clarke J A, Johnstone C, Kim J and Tuohy P (2007) 'Energy and carbon performance of housing: upgrade analysis. Energy Labelling and National Policy Development', *Proc. 28th AIVC Conf.*, Crete.

Crawley D B (2008) 'Building performance simulation: A tool for policymaking', *PhD Thesis*, ESRU, University of Strathclyde.

Crawley D B (2019) 'Building simulation for policy support', In Hensen J L M and Lamberts R (eds.), *Building Performance Simulation for Design and Operation* (2nd Edn.), Chapter 12, pp. 384–395, Routledge, ISBN 9781138392199.

Darby S (2000) 'Making it obvious: designing feedback into energy consumption', *Proc. 2nd Int. Conf. on Energy Efficiency in Household Appliances and Lighting*, Italian Association of Energy Economists, EC-SAVE Programme.

DTI (2003) *Energy White Paper: Our Energy Future – Creating a Low-Carbon Economy*, Department of Trade and Industry, Report URN 06/660, Stationery Office, London.

DUKES (2023) *DUKES chapter 7: statistics on the contribution made by Combined Heat and Power (CHP) to the UK's energy requirement,* https://gov.uk/government/statistics/combined-heat-and-power-chapter-7-digest-of-united-kingdom-energy-statistics-dukes/.

EA Workshop (1999) 'Network Connection of Photovoltaic Systems', EA Technology Limited, UK.

EAS (2023) *Fuel Poverty across the UK, Energy Action Scotland, https://nea.org.uk/wp-content/uploads/2024/01/NEA-Fuel-Poverty-Monitor-FULL-REPORT-FINAL.pdf/.*

EC (2002) *On the Energy Performance of Buildings*, European Commission, Directive 2002/91/EC of the European Parliament.

Electricity North West (2020) 'Measures to reduce CO_2 emissions for offices', *https://enwl.co.uk/globalassets/future-energy/net-zero/reducing-carbon-for-businesses/office/measures-to-reduce-emissions-offices.pdf/.*

Energy Guide (2022) *Average Carbon Footprint per House in the UK*, https://energyguide.org.uk/average-carbon-footprint-uk/.

Environmental Change Institute (2005) 40% House, Environmental Change Institute, Oxford.

ESNZ (2023) *Energy Trends: UK renewables*, Department for Energy Security and Net Zero, Stationary Office, London.

Field E and Ghosh A (2023) 'Energy assessment of advanced and switchable windows for less energy-hungry buildings in the UK', *Energy*, 283.

FREE (2018) *Future of Rural Energy in Europe*, https://rural-energy.eu/country-data/united-kingdom/.

Girard F, Toublanc C, Andres Y, Dechandol E and Pruvost J (2023) 'System modeling of the thermal behavior of a building equipped with facade-integrated photobioreactors: Validation and comparative analysis', *Energy and Buildings*, 292.

Glicksman L R (2008) 'Energy efficiency in the built environment', *Physics Today*, 61(7), pp. 35–40.

Gosnell G and McCoy D (2023) 'Market failures and willingness to accept smart meters: Experimental evidence from the UK', *Journal of Environmental Economics and Management*, 118.

Harvey L D D (2006) *A Handbook on Low-Energy Buildings and District-Energy Systems*, Earthscan, London.

Hong T, Chou S K and Bong T Y (2000) 'Building simulation: an overview of developments and information sources', *Building and Environment*, 35(1), pp. 347–361.

IEA (2023) *Report of the IEA's Energy Technology Systems Analysis Program (ETSAP)*, https://iea-etsap.org/index.php.

IEA (2023a) *Global EV Outlook 2023*, International Energy Agency, www.iea.org

Institute for Fiscal Studies (2021) 'Tax policies to help achieve net zero carbon emissions', https://ifs.org.uk/sites/default/files/output_url_files/8-Tax-policies-to-help-achieve-net-zero-carbon-emissions-.pdf/.

Johnson S (1993) *Greener Buildings: The Environmental Impact of Property*, MacMillan Press, Hong Kong.

Kalantar N and Borhani A (2017) 'Breathable Walls – Computational Thinking in Early Design Education', *Proc. 22nd CAADRIA Conference*, Xi'an Jiaotong-Liverpool University, Suzhou, China, 5–8 April, pp. 377–386.

Kelly N J (2006) 'The vital role of demand reduction in reducing Scotland's CO_2 emissions', *Energy Policy*, 34(18), pp. 3505–3515.

Kokogiannakis G (2008) 'Support for the integration of simulation in the European Energy Performance of Buildings Directive', *PhD Thesis*, ESRU, University of Strathclyde.

Liu Y, Dong B, Hong T, Olesen B, Lawrence T and O'Neill Z (2023) 'ASHRAE URP-1883: Development and analysis of the ASHRAE global occupant behavior database', *Science and Technology for the Built Environment*, 29(8), pp. 749–781.

Loulou R, Goldstein G and Noble K (2004) *Documentation for the MARKAL Family of Models*, IEA-ETSAP.

Loulou R, Remne U, Kanudia A, Lehtila A and Goldstein G (2005) *Documentation for the TIMES Model - PART I* 1–78.

Maver T W (1985) 'CAAD: A mechanism for participation', *Proc. Design Coalition Team Conf.* (Ed: Beheshti M), vol. 2, pp. 89–105, Eindhoven, https://papers.cumincad.org/cgi-bin/works/Show?0c3c/.

McElroy L B, Clarke J A, Hand J W and Macdonald I A (2001) 'Delivering energy simulation to the profession: the next stage', *Proc. Building Simulation'01*, Rio de Janeiro.

Mohammad A K and Ghosh A (2023) 'Exploring energy consumption for less energy-hungry building in UK using advanced aerogel window', *Solar Energy*, 253, pp. 389–400.

ONS (2019) Percentage of households with central heating systems in the United Kingdom from 1970 to 2018, Office for National Statistics, https://statista.com/statistics/289137/central-heating-in-households-in-the-uk/.

ONS (2023) *Families and households in the UK: 2022*, Office for National Statistics, https://ons.gov.uk/peoplepopulationandcommunity/birthsdeathsandmarriages/families/bulletins/familiesandhouseholds/2022/.

Passive House Plus (2017) 'UK may deliver EU sustainable building targets in spite of Brexit – while Scotland & Wales commit', *https://passivehouseplus.ie/news/government/exclusive-uk-may-deliver-eu-sustainable-building-targets-in-spite-of-brexit-while-scotland-wales-commit/*.

Peuportier B and Michel J (1995) 'Comparative analysis of active and passive solar heating systems with transparent insulation', *Solar Energy*, 54(1), pp. 13–1 8.

Piddington J, Nicol S, Garret H and Custard M (2017) *The Housing Stock of the United Kingdom*, Building Research Establishment, Garston.

Pugsley A, Zacharopoulos A and Chemisana D (2022) 'Chapter 10 - Polygeneration systems in buildings', *Polygeneration Systems*, Academic Press, pp. 351–410.

RAE (2021) 'Construction sector must move further and faster to curb carbon emissions, say engineers', Royal Academy of Engineering, *https://raeng.org.uk/news/construction-sector-must-move-further-and-faster-to-curb-carbon-emissions-say-engineers*.

Scottish Statutory Instruments (2023) *The Building (Scotland) Amendment Regulations 2023*, https://legislation.gov.uk/ssi/2023/177/contents/made/.

SG (2017) *Draft Energy Strategy*, Scottish Government, https://consult.gov.scot/energy-and-climate-change-directorate/draft-energy-strategy/.

SG (2022) *Promoting Net Zero Carbon and Sustainability in Construction*, Scottish Government, https://assets.publishing.service.gov.uk/media/631222898fa8f54234c6a508/20220901-Carbon-Net-Zero-Guidance-Note.pdf/.

SG (2023) *Draft Energy Strategy and Just Transition Plan*, Scottish Government, https://gov.scot/publications/draft-energy-strategy-transition-plan/pages/11/.

Shorrock L D and Utley J I (2003) *Domestic Energy Fact File* 2003, Building Research Establishment, Watford.

Smart Energy Code (2023) *Smart Metering Implementation Programme*, https://smartenergycodecompany.co.uk/smip/.

Spitler J D (2006) 'Building performance simulation: the now and the not yet', *HVAC&R Research*, 12(3a), pp. 549–551.

Tuohy P G, Strachan P A and Marnie A (2006) 'Carbon and energy performance of housing: a model and toolset for policy development applied to a local authority housing stock', *Proc. Eurosun*, Glasgow.

UK Statutory Instruments (2019), *The Climate Change Act 2008 (2050 Target Amendment) Order 2019*, https://legislation.gov.uk/uksi/2019/1056/contents/made/.

UK Statutory Instruments (2023) *The Building Regulations etc. (Amendment) (England) Regulations 2023*, https://legislation.gov.uk/uksi/2023/911/contents/made/.

Wales Statutory Instruments (2022) *The Building (Amendment) (Wales) (No. 2) Regulations 2022*, https://legislation.gov.uk/wsi/2022/993/contents/made/.

Wang X, Xi Q and Ma Q (2012) 'A review of current work in research of Trombe walls', *Proc. E3S Web Conference*, 248, 03025.

Wilkinson P, Smith K R, Beevers S, Tonne C and Oreszczyn T (2007) 'Energy, energy efficiency, and the built environment', *The Lancet*, 370(9593), pp. 1175–1187.

Zhu H, Zhang D, Goh H H, *et al.* (2023) 'Future data center energy-conservation and emission-reduction technologies in the context of smart and low-carbon city construction', *Sustainable Cities and Society*, 89, 104322.

Chapter 2

Building performance simulation

This chapter summarises the BPS+ tool functionality required to address the issues introduced in Chapter 1 and the actions required to apply the approach effectively in practice.

Designers have traditionally relied on a variety of simplified calculation methods to size building and plant components at the design stage. Such methods were mostly confined to the steady-state calculation domain and were focused on independent issues such as construction heat loss, warm weather overheating, boiler/radiator sizing, indoor daylight level estimation, and the like. Given the relative ease of use of such simplified tools, is it any wonder that practitioners have been inured to their low integrity shortcomings (Section 4.9 elaborates on an approach to simplified tool development that avoids over-simplification). In real buildings, complexities exist that expose the deficiencies of these methods, including, inter alia, time-varying boundary conditions, complex heat and mass transfers, and stochastic occupant behaviour. BPS+ programs employ simulation methods that are able to address such issues whilst, as summarised in Figure 2.1, coupling the different domains that interact to influence the overall solution.

Here, the three principal physical domains are evolved under control system action, with the underlying processes solved in an integrated manner. Also represented are the points of interaction between these domains and the extraction of quantities to enable the determination of embodied energy.

Applying BPS+ in practice requires new activities, not least input model planning (focused on the issues at hand and not overly complicated), simulation planning (to ensure that the issues likely to be encountered in use are covered), and a systematic approach to the interrogation of large results datasets.

The following description relating to the ESP-r system is intended to give an insight into the modus operandi of a numerical simulation program after receiving an input model.

The defined physical system is made discrete by the placement of 'nodes' at preselected points of interest. These nodes represent small homogeneous or non-homogeneous volumes corresponding to building parts (internal air, opaque and transparent boundary surfaces, constructional elements, contents, etc.), plant parts (boiler, chiller, radiator, heat exchanger, pipes, ducts, etc.), renewable energy system parts (solar thermal/PV panels, ASHP components), and control system parts (room temperature sensors, radiator TRVs, photocells, PIR sensors, motorised

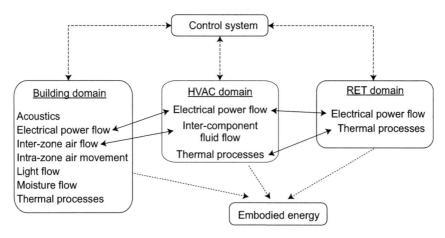

Figure 2.1   Coupled domains within BPS+

valves, dampers, etc.). As an example, Figure 2.2 shows the discretisation of a multi-layered construction and water/air solar collector systems. A simulation would proceed as follows.

- For each node in turn, and in terms of all surrounding nodes representing regions in thermodynamic contact, conservation equations are established to represent the nodal condition (temperature, pressure, voltage, species concentration, etc.) and the inter-nodal transfers of energy, mass, and momentum.
- The entire equation set is solved simultaneously at successive time steps to obtain the time series of nodal state variables as a function of the changing influences from weather, occupant behaviour, and control action.
- The time series of nodal state variables and internode flux transfers is made available for interrogation by the user to obtain performance metrics in support of design decision-making relating to occupant comfort, indoor air quality, energy consumption, plant efficiencies, delivered renewable energy, and so on.

The approach as implemented in ESP-r is summarised in Appendix A and explained in detail elsewhere (Clarke 2001).

Table 2.1 summarises the evolution of BPS+ tools from the early manual methods that were ubiquitous in the 1970s, to the present-day integrated performance simulators and beyond. As indicated, the tendency over time is for tools to become more refined in how they represent the underlying physics, with improved graphics and CAD interoperability to enhance the human-computer interface.

The adoption of simulation procedures in the 3rd generation represented a profound shift in assessment rigour. To avoid confusion when applying BPS+, a clear distinction must be drawn between the terms *modelling* and *simulation*. Becker and Parker (2009) have stated that

"... it is common to see the words simulation and modelling used as synonyms, but they are not really the same thing; at least, not to those in

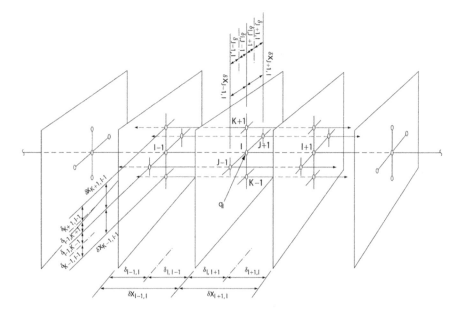

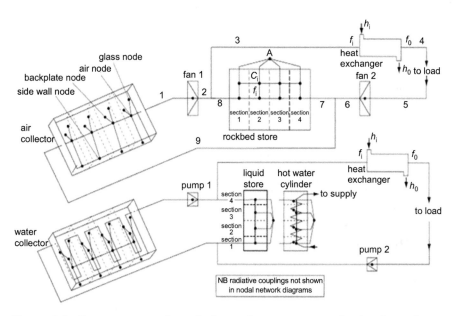

Figure 2.2   Discretisation of a multi-layered construction and solar thermal systems

Table 2.1   The evolution of BPS+ tools

1st Generation	Codification of simplified, manual methods Limited application range and depth Weak correspondence with reality Piecemeal treatment of design issues	Increasing integrity vis-à-vis the real world,
2nd Generation	Fabric dynamics emphasised Some simplifications removed Move to personal computers Remains piecemeal in application	↓ ↓ ↓ ↓
3rd Generation (current)	Simulation methods to handle dynamics Progress toward holistic evaluations Improved but still bespoke user interfaces Partial CAD integration	leading gradually to design tools that are ↓
Future Generations	Extension of application scale and scope Full CAD interoperability Unified user interfaces based on BIM Fully holistic simulation capability Move to operational resilience testing	↓ ↓ holistic, generalised, and easy to apply.

the field bearing those words in its name. To be precise in terminology, a simulation enacts, or implements, or instantiates, a model. A model is a description of some system that is to be simulated, and that model is often a mathematical one. A system contains objects of some sort that interact with each other. A model describes the system in such a way that it can be understood by anyone who can read the description and it describes a system at a particular level of abstraction."

This chapter summarises the BPS+ state-of-the-art in terms of the modelling capabilities of contemporary simulation tools and the general developments underway to extend this capability to deepen analysis possibilities and widen tool applicability.

2.1   Required functionality

The drive towards a sustainable built environment raises challenges for built environment practitioners. These stem from the need to reduce energy consumption, integrate cleaner sources of energy, and meet expectations for human well-being and environmental protection.

In relation to reducing energy consumption and the related environmental emissions, the dual causal factors of population increase and growing consumption present intractable problems: population control is not a moral option, whilst the punitive actions necessary to bring about the required degree of lifestyle change are misaligned with the economic growth objectives of present political systems. That leaves technology intervention, with its plethora of options, as the best and only contender for positive action. Appraising the impact of possible interventions in the built environment in a timely and cost-effective manner is the function of BPS+.

On the supply side, three options are identifiable: decarbonising/replacing fossil fuels, new nuclear power, and new and renewable technologies when distributed nationally or embedded within the urban environment. It is the second part of the last option, embedding supply alongside demand, that provides the greatest application potential for BPS+, with the implication that various scales must be accommodated from a single building, through a large estate, to a whole city.

Ensuring acceptable overall performance of the built environment is a task made complex by the presence of interacting technical domains, diverse performance expectations, and pervasive uncertainties. BPS+ provides a means to accommodate this complexity whilst allowing exploration of the impact of design parameters on solutions intended to provide the required life cycle performance at an acceptable cost. Various BPS tools exist (BEST 2023) that collectively provide practitioners with the means to simulate the flow of heat, air, moisture, light, sound, electricity, pollutants, and control signals within buildings and thereby nurture systematic performance improvement. Crawley and Hand (2008) contrasted the capabilities of several of these tools. The BPS+ acronym used hereafter signals an extension of these capabilities to the urban scale and beyond.

Such tools help practitioners bring balance to the competing aspects of performance whilst addressing the underlying technical complexities that are difficult to handle by any other means. They can be used (separately or together) to:

- evaluate new building components and systems;
- ensure requisite levels of thermal/visual/acoustic comfort and indoor air quality;
- devise energy efficiency and demand management solutions;
- embed new and renewable energy technologies;
- lessen environmental impact;
- ensure conformance with legislative requirements;
- participate in smart control; and
- help formulate energy action plans at various scales.

Such targets can be pursued building-by-building by design practitioners or applied on a larger scale by policymakers. This functionality defines a best practice approach to building design and refurbishment because it respects temporal and spatial interactions, integrates all performance domains, supports cooperative working, and links life cycle performance to health and environmental impact. The approach is rational because it allows the gradual evolution of a design hypothesis, with incremental performance outputs informing the actions to be taken at progressive stages based on models of growing fidelity. The challenge for software developers is to ensure that the technology continues to evolve to adequately represent the changing built environment and its myriad supply technologies in terms of overall performance, impact, and cost.

In many regions throughout the world, sustainable energy solutions are being driven by schemes intended to bring about high-performance buildings through a rigorous approach to performance appraisal. These schemes, in general, encapsulate or encourage simulation-based appraisal. Examples include ASHRAE's Standard 189 (ASHRAE 2023), CIBSE's Application Manual 11 (CIBSE 2015), the European Energy Efficiency Directive (EEED 2023), and rating schemes such as LEED[1], Nabers[2], and BREEAM[3]. Whilst recognising the aspirational intention of such schemes, they can exacerbate the performance gap by enabling simplified methods rather than mandating a multi-physics representation of the problem domain.

Since its inception, BPS+ has been on a development trajectory towards a high-resolution, holistic representation of problems at various scales – from the features comprising a low-energy building, to the complexities underlying a community with disparate building types and extraneous or embedded means of energy production (Clarke and Hensen 2015). Supporting such representations accurately requires modelling capabilities relating to aspects such as hybrid system operation, smart grid interaction, stochastic occupant behaviour, and energy demand manipulation by external, progressively AI-based agents.

New approaches to building environmental control are driving the need for a closer coupling between building and plant models (here, the term 'plant' refers to both the equipment that generates/captures the required energy and the associated distribution system). In the early days of BPS+ development, the building and plant domains were processed independently. Whilst refinements have been introduced over time to address the deficiencies of this decoupled approach, most BPS+ tools still have a degree of decoupling and operate with manufacturers' data corresponding to standard test conditions that invariably do not occur in practice. Indeed, it is normal for a BPS+ tool to employ different mathematical approaches to the building and plant domains. Whilst the former is treated dynamically, albeit with some simplifications imposed (e.g., one-dimensional conduction and fully mixed zone air), the latter is often treated as a succession of steady-state solutions with time-invariant component parameters. Whilst this approach captures the logic of plant behaviour, it does not represent the plant objects themselves, as is done with building-side objects. This has been justified on the basis that thermal capacity effects in plant components are at least an order of magnitude smaller than within the building fabric. This is questionable when considering technologies such as thermal storage for renewable energy supply/demand matching, ground source heat pumps, solar thermal systems, and control action imposed on highly coupled systems to effect demand management.

This disparity of treatment between the building and plant domains is not an intrinsic characteristic of the two subsystems but a legacy from a time of low computing power. It is possible and helpful to apply the same mathematical methods, as used to represent a building as a dynamic, non-linear, systemic, and stochastic problem domain, to also model plant systems. Likewise, dynamic models

[1] https://usgbc.org/leed/
[2] https://nabers.gov.au/about/
[3] https://vts.com/blog/leed-vs-breeam-understanding-the-differences/

can be introduced to represent diverse aspects such as building contents, stochastic occupant behaviour, thermal bridges, and local electricity network interactions.

Although there has been good progress with fundamental physical representation[4], little formal research has yet been undertaken into levels of problem abstraction corresponding to the many possible performance appraisal tasks – it is clearly unhelpful to require a complete and detailed building description to enable every analysis. Full design process integration remains a work in progress, although integrative mechanisms have been demonstrated based on building information and workflow models that can accommodate the different skill levels and conceptual outlooks of those who collaborate in the design process (Clarke and Mac Randal 1991). The aim is to allow BPS+ programs to interoperate with other applications such as CAD and software for structural analysis, regulation compliance, cost estimating, pipe/wiring layout, workflow scheduling, and the like. In this way, much of the required data model for a specific tool may be automatically generated and workflows pre-defined to allow BPS+ to be seamlessly embedded within the design process (Augenbroe 1995, Chen *et al.* 2022).

To summarise: buildings are complex systems because their energy use and indoor conditions vary dynamically under the stochastic influence of weather, occupancy, and equipment erraticism. Accommodating this complexity has been the principal driver of the evolution of performance simulation tools since the beginning of the personal computer era in the 1970s. Despite the significant progress made, it is evident that a gap is growing between tool capability and the expanding demands of the clean energy transition as it affects buildings and cities. This unwelcome situation stems from the growing pressure to radically reduce city energy demand, integrate cleaner sources of energy supply, ensure that indoor/outdoor spaces foster wellbeing, and mitigate local/global environmental impacts; all whilst addressing interacting technical domains, diverse performance expectations, and pervasive uncertainties.

It is here contended that the ultimate goal of BPS+, when applied at whatever scale, is to provide practitioners with the means to emulate reality in a manner that tests operational resilience and thereby renders it more likely in reality. Such a capability portends a future in which the conjugate heat, air, moisture, light, sound, electricity, pollutant, and control signal flows are simulated in an integrated manner based on high-resolution descriptions of proposed schemes subjected to industry standard (and thereby widely understood) performance assessment procedures. The merits of an approach that enables whole system, multi-variate performance appraisal under realistic operational scenarios cannot be understated. The challenge is to ensure that performance simulation tools evolve to provide the required functionality. Young researchers take note! Creating such highly functional software tools and embedding them within the design process is a task that is hindered by the present situation, where the development community encompasses diverse technical and business interests and has yet to evolve mechanisms by which long-term development goals can be agreed upon and pursued collectively and collaboratively.

---

[4]https://publications.ibpsa.org/building-simulation-conference-proceedings/

2.2   Program development

All widely used BPS+ programs are based on high-level programming languages such as C and FORTRAN. Examples of such programs include EnergyPlus[5], ESP-r[6], IDA-ICE[7], IES VE[8], TAS Engineering[9], and TRNSYS[10] among others[11]. These programs have evolved over several decades, with evolution dictated by deficiencies exposed through use and evolving user demand. The development of ESP-r, for example, commenced in 1974 and continued unabated to the present day. Over that period, around 50 individuals contributed at the source code level[12].

In general, code development, maintenance, and user support are serviced by a dedicated organisation comprising individuals possessing expertise in building physics, numerical methods, programming, and application support. In some cases, these organisations offer program licences and related services on a commercial basis. In other cases, especially where an organisation is academic, programs are made available at no cost under an open-source licence (such organisations then benefit by leveraging industry collaboration to attract research funding).

The lack of development task sharing has introduced application problems, not least the proliferation of semantic differentials, incompatible input models, and bespoke user interfaces. For these reasons, researchers have pursued generic program-building environments, with notable examples including the work of IEA Annex 60[13], IBPSA Project 1[14], and the UK EPSRC Energy Kernel System[15] (Charlesworth *et al.* 1991). Whilst no program has yet emerged from these generic environments that can compete with the functionality of existing whole-building simulators, some partial prototypes have been demonstrated and innovative support tools have been developed – such as the Modelica Buildings Library[16] that offers models for the rapid prototyping of buildings, district energy, and control system simulation.

2.3   Model data types and sources

The data comprising a CAD model can be a subset of a BPS+ model because the latter requires 3D geometry and much more. Whilst it is possible that CAD tools will control the BPS+ application process in the future, an extension to present

---

[5]https://energyplus.net/
[6]https://esru.strath.ac.uk/applications/
[7]https://equa.se/en/ida-ice/
[8]https://iesve.com/software/virtual-environment/
[9]https://edsl.net/tas-engineering/
[10]https://trnsys.com/demo/
[11]https://ibpsa.us/best-directory-list/
[12]https://esru.strath.ac.uk/Downloads/ESP-r/Contributors/list.html
[13]https://iea-annex60.org/
[14]https://ibpsa.github.io/project1/
[15]https://esru.strath.ac.uk/Applications
[16]https://simulationresearch.lbl.gov/modelica/

building information model (BIM) schemas (Eastman *et al.* 2008, Uddin *et al.* 2021) will be required to cover all performance domains possible in BPS+. There exists a plethora of BIMs and their related construction platforms (e.g., Revit; Syed and Manzoor 2022), with some attempts to extend the data model to include workflow management. The core issue is how to transfer information between tools without needing to access different BIMs. There have been several initiatives in this regard. For example, Bazjanec (2002) developed a data model based on the concept of 'Industry Foundation Classes' (IFC 2012), which provide an open, international standard (ISO 16739-1: 2018) and cover aspects such as building geometry, structure, plant, electrical services, and facilities management. For example, this has enabled Blender[17], an open-source 3D modelling program, to read, modify, and export IFC models to support information transfer between tools. This is achieved via an add-on feature, named BlenderBIM (2023). Likewise, Green Building XML[18] (gbXML 2023) is a separate open-source BIM schema covering the geometry and some energy-related aspects of a building; the schema supports the transfer of information between CAD models and a variety of engineering analysis tools.

Along complementary lines, Augenbroe (1995) led a project to develop a prototype data exchange environment in which applications can share data under the control of a formally declared design process model. This work was extended within a 'Design Analysis Integration Initiative' (Augenbroe *et al.* 2004) that included workflow management whilst recognising that much of the available information is loosely structured and available in incompatible forms.

Although the BIM approach has been widely adopted by industry and can be regarded as a de facto standard, further development is required to support information exchange between CAD systems that satisfy the complete requirements of BPS+ applications. The current gbXML schema (Version 6.01 of January 2017), for example, comprises 122 *SimpleTypes* and 255 *ComplexTypes*, whilst an updated schema (Version 7.02 of June 2023) has been proposed that adds further elements and enumeration parameters, bringing the totals to 129 *SimpleTypes* and 264 *ComplexTypes* (gbXML 2017). However, many of these entities are empty containers that need to be evolved if the needs of BPS+ are to be served (the same is true of the IFC schema (Mediavilla *et al.* 2023)). For example, the gbXML *equipmentTypeEnum* offers 19 choices, whilst *systemTypeEnum* offers 49 choices, but all are shallow lists with only a name and a brief description but no describing data. The *AirLoopEquipment* and *HydronicLoopEquipment* types have a number of general attributes but lack the specific data definitions that could be used to encapsulate actual representations within a BPS+ model. Many domains – renewable energy systems, low voltage electrical networks, constructional moisture flow, air movement, and embodied energy – are not addressed. Likewise, the performance side is superficially represented: the *resultsTypeEnum* includes 21 items, some of which are dimensional wrappers (e.g., *FootCandles*) and some lack specificity (e.g., *DemandCost, Flow*). Concepts such as

[17]https://blender.org
[18]https://www.gbxml.org/Schema_Current_GreenBuildingXML_gbXML

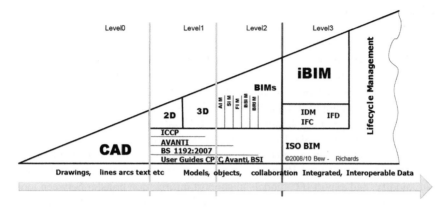

Figure 2.3   The Bew–Richards BIM Maturity Model (source: BIM Industry Working Group 2011)

glare, local draught, contaminant concentration, condensation, thermal comfort, acoustics, indoor air quality, electrical power factor, emissions, and so on are absent, as is plant component performance, making it impossible to represent issues such as evaporator defrosting, flue gas heat recovery, boiler firing rates, and damper control response. In addition, the constituents of energy and mass balances, which are often employed to investigate the cause of poor performance, are missing.

Although BIM mainly focuses on geometrical and constructional aspects of a building, future developments will need to bring extensions to cover all technical domains and performance outputs. Perhaps Level 3 BIM as shown in Figure 2.3 will include such extensions.

The BIM levels shown are as follows.

1. Geometry in 2D format with files shared as separate sources of information supporting low collaboration.
2. A common data model with a mix of 2D and 3D information supporting partial collaboration.
3. 3D model and project data shared via a BIM file encoding to support full collaboration.
4. A unified, cloud-based BIM model with extensions for life cycle management.

In any event, the required data types are already expressed within BPS+ and there follow brief descriptions of four technical domains within ESP-r that are likely to add significant value to an existing BIM schema in terms of the new appraisal functionality then supported: network airflow, computational fluid dynamics (CFD), lighting systems, and electrical networks.

Network fluid flow

Modelling air movement requires descriptive parameters such as room volume, leakage distribution, control valve response rate, and the like. Many of these parameters are difficult to measure, are dependent on the quality of construction work, and only become apparent when the building is in use. Such issues remain largely beyond the scope of BIM.

Thermodynamic forces cause energy to flow throughout a building, and through conduction, convection, and radiation, energy is gained and lost to the external environment. Dynamic simulation programs replicate these energy flows by tracking the underlying thermodynamic processes. Air within a building is often treated as a non-dynamic element, with prescribed flow rates and losses specified in terms of scheduled air change rates for infiltration and natural ventilation. This approach is a throwback to manual calculation methods and finds its way into various BPS+ tools as a means to avoid the necessity of constructing data models that describe distributed airflow paths. The result is a significant loss of similitude to real buildings. Air flows result from the interactions between external wind pressure, buoyancy due to temperature differences, the operation of plant components (e.g., ventilation fans), and the interaction of occupants with ventilation control devices. These forcing effects can have high temporal variability, and the resulting building behaviour often diverges markedly from predictions based on over-simplification. Most BPS+ programs offer approaches that avoid this situation. Within ESP-r, for example, the data required to model dynamic airflow falls into four categories, as follows.

- Boundary condition determinants – e.g., external wind speed, wind direction, and position of obstructions. Weather data and environment descriptors can be used to create suitable driving force information, whilst pressure coefficient data for typical building shapes are included in a support database. In the past, complex geometries required wind tunnel studies to be undertaken to identify surface pressure coefficients, but now computational fluid dynamics (CFD) can be used for this purpose.
- Generic flow resistances – e.g., cracks around windows, doorways, fixed louvres, ductwork, and fittings. Formulae describing the non-linear behaviour of such components along with coefficient values for specific cases are generally available in the literature and contained within a support database.
- Ventilation control devices – e.g., fans and adjustable louvres. Manufacturers of such products usually provide the necessary flow relationships, and these underpin BPS+ algorithms.
- Occupant behaviour descriptions – e.g., window opening patterns, which may be prescribed or delivered from a model of occupant behaviour at runtime.

Dynamic airflow modelling, integrated into the overall building thermal simulation as depicted in Figure 2.4, allows typical building performance behaviours to be investigated. Any building that is naturally ventilated, or relies partially on natural ventilation (so-called mixed-mode) to achieve performance objectives cannot be simulated reliably without integrated airflow modelling. That applies especially where reliance on buoyancy forces, or on natural wind effects, is designed to achieve specific performance objectives. Extreme conditions, such as warm weather overheating, cannot be reliably predicted by simplified means.

Overheating control strategies, such as overnight free cooling, depend on an accurate assessment of the energy flows into and out of fabric thermal storage, with

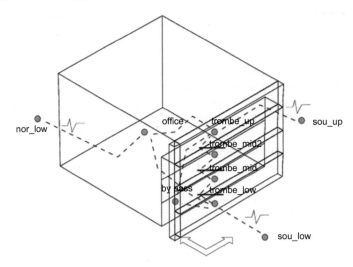

Figure 2.4   An ESP-r conflated thermal and airflow model representing a Trombe-Michel wall

the ventilation air acting as a transport medium. This requires the simultaneous processing of the thermal and airflow domains. Active facade performance simulation, where solar inputs, thermal transfers, and airflow interact in a dynamic manner, is another example of the power of the integrated modelling approach.

Once established, a network airflow model allows the appraisal of ventilation efficacy and controllability, the impact of infiltration on energy use, heat recovery potential, approaches to draught proofing, and the contribution of draught to local thermal comfort. Table 2.2 lists the specific data requirements of a fluid flow model.

Computational fluid dynamics

The airflow network method is appropriate where there is a strong coupling between thermal zones and the distributed airflow associated with building leakage pathways, and HVAC plant: temperature and mass flow data can be readily passed between the two domains. On the other hand, the approach makes simplifying assumptions: that mass flow is a non-linear function of pressure difference only, that the air within a zone is well mixed and characterised by a single temperature and pressure, and that flows are realised instantaneously. Where such assumptions are unacceptable, a CFD model may be established. Air movement is then determined from the solution of mass, energy, and momentum balance equations at points throughout a discretised zone. Because the CFD domain is linked to the zone thermal and network airflow domain models, the CFD boundary conditions (temperature, mass/momentum inputs/extracts) and source terms (heat, contaminants) can be varied throughout the simulation.

Within ESP-r, a typical application of the approach is to use one or more flow networks to represent the macroscopic aspects of overall building/plant fluid flow

Table 2.2   Data requirements of an ESP-r fluid flow model

- Fluid type (air, water, etc.).
- For each internal network node:
  - user-specified pressure and/or temperature, or
  - location of equivalent building domain node (to allow imposition of temperature prior to solution for pressure).
- For each external network node:
  - user-specified pressure and/or temperature, or
  - related surface azimuth, height above datum, and related pressure coefficient set (temperature assigned to prevailing external air temperature, pressure established from pressure coefficient based on wind velocity adjusted for height).
- For each flow component:
  - type (pump, fan, restrictor, tee, crack, doorway, opening, etc. – automatically establishes an empirical mass flow model);
  - model parameters (area, diameter, equation coefficients/exponents, etc.).
- For each inter-node connection:
  - node on the 'positive' side (arbitrary designation) of connection;
  - node on the 'negative' side of the connection;
  - height of connection link relative to the positive side node;
  - height of connection link relative to the negative side node;
  - list of linked components (e.g., vent, door, fan, etc.).
- Solver parameters:
  - type (Newton–Raphson, Gauss–Seidel, etc.);
  - maximum number of iterations;
  - largest allowable cell residual;
  - pressure correction relaxation factor.

---

and CFD to represent the microscopic aspects of air movement within zones over periods of interest. Temperature, momentum, and heat flux boundary conditions are then imported from the thermal/airflow simulation.

When defining a CFD domain in ESP-r, a large amount of input data is required to describe the problem being addressed and the parameters that control the solution: mesh size and location, treatment of turbulence, near-wall conditions, convergence criteria, source terms, and solver-related directives. Because a building may be characterised as an unsteady, low Reynolds Number flow problem, the CFD domain is re-established at each computational time-step based on an exchange of variables with coupled domains. The simplest approach is the one-way transfer of information from the thermal and airflow domains. For a more realistic 'digital twin' simulation, a two-way transfer of information is required, whereby thermal, airflow, and CFD solvers search for mutually converged solutions. This is done by initialising each domain with information from the others and iterating until mutual convergence is achieved. Adaptive mechanisms are available that readjust heat transfer coefficients, near-wall functions, mesh geometry, turbulence model parameters, buoyancy effects, and heat and mass sources based on the prevailing thermodynamic states. For cases with convergence problems, coupling parameters can be regulated to aid convergence.

Table 2.3   Data requirements of an ESP-r CFD model

- Number of grid lines along x-, y- and z-axes and power law coefficients to dictate mesh reduction at near-surface locations.
- Activation of optional equations for solution (buoyancy, contaminant concentration).
- For each boundary opening:
  - type (pressure, mass/volume flow, zero velocity gradient, etc.);
  - location (East, West, North, South, High, Low);
  - initial and final cells in x-, y- and z-direction.
- For each solid boundary:
  - type (temperature, heat flux, symmetry plane);
  - location (East, West, North, South, High, Low);
  - initial and final cell in x-, y- and z-direction.
- For each heat/contaminant source:
  - type (heat flux, contaminant flux);
  - initial and final cell in x-, y- and z-direction;
  - blocked cell identification.
- Solver parameters:
  - initialisation of cell pressure, temperature, and flow rates based on previous time step values;
  - maximum number of iterations;
  - number of solution sweeps;
  - largest allowable cell residual;
  - relaxation factors per parameter.

Once established, the CFD domain can provide an impressive level of detail on critical aspects such as temperature, humidity and contaminant distribution, local air movement (draught), the nature of flows (laminar, turbulent, buoyant), the mean age of air and air quality, ventilation system effectiveness, cross contamination potential, stratification, and local comfort levels. Table 2.3 lists the specific data requirements of a CFD model.

Lighting simulation

The way buildings are used has a significant impact on the magnitude and distribution of internal heat gains, occupants' perceptions of their environment, and worker productivity. Decisions about the provision and control of lighting are a significant issue for design teams, as is code compliance and adherence to points-based standards such as the LEED[19] (Clay *et al.* 2023) and RELi[20] (Siu *et al.* 2023) rating schemes. As with other aspects of simulation, model definition involves selecting from several possible levels of resolution. The subsequent understanding of lighting performance is predicated on clear performance indicators that can vary depending on the stage of the design process and the specific performance issue being addressed. Lighting functionality follows closely from the resolution level.

[19]https://usgbc.org/leed/
[20]https://www.resiliencerisingglobal.org/resilience-toolbox/reli/

Table 2.4 Data requirements of an ESP-r lighting model

- Full geometrical description (with higher resolution at the facade than for thermal models).
- Geometry and location of zone furniture and fittings.
- Surface optical properties for opaque elements.
- Optical transmission properties for transparent elements.
- Optical properties and control states for blinds.
- Switching characteristics of electro-, thermo- or photo-chromic glazing.
- Position, viewing direction, and response characteristics of photocells.
- Position and distribution characteristics of luminaires, use schedules if control is not automated.
- Sky luminance distribution.
- Ground topography and reflectance.
- Geometry and surface properties of adjacent buildings and natural features.

At a low level of resolution, only the heat generation properties of luminaires are of interest, and thus schedules of sensible gain magnitudes are required along with heat distribution fractions to zone air and surfaces via radiation. Control may be based on concepts of daylight factors or radiation levels incident on the facade. Where the design question is related to daylight utilisation as a means to displace the electricity used for artificial lighting, the resolution level must be increased by including the characteristics of photocells and the related control logic (Doulos *et al.* 2019). Where facades include controllable shading, the different possible deployment states of shading devices also need to be included. A simple control actuator may be employed to alter the optical characteristics of the facade in response to some excitation. For glazing, angle-dependent reflectivity, absorptivity, and transmissivity data are required (bi-directional reflectivity in the case of holograms). Such data might be obtained from fenestration suppliers, from physical measurements, or from ray-tracing computations.

Much research has focused on occupant control of lighting and blinds. In actual buildings, occupants move around and have different perceptions and likely responses to perceived changes in their visual field. In this case, the input model needs to hold information about temporal movement and the range of preferences. Table 2.4 lists the specific data requirements of a lighting model.

Electrical networks and renewable energy systems

An electrical network may be established to represent alternating and/or direct current systems, with time-varying, multi-phase, real and reactive power flows at multiple voltage levels. In multi-phase mode, each phase of the network is fully elaborated using a similar approach to that outlined by Kersting (2017) among others. As with a fluid flow network, the electrical network is represented as a series of nodes and arcs, the former being points of interest, such as where power is withdrawn or supplied, or where two conductors meet. The arcs are routes of electrical conduction and can represent electric cabling, power electronics (e.g., inverters), or transformers. The data structure for each node includes its voltage, real and reactive power supplied, real and reactive power drawn, and real and reactive power transmitted to other nodes. The connector data structure typically only includes impedance data in the form of resistance and reactance.

Within ESP-r, the electrical network is integrated with the other domain models (Kelly 1998), using real and reactive power demands or supplies calculated within other technical subsystems of the integrated model: e.g., PV (renewable devices subsystem), lighting (internal schedules and control subsystems), and combined heat and power (CHP) (plant subsystem). This information is used as the boundary condition for the solution of the electrical network, yielding voltage levels and internal and boundary power flows on a time-step basis. As with the other domains within ESP-r, the electrical network model can be developed to represent power flows in more or less detail. Figure 2.5 summarises part of a model where heat is recovered from the cavity of a hybrid PV facade and the output electricity is used locally or exported to the local electricity network.

At the simplest level, the network can be used to record electrical real and reactive power supply and demand, including import and export from the grid and supply from connected low-carbon technologies. This requires the definition of a single, fixed voltage node and the connection of other ESP-r entities, such as lighting loads, PV, or CHP generation, to it. Adding a single-phase electrical network topology allows the model to track real and reactive power flows within the building's distribution system and to calculate voltages at critical points. This allows the identification of phenomena such as cable losses, possible overloading, and low and high voltage levels. Such a network requires the definition of an electrical network topology (comprising multiple nodes and their connectivity), the provision of live and neutral conductor impedance data, and the connection of power-consuming or generating devices (as described elsewhere in an ESP-r input model).

At higher resolution, a fully elaborated three-phase network topology can be defined. This requires all of the information outlined above, along with details on

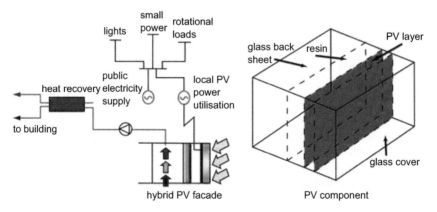

Figure 2.5   A facade-integrated PV component with local power use

the mutual magnetic and electrical couplings between conductors occupying the different phases in an electrical component such as a cable or transformer; these are defined in terms of coupling impedances and would typically be derived from a test on electrical equipment. A full three-phase model allows the electrical network to track real and reactive power flows, power factors, and voltages through the individual phases of a typical power distribution system such as that found in large buildings or serving communities. This enables the identification of power quality problems such as current and voltage imbalances in conductors and devices.

The ESP-r electrical model can be interrogated at different levels, with many of the quantities mentioned above aggregated for the entire network (e.g., total renewable electricity contribution) or reported at the level of individual components. Table 2.5 lists the specific data requirements of an electrical model.

Table 2.5 Data requirements of an ESP-r electrical model

- For the whole network:
  - description;
  - type (a.c., d.c., single-phase, multi-phase, mix of previous).
- For each network node:
  - description;
  - type (variable voltage, fixed voltage);
  - current (a.c., d.c.);
  - single-phase or multi-phase.
- For each electrical load:
  - description;
  - type (lighting, fan, pump, etc.);
  - building location (zone, plant component, etc.);
  - load model;
  - model data (position, associated building component, etc.).
- For each generator:
  - description;
  - type (PV, CHP, etc.);
  - building location (facade-integrated, roof-mounted, free-standing, associated with an HVAC component, etc.);
  - generator model;
  - model data (position, associated plant component, etc.).
- For each connector component:
  - description;
  - type (cable, transformer, etc.);
  - start and end node (or nodes if multi-phase);
  - phase and neutral conductor impedance data;
  - mutual impedances between conductors;
  - connector model;
  - model data.
- Solver parameters:
  - type (Newton–Raphson, Gauss–Seidel, etc.);
  - voltage convergence criteria;
  - apparent power flow convergence criteria;
  - maximum voltage change/iteration.

When using a BPS+ tool to define such models, the user will typically be guided through the process and offered pre-constructed components at key points. The tool might also be able to infer data entries based on prior user input and offer these as default answers. In this way, the input process need not be overly onerous.

2.4 Data representation

The format of the input data model required by BPS+ tools is program-specific and mostly hidden from the user. Appendix B gives an insight into the content and organisation of an ESP-r input model. Table 2.6 lists portions of an EnergyPlus and ESP-r top-level file.

These files comprise sets of keywords and related alphanumeric data. They reflect developer decisions about how model data is held. For example, an EnergyPlus model is held in a single file, whilst an ESP-r 'configuration' file references other files and databases held elsewhere in a project folder.

Both approaches employ keywords to identify specific topics and entities, and these are optionally followed by attributes that are checked for syntax and value when the model is read. Users familiar with the tool are able to parse these files without the help of a program interface. EnergyPlus attributes are comma-separated and a ';' marks the end of the entity or topic. A '!' marks a comment. The 'SimulationControl' entry in Table 2.6 could be concatenated into a single line if the order and number of attributes are correct. The ordering of topics and entities is not significant.

With ESP-r, a '#' marks the start of a comment. Keywords such as *clm and *stdclm signal that a local database is being used in the first case and a standard one as distributed with a release in the last case. Paths to model folders are initially created by the tool but can be changed by the user as required.

EnergyPlus supports multiple representations of surfaces, whilst ESP-r includes lists of vertices, edge-ordering lists, and surface attributes. EnergyPlus punctuation is strict, whilst for ESP-r, commas are only required where tokens such as entity names could have spaces. As described in Section 4.5, productivity tools can be created to manipulate model files, and these will need to perform syntax and relational checks due to bypassing the normal program interface.

Data on model entities such as surfaces are shown in Table 2.7 for both programs.

An EnergyPlus IDF file is paired with a data dictionary, which is included in a distribution. This specifies for each key phrase (e.g., 'BuildingSurface:Detailed') the number, type, and order of the tokens and attributes that follow. Normally, an interface such as IDF editor[21] or Simergy[22] manages the model definition and assessment process, whilst commercial tools such as Designbuilder[23] are able to generate

---

[21]https://bigladdersoftware.com/epx/docs/8-3/getting-started/idf-editor-brief-introduction.html
[22]https://energy-models.com/software/simergy
[23]https://designbuilder.co.uk/35-support/tutorials/96-designbuilder-online-learning-materials

Table 2.6   Extracts of top-level EnergyPlus and ESP-r model files

---

EnergyPlus 'idf' file

Version,9.2;
Timestep,6;
SimulationControl,
Yes,   *!- Do Zone Sizing Calculation*
Yes,   *!- Do System Sizing Calculation*
Yes,   *!- Do Plant Sizing Calculation*
Yes,   *!- Simulate for Sizing Periods*
No,   *!- Simulate Weather File Run Periods*
RunPeriod,
Run Period 1,   *!- Name*
1,   *!- Begin Month*
1,   *!- Begin Day of Month*
,   *!- Begin Year*
12,   *!- End Month*
31,   *!- End Day of Month*
,   *!- End Year*
Sunday,   *!- Day of Week for Start Day*
No,   *!- Assume Holidays and Special Days*
No,   *!- Assume Daylight Saving Period*
No,   *!- Apply Weekend Holiday Rule*
Yes,   *!- Use Weather File Rain Indicators*
Yes;   *!- Use Weather File Snow Indicators*
SurfaceConvectionAlgorithm:Inside,TARP;
SurfaceConvectionAlgorithm:Outside,DOE-2;
HeatBalanceAlgorithm,
ConductionTransferFunction,200.0;
ZoneAirHeatBalanceAlgorithm,
AnalyticalSolution;   *!- Algorithm to use*
. . .

. . .

ESP-r system configuration file
*configuration 4.2   *# system configuration file version*
*date Mon Nov 6 12:04:48 2023   *# latest modification*
*base_name projectX   *# base name for model files*
*indx 1   *# simulation level (building, plant, both)*
 # —— model folders ——
*zonpth ../zones   *# path to zone files*
*netpth ../nets   *# path to network files*
*ctlpth ../ctl   *# path to control files*
*mscpth ../msc   *# path to miscellaneous files*
*radpth ../rad   *# path to Radiance files*
*imgpth ../images   *# path to project images*
*docpth ../doc   *# path to project documents*
*dbspth ../dbs   *# path to project-specific databases*
*tmppth ../tmp   *# path to results files*
 # —— databases ——
*stdmat material.db   *# '*std' indicates that the*
*mlc ../dbs/constr.db1   *# database was included*
*stdcfcdb cfc.db   *# in a distribution else it is*

(Continues)

Table 2.6   (Continued)

stdopt optics.db	# project-specific
stdprs pressc.db	
stdclm kew67	
stdmld mould.db	
stdpdb plantc.db	
stdpdf ../dbs/predef.db	
# —— documentation and images ——	
notes ../doc/office_natvent.log	# project notes
qarep ../doc/qa.txt	# QA report
# —— year and seasons ——	
year 1997	# assessment year
. . .	
. . .	

EnergyPlus models and use its API. All ESP-r files are managed by a Project Manager (PM), which is included in an ESP-r distribution. Another option is, a CAD interface that is called by the PM to handle geometry definition.

Because the model files are held in ASCII format, manual editing is possible, although this requires knowledge of the respective data models and syntax rules. Alternatively, specialist software agents can be used to manipulate model files. The ability to deploy software agents to manipulate model files as part of application automation has a long history. Where a program holds its input data in binary form, the option for automated manipulation is constrained unless the hidden format is published.

In some cases, it is possible to import data from other tools, especially CAD, and reformat the input model as required for delivery to another program. ESP-r, for example, has the facility to import/export models in EnergyPlus IDF format, although it is stressed that such facilities can easily age and fail to keep up with tool changes.

The first step in developing interoperable design tools is to identify the input data requirements and the performance outputs for the various domains, as typified in the previous section. The second step is to develop a structure for describing the relationships between the data types, constraints on these relationships, and essential validation checks. This requires a data model schema that is neutral in relation to the many user applications – an early example is the Neutral Model Format proposed by Sahlin (1992).

A novel schema for building performance modelling was established within the COMBINE project (Augenbroe 1995, Clarke *et al.* 1998). Graphical entity-relationship modelling was used to describe the relationships between the data describing the building context, fabric, airflow networks, and system control. This used ATLIAM representations (Vogel 1991, Clarke *et al.* 2012), an extension of the Natural Language Information Analysis Method (NIAM; Nijssen and Halpin 1989, Mascle 2013). The graphical data can then be parsed into the EXPRESS language

Table 2.7   *Extracts of surface attributes in EnergyPlus and ESP-r model files*

EnergyPlus IDF
<u></u>

...

BuildingSurface:Detailed,
  25C85A,   *!- Name*
  Wall,   *!- Surface type*
  Standard_Int-Wall,   *!- Construction name*
  Floor 1 Café,   *!- Zone name*
  Adiabatic   *!- Outside boundary condition*

  ,
Condition Object
  NoSun,   *!- Sun exposure*
  NoWind,   *!- Wind exposure*
  AutoCalculate   *!- View factor to ground*
  4,   *!- Number of vertices*
  0.0000,66.1416,3.0480,   *!- X,Y,X of vertex 1*
  0.0000,66.1416,0.0000,   *!- X,Y,Z of vertex 2*
  -14.9352,66.1416,0.0000,   *!- X,Y,Z of vertex 3*
  -14.9352,66.1416,3.0480;   *!- X,Y,Z of vertex 4*
Wall:Exterior,
  Zn002:Wall001,   *!- Name*
  EXTERIOR,   *!- Construction name*
  ZONE 2,   *!- Zone name*
  180,   *!- Azimuth angle*
  90,   *!- Tilt angle*
  0,   *!- Starting X coordinate*
  0,   *!- Starting Y coordinate*
  0,   *!- Starting Z coordinate*
  20,   *!- Length*
  10,   *!- Height*
  ...
  ...

ESP-r zone geometry file
<u></u>
 # tag, version, format, name
*Geometry 1.1,GEN,Meeting
 # brief description
Meeting room in building core
 # Vertex coordinates
 # tag, id, X, Y, Z
*vertex,1,33.60,34.60,0.0
*vertex,2,29.43,33.40,0.0
*vertex,3,31.95,24.60,0.0
*vertex,4,36.14,25.81,0.0
*vertex,5,33.61,34.61,2.8
*vertex,6,29.43,33.41,2.8
*vertex,7,31.95,24.61,2.8
*vertex,8,36.14,25.81,2.8
 # Surface edges
 # tag, id, no. vertices, vertex list
*edges,1,4,1,2,6,5

(Continues)

Table 2.7   (Continued)

---

*edges,2,4,2,3,7,6
*edges,3,4,3,4,8,7
*edges,4,4,4,1,5,8
*edges,5,4,5,6,7,8
*edges,6,4,1,4,3,2
 # Surface attributes
 # tag, id, name, orientation, construction, transparency, boundary condition (+ 2 items)
*surf,1,ptn_core,vert,wall_core,opaq,another,13,3
*surf,2, ptn_core,vert,wall_core,opaq,similar,0,0
*surf,3, ptn_TP,vert,wall_core,opaq,another,20,2
*surf,4,ptn_coffee,vert,wall_core,opaq,another,16,35
*surf,5,ceil_mtg,hor_d,ceiling,opaq,another,22,33
*surf,6, floor_mtg,hor_u,floor,opaq,another,24,11
. . .
. . .

(Spiby 1991, Jiang *et al.* 2022) for neutral format distribution. An example ATLIAM diagram describing the data representation of a fluid model network is shown in Figure 2.6 (Hand and Strachan 1998).

Some elements in this diagram, for example *flow_node*, are further decomposed in linked diagrams. Translators were developed to map the EXPRESS format into ESP-r input files and vice versa. These translators needed to be sophisticated enough to allow the application of constraints such as the thermodynamic range of validity.

The advantage of using a schema representation model such as NIAM is that it provides a conceptual framework for describing the data model and allows related items to be stored and accessed in an effective way. A more up-to-date variation of NIAM is the Natural Object Role Modelling Architect (NORMA 2023), which is an open-source plug-in to Microsoft Visual Studio. With this, it is possible to transform a NIAM schema into an XML schema.

Currently, all BIMs are partial, many are proprietary, and there is scant support for issues such as problem decomposition, the changing nature of the information available throughout the design process, and the differing conceptual outlooks of users. Application interoperability in the context of the multi-actor, temporally evolving, semantically diverse building design process remains an elusive goal. True performance optimisation must recognise the complex and multi-actor nature of the building design activity. This requires consideration of non-traditional aspects such as problem decomposition, the temporal aspects of design, model quality assurance, tool interoperability, the adoption of standardised assessment methods, judging performance in terms of diverse criteria, and rules for the mapping of simulation outcomes to design intervention.

In the early days of BPS+ development, users were likely to be building services engineers concerned with evaluating the impact of energy efficiency

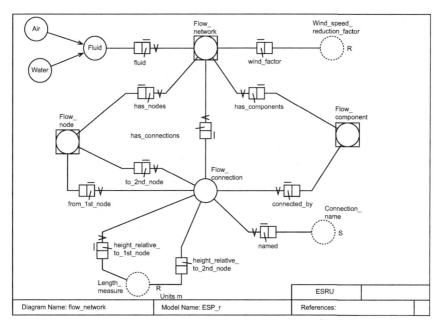

Figure 2.6   ATLIAM diagram for an ESP-r fluid flow network

measures and the sizing of HVAC equipment. The functionality of available tools was therefore limited, and, even now, their use is constrained by a paucity of challenging demands from users, who are preoccupied with regulation compliance and attaining net-zero targets. The application of BPS+ tools should be driven by multiple societal concerns such as occupant wellbeing, fuel poverty alleviation, environmental protection, gradual fossil fuel replacement, security of supply, and cost minimisation. Realising this potential requires that attention be given to several enabling needs: support for diverse user types and applications; the upward and downward extension of the application scale; the linking of energy, environment, wellbeing, and productivity; the imposition of uncertainty and risk; consideration of life cycle performance; and support for both design and policy objectives. Further, BPS+ is evolving to be more than a building design tool by encapsulating urban scale aspects such as demand/supply matching, community energy schemes, aggregate demand management, smart grid operation, renewable technology integration, and electric vehicle participation.

Another vexing issue is the form of the human-computer interface. Why do practitioners accept programs that offer bespoke interfaces to ad hoc and partial models of performance? What is needed is a way to tackle the two fundamental problems that underlie the use of BPS+ programs: the quantity and nature of the data being manipulated and the expertise and conceptual outlook of users. By constructing an industry-standard user interface that incorporates a significant level of knowledge in relation to users, the domain, and the different programs, a more cooperative dialogue can be enabled. This, in turn, would allow designers to more easily abstract a proposal into a form suitable for computer manipulation, and then initiate and control the required performance assessments. Whilst there have been attempts at creating front-end authoring environments to construct interfaces suited to the different conceptual outlooks, technical capabilities, and computer aptitudes of users (e.g., Clarke and Mac Randal 1991), these have not been taken up for two principal reasons. First, the absence of a shared vision of user requirements and, second, the marketing stance of tool vendors that downplay ease-of-use problems. Appendix E summarises a project that established a prototype BPS+ front-end authoring facility that separates user dialogue handling, application coordination, data manipulation, and appraisal procedures, and places these under the control of knowledge predicates corresponding to the facts and rules of building performance assessment. The approach supports the creation of knowledge-based user interfaces that are tailored to the conceptual outlooks of different user types.

Notwithstanding this non-optimum situation, BPS+ tools exist that offer helpful, if partial solutions to the performance assessment requirement of practitioners (Østergård *et al.* 2016, 2020). The Building Energy Software Tools directory (BEST 2023) provides links to a variety of application types and capabilities, both free to use and commercial. On an optimistic note, there are two distinct tool evolution trends. Greater modelling realism is being introduced by providing the means to model ever more complex phenomena. In addition, the tool scope is being expanded to include all technical domains that affect overall performance including, but not limited to, air quality, comfort, energy (required and embodied), environmental emissions, and operational resilience.

2.5   The state-of-the-art

BPS+ tools have traditionally been simplified to a level that preserves the essence of some domains whilst excluding others. Mahdavi (2003) and Mazuroski *et al.* (2018) have pointed out that such simplifications are sometimes necessary to facilitate more effective performance explorations and inter-person communication. Although contemporary programs are highly functional, deficiencies remain vis-à-vis designs that embody non-trivial phenomena such as thermal bridges, conjugate heat and moisture flow, geometrically complex shading devices, embedded renewable energy systems, stochastic occupant interactions, changing thermo-physical properties, and smart control. Addressing inadequacies exposed through use appears to be the main driver for tool refinement, and it is apparent that this responsive development approach is leading to 'over-engineered' programs whereby particular phenomena may be modelled in alternate ways. Airflow, for example, can be represented by prescribed schedules, and/or modelled via distributed leakage descriptions (network airflow) and/or room discretisation (zonal method and computational fluid dynamics), with the user left to decide on suitable approaches. Such alternative abstractions will typically exist for all modelled constructs relating to form, fabric, technical systems, occupants, and control. This theoretical pluralism stems from a perceived need to accommodate different user viewpoints whilst integrating state-of-the-art modelling techniques: a form of theoretical backward compatibility.

Some researchers have attempted to assist users in making context-specific model selections (e.g., Goncalves *et al.* 2020 and Djunaedy *et al.* 2003 in the case of airflow modelling) although progress is often exacerbated by an inappropriate emphasis on the software engineering aspects of a problem at the expense of an evolution of the underlying physical model. Independent but often duplicated work has been undertaken to evolve models for particular processes:

- air movement models to represent the low Reynolds Number, non-steady flow regimes occurring in buildings;
- light flow models to assess visual comfort and the contribution of daylight;
- moisture flow models to assess impacts on materials and air quality;
- occupant models to represent human anatomy and adaptive building interactions;
- fuzzy logic models to accommodate subjective human perception by representing imprecise concepts;
- exergy models to assess the quality of energy sources in addition to quantity;
- uncertainty assessment models to determine the impact of variations in design parameter values;
- supply models to represent new and renewable energy systems;
- control models to regulate energy systems and coordinate hybrid schemes;
- micro-grid models to enable load switching within the context of local renewable power trading;
- material models to support adaptive behaviour (e.g., phase change materials, photovoltaics, and switchable glazing);
- efficient equation solvers to reduce simulation times and move towards real-time design support; and
- enhanced geometry models to represent complex shading devices, solar tracking, thermal bridges, and the like.

Because such developments are pursued in the absence of an overarching framework for BPS+ development, it has proved difficult to share and systematically integrate new developments within existing software tools. One way to alleviate this problem is to find a way to link open source and commercial developments so that users of the latter benefit from the rapid response and collaborative capability of the former.

A particular need is to integrate the different performance domain models in order to represent the interactions and conflicts that occur between problem parts. It is then possible to make explicit the trade-offs that underlie acceptable design solutions – between improved energy efficiency but increased local air pollution in the case of micro-CHP; or between daylight capture (to displace electricity use for lighting) and its exclusion in the case of a photovoltaic facade to generate electricity. Such solutions require that several domain models be able to interact at an appropriate resolution. For example, ESP-r has demonstrated a 3D CFD simulation solved in step with zone, plant, and nodal-network mass flows (Negrao 1995, Beausoleil-Morrison 2000, Bartak *et al.* 2002). Within a time step, the calculated velocities from the CFD analyses are used to determine surface convective heat transfer coefficients, while the calculated pressures influence the overall building flow network solution. In turn, the temperature boundary conditions for the CFD analysis are provided dynamically from the building-side solution, with domain momentum inputs from the flow network. This approach allows for the granularity of the flow description to be closely matched to the problem requirements – a network approach to represent flows at the whole-system level, with one or more CFD domains activated where and when local detail is required.

Whilst some principal domain couplings have been established – e.g., between the thermal and lighting domains, and between the heat and airflow domains – progress is modest, and approaches to integrative modelling have yet to be standardised. Because of this, the impact of domain interactions on problem parameters is poorly understood and essentially excluded from performance appraisals. There is, however, a distinct trend in BPS+ program evolution: new domain models are initially simplified (presumably, to provide a rapid response to user needs or reflect a limited understanding of the domain being addressed) and subsequently refined to remove limitations exposed through use. For example, a 3D construction heat flow model might typically be blended with a 1D moisture flow model (e.g., Nakhi 1995, Clarke 2013), or a dynamic building model might operate alongside a steady-state plant representation (e.g., Ihm *et al.* 2003, Schweiger *et al.* 2018). This has led to the situation where most simulators are hybrid, with detailed models of construction heat transfer, inter- and intra-zone airflow, and radiation exchange co-existing with simplified models of conventional and renewable supply-side components, emissions, and human comfort. It is likely that the simplified models will be refined over time and new domain models added (e.g., for life cycle impact assessment; Citherlet 2001, Casini 2022) in order to address performance trade-offs where designs aim to reduce/reshape energy demand, mitigate environmental impact, improve indoor conditions, and accommodate low carbon supply solutions, all at an acceptable cost.

The ultimate aim of BPS+ is to respect thermodynamic integrity and target reality emulation – as opposed to attempting to predict the future – to confirm operational resilience. With reference to Figure 2.7, this requires due consideration of four intrinsic characteristics of a building when viewed as an energy system.

First, all energy processes evolve in a dynamic manner, requiring the solution (however achieved) of sets of partial differential equations corresponding to the conservation of mass, momentum, and energy. Second, the defining parameters are non-linear in that their values vary as a function of the evolving variables of the state. Third, the overall problem is systemic in that anything that affects one performance aspect will likely affect the others. Fourth, extraneous variables often evolve in a stochastic manner.

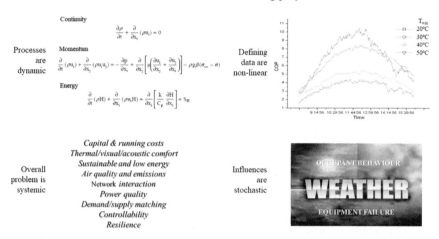

Continuity

$$\frac{\partial \rho}{\partial t} + \frac{\partial}{\partial x_i}(\rho u_i) = 0$$

Processes are dynamic

Momentum

$$\frac{\partial}{\partial t}(\rho u_i) + \frac{\partial}{\partial x_j}(\rho u_i u_j) = -\frac{\partial p}{\partial x_i} + \frac{\partial}{\partial x_j}\left[\mu\left(\frac{\partial u_i}{\partial x_j} + \frac{\partial u_j}{\partial x_i}\right)\right] - \rho g \beta(\theta_{\infty} - \theta)$$

Energy

$$\frac{\partial}{\partial t}(\rho H) + \frac{\partial}{\partial x_i}(\rho u_i H) = \frac{\partial}{\partial x_i}\left[\frac{k}{C_p}\frac{\partial H}{\partial x_i}\right] + S_H$$

Defining data are non-linear

Overall problem is systemic

Capital & running costs
Thermal/visual/acoustic comfort
Sustainable and low energy
Air quality and emissions
Network interaction
Power quality
Demand/supply matching
Controllability
Resilience

Influences are stochastic

OCCUPANT BEHAVIOUR

WEATHER

EQUIPMENT FAILURE

Figure 2.7   Four intrinsic characteristics of a building as an energy system

As an example of integrated simulation, consider an ESP-r application when the input model adheres to a whole-system, high-resolution approach. At simulation time, the overall model is subdivided into finite volumes, each represented by conservation equations relating to energy, mass, and momentum exchange, as appropriate. These equations are solved simultaneously and repetitively by numerical methods to obtain the spatial distribution of state variable values and their variation over time. There is no differentiation between the mathematical treatment of the building and the plant. Each has interacting energy flows associated with material conduction and moisture flow, surface convection, inter-surface radiation, intra-space fluid movement, and electricity flow. For example, the rooms within the dwelling and the boiler combustion chamber are thermodynamically equivalent, differing only in the nature of the heat source – a radiator, the sun, or casual gain in the former case and by the rapid oxidation of fuel in the latter case. Further, and significantly, control action influences the whole-system model by driving parameter adaptation in response to changing system states and linking hitherto disconnected problem parts (e.g., the power from a PV array is delivered to the electrical network model while the generated heat is absorbed by the building fabric or a ventilation heat recovery device).

By degrading this high-resolution model (e.g., by imposing published boiler/ PV efficiency factors or removing content thermal capacity), it is possible to emulate the lower-order modelling approaches and thereby characterise the inherent loss of appraisal capability and realism. Since this loss has been demonstrated to be significant, the implication is that tool users require better support for plant component representation with equipment efficiency delivered as an output from simulations rather than required as an input corresponding to performance data produced under test conditions that will rarely prevail in practice.

Whether it is best to integrate domain mathematical models within a single program or within separate, cooperating programs is a matter of preference. The former approach supports high-integrity modelling because all domain models have access to the same data model and can therefore interact at the solver level. The latter approach implies that a problem can be decomposed into a hierarchy of sub-problems, each represented by a different tool – for comfort, HVAC, airflow, lighting, etc. – with an overarching coordination mechanism to ensure that domain interactions are translated to individual input model reconfigurations as a simulation progresses. Cóstola *et al.* (2009) demonstrated the latter approach using separate programs for building energy simulation and construction heat and mass transfer, whilst other researchers have applied the approach to simplified design problems (Klemm *et al.* 2000, Choudhary *et al.* 2003, Santos *et al.* 2017). This is the approach presently favoured by CAD vendors because it effortlessly adds energy modelling functionality to their existing systems. The barriers resulting from weak tool interoperability are expected to be mitigated by adherence to a common data exchange schema that supports a shared semantic.

Several general observations may be made on the state-of-the-art of BPS+ as follows.

- Most tool features remain program-specific, with slow progress on regularisation and standardisation in order to facilitate feature sharing and allow users to effortlessly access multiple tools. For example, databases relating to material properties, weather data, modelling entities (plant components, control systems, construction types, etc.), past solutions, and exemplar models are typically ad hoc, held in proprietary format, have no central repository, and offer no accompanying evidential basis. Interfaces are likewise ad hoc, adhere to weak human-computer interface principles, and show little sign of evolution based on formal requirements analysis and industry buy-in. In addition, approaches to model-making cover a wide spectrum, employ dissimilar syntax, and require procedures that imply radically different semantics.
- Program outputs typically relate to different subsets of overall performance, with interpretation confounded by the use of alternative performance criteria expressed in a variety of styles.
- The time cycle of program application remains unacceptable in relation to the requirements of design practice. Attempts to address this by simplification of the theoretical basis and/or reduction in domain representation are ill-founded and unlikely to evolve the state-of-the-art in the direction required.
- Progress with the conflation of existing domain models – thermal, air/light/ moisture flow, plant, etc. – is tentative and, where attempted, there is often an inappropriate balance of model resolution. This situation is often the result of an inappropriate emphasis on the software engineering aspects of the problem at the expense of the evolution of the underlying theory.
- Whilst specialist tools exist for significant non-thermal domains such as acoustic comfort, indoor air quality, embodied energy, renewables, and economic aspects, slow progress is being made on BPS+ integration. The principal

barrier here is the absence of an overarching data model, which necessitates inelegant adaptations to specific tools.

- Despite its core importance, support for the mapping of simulation outcomes to design interventions is nebulous. Likewise, there is little acceptance of, and agreement on the procedures required to quality-assure models, coordinate BPS+ tool applications, and train/support users.
- User interfaces tend to coerce users to follow an established path when creating tool-specific input models, as opposed to helping users identify the level of detail required to answer particular design questions. There are many actors in this 'conspiracy': tool developers, who hide complex choices from users; limited appreciation of the underlying physics within the user community; and a lack of observational skills about the temporal nature of building performance (to name but three from many).

Notwithstanding such deficiencies, BPS+ tools offer a tentative mechanism to meet the challenges posed by increased user expectations and legislation. Good indoor air quality, for example, is a fundamental prerequisite of a health-promoting building. Numerous issues can be addressed through the application of macroscopic and microscopic airflow modelling or a combination of the two. Examples of where airflow modelling may be deployed include (in a domestic context) the prediction of condensation, dampness, mould growth, and the transport of harmful pollutants such as mould spores, carbon monoxide, and volatile organic compounds. In commercial and industrial contexts, airflow modelling can address some of the issues underlying sick building syndrome and low occupant productivity. Furthermore, the modelling of air quality can address safety issues such as the removal of smoke and hazardous contaminants. Whilst the application of CFD to the above issues is not new, the embedding of CFD within BPS+ enables the reconfiguration of turbulence models and boundary conditions as a simulation proceeds as a means to enhance modelling fidelity.

The prediction of surface condensation and mould growth is an area where the CFD and fabric moisture domains can be applied jointly. These domain models can cooperate with the building thermal model to assess near-wall and surface conditions – outputs required for the prediction of phenomena such as mould growth include the temperature and free moisture at the wall surface.

At lesser granularity, a network airflow model may be used in the analysis of contaminant dispersal throughout a building. An example of this approach is the modelling of the diffusion of combustion-based contaminants through a naturally ventilated building close to a busy road. If more detail is required about specific localised concentrations, the predictions of the network airflow model can be used as boundary conditions for the CFD domain. This demonstrates an advantage of the modular solution approach in that the same phenomena can be examined at different levels of detail depending upon the context of the analysis.

The ability to couple an HVAC domain to other domains enables it to be used in the modelling of low-carbon energy systems. For example, a ventilated photovoltaic facade can be modelled using a combination of an HVAC network, an airflow network, and a 3D model of the facade. In this case, the photovoltaic elements are explicitly modelled as part of the building construction, allowing alteration to local material properties that are temperature-dependent and control volume source terms to be adjusted to reduce the absorbed solar radiation to account for its conversion to electricity. The photovoltaic elements can themselves be coupled to an electrical network. These issues are demonstrated in the application examples in Chapter 6.

2.6 Program validation

Given the complexity of the building physics and design domains, assessing the predictive veracity of a BPS+ program is a non-trivial task. Where an aspect of a program is simplified, its parameters will be non-physical. Where a program is comprehensive, there will be numerous interacting parameters. A recurring finding from validation studies has been the need to improve the representation of the underlying building physics – notable examples include the treatment of inter-surface longwave radiation exchanges (Moore and Numan 1982) and surface convection heat transfer (Beausoleil-Morrison 2001). Researchers responded to this problem by establishing validation procedures within various projects:

- the International Energy Agency's Annex 1 (Irving 1982), 4 (IEA4 1980), 21 (Lomas *et al.* 1994) and 58 (Janssens 2014, Kersken *et al.* 2015) of the Buildings and Community Systems Programme; and Annex VIII of the Solar Heating and Cooling Programme (Bloomfield 1988);
- the European Commission's PASSYS distributed test cell programme (Jensen 1995);
- national efforts in the USA (Judkoff and Neymark 2006, Faulkner *et al.* 2023, ASHRAE 2023) and the UK (Hand *et al.* 1998, Bloomfield *et al.* 1988, Kersken *et al.* 2020, Ferreira *et al.* 2023); and
- program-specific developments undertaken by developers and user groups (e.g., Strachan and Baker 2008, Queiroz *et al.* 2020).

These activities have culminated in the development of robust validation procedures comprising a formal review of theory/algorithms, code checking, analytical testing, inter-program comparison, parametric sensitivity analysis, and empirical testing. Most BPS+ programs have been subjected to these procedures (e.g., Strachan *et al.* 2015 in the case of ESP-r) and some tools have incorporated them into standard tests for invocation by vendors and users following software updates. A recurring insight is the need to eliminate user input error through training and better model calibration and quality assurance procedures. A rational approach to program validation will involve the following procedures.

Review of theory/algorithms

This element is needed to ensure that tools adhere to state-of-the-art algorithms. The activity is well supported within collaborative projects where expertise is available across all aspects of building physics. Issues stemming from algorithmic shortcomings are particularly pernicious and, therefore, difficult to detect. Consequently, a BPS+ tool's source code must have strict versioning control and, most helpfully, be made available for third-party scrutiny.

Code checking

Code checkers exist to trap programming errors and are an essential part of professional quality control procedures. Third-party checking is enabled where source code is available (e.g., through an open-source licence) but frustrated where the code is proprietary.

Analytical testing

Although restricted to simple cases for which an analytical solution is possible (e.g., wall conduction under steady-cyclic boundary conditions), this technique is powerful because it provides an unambiguous truth model. It can be creatively applied to many of the technical subsystems comprising a BPS+ program. For example, a complex flow network consisting of many but simple connected resistances can be transformed into a single equivalent resistance and solved manually for a given pressure difference. This can then be compared with the numerical solution that emerged for the original network.

Inter-program comparison

This validation element helps a new entrant program to align with existing programs that have been previously well-tested and applied in practice. Where alignment is poor, this will lead to an appropriate development of the new entrant or, perhaps, a general adjustment across the community.

Empirical testing

This element requires a detailed procedure for experimental design, data processing, and result analysis. Several significant datasets exist, along with detailed descriptions of the related experimental setup (e.g., Mahdavi and Tahmasebi 2015, Vazifeh *et al.* 2015, Mantesi *et al.* 2019). Empirical testing can be difficult to implement as errors from various sources can occur simultaneously (Judkoff *et al.* 2008, Chong *et al.* 2021); some significant sources include differences between real and simulated weather data, between real and simulated occupancy behaviour, and between real and modelled building properties.

Parametric sensitivity analysis

This element helps with the understanding of error sources in situations where a program does not align with expectations. It can also be used to characterise aspects of a BPS+ input model, as demonstrated by Monari and Strachan (2016) in the context of establishing an airflow network as routinely used with BPS+.

Repeated application

Although the above elements have evolved to a high degree of refinement, they remain difficult to apply in practice for several reasons.

- Validation is a continuing process with repeated application required to maintain confidence in predictions. Conversely, major conclusions regarding a program's validity cannot be made from individual tests since these will not cover all use cases.
- With analytical, inter-program, and empirical testing, it is usually necessary to reduce the complexity of the test case to control parameter uncertainties and enable input model equivalency.
- Significant resources are required to establish an appropriately instrumented test facility and quality assure monitored data. Where such data is already available, it invariably transpires that the data required to create an input model is partial.

For these reasons, most BPS+ tools are evaluated using the Building Energy Simulation Test (BESTEST) procedure as developed within the International Energy Agency's Solar Heating and Cooling Programme Task 12 and Annex 21 (Judkoff and Neymark 2006, Ohlsson and Olofsson 2021). BESTEST assesses a program's ability to produce outputs that are acceptably aligned with other programs of the same type. It offers a number of test cases that progress from simple to realistic, with the outputs from a tested program evaluated according to diagnostic logic to determine the algorithms responsible for predictive shortcomings. Since first published in 1995, BESTEST has been adopted by certifying bodies and research groups, e.g., the procedure underpins ASHRAE Standard 140 (ASHRAE 2004) for BPS+ program testing, which at the time of writing is in further development (ASHRAE 2023).

Regression testing

As BPS+ tools are continually evolving, there is a need to ensure that changes do not adversely affect predictions. One approach is to include a suite of representative models and a means to process them automatically to identify if, when, and where performance predictions diverge between releases. Such a feature can be used by developers to scrutinise updates prior to each release, and by users to help them decide if they should move to a new release mid-project or defer until the implications of a discrepancy are fully explored.

For example, ESP-r includes a suite of around 200 test models, and some of these can be invoked and the outcome compared with results obtained from a prior version. The comparison presents sufficient information to help isolate the cause of discrepancies that fall outside a defined tolerance. In addition, users can invoke the BESTEST procedure to determine if the program remains compliant.

Experimental support for empirical validation

Where empirical validation is carried out in a research context, substantial resources are normally available. The following procedure was developed within the EC-funded PASSYS project (Gicquel 1988), which utilised identical test cells (BBRI 1990, Strachan and Baker 2008) sited at locations throughout Europe as indicated in Figure 2.8.

The project developed standardised experimental procedures for building component testing and used these to capture datasets for empirical validation (Jensen 1995). A validated program could then be used to extrapolate the results to other building types and climates.

Prior to any experiment, an estimate of the likely principal factors is obtained to ensure that the measured data's resolution is matched to the program being tested. Where possible this assessment is carried out by the target program so that its inherent sensitivities become the focus of the experiment. The use of a simulation program in this way requires:

- an accurate description of the experimental configuration;
- a high-quality experimental facility – an outdoor test cell in the case of the PASSYS project – with a high fidelity monitoring capability; and
- a sensitivity analysis to assess the influence of the program's input parameters on predictions in order to determine sensor and accuracy requirements.

Experiment implementation now follows and adheres to the requirements and constraints identified in the preceding step. The final dataset requires careful checking and documenting if it is to be considered high-quality and useful to others.

Program/data comparisons can now proceed and will entail:

- an initial blind run of the program using a carefully formed input model and measured weather data;
- goodness-of-fit is assessed by means of parametric sensitivity analysis, used to estimate the uncertainty bands associated with the predicted time series; and
- for cases in which dynamic aspects dominate and/or where more information is required on the cause of poor agreement, a statistically-based approach (Palomo and Tellez 1991) can be employed, which is based on an analysis of

Figure 2.8   PASSYS test cells at the University of Strathclyde and other locations

residuals (the difference between measurements and predictions). This entails estimating the autocorrelation function and power spectrum of the residuals and determining the cross-correlation functions between program inputs and these residuals in the time and frequency domains. Tests applied to these data will yield information on which program inputs are mainly responsible for the residuals. In this way, some insight is obtained into which physical processes are not being well represented by the program.

Where program deficiencies are exposed, remedial action is taken. Where deficiencies are not excessive, a pragmatic alternative is to employ the monitored data in model calibration mode (Clarke *et al.* 1993). By this means, a program can be tuned to represent a system over a realistic range of operating conditions. The calibrated program can then be used, with caution, to extrapolate performance to other contexts by means of scaling and replication procedures (Strachan and Guy 1991).

As simulation becomes more widely used as the basis of design tools, the need for program accreditation will grow. Validation procedures will be an essential part of any accreditation procedure, acting to ensure that, for a limited number of cases at least, the predictions from candidate programs are acceptable.

2.7 Model quality assurance

Given the number of issues at play within the design process, errors and omissions are natural artefacts of any model creation process. BPS+ tools will typically apply consistency checks to such models prior to use. Some checks can be applied automatically, e.g., to ensure that zones are bounded, that windows lie within the bounds of their parent wall, and that thermophysical properties are reasonable. Figure 2.9 presents two cases of illegal geometry: a surface topology problem as indicated by the inward-facing normal on the left (the surface vertex ordering direction is incorrect); and a missing surface highlighted on the right (a frequent occurrence after CAD import).

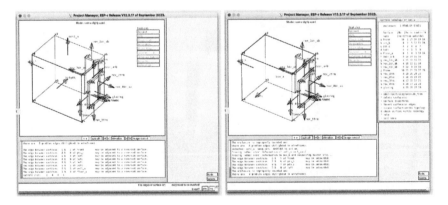

Figure 2.9   ESP-r model geometry warnings

Table 2.8   Extract from an ESP-r model contents report relating to zone control

Zone	Description
O_BC	Base case office.
O_PI	Office with PI control.
O_P	Office with P control.
O_3pos	Office with a 3-position controller.
O_PID	Office with PID control.
O_PD	Office with PD control.
Corid	Passage adjacent to offices.
Void	Ceiling void.
Master	Office with master controller.

Period	Proportional control parameters
00:00	Max. heating 2.5 kW, min. heating 0.2 kW, heating set point 15 °C, throttling range 3 °C. Max. cooling 3.0 kW, min. cooling 0.3 kW, cooling set point 26 °C, throttling range 3 °C.
06:00	Max. heating 2.5 kW, min. heating 0.2 kW, heating set point 20 °C, throttling range 3 °C. Max. cooling 3.0 kW, min. cooling 0.3 kW, cooling set point 24 °C throttling range 3 °C.
18:00	Max. heating 2.5 kW, min. heating 0.2 kW, heating set point 15 °C, throttling range 3 °C. Max. cooling 3.0 kW, min. cooling 0.3 kW, cooling set point 26 °C, throttling range 3 °C.

Likewise, all model data are range-checked. To trap data that are plausible but incorrect, the user can request a tidy statement of any part of the problem and correct detected errors. Table 2.8 shows an extract from a model content report as generated by ESP-r. Here, the input model has been established to examine the impact of different control types.

The first part is a terse zone summary (the verbosity level can be activated by topic). This echoes the user-supplied zone names and descriptions of the assigned control function. The second part is a synopsis of one control function indicating heating and cooling inputs within three periods. This allows the user to check that the model remains aligned with an evolving project brief and can assist in troubleshooting in the case of an unexpected performance outcome.

Appendix B shows a more complete description of a model content report as generated to assist with quality assurance.

2.8   Support for teaching, design, and research

Modelling and simulation have for many years contributed to teaching and learning activities at the undergraduate, postgraduate, and continuing professional development levels, and on a range of topics underpinning energy systems and the environment. A feature of the technology is its ability to amplify the learning experience by enabling taught concepts to be immediately applied and explored via pre-constructed models. An increasing number of practical courses are assisting students and practitioners to refine their application skills. Augenbroe (2008) has argued that effective application requires knowledge of the underlying theory as well as the model manipulations supported at the interface. Conversely, Göçer and Dervishi (2015) have called for tools whose use is not predicated on theoretical insight to enable application at the concept design stage (where presumably many of the model parameters are assigned default values). The approach taken at the authors' institution is to provide pre-formed exemplar models and require students to undertake model exploration followed by simulations of incremental complexity matched to corresponding theory classes[24]. Other researchers have established internet-based classes for students and practitioners (Hensen *et al.* 1998). An exciting future prospect is to utilise the BPS+ approach early in the educational journey (e.g., university first year) to embed an appreciation of cause and effect prior to elaborating the underlying principles and theories in later years.

When used in a design context, the BPS+ approach has the potential to be better, quicker, and cheaper than traditional methods, but only if application procedures are well managed. Realising this potential will require a change to business work practices and innovative company support services, such as those espoused by Macdonald *et al.* (2005), who operated an industry club that delivered regular training courses accompanied by 'supported technology deployments' whereby modelling specialists were seconded to a design team utilising BPS+ for the first time.

The paradigm shift implied by the use of BPS+ in a design context is that the purpose has less to do with sizing components and more to do with ensuring that selected components will provide the desired outcome under realistic operating conditions. An enticing prospect is to extend the design team to include non-traditional users such as building occupants, facility managers, community associations, trade bodies, and the like as a means to ensure full participation in the design process (Maver 1985).

When used in a research context the power of simulation becomes most apparent. Researchers generally have the time and facilities to create calibrated input models and undertake multiple simulations to accommodate parameter uncertainty. Where such applications lead to theoretical refinements that are then encapsulated in the tools used, this is surely a better way to deliver new knowledge than to only publish a description of the advance in a journal or conference paper. The need here is to require that new methods emerge from research projects be made available in a ready-to-use software format.

2.9   Required tool refinements

New and renewable energy (NRE) systems have typically been pursued at the strategic level, with distributed hydroelectric stations, biogas plants, and wind farms being connected to the electricity network at the transmission or distribution levels. To avoid problems with fault clearance, network balancing, and power quality, it has been estimated that the deployment of systems with limited control possibilities should be restricted to around 25% of the total installed capacity

---

[24]For example, see www.esru.strath.ac.uk/Courseware/Class-ME404/index.htm

(Jani *et al.* 2022). This limitation is due to the intermittent nature of NRE sources, which require controllable, fast-responding reserve capacity to compensate for fluctuations in output, and energy storage to compensate for non-availability. To achieve a greater penetration level, NRE systems can be embedded within the built environment, where they serve as a demand reduction device (as seen from the electricity network).

Embedded generation, a concept summarised in Figure 2.10, requires the matching of local supply potentials to optimised demands arrived at by the application of demand reshaping measures. For example, passive solar devices, adaptive materials, heat recovery, and/or smart control may be used to reduce energy requirements, and NRE systems – any mix of combined heat and power (CHP), heat pumps (HP), small gas turbines (GT), photovoltaic (PV) components, fuel cells (FC), and ducted wind turbines (DWT) – used to meet a significant portion of the residual demand. Any energy deficit is met from the public electricity supply operating in cooperative mode. For the embedded approach to be successful, energy reduction and management measures must be deployed alongside the NRE systems. This requires BPS+ application, by which appropriate technology matches may be identified and specific schemes assessed in terms of criteria relating to air quality, human comfort, energy use, environmental impact, and cost (see Section 7.7).

A potential paradigm for future energy systems is the concept of the microgrid, where small power generators cooperate in a local electricity network. BPS+ tools equipped with the appropriate domain models are well suited to the analysis of such systems where the production of heat and electricity is inherently linked to the time-varying building loads. In addition, the approach may be extended to the modelling of communities as opposed to individual dwellings. In this regard, some challenges remain to be resolved, including the development of demand management algorithms that may be applied to switch certain loads within the context of renewable power trading. Such a facility would primarily interact with the building's thermal, HVAC, and electrical domains by acting to reschedule heating/cooling system set-point temperatures and withholding/releasing power-consuming appliances where acceptable.

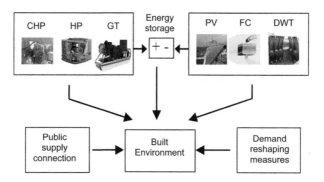

Figure 2.10 Possible elements of an embedded generation solution

As they evolve, BPS+ tools will encompass a widening range of problems. The pertinent question is whether the underlying modelling approach can accommodate future requirements. Sometimes a new feature will require only a minor code change: e.g., the modelling of phase change material can be implemented in a numerical code via a temporally varying conservation-equation source term and material heat capacity. In other cases, the required change will be non-trivial.

The following summaries relate to current and potential developments that will help to bridge the gap between current BPS+ capabilities and future needs.

- Buildings are complex 3D entities and should be treated as such without exception. There is presently a gap between the representations of building geometry in generalised BPS+ tools and specialised facade modelling tools such as WUFI and Radiance concerned with construction heat and mass transfer and lighting simulation, respectively. The elimination of this gap is presently being tackled from both ends: enhanced representations are being added to the BPS+ tools and functional generalisations are being introduced to the specialised tools.

- The network airflow modelling approach (Cockroft 1979, Walton 1989, Hensen 1991, Abbas *et al.* 2023) is now widely implemented in BPS+ and the present need is to extend the scope and robustness of component models defining the fluid flow as a function of pressure difference. Other possible improvements include the introduction of a pressure capacity term into the mass balance equations to facilitate the modelling of compressible flows, and the introduction of transport delay terms to the network connections: where the fluid velocity and geometrical characteristics of a particular connection are known, a transport delay can be calculated automatically.

- Because buildings define a low Reynolds Number, non-steady flow problem, work has been undertaken to embed the computational fluid dynamics technique within BPS+. Beausoleil-Morrison (2000), for example, implemented a conflation controller in ESP-r whereby the turbulence model is reconfigured at each time step to accommodate changing flow regimes resulting from occupant and control system actions.

- In the context of health and comfort, it is inappropriate to couple a detailed surface model to a lumped air volume, as the detail of the surface model predictions would be lost. An adequate representation of zone vapour distribution requires that a CFD domain be coupled to a heat and mass transfer construction model. This requires extensions to the linkages between the building thermal/ moisture and CFD domains to handle the possible surface/zone grid coupling cases: 1D/3D, 2D/3D, and 3D/3D (Amissah 2005).

- The modelling of fire/smoke requires that extra equations be added to handle combustion reactions and the transport of combustion products (e.g., via the implementation of mixture fraction or grey gas radiation transport models). Such adjustments can be readily implemented within existing numerical models since the governing equations are of the diffusive type and can be treated in the same way as the energy and species diffusion equations. In conjunction with the network flow model, the distribution of smoke can then be applied as a boundary condition for the prediction of the movement of occupants during a fire via a Markov Chain model.

- A further CFD refinement is to enable modelling of small solid particulates that have appreciable mass within the main air stream. Two particle dispersion/ deposition modelling methods have been considered for implementation: treating the particles as a continuum or particle tracing using Lagrange coordinates. A prototype continuum-based model has been implemented in ESP-r to enable the modelling of very small particles in flows with Stokes number significantly less than unity (Kelly and Macdonald 2004).

- Between the techniques of CFD and network airflow modelling, there exists a technique of intermediate complexity termed zonal modelling (Inard *et al.* 1996, Fornari *et al.* 2023). The advantage of the approach is its ability to represent intra-zone air movement without the CPU-intensive need to solve the Navier-Stokes equations as is done in CFD. An issue with the approach is the need to partition rooms in a manner that represents the possible flow regimes (plumes, jets, and boundary layers). Several researchers have devised empirical rules to automate such partitioning (e.g., Gagneau and Allard 2001, Guernouti *et al.* 2003, Fornari *et al.* 2023).

- In the context of daylight simulation, there is a need to model the rapidly changing sky brightness distribution whilst eliminating the need to reinitiate an entire daylight simulation at each time step. Typical approaches to the calculation of indoor illuminance distribution are to categorise daylight conditions as a function of sun position (Herkel and Pasquay 1997, Mazzeo *et al.* 2020) or to treat real skies as blends of clear and fully overcast states (Erhorn *et al.* 1998). Such approaches are computationally efficient but accuracy-constrained. To overcome the latter deficiency, numerical models have been pursued (Reinhart and Herkel 2000, Liu and Wu 2022); the issue then is how best to reduce the computational burden. One approach is to operate in terms of pre-computed daylight coefficients based on sky patch pre-processing (Tregenza and Waters 1983, Ayoub 2019). To process a year at hourly time steps would require 'n' sky patch simulations rather than 8760 hourly ones, where 'n' is typically 145. Geebelen and Neuckermans (2003) demonstrated that further reductions in the computational time are possible by increasing the discretisation mesh size within a radiosity algorithm: for rooms with common office proportions, a mesh dimension of one-third to a half of the smallest room dimension performed well in terms of acceptable accuracy and practical computation time. Of course, the computational burden will rise in cases where the model itself is time-varying due, for example, to the operation of window shading devices.

- Whilst component models exist to represent most HVAC systems, many of these relate to pseudo-steady-state operation with little or no account of dynamic response. Work is therefore required to introduce dynamic considerations into these models. The best way to do this is to treat plant components in the same manner as building zones by establishing a connected set of conservation equations corresponding to a discretised digital twin (Clarke 2001).

The extensibility of the approach is essentially unlimited: new component models may be implemented as new products emerge. The major issue confronting BPS+ is the generation of the component models in the first place and the combination of the selected components to form a hybrid supply system. To this end, two issues need to be addressed: the synthesis of component models from basic heat transfer/flow elements and the automatic linking of the resulting models. The former issue has been addressed by the 'primitive parts' technique (Chow 1995) whereby a new component model may be synthesised from pre-constructed models representing individual heat and mass transfer processes. The merit of the approach is that arbitrarily complex models may be rapidly configured for particular problems. The latter issue, automatic linking, might build upon previous research into object-oriented HVAC modelling (Tang 1985, Sowell and Moshier 1995, Ayoub 2019).

The main aim of building design is to provide satisfactory comfort conditions for occupants. The adaptive thermal comfort approach (Humphreys and Nicol 1998, Liu *et al.* 2020) recognises that occupants will actively change their environment to secure comfort. This implies that thermal comfort should be considered a self-regulating system, incorporating the heat exchange between the occupant and environment, the physiological/behavioural/psychological responses of the occupant, and the control opportunities afforded by the building design. Within BPS+, this requires maintaining comfort-related parameters within certain limits. Such parameters must be chosen with respect to the control strategy being implemented, the technical feasibility of parameter measurement, and economic considerations. Some principal control system terms include:

- controlled parameter – the parameter that has to be kept constant or within a fixed range (e.g., indoor air temperature);
- control parameter – the parameter that can be modified by the control system in order to move the controlled parameter close to its set point (e.g., the position of a valve or the status of a pump);
- set point – the value at which the controlled parameter has to be maintained (e.g., 21 °C in an occupied space during winter).
- logic control – a technique based on 'if-then-else' statements, used by the first generation of intelligent building energy management systems that employed hierarchical conditional statements; and
- time programming – a control function implemented in modern buildings in order to programme the operation of equipment or the value of control parameters as a function of time.

A building control system comprises three elements: sensors to measure the value of the controlled parameters, actuators to modify equipment according to a control parameter's value, and a controller to set the value of control parameters as a function of controlled parameters. There are two main categories of control techniques, as follows.

- Open loop control – where the controlled parameter is not explicitly taken into account and inputs to the controller are one or more disturbance parameters. The technique is also called feed-forward control. A well-known example is the determination of the mixed flow water temperature as a function of ambient temperature in a heating system with a boiler and radiant panels.
- Close loop control – where the controlled parameter is an input to the control system and is compared to the set point in order to determine the value to be given to the control parameter. This is also known as feedback control. The most common closed-loop control systems are on/off, proportional (P), proportional-integral (PI), and proportional-integral-differential (PID) control. This last type takes into account the value of the controlled parameter, the difference between the controlled parameter's value at the current time step and the previous time step, and the previous values of the controlled parameter.

Unfortunately, controllers that implement the above techniques do not assure optimal control of all controlled variables. To address this issue, more advanced techniques have been developed, as follows.

Fuzzy logic

Given that human perception is difficult to model mathematically because of its subjective nature (different individuals will respond differently to the same environmental stimuli), the fuzzy logic technique allows imprecise concepts (sets) such as 'cold', 'cool', 'neutral', 'hot', and 'warm' (or low, neutral, high) to be represented. The values of the set members range, inclusively, between 0 and 1, with the possibility of these members being shared between sets. In relation to thermal comfort, set members will include principal environmental stimuli, such as dry bulb and mean radiant temperatures, relative humidity, and local airspeed, and personal parameters such as metabolic rate and clothing level. Thus, a 21 °C dry bulb temperature, for example, may have a member value of 0.6 when associated with the neutral category but 0.0, 0.3, 0.1, and 0.0 when associated with cold, cool, warm, and hot, i.e., a 21 °C stimulus can give rise to one of three perceived responses. Fuzzy rules are then applied to all environmental stimuli (ES) and personal factors (PF) to obtain an overall fuzzy prediction: for example: if <dry bulb temperature> is neutral; <mean radiant temperature> is neutral; <relative humidity> is high; <airspeed> is low; <metabolic rate> is high; and <clothing level> is low; then HOT. Because each ES or PF can simultaneously be a member of more than one class, all possible if-then rules must be evaluated to give the distribution of possible outcomes. Such controllers are suitable for the control of imprecise and uncertain parameters (such as comfort indices) and the incorporation of more than one control parameter (multi-criteria control).

Neural networks

These have been used for system identification purposes, especially pattern recognition (Banihashemi *et al.* 2022). Their structure corresponds to a simple replication of the human brain. The controller is divided into several layers, with the first one being the inputs and the last one being the outputs of the controller. A neural network is, therefore, a black box, giving no insight into the operation of the system being controlled. Whilst it is well suited to the control of non-linear systems, the method requires time-consuming tuning of the hidden layers. It is therefore rarely implemented in building energy management systems but might become more prevalent as AI systems proliferate.

Stochastic control

An early example of this control type is the pioneering algorithm by Hunt (1979), who developed a relationship between the lighting conditions in offices and the probability that occupants would switch on lights on arrival. Other examples are the model of occupant presence by Page *et al.* (2007) and the work of IEA Annex 66 project[25] to develop a toolkit for occupant behaviour simulation (Yan *et al.* 2017).

Simulation-assisted control

This approach can enact trade-offs between different sub-systems (e.g., between HVAC and lighting systems). Clarke *et al.* (2002), for example, established a prototype simulation-assisted controller, in which ESP-r was embedded in a real-time control system. Results from experiments demonstrated the feasibility of the approach. Other researchers have demonstrated that the use of embedded simulation to rank alternative control options could be a resilient approach to control despite moderate forecasting errors (Mahdavi *et al.* 2009, Baranski *et al.* 2018).

Occupant-responsive control

Most interactions between buildings and occupants are two-way processes. A poorly designed building may give rise to overheating, resulting in occupant responses that worsen the indoor condition. No matter how well the technical domains of the building are modelled, the effects of occupancy can vastly alter the physical behaviour and ultimately the predicted performance. These effects arise from two aspects: behaviour (e.g., the occupants' response to window opening) and attitude (e.g., the rejection of facilities on other than performance grounds). Two approaches to occupancy modelling are possible. Typical interactions may be included within a controller that has the authority to adapt the parameters of the affected domain models prior to the solution. More realistically, an explicit model of occupant behaviour may be introduced (as elaborated in Section 11.2) by which the physiological and psychological responses to stimuli are explicitly represented. In both cases, the aim is to address the distributed impacts of occupant actions (e.g., the impact of window opening on the network flow, CFD, thermal, and lighting domains).

Mahdavi and Pröglhöf (2009) reviewed developments in the field and concluded that control-related behavioural trends and patterns for groups of building occupants can be extracted from long-term observational data. To this end, Rijal *et al.* (2007) used results from field surveys to formulate an adaptive model of window opening by occupants in response to indoor/outdoor conditions in the

---

[25]https://annex66.org

context of offices. Fiala *et al.* (2003) developed an explicit model of occupant geometry and thermoregulatory response. Conejo-Fernández *et al.* (2021) added to this approach by developing a method to predict the effective radiation area of the human body for any posture, thus allowing the determination of the apparent mean radiant temperature.

Uncertainties abound in the real world, and it is important to allow for these at the design stage. There are essentially two ways to do this: by undertaking multiple simulations while perturbing model parameters to reflect expected variations; or by embedding a model of uncertainty within a program's algorithms. The former approach has been widely applied within program validation and applicability studies in the form of differential, factorial, and Monte Carlo methods (Lomas and Eppel 1992, Silva & Ghisi 2020) depending on whether the aim is to determine overall uncertainty, the contribution to uncertainty of individual parameters, or both. With this approach, the BPS+ program remains unaltered, but the computational burden rises because of the need to undertake multiple simulations. The attractiveness of the latter approach lies in its ability to identify individual and overall uncertainties based on a single simulation. Macdonald (2002), for example, replaced uncertain parameters within ESP-r's energy conservation equations for room air, surfaces, and intra-construction energy balances with affine representations (Figueiredo and Stolfi 2004) whereby single parameter values are replaced with a first-order polynomial comprising the mean value of the parameter and individual uncertainties represented by interval numbers (Neumaier 1990). The solution of the conservation equations based on affine operations then leads to an affine number representation of state variable uncertainties (as opposed to a single, definite value). In addition to computational efficiency, the approach facilitates simulation-time control based on uncertainty considerations: for example, a simulation might be terminated or the imposed control modified if the ability to maintain thermal comfort becomes ambiguous.

Some developments have focused on the technique of exergy analysis to allow the quality (usefulness) of energy to be assessed in addition to the quantity consumed: a likely important attribute of future BPS+ tools when applied to hybrid conventional and renewable energy schemes in an urban context. The objective is to attain a solution in which the lowest quality energy (i.e., low exergy) is harnessed, such as a low-temperature water source being used for space heating rather than electricity. The transformation of conventional (first law) building simulation theory to a (second law) exergy theory requires that the individual sources within the building's energy balance be referred to a reference state that defines an acceptable datum. Deciding on these reference states and interpreting the results from exergy simulations of real systems will likely prove challenging (Asada and Boelman 2003).

Despite advances in computational power, state-of-the-art simulation stubbornly defies real-time design application. The main reason for this situation is that domain integration (e.g., the embedding of CFD within the building thermal/lighting model) is computationally demanding, typically exceeding processing power capacity enhancement by several orders of magnitude. In an effort to address this problem, attempts have been made to reduce the computational demand by introducing equation solver refinements targeted on the sparse system of equations that characterise the building problem domain. ESP-r, for example, implements a matrix partitioning technique whereby the sparsity of the multi-zone, energy balance, matrix equation is eliminated and individual matrix partitions are processed at different frequencies depending on the time constant of the related building part. This results in a substantial reduction in computational requirements relative to sparse matrix processing techniques. With multi-core processors, it is also possible to reduce simulation times significantly by dividing long-term simulations into parts and then assigning each part to a different processor. The results for each part are then concatenated with suitable adjustments at the interfaces. Another approach is to apply a model reduction technique to construct a lower-order model via a projection on a state space of lower dimension (Van Dooren 1999). Berthomieu and Boyer (2003), for example, have applied a technique from control theory to a building thermal model. The approach balances so-called controllability and observability Gramian matrices (Moore 1981), which embody information on the input-output behaviour of the system and characterise system stability. In this way, they obtain a model that is smaller in size than the original. When applied to a typical dwelling, computing time was reduced by a factor of 3, whilst the error associated with the model reduction was constrained to less than $0.2\,°C$.

In the longer term, solver developments may be implemented to bring about computational efficiencies and thereby assist with the translation of simulation to the early design stage. For example:

- additional, context-aware solution accelerators may be embedded within the solvers to control their appropriate invocation;
- parallelism may be introduced to allow the different domains to be established and solved in tandem to reduce simulation times; and
- cloud computing could be exploited to allow different aspects of the same problem to be pursued at different locations as an aid to teamwork.

Such developments might well be built upon entirely new methods such as 'intelligent matrix patching' whereby the coupling information between domain models is stored in a 'patch matrix' allowing the numerical model of the coupling components to be activated only when the actual coupling takes place. Further, a greater level of coefficient management may be introduced to ensure that the matrix coefficients are only updated when required and otherwise never reprocessed. Such smart devices would lead to significant reductions in computing times.

The future is likely to be characterised by a significant utilisation of new and renewable energy technologies. When deployed at the local level, it will be beneficial to group building types and implement load control strategies so that the aggregate load profile can be adapted to the uncertain heat and power variations associated with renewable energy supplies. Modelling such schemes will require the development of demand management algorithms that can switch certain loads within the context of renewable power trading. Such a facility will primarily interact with the building heat and electrical domains by acting to reschedule heating/cooling system set-point temperatures and restricting power-consuming appliances where acceptable.

Another major requirement will be to extend current plant simulation capabilities by adding models for new and renewable energy technologies – heat pumps, combined heat and power, solar thermal/electric, fuel cells, wind/tidal turbines, district heating, etc. BPS+ has already proven itself amenable to the modelling of passive and active renewable energy systems. Grant and Kelly (2003), for example, developed a model of a ducted wind turbine, which may be deployed around roof edges to convert the local wind energy to electrical power. In one implementation of the technology (McElroy and Kane 1998), an air spoiler was incorporated to increase the air speed through the turbine. By covering the spoiler with photovoltaic cells, the device was able to convert solar energy into electrical power, thus improving the uniformity of the electrical output throughout the year. Other development examples relate to ground and air source heat pumps (Andresen *et al.* 2022) and Stirling engines for combined heat and power (Ferguson and Kelly 2006).

A core issue to be addressed derives from the fact that renewable energy component models can span several domains. For example, a hybrid photovoltaic model has thermal, airflow, electrical, and control constituents: the model is essentially the link point between the domains, interchanging key coupling variables as the simulation progresses. This issue is further exemplified by a dynamic fuel cell model (Carl *et al.* 2010) where control volumes were used to represent the fuel cell plates, gas channels, and balance of plant such as the gas desulphuriser and reformer. The equation set describing the fuel cell includes those for 3D heat conduction in the stack, gas dynamics, and electrochemical reactions, whilst the boundary conditions for these equations were supplied from other domains: environment temperature, relative humidity, electrical demand, etc.

Developments are also required in relation to the modelling of micro-generation systems that will often be subjected to control actions based on electrical as opposed to thermal criteria. Examples include voltage regulation, network stability, and phase balancing. This requires the creation of high-level controllers to iteratively couple the electrical and associated domains. Such controllers will need to be intelligent enough to balance the conflicting demands of local comfort and community benefit. This type of control will likely include some form of finance-based decision-making.

Result interpretation is the element that differentiates most between the requirements of the novice and the expert. The expert will be trying to detect patterns in, and relationships between, the different building and performance parameters in an attempt to isolate the dominant causal factors. To do this, all the data generated by the program have to be available. The novice, on the other hand, merely requires a concise summary of performance, preferably in terms of those parameters that are most meaningful to the design team and client. Abstracting program outputs into relevant performance summaries are difficult and may well require the development of smart analysis features. One approach to this issue is to develop a mechanism that ensures that the user's attention is focussed on the issue of greatest impact. This can be done through the technique of 'causal energy breakdown tracking' as elaborated in Section 4.6.

An accurate estimate of the cost of building retrofits, including the introduction of low-carbon energy systems, is characterised by many uncertainties involving interrelated factors that are difficult to assess. Most simulation models exclude cost considerations, thereby supporting design comparisons on performance grounds alone. A promising development would be to align BPS+ with rules-based expert systems that address the complex, cost estimation problem.

At some point in the future, BPS+ will reach a stage where performance prediction (i.e., user reward) follows problem description (i.e., user effort) with an imperceptible time delay. As the level of descriptive detail increases, the underlying simulation engine is able to provide progressive performance insight. For example, Figure 2.11 summarises a scenario in which descriptive detail is progressively added to a model to enable more focussed performance appraisals as a design evolves. Assuming the five columns in this figure are labelled 'a' through 'e' and the four rows '1' through '4', a rational performance assessment might proceed as follows (example taken from ESP-r).

a1 A Project Manager (Hand *et al.* 1998) gives access to support databases, model definition tools, a simulation engine, and performance assessment tools. Its role is to coordinate problem definition and pass data models between applications as a problem is incrementally evolved.

b1 Projects commence by making ready system databases. These typically include hygro-thermal, embodied energy, and optical properties for constructions, typical occupancy profiles, pressure coefficient sets, plant components, and weather collections representing different locations and severity. Embedded within such databases is the knowledge that supports design conceptualisation. For example, the construction elements database contains derived properties from which behaviour may be deduced (e.g., thermal diffusivity to characterise a construction's rate of response or thermal transmittance to characterise its rate of heat loss).

c1 It is common to commence problem definition with the specification of a building's geometry using an external CAD tool or in-built equivalent.

d1 Simple wire-line or false-coloured images can be generated as an aid to the communication of design intent or the study of solar/daylight access.

e1 Constructional and operational attribution is achieved by selecting products (e.g., wall constructions) and entities (e.g., occupancy profiles) from the support databases and associating these with the problem geometry. Embodied energy determination is then enabled.

a2 Temperature, wind, radiation, and luminance boundary conditions, of the required severity, are now associated with the model to enable an appraisal of 'no-plant' environmental performance (e.g., thermal and visual comfort levels).

b2 As required, geometrical, constructional, or operational changes can be applied to the model in order to determine the impact on performance. For example, alternative constructional systems might be investigated or different approaches to daylight utilisation assessed along with the extent and location of glare, as shown here in the case of an office with a light shelf.

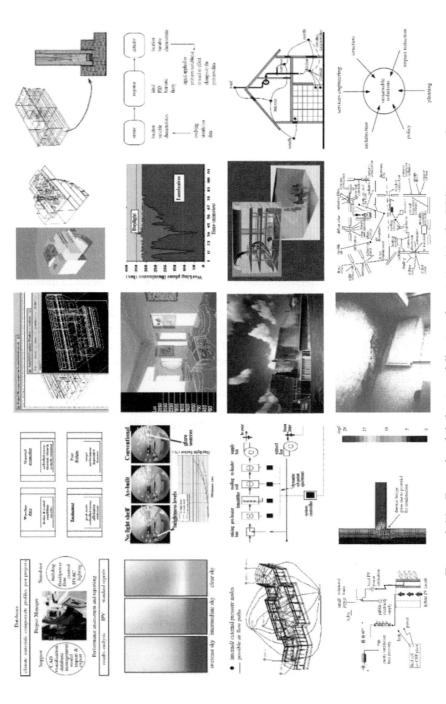

Figure 2.11   Example of 'behaviour following description' in the BPS+ approach

c2 Indoor visual comfort and daylight utilisation can be studied by investigating the impact of facade features and internal finishes on luminance distribution.

d2 To access the energy displacement potential of daylight, a luminaire control system might be introduced, comprising photocells linked to a dimming device. Simulations can then be undertaken to optimise the parameters of this control system in order to minimise the use of electricity for artificial lighting whilst maintaining visual comfort.

e2 The issue of integrated environmental control can now be explored by establishing a control system to dictate the availability of heating, cooling, ventilation, and lighting and so resolve conflict between these delivery systems.

a3 To study ventilation, a flow network can be associated with the building model so that dynamic interactions are explicitly represented. The control definition may then be extended to apply to the components of this network, e.g., to emulate window opening or fan operation.

b3 Where mechanical intervention is necessary, a component network can be defined to represent an HVAC system for association with both the building model and any active flow network.

c3 Special facade systems might now be considered: PV components to transform part of the solar power spectrum into electricity (and heat) or electro-, photo- or thermo-chromic glazing to control glare and/or illuminance distribution.

d3 To examine indoor air quality, one or more spaces within the model can be further discretised to enable the application of CFD in order to evaluate intra-space air movement and the distribution of temperature, humidity, and contaminants.

e3 Whilst the components of a model – the building, flow and HVAC networks, and CFD domain(s) – may be processed independently, it is usual to subject them to an integrated assessment whereby dynamic interactions are included. In the example shown here, a house model has been assigned a flow network to represent natural ventilation, an HVAC network to represent ventilation heat recovery, a CFD domain to enable an analysis of air quality, and a moisture flow model to allow an assessment of interstitial condensation.

a4 A further network might now be added to represent electrical power circuits for use in conjunction with previously established models for PV, µCHP, fuel cells, and the like to study the utilisation of outputs from building-integrated RE components, cooperative switching with the public electricity supply, and approaches to demand management.

b4 For specialist applications, the resolution of parts of the model can be enhanced. For example, a portion of a multi-layered wall might be finely discretised to enable the identification of possible thermal bridges, or a moisture flow network might be added to support an assessment of the potential for interstitial or surface condensation.

c4 By associating the time series pairs of near-surface temperature and relative humidity with growth limit data for different mould species, it is possible to determine the risk of mould growth.

d4 The message is that any problem, from a single space to an entire building with systems, distributed control, and enhanced resolutions, can be passed to the Simulator where its multi-variate performance is assessed and made available to inform design evolution.

e4 Integrated modelling supports teamwork because it provides a mechanism whereby the different professional viewpoints can come together and contribute to the eventual outcome.

There are, however, formidable barriers to the attainment of such functionality within a resource-constrained design process. Many modelling features are tool-specific, with scant commonality of approach across tools. Tool outputs invariably relate to different performance aspects or, worse still, different flavours of the same performance aspect, with interpretation confounded by the use of alternative criteria expressed in a variety of styles. Support databases are typically held in proprietary format and have a weak evidential basis in relation to data sources. Human-computer interfaces correspond to bespoke designs that are tailored to each tool's specific targets (as opposed to a general-purpose description of a building, for example). In addition, approaches to problem definition cover a wide spectrum, employ dissimilar syntax, and imply radically different semantics. The power of BPS+ would increase exponentially if such issues could galvanise developments that were pursued collaboratively. This will require that the industry, through its professional bodies, adopt a more proactive stance.

2.10 Chapter summary

There are many issues confronting the transition to clean energy systems, and these give rise to challenges for modelling and simulation tools. Examples include the matching of supply to demand at various scales, intelligent control within individual buildings and at the community level, ensuring acceptable trade-offs between disparate performance expectations, and avoiding unintended consequences.

Integrated performance simulation provides a means to address such issues. Effective application requires high resolution, feature-rich input models, standardised performance assessment procedures, and multi-variate views of performance. Support for these aspects is being incorporated increasingly in BPS+ programs and such developments would be accelerated if the industry were to take the lead in prescribing the required functionality and bringing forward standard application procedures.

References and further reading

Abbas G M, Dino I G and Percin M (2023) 'An integrated pipeline for building performance analysis: daylighting, energy, natural ventilation, and airborne contaminant dispersion', *Building Engineering*, 75.

Amissah P A (2005) 'Indoor air quality - combining air humidity with construction moisture', *PhD Thesis*, ESRU, University of Strathclyde.

Andresen I, Trulsrud T H, Finocchiaro L, *et al.* (2022) 'Design and performance predictions of plus energy neighbourhoods – case studies of demonstration projects in four different European climates', *Energy and Buildings*, 274, 112447.

Asada H and Boelman E (2003) 'Exergy analysis of a low temperature radiant heating system', *Proc. Building Simulation'03*, 1, pp. 3–17, Eindhoven.

ASHRAE (2004) Standard Method of Test for the Evaluation of Building Energy Analysis Computer Programs, ANSI/ASHRAE Standard 140-2004.

ASHRAE (2023) *Standard for the Design of High-Performance Green Buildings*, ANSI/ASHRAE *Standard 189.1-2020*, https://ashrae.org/technical-resources/standards-and-guidelines/read-only-versions-of-ashrae-standards/.

Augenbroe G (2008) 'Lessons from an advanced building simulation course', *Proc. SimBuild'08*, Berkeley.

Augenbroe G L M (Ed) (1995) Computer models for the building industry in Europe, Second Phase (COMBINE 2), *Final report for EC contract JOU2-CT92-0196*, Delft University of Technology, Faculty of Civil Engineering.

Augenbroe G, de Wilde P, Moon H J and Malkawi A (2004) 'An interoperability workbench for design analysis integration', *Energy and Buildings*, 36(8), pp. 737–748.

Ayoub M (2019) '100 Years of daylighting: a chronological review of daylight prediction and calculation methods', *Solar Energy*, 194, pp. 360–390.

Banihashemi F, Weber M and Lang W (2022) 'Model order reduction of building energy simulation models using a convolutional neural network autoencoder', *Building and Environment*, 207.

Baranski M, Fütterer J and Müller D (2018) 'Distributed exergy-based simulation-assisted control of HVAC supply chains', *Energy and Buildings*, 175, pp. 131–140.

Bartak M, Beausoleil-Morrison I, Clarke J A, *et al.* (2002) 'Integrating CFD and building simulation', *Building and Environment*, 37(8–9), pp. 865–871.

Bazjanec V (2002) 'Early lessons from development of IFC compatible software', *Proc. 4th European Conf. on Product and Process Modelling*, Portoroz, ISBN 90-5809-507-X.

BBRI (1990) *The Passys Test Cells: A Common European Outdoor Test Facility for Thermal and Solar Building Research*, Edited by the Belgium Building Research Institute, Publications Office of the European Union, Catalogue Number EU-NA-12882-EN-C.

Beausoleil-Morrison I (2000) 'The adaptive coupling of heat and air flow modelling within dynamic whole-building simulation', *PhD Thesis*, ESRU, University of Strathclyde, Glasgow.

Beausoleil-Morrison I (2001) 'An algorithm for calculating convective coefficients for internal building surfaces for the case of mixed flow in rooms', *Energy and Buildings*, 33(4).

Becker K and Parker J R (2009) 'A simulation primer', In Gibson D and Baek Y (eds.), *Digital Simulations for Improving Education: Learning through Artificial Teaching Environments*, Hershey, PA: IGI Global.

Berthomieu T and Boyer H (2003) 'Time-varying linear model approximation: application to thermal and airflow building simulation', *Proc. Building Simulation'03*, 1, pp. 101–106, Eindhoven.

BEST (2023) *Building Energy Software Tools directory*, https://buildingenergy-softwaretools.com/.

Blender BIM (2023) https://blenderbim.org/.

Bloomfield D, Bland B and Gough M (1988) 'An Investigation into analytical and empirical validation techniques for dynamic thermal models of buildings', *Final Report of an BRE/SERC Collaborative Research Project*, Building Research Establishment, Garston.

Carl M, Djilali N and Beausoleil-Morrison I (2010) 'Improved modelling of the fuel cell power module within a system-level model for solid-oxide fuel cell cogeneration systems', *Power Sources*, 195(8), pp. 2283–2290.

Casini M (2022) 'Chapter 5 - building performance simulation tools', *Civil and Structural Engineering*, pp. 221–262.

Charlesworth P, Clarke J A, Hammond G, *et al.* (1991) 'The energy Kernel system', *Proc. Building Simulation'91*, Nice, pp. 313–322.

Chen X, Abualdenien J, Singh M M, Borrmann A and Geyer P (2022) 'Introducing causal inference in the energy-efficient building design process', *Energy and Buildings*, 277.

Chong A, Augenbroe G and Yan D (2021) 'Occupancy data at different spatial resolutions: building energy performance and model calibration', *Applied Energy*, 286.

Choudhary R, Malkawi A and Papalambros P Y (2003) 'A hierarchical design optimisation framework for building performance analysis', *Building Simulation'03*, 1, pp. 179–186, Eindhoven.

Chow T T (1995) 'Atomic modelling in air-conditioning simulation', *PhD Thesis*, University of Strathclyde, Glasgow.

CIBSE (2015) *Applications Manual 11: Building performance modelling*, https://cibse.org/knowledge-research/knowledge-portal/applications-manual-11-building-performance-modelling-2015/.

Citherlet S (2001) 'Towards the holistic assessment of building performance based on an integrated simulation approach', *PhD Thesis*, Swiss Federal Institute of Technology.

Clarke J A (2001) *Energy Simulation in Building Design*, Butterworth-Heinemann.

Clarke J A (2013) 'Moisture flow modelling within the ESP-r integrated building performance simulation system', *Building Performance Simulation*, 6(5), pp. 385–399.

Clarke J A and Hensen J (2015) 'Integrated building performance simulation: progress, prospects and requirements', *Building and Environment*, 91, pp. 294–306.

Clarke J A and Mac Randal D F (1991) 'An intelligent front-end for computer-aided building design', *Artificial Intelligence in Engineering*, Computational Mechanics Publications, 6(1), pp. 36–45.

Clarke J A, Cockroft J, Conner S, *et al.* (2002) 'Simulation-assisted control in building energy management systems', *Energy and Buildings*, 34(9), pp. 933–40.

Clarke J A, Hand J W, Kelly N, *et al.* (2012) 'A data model for integrated building performance simulation', *Proc. Building Simulation and Optimization Conference*, Loughborough, UK, pp. 340–347.

Clarke J A, Hand J, Mac Randal D F and Strachan P A (1998) 'Design tool integration within the COMBINE Project', *Proc. 2nd European Conf. on Product and Process Modelling in the Building Industry*, BRE, Garston, 19–21 Oct.

Clarke J A, Strachan P A and Pernot C (1993) 'An approach to the calibration of building energy simulation models', *ASHRAE Transactions*, 99(2), pp. 917–927.

Clay K, Severnini E and Sun X (2023) 'Does LEED certification save energy? Evidence from retrofitted federal buildings', *Journal of Environmental Economics and Management*, 121.

Cockroft J P (1979) 'Heat transfer and air flow in buildings', *PhD Thesis*, University of Glasgow.

Conejo-Fernández J, Cappelletti F and Gasparella A (2021) 'Including the effect of solar radiation in dynamic indoor thermal comfort indices', *Renewable Energy*, 165, pp. 151–161.

Cóstola D, Blocken B and Hensen J L M (2009) 'External coupling between BES and HAM programs for whole-building simulation', *Building Simulation '09*, Glasgow.

Crawley D B and Hand J (2008) 'Contrasting the capabilities of building energy performance simulation programs', *Building and Environment*, 43, pp. 661–673.

de Figueiredo L H and Stolfi J (2004) 'Affine arithmetic: concepts and applications', *Numerical Algorithms*, 37(1–4), pp. 147–158.

Djunaedy E, Hensen J L M and Loomans M G L C (2003) 'Development of a guideline for selecting a simulation tool for airflow prediction', *Building Simulation'03*, 1, pp. 267–274, Eindhoven.

Doulos L T, Kontadakis A, Madias E N, Sinou M and Tsangrassoulis A (2019) 'Minimizing energy consumption for artificial lighting in a typical classroom of a Hellenic public school aiming for near Zero Energy Building using LED DC luminaires and daylight harvesting systems', *Energy and Buildings*, 194, pp. 201–217.

Eastman C, Teicholz P, Sacks R and Liston K (2008) *BIM Handbook: A Guide to Building Information Modeling for Owners, Managers, Engineers and Contractors*, John Wiley & Sons Ltd., New York.

EEED (2023) *European Energy Efficiency Directive*, https://energy.ec.europa.eu/topics/energy-efficiency/energy-efficiency-targets-directive-and-rules/energy-efficiency-directive_en/.

Erhorn H, de Boer J and Dirksmoller M (1998) 'ADELINE - An integrated approach to lighting simulation', *Proc. Daylighting'98*, pp. 21–28, Ottawa.

Faulkner C A, Lutes R, Huang S, Zuo W and Vrabie D (2023) 'Simulation-based assessment of ASHRAE Guideline 36, considering energy performance, indoor air quality, and control stability', *Building and Environment*, 240, 110371.

Ferguson A and Kelly N (2006) 'Modelling building-integrated Stirling CHP systems', *Proc. eSim'06*, pp. 94–101, Toronto.

Ferreira A, Pinheiro M D, de Brito J and Mateus (2023) 'A critical analysis of LEED, BREEAM and DGNB as sustainability assessment methods for retail buildings', *Building Engineering*, 66.

Fiala D, Lomas K J and Stoher M (2003) 'First principles modelling of thermal sensation responses in steady state and transient conditions', *ASHRAE Trans.*, 109, pp. 179–186.

Fornari W, Grozman G, Wikström M and Sahlin P (2023) 'Development of a BLOCK zonal model in a BPS software to predict thermal stratification and air flows', *Energy and Buildings*, 294.

Gagneau S and Allard F (2001) 'About the construction of autonomous zonal model', *Energy and Building*, 33, pp. 245–250.

gbXML (2017) https://gbxml.org/schema_doc/6.01/GreenBuildingXML_Ver6.01.html/.

gbXML (2023) https://gbxml.org/schema_doc/7.02/GreenBuildingXML_Ver7.02.html/.

Geebelen B and Neuckermans H (2003) 'Optimizing daylight simulation for speed and accuracy', *Proc. Building Simulation'03*, 1, pp. 379–386, Eindhoven.

Gicquel R (1988) 'The Project PASSYS', In Steemers T C (ed.), *Solar Energy Applications to Buildings and Solar Radiation Data. Solar Energy Development — Third Programme, 4*, Springer, Dordrecht, ISBN 978-94-010-7831-3.

Göçer O and Dervishi S (2015) 'The use of building performance simulation tools in undergraduate program course training', *Proc. Building Simulation '15*, Hyderabad.

Goncalves J E, van Hooff T and Saelens (2020) 'A physics-based high-resolution BIPV model for building performance simulations', *Solar Energy*, 204, pp. 585–599.

Grant A and Kelly N J (2003) 'The development of a ducted wind turbine simulation model', *Proc. Building Simulation'03*, 1, pp. 407–414, Eindhoven.

Guernouti S, Musy M and Hegron G (2003) 'Automatic generation of partitioning and modelling adapted to zonal method', *Proc. Building Simulation'03*, 1, pp. 427–434, Eindhoven.

Hand J W and Strachan P A (1998) 'ESP-r data model decomposition', Project Report TR 98/9, ESRU, University of Strathclyde.

Hand J W, Irving S J, Lomas K J, *et al.* (1998) 'Building energy and environmental modelling', *Applications Manual AM11*, CIBSE, London.

Hensen J L M (1991) 'On the thermal interaction of building structure and heating and ventilating systems', *PhD Thesis*, Eindhoven University of Technology.

Hensen J L M, Janak M, Kaloyanov N G and Rutten P G S (1998) '*Introducing building energy simulation classes on the Web'*, *ASHRAE Transactions*, 104(1).

Herkel S and Pasquay T (1997) 'Dynamic link of light and thermal simulation: on the way to integrated planning tools', *Proc. Building Simulation '97*, Prague.

Humphreys M A and Nicol F (1998) 'Understanding the adaptive approach to thermal comfort', *ASHRAE Transactions*, 104(1), pp. 991–1004.

Hunt D (1979) 'The use of artificial lighting in relation to daylight levels and occupancy', *Building and Environment*, 14, pp. 21–33.

IEA4 (1980) 'IEA Annex 4: computer modelling of building energy performance: results and analysis of Avonbank building simulation', Oscar Faber, St. Albans.

IFC (2012) https://buildingsmart-tech.org/specifications/ifc-releases/ifc2x4-release/rc3-release/rc3-release-summary/.

Ihm P, Krarti M and Henze G (2003) 'Integration of a thermal energy storage model within EnergyPlus', *Proc. Building Simulation'03*, 2, pp. 531–538, Eindhoven.

Inard C, Bouia H and Dalicieux P (1996) 'Prediction of air temperature distribution in buildings with zonal model', *Energy and Buildings*, 24, pp. 125–132.

Irving S (1982) 'Energy program validation: conclusions of IEA Annex 1', *Computer-Aided Design*, 14(1), pp. 33–38.

Jani H K, Kantipudi M V, Nagababu G, Prajapati D and Kachhwaha S S (2022) 'Simultaneity of wind and solar energy: a spatio-temporal analysis to delineate the plausible regions to harness', *Sustainable Energy Technologies and Assessments*, 53C.

Janssens A (2014) 'IEA Annex 58 Report: State of the Art of Full Scale Test Facilities for Evaluation of Building Energy Performances', https://kuleuven.be/bwf/projects/annex58/index.htm/.

Jensen S (1995) 'Validation of building energy simulation programs: a methodology', *Energy and Buildings*, 22(2), pp. 133–144.

Jiang L, Shi J and Wang C (2022) 'Multi-ontology fusion and rule development to facilitate automated code compliance checking using BIM and rule-based reasoning', *Advanced Engineering Informatics*, 51.

Judkoff R and Neymark J (2006) 'Model Validation and Testing: The Methodological Foundation of ASHRAE Standard 140', *Proc. ASHRAE 2006 Annual Meeting*, Quebec City, Canada, June 24–29.

Judkoff R, Wortman D, O'Doherty B and Burch J (2008) 'A methodology for validating building energy analysis simulations', *Technical Report NREL/TP-550-42059*, National Renewable Energy Laboratory, Boulder.

Kelly N J (1998) 'Towards a design environment for building-integrated energy systems: the integration of electrical power flow modelling with building simulation', *PhD Thesis*, ESRU, University of Strathclyde.

Kelly N J and Macdonald I (2004) 'Coupling CFD and visualisation to model the behaviour and effect on visibility of small particles in air', *Proc. eSim*, pp. 153–160, Vancouver, 9–11 June.

Kersken M, Heusler I, Strachan P and Sinnesbichler H (2015) 'Introduction of a new validation scenario for building energy simulation tools based on measurement data', *Bauphysik*, 37(3), pp. 153–158.

Kersken M, Strachan P, Mantesi E and Flett G (2020) 'Whole building validation for simulation programs including synthetic users and heating systems: experimental design', *E3S Web of Conferences*, 172, 22003.

Kersting W H (2017) 'Distribution System Modeling and Analysis', *CRS Press*.

Klemm K, Marks W and Klemm A J (2000) 'Multi-criteria optimization of the building arrangement with application of numerical simulation', *Building and Environment*, 35, pp. 537–544.

Liu S, Kwok Y T, Lau K K, Ouyang W and Ng E (2020) 'Effectiveness of passive design strategies in responding to future climate change for residential buildings in hot and humid Hong Kong', *Energy and Buildings*, 228.

Liu X and Wu Y (2022) 'Numerical evaluation of an optically switchable photo-voltaic glazing system for passive daylighting control and energy-efficient building design', *Building and Environment*, 219.

Lomas K J and Eppel H (1992) 'Sensitivity analysis techniques for building thermal simulation programs', *Energy and Buildings*, 19, pp. 21–44.

Lomas K J, Eppel H, Martin C and Bloomfield D (1994) 'Empirical validation of thermal building simulation programs using test room *data', Final Report for IEA Annex 21/Task 12*, 1–3.

Macdonald I A (2002) 'Quantifying the effects of uncertainty in building simula-tion', *PhD Thesis*, Energy Systems Research Unit, University of Strathclyde.

Macdonald I A, McElroy L B, Hand J W and Clarke J A (2005) 'Transferring simulation from specialists into design practice', *Proc. Building Simula-tion'05*, Montreal, 15–18 August.

Mahdavi A (2003) 'Computational building models: theme and four variations', *Proc. Building Simulation'03*, 1, pp. 3–17, Eindhoven.

Mahdavi A and Pröglhöf C (2009) 'Toward empirically-based models of people's presence and actions in buildings', *Proc. Building Simulation'09*, Glasgow.

Mahdavi A and Tahmasebi F (2015) 'Predicting people's presence in buildings: an empirically based model performance analysis', *Energy and Buildings*, 86, pp. 349–355.

Mahdavi A, Orehounig K and Pröglhöf C (2009) 'A simulation-supported control scheme for natural ventilation in buildings', *Proc. Building Simulation '09*, Glasgow.

Mantesi E, Hopfe C J, Mourkos K, Glass J and Cook M (2019) 'Empirical and computational evidence for thermal mass assessment: the example of insu-lating concrete formwork', *Energy and Buildings*, 188–189, pp. 314–332.

Mascle C (2013) 'Design for rebirth (DFRb) and data structure', *Production Eco-nomics*, 142(2), pp. 235–246.

Maver T W (1985) 'CAAD: a mechanism for participation', *Proc. Design Coalition Team Conf.* (Ed: Beheshti M), vol. 2, pp. 89–105, Eindhoven, https://papers.cumincad.org/cgi-bin/works/Show?0c3c/.

Mazuroski W, Berger J, Oliveira R C and Mendes N (2018) 'An artificial intelligence-based method to efficiently bring CFD to building simulation', *Building Performance Simulation*, 11(5), pp. 588–603.

Mazzeo D, Matera N, Cornaro C, Oliveti G, Romagnoni P, and De Santoli L (2020) 'EnergyPlus, IDA ICE and TRNSYS predictive simulation accuracy for building thermal behaviour evaluation by using an experimental campaign in solar test boxes with and without a PCM module', *Energy and Buildings*, 212.

McElroy L B and Kane B (1998) 'An integrated renewable project for Glasgow - City of Architecture and Design 1999', *Proc. 5th European Conf. Solar Energy in Architecture and Urban Planning*, Bonn.

Mediavilla A, Elguezabal P and Lasarte N (2023) 'Graph-based methodology for multi-scale generation of energy analysis models from IFC', *Energy and Buildings*, 282, 112795.

Monari F and Strachan P (2016) 'Characterization of an airflow network model by sensitivity analysis: parameter screening, fixing, prioritizing and mapping', *Building Performance Simulation*, 10(1), pp. 17–36.

Moore B C (1981) 'Principal component analysis in linear systems: controllability, observability, and model reduction', *IEEE Trans. Autom. Control*, 26(1), pp. 17–32.

Moore G and Numan M Y (1982) 'Form factors: the problem of partial obstruction', *Martin Centre Report, University of Cambridge*.

Nakhi A E (1995) 'Adaptive construction modelling within whole building dynamic simulation', *PhD Thesis*, ESRU, University of Strathclyde.

Negrao C O R (1995) 'Conflation of computational fluid dynamics and building thermal simulation', *PhD Thesis*, ESRU, University of Strathclyde, Glasgow.

Neumaier A (1990) *Interval Methods for Systems of Equations (No. 37)*, Cambridge University Press.

Nijssen G M and Halpin T A (1989) *Conceptual Schema and Relational Database Design: A Fact Oriented Approach*, Prentice-Hall, Australia, ISBN 978-0131672635.

NORMA (2023) 'Natural object-role modeling architect', https://github.com/orm-solutions/NORMA/.

Ohlsson K E and Olofsson T (2021) 'Benchmarking the practice of validation and uncertainty analysis of building energy models', *Renewable and Sustainable Energy Reviews*, 142.

Østergård T, Jensen R L and Maagaard S E (2016) 'Building simulations supporting decision making in early design – a review', *Renewable and Sustainable Energy Reviews*, 61, pp. 187–201.

Østergård T, Jensen R L and Mikkelsen F S (2020) 'The best way to perform building simulations? One-at-a-time optimization vs. Monte Carlo sampling', *Energy and Buildings*, 208.

Page J, Robinson D, Morel N and Scartezzini J-L (2007) 'A generalised stochastic model for the simulation of occupant presence', *Energy and Buildings*, 40, pp. 83–98.

Palomo E and Tellez F M (1991) 'PAMTIS – version 1.0 – Package to analyse multivariate series', *Report 194-91-PASSYS-MVD-WD-216, IER*, Spain.

Queiroz N, Westphal F S and Pereira F O R (2020) 'A performance-based design validation study on EnergyPlus for daylighting analysis', *Building and Environment*, 183.

Reinhart C and Herkel S (2000) 'The simulation of annual daylight illuminance distributions – a comparison of six radiance-based methods', *Energy and Buildings*, 32, pp. 167–187.

Rijal H B, Tuohy P, Humphreys M A, Nicol J F, Samuel A and Clarke J A (2007) 'Using results from field surveys to predict the effect of open windows on thermal comfort and energy use in buildings', *Energy and Buildings*, 39(7), pp. 823–836.

Sahlin P (1992) *The Neutral Model Format for Building Simulation, motivation and syntax*, Swedish Institute of Applied Mathematics, Stockholm.

Santos L, Schleicher S and Caldas L (2017) 'Automation of CAD models to BEM models for performance based goal-oriented design methods', *Building and Environment*, 112, pp. 144–158.

Schweiger G, Heimrath R, Falay B, *et al.* (2018) 'District energy systems: modelling paradigms and general-purpose tools', *Energy*, 164, pp. 1326–1340.

Silva A S and Ghisi E (2020) 'Estimating the sensitivity of design variables in the thermal and energy performance of buildings through a systematic procedure', *Cleaner Production*, 244, 118753.

Siu C Y, O'Brien W, Touchie M, *et al.* (2023) 'Evaluating thermal resilience of building designs using building performance simulation – a review of existing practices', *Building and Environment*, 234.

Sowell E F and Moshier M A (1995) 'HVAC component model libraries for equation-based solvers', *Proc. Building Simulation '95*, Madison, Wisconsin.

Spiby P (Ed) (1991) *EXPRESS Language Reference Manual*, ISO TC184/SC4/WG5, Document N14.

Strachan P and Baker P (2008) 'Outdoor testing, analysis and modelling of building components', *Building and Environment*, 43(2), pp. 127–128.

Strachan P and Guy A (1991) 'Modelling as an aid in the thermal performance assessment of passive solar components', *Building Environmental Performance '91*, Canterbury, April, pp. 78–88.

Strachan P, Svehla K, Heusler I and Kersken M (2015) 'Whole model empirical validation on a full-scale building', *Building Performance Simulation*, 9(4), pp. 331–350.

Syed E U and Manzoor K M (2022) 'Analysis and design of buildings using Revit and ETABS software', *Materials Today: Proceedings*, 65(2), pp. 1478–1485.

Tang D (1985) 'Modelling of heating and air-conditioning systems', *PhD Thesis*, ESRU, University of Strathclyde.

Tregenza P and Waters I (1983) 'Daylight coefficients', *Lighting Research & Technology*, 15(2), pp. 65–71.

Uddin M N, Wang Q, Wei H H, Chi H L and Ni M (2021) 'Building information modeling (BIM), System dynamics (SD), and Agent-based modeling (ABM): Towards an integrated approach', *Ain Shams Engineering Journal*, 12(4), pp. 4261–4274.

Van Dooren P (1999) 'Gramian based model reduction of large-scale dynamical systems', In Griths and Watson (eds.), *Numerical Analysis*.

Vazifeh E, Schuß M and Mahdavi A (2015) 'Radiometric boundary condition models for building performance simulation: an empirical assessment', *Energy Procedia*, 78, pp. 1775–1780.

Vogel T (1991) 'Configurable graphical editor: users guide', *Report 91-ITI-382*, TNO Institute for Applied Computer Science, Delft, the Netherlands.

Walton G N (1989) 'Airflow network models for element-based building airflow modelling', *ASHRAE Trans.*, 95(2), pp. 613–620.

Yan D, Hong T, Dong C, *et al.* (2017) 'IEA EBC Annex 66: definition and simulation of occupant behavior in buildings', *Energy and Buildings*, 156, pp. 258–270.

Chapter 3

Performance assessment requirements

Whilst it has long been recognised that building performance is a multi-variate problem domain (Markus *et al.* 1972), it is only with the advent of performance simulation that the issue can be adequately addressed at the design stage.

This chapter describes some of the assessments and ratings that are used to evidence acceptable performance and demonstrate compliance with building regulations. In some cases, equations are given to enable BPS+ outputs to be mapped to indices that are not included in a specific program or to support a comparison with what might have been calculated otherwise.

An issue with current BPS+ tools is the pluralism of output in terms of style and substance. The suggestion here is not that programs should be mandated to deliver prescribed outputs but that they should be able, on request, to deliver outputs that conform to some industry standard. In this way, the outputs from different programs would be both comprehensible and comparable. Devising standard output templates will require agreement on performance metrics and assessment criteria.

All building designs aim, among many aspirations, to minimise energy consumption and optimise indoor conditions. This means that BPS+ must address multiple performance domains, most notably:

- thermal comfort, including the effects of global and local parameters;
- visual comfort in terms of adequate illumination and the avoidance of discomfort and disability glare;
- indoor air quality as associated with the ventilation system and influenced by thermal comfort and energy efficiency measures;
- energy performance, both operational and embedded; and
- environmental impact, both local and global.

At the same time, buildings and their contexts are unique (each design is its own prototype), implying that BPS+ applications are context-sensitive and must be adaptable to specific system configurations. Further, outcomes will depend on the spatial and temporal variation of multiple physical variables that must be somehow combined and presented to different stakeholders (client, architect, engineer, occupant, researcher, and student) in a semantically acceptable manner that supports understanding and translation to action.

3.1   Performance parameters and assessment criteria

The assessment of performance in practice is compounded by the existence of alternative definitions and adherence to different targets and benchmarks (see de Wilde (2018) for a comprehensive discussion on this topic). Control strategies normally target indoor comfort parameters to regulate room conditions (internal temperature, humidity, etc.), and external weather parameters to regulate central heating and air conditioning plant. Internal control parameters are measured by sensors located throughout the building, whereas external parameters may be measured in the immediate vicinity or obtained from outside sources, such as a nearby weather station. It is also sometimes necessary to allow direct regulation of appliances by users in order to accommodate divergent occupant behaviour and levels of technical understanding.

The following subsections cover some of the important parameters that can be used to quantify the performance of a building. These parameters can usually be extracted from BPS+ simulations and, in many cases, are already mapped to higher-level performance indices, e.g., PPD or JPPD in the case of thermal and visual comfort, respectively. In other cases, BPS+ output will need to be converted to a prescribed form.

3.1.1   Thermal comfort

A combination of environmental and personal parameters influences the perceived thermal comfort in the indoor environment (Fanger 1972, van Treeck and Wölki 2019). Air temperature is the principal parameter to control thermal comfort; indeed, this is the parameter used most as a thermal comfort index within control systems (where a room sensor is aspirated, for example). However, control of air temperature alone does not ensure thermal comfort. The theory implies that four physical and two personal parameters are needed to describe a person's interaction with their environment:

- air temperature (T_a);
- radiant temperature (T_r);
- air velocity (V_a);
- water vapour pressure (P);
- activity level $(-)$; and
- clothing level $(-)$.

To complicate matters, thermal comfort (like visual comfort) is position-specific and dependent on attributes that may not be part of a routine performance assessment. The effort expended to ensure optimal conditions in an operating theatre is rarely applied in the case of an office environment, even though BPS+ can deliver a comfort assessment for individual occupants.

Conventional comfort theory evaluates the heat exchanges that occur between a person and their thermal environment, as well as the physiological conditions that are needed for human comfort. From this analytical approach, rationally based indices of thermal comfort have emerged, such as Predicted Mean Vote (PMV; Fanger 1972) and Standard Effective Temperature (SET; Gagge *et al.* 1986) based on a two-node model to represent the thermal regulation process of the human body. In addition, field studies have led to statistically based thermal comfort indices. Significant differences therefore exist between indices, and such discrepancies have led to the view that thermal comfort is part of a self-regulating system (Humphreys and Nicol 1998, Brager and de Dear 1998), an approach generally known as adaptive comfort.

Simulation yields temporal profiles of air and radiant temperature, humidity, and air movement (draught) at various locations or spatially averaged, which can be combined to give a PMV value. It is important to note that PMV outputs should only be requested for periods where indoor conditions have stabilised in order to comply with the constraints of the underlying theory. Of course, it is possible to scan simulation results and then request PMV values for those periods where the internal conditions are stable.

PMV is the mean vote obtained by averaging the thermal sensation votes of a large group of people in a given environment. It is a function of the imbalance in the heat equation of the human body under comfort conditions, L (W/m^2), and of the metabolic rate, M (W/m^2), both being related to the surface area of the human body:

$$PMV = [0.303 \ e^{-0.036M} + 0.0275] \ L$$

$$L = (M - V) - 3.05^{-3} \ [5733 - 6.99(M - W) - P_v]$$

$$-0.42[(M - W) - 58.15] - 1.7^{-5}M(5867 - P_v) - 0.0014M(34 - T_a)$$

$$-3.96^{-8}f_{cl}[(T_{cl} + 273)^4 - (T_r + 273)^4] - f_{cl}h_c(T_{cl} - T_a)$$

$$T_{cl} = 35.7 - 0.028(M - W)$$

$$-0.155 \ I_{cl}\{3.96^{-8}f_{cl}[(T_{cl} + 273)^4 - (T_r + 273)^4] + f_{cl}h_c)(T_{cl} - T_a)\}$$

$$h_c = \max[2.38(T_{cl} - T_a)^{0.25}, 12.1(V_{ar})^{0.5}]$$

$$f_{cl} = 1.0 + 1.29I_{cl} \text{ for } I_{cl} \leq 0.078$$
$$= 1.05 + 0.645 \text{ for } I_{cl} > 0.078$$

where,

W is the external work (W/m^2) for the activities (metabolic rate),
I_{cl} clothing insulation (m^2.K/W),
f_{cl} ratio between the clothed body area and the body area,
T_a air temperature (°C),
T_r mean radiant temperature (°C),
T_{cl} clothing temperature (°C),
V_{ar} relative air velocity near the external clothing surface (m/s),
P_v partial vapour pressure in the air (Pa), and
h_c convective heat transfer coefficient.

The most commonly used index to express the correspondence between the thermal sensation and the percentage of people expressing discomfort is Predicted Percentage Dissatisfied (PPD), which is directly related to PMV:

$$PPD = 100 - 95 \, e^{(-0.03353 \; PMV^4 + 0.2179 \; PMV^2)}$$

ISO Standard 7730[1] gives specifications of the conditions for thermal comfort in moderate thermal environments based on the PMV/PPD model. The imbalance of a steady-state, one-node energy balance is related to the PMV index, which gives the average response of a population according to the ASHRAE thermal sensation scale, which varies from -3 (cold) to $+3$ (hot), 0 being neutral. Environmental and personal variables are measured or quantified in order to assess, through the PMV and PPD values, the thermal comfort conditions for the environment. To assess these conditions, environmental variables have to be maintained within a range defined by the standard. This range corresponds to the interval -0.5 to 0.5 for offices in which a relatively high standard of comfort is required, although in many cases a wider range is acceptable.

In the real world, there is a low possibility of basing a control strategy on PMV/PPD because of the many parameters involved, and so control is usually achieved using a few environmental parameters, mainly temperature. The efficiency of a control strategy for thermal comfort is closely related to the measurement accuracy of the environmental variables, and this accuracy can only be quantified for specific cases. For example, in some room configurations, the globe temperature gives a reasonable approximation of the mean radiant temperature of the body. If a complex comfort index is selected as a control parameter, then indoor temperature, humidity, and air velocity must be measured. When not available, a constant value of 0.1 m/s is often assumed for the air velocity, and the mean radiant temperature is set equal to the air temperature. Typical and constant values are often assumed for the personal parameters.

Returning to the concept of adaptive comfort. If conditions are allowed to vary and subjects dress and behave as normal, then the measured physical variables can be related to the subject's feeling of warmth, which is termed a 'Comfort Vote'. It is then possible to form a comfort index, *TPV* (Thermal Preference Vote), by using globe temperature, T_g, water vapour pressure, P_v, and relative air velocity, V_{ar}, to establish a comfort vote regression equation:

$$TPV = 0.186 \, T_g - 0.032 \, P_v - 0.366 \, V_{ar}{}^{1/2} - 0.820.$$

Comfort standards based on adaptive assumptions are more than just a temperature to target. The comfort temperature is defined as the temperature at which there is the least probability of discomfort. Its value varies with the climate and season. TPV is therefore a parameter that is suited to building performance simulations.

---

[1]https://iso.org/standard/39155.html

The results obtained from comfort surveys performed during the Smart Controls and Thermal Comfort (SCATS) project (Wagner 2007) were used to develop an algorithm for comfort temperature in terms of outdoor temperature. Researchers from five countries (France, Greece, Portugal, Sweden, and the UK) conducted surveys at two levels, differentiated by the detail of the environmental monitoring. The study was based on the assumption that the comfort temperature changes with time in a way that is related to the outdoor temperature.

To decide on the appropriate measure of outdoor temperature, the rate at which the comfort temperature, T_c, changes must be characterised. A common measure of outdoor temperature used as a predictor for indoor comfort temperature is the exponentially weighted running mean of the daily mean outdoor temperature, which is straightforward to track in a simulation:

$$T_{rm} = (1 - a) \left[T_{od-1} + a T_{od-2} + a^2 T_{od-3} ... \right]$$

where,

T_{rm} is the running mean temperature,
T_{od} the daily mean outdoor temperature (when T_{rm} is the running mean temperature for a particular day),
T_{od-1} the daily mean outdoor temperature for the previous day, and
'a' a constant between 0 and 1 that defines the speed at which the running mean responds to the outdoor temperature and the characteristic time period for the relationship.

This equation reduces to $T_{rm}^{n} = (1 - a)\, T_{od}^{n-1} + a\, T_{rm}^{n-1}$ where T_{rm}^{n} is the running mean temperature for day 'n'.

When the running mean has been calculated for one day, it can be calculated for the next day and all following days from this value and the daily mean outdoor temperature. The best value for constant 'a' was determined as 0.80. T_{rm80} is then the running mean temperature, with $a = 0.80$. Taking account of the above considerations, adaptive algorithms to estimate T_c have been suggested for both monitoring levels, as given in Table 3.1.

The two surveys gave rise to different recommended temperatures. The Level I results are considered more applicable since the measurements were more accurate.

Table 3.1 National T_c calculation equations ($T = T_{r80}$)

	Level I		Level II	
	$T_{r80} < 10$	$T_{r80} > 10$	$T_{r80} < 10$	$T_{r80} > 10$
France	0.049T+22.58	0.206T+21.42	0.041T+21.59	0.188T+20.10
Greece	Not applicable	0.205T+21.69	Not applicable	0.244T+18.99
Portugal	0.381T+18.12		0.452T+16.37	
Sweden	0.051T+22.83		0.061T+23.03	0.084T+22.24
UK	0.104T+22.58	0.168T+21.63	0.047T+21.10	0.188T+19.55
All	22.88	0.302T+19.39	21.61	0.267T+18.88

It should be noted that people in different European countries could be expected to have a different concept of thermal comfort, with people located at Northern latitudes feeling comfortable at lower temperatures. Note also that the above correlations are based on values of T generally lower than 30 °C. These algorithms can be used to define comfortable indoor conditions, control air conditioning systems, or assess whether naturally ventilated buildings will provide acceptable indoor temperatures.

3.1.2   Visual comfort

The role of lighting is to produce an adequate visual environment that ensures visual comfort and serves the intended visual tasks. The following characteristics are generally specified to define an acceptable visual environment:

- average illuminance;
- uniformity;
- ratios of luminance;
- allowable glare level;
- light direction and effect of shadows;
- colour temperature; and
- colour rendering.

The most important issue in assessing visual comfort and illuminance quality is glare and different indexes have been proposed:

- Bodmann-Sollner – for different activities, a minimum viewing angle is defined for light sources in order to avoid glare (Bodmann and Söllner 1965);
- Unified Glare Rating (UGR; ICI 2019) – this is a measure of discomfort glare and is calculated using visual parameters in the field of view, with discomfort evaluated in terms of the position on a scale of discomfort:

$$ UGR = 8 \ \log_{10} \left[\sum \frac{0.25}{L_b} \ \frac{L_s^2 \omega}{p^2} \right] $$

where L_b is the background luminance, L_s is the luminance of each source, ω the solid angle, and p the Guth position index, which relates to the displacement of each luminaire from the line of sight (Guth 1966); and

- Visual Comfort Probability (VCP; Guth 1963) – this represents the percentage of people who probably will not complain about glare:

$$ VCP = \frac{100}{\sqrt{2\pi}} \int_0^{6.374-1.3227\ln(UGR)} \exp\left(-\frac{t^2}{2} \right) dt. $$

Professional associations such as the Illuminating Engineering Society[2] and the Association Francais d'Eclairage[3] have provided recommended illuminance

---

[2]https://ies.org
[3]https://afe-eclairage.fr

values for lighting design. Tables 3.2–3.4 give categories and values corresponding to different activities.

Lighting should not be excluded as a contributor to energy conservation. Lighting controls are used to manage energy use more effectively and efficiently inside buildings. They can be used to:

Table 3.2   Illuminance values for various activities

Activity	Illuminance category	Range of illuminances (lux)
Public spaces with dark surroundings.	A	20–50
Simple orientation for a short temporary visit.	B	50–100
Working spaces where visual tasks are only occasionally performed.	C	100–200
Visual tasks of high contrast or large size.	D	200–500
Visual tasks of medium contrast or small size.	E	500–100
Visual tasks of low contrast of very small size.	F	1000–2000
Visual tasks of low contrast and very small size over a prolonged period.	G	2000–5000
Very prolonged and exacting visual tasks.	H	5000–10,000
Very special visual tasks or extremely low contrast and small size.	I	10,000–20,000

Table 3.3   Lighting levels for workspace types

Task group and typical interior	Illuminance (lux)
Minimum for a visual task.	200
Rough work, assembly, writing, reading.	300
Routine work, offices, control room.	500
Drawing office.	750
Fine work, machining, and inspection.	1000
Very fine work.	1500
Very demanding task, industry/laboratory.	2000

Table 3.4   Illuminance categories

Indoor activity	Category
Residences	
General lighting	B
Specific visual tasks	
Dining	C
Ironing	D
Laundry	D
Reading	
book	D
poor copy	E

- reduce lighting during unoccupied periods;
- switch off or time-set the lights in daylight areas;
- compensate for lumen depreciation;
- adjust light levels according to local tasks; and
- adjust light levels to suit visual adaptation.

There are several control types:

- dimmers;
- on/off switching;
- differential switching control;
- photoelectric switching with a time delay;
- solar reset;
- photoelectric dimming; and
- manual switching.

The selection of effective lighting control may be determined by any mix of the factors presented below:

- size of the building;
- lighting system type;
- daylight availability;
- building usage;
- budgetary constraints; and
- dimming requirements.

Visual comfort is location-specific and even includes the orientation of the head. This presents a dichotomy between spatially aggregated measures (e.g., average daylight factor) and assessments of glare affecting an individual. BPS+ can go some way towards resolving this issue.

3.1.3   Indoor air quality

Every building has a number of potential contaminant sources. These are related to building materials and furnishings (continuous release), and cooking, smoking, solvents, paints, and cleaning products (intermittent release). The most significant pollutant sources are human and animal metabolism, occupant activities, building materials, and equipment. The main air pollutants are carbon dioxide, carbon monoxide, tobacco smoke, formaldehyde, moisture, odour, ozone, particulate matter, and volatile organic compounds.

A review of international standards has been undertaken by the International Energy Agency. This included World Health Organisation standards in which three different concentration levels are listed:

- Maximum Allowable Concentration (MAC) at the work space for an 8 hour period (occupational health criteria);
- Maximum Environmental (ME) value;
- Acceptable Indoor Concentration (AIC) – for concentrations below the AIC, the negative health effects are either negligible or, if no threshold is known, are at least tolerable.

Tables 3.5–3.7 list the different national limits for each of the major pollutants found in the indoor environment.

The measurement of indoor air pollutants in buildings is difficult for technical and financial reasons, whilst the corresponding simulation of pollutant concentration can be a non-trivial task. The following indexes may be considered representative of the quality of the indoor air.

- CO_2 level may be considered an indicator of the pollutants directly emitted by occupants in non-smoking areas. It is especially useful for variable occupancy zones and is a commonly used index, with concentration values between 500 and 1,000 ppm.
- CO can be a useful index where tobacco smoke is predominant (or in conjunction with an index such as CO_2). In particular, it can be employed as a security indicator to avoid combustion-related poisoning from heating appliances.

Table 3.5 International standards for indoor CO_2 concentration (ppm)

Country	MAC	Peak limit	AIC
Canada	5000		1000–3500
Germany	5000	2 × MAC	1000–1500
Finland	5000	5000	2500
Italy			1500
Netherlands	5000	15,000	1000–1500
Norway	5000	MAC+25%	
Sweden	5000	10,000	
Switzerland	5000		1000–1500
UK	5000	15,000	
USA	5000		1000

Table 3.6 International standards for indoor CO concentration (ppm)

Country	MAC	Peak limit	ME value	AIC	Remark
Canada	50	400		9	
Germany	30	2 × MAC	8–43	1–18	Depends on duration and room type.
Finland	30	75		8.7–26	Depends on duration.
Italy	30				
Netherlands	25	120		8.7–35	Depends on duration.
Norway	35	+50%			
Sweden	35	100		12	
Switzerland	30		7	1000–1500	
UK	50	400			
USA	50	400		9	Depends on duration.
WHO		9–87			

- H_2O, though not an air pollutant, is frequently used to control airflow rates in order to avoid excessive relative humidity levels.
- Occupancy may be employed as an air quality indicator where occupants represent the main source of pollution, as it is easier to measure presence than pollutant concentration.

Ventilation is the process of supplying and removing air to and from spaces and can be achieved by natural or mechanical means. Hybrid ventilation combines both ventilation modes. In a natural ventilation scheme, outdoor air moves through building openings such as doors, windows, and cracks. It can be used to provide fresh air for occupants and to provide cooling when external weather conditions allow it. With mechanical ventilation, the air is supplied by electrical fans, which may be part of an HVAC system used to heat, cool, and filter the air. There are many techniques for measuring airflows and ventilation rates:

- tracer gas decay;
- tracer gas constant injection;
- tracer gas constant concentration; and
- the multi-tracer gas method.

With these techniques, the tracer gas has desirable properties: it is non-toxic and non-reactive, measurable at low concentrations, not a normal constituent of ambient air, and has a similar molecular weight to the normal constituents of ambient air. SF_6 closely matches these requirements and is the most widely used tracer gas.

An alternative method is based on occupant-exhaled CO_2. The two most common methods of CO_2 concentration analysis are the decay and constant injection methods. The critical parameter in both cases is the CO_2 generation rate, which cannot be measured accurately because it depends on many characteristics. The advantages of this technique are that there is no need to release a tracer gas in the building, and the equipment for sampling CO_2 is relatively inexpensive and simple to operate.

Table 3.7   International standards for indoor NO_2 concentration level (ppm)

Country	MAC	Peak limit	ME value	AIC	Remark
Canada	3	5		0.3	Offices, homes.
				0.052	
Germany	5	2 × MAC	0.05–0.1		
Finland	3	6		0.08	Daily average,
				0.16	hourly average.
Netherlands	2			0.08–0.16	Depends on duration.
Sweden	2	5		0.15–0.2	
Switzerland	3		0.015–0.04	1000–1500	
UK	3	5			
USA	3	5	0.16	0.3	
WHO		9–87		0.08–0.21	

It is important to carry out regular air quality monitoring to detect contaminants that are capable of generating adverse health effects. Monitoring might aim at comparing indoor and outdoor conditions and recording gas concentrations over various periods. The levels recorded indoors are compared with available national air quality guidelines. Building- and room-level air change rates should be periodically monitored to verify ventilation effectiveness.

3.1.4   *Energy use*

The assessment of energy consumption in a building is required to invoice delivered energy, reduce peak load, analyse system energy behaviour, support financial management, and optimise the energy tariff. In order to reach these objectives, different values may be measured: delivered thermal and electrical power and/or fossil fuel usage. Moreover, the power metering may be organised hierarchically per building, per flat, per zone, etc. Considering energy tariffs is essential when analysing and managing energy consumption. In order to evaluate the energy performance of existing buildings, methods have been developed to assess the total energy consumption, which may be estimated by two distinct approaches, black box and analytical as follows. These methods may be readily emulated by a BPS+ tool or evaluated separately based on simulation outputs. It may even be possible to use simulation results in place of measured data.

In the *black box* approach, the weather normalised energy performance can be deduced from a statistical analysis of actual energy consumption and a control parameter such as outdoor temperature. Weather normalisation techniques are based on the assumption that energy consumption is composed of both weather- and non-weather-dependent components. The weather-dependent component varies linearly with weather (usually expressed as heating/cooling degree-days), whilst the non-weather component is due to local equipment, lighting, etc.

The Princeton Scorekeeping Method (PRISM)[4] is a weather normalisation technique that assumes a linear relationship between energy consumption and heating/cooling degree-days. The normalised annual energy consumption (NAC) is given by the weather-dependent term that is calculated for reference or long-term annual average conditions:

$$NAC = a + b \cdot f \text{(climate conditions)}.$$

The daily average energy consumption ($kWh/m^2.d$) is calculated from

$$E_i = a + b \cdot H_i(T_{ref})$$

where E_i is calculated by dividing the total amount entered on the utility bill by the total floor area and by the number of days, 'a' is the non-weather-dependent daily energy consumption, 'b' the slope of the weather-dependent energy consumption, and $H_i(T_{ref})$ the number of degree-days computed for a reference temperature T_{ref}. The following weather data are necessary to apply the method:

---

[4]https://marean.mycpanel.princeton.edu/Details.html

- average temperatures;
- long-term heating degree-days for several reference temperatures; and
- long-term cooling degree-days for several reference temperatures.

If T_{ref} is chosen arbitrarily, 'a' and 'b' must be estimated using least squares techniques. T_{ref} can subsequently be estimated according to the criterion of the highest coefficient, R^2. Finally, normalised annual energy consumption, NAC (kWh/m^2), is calculated using the relation

$$NAC = 365 \cdot a + b \cdot H_o(T_{ref})$$

where H_o is the annual heating degree-days.

The Zmeureanu method (Zmeureanu 1992) is another weather normalisation method that can be applied to buildings using gas or fuel for heating and electricity for cooling, lighting, and equipment. The method uses the term 'building energy signature' to define the linear relation between daily average consumption, E_i (kWh/m^2.d), and the daily average temperature, T_i, for each meter-reading interval:

$$E_i = a + b \cdot T_i$$

where 'a' is the base level of non-weather-dependent daily energy consumption and 'b' is the slope of the weather-dependent energy consumption corresponding to the ratio between the heat loss rate and the efficiency of the HVAC system. The building consumes energy above the base load only when the outdoor temperature drops below the reference temperature, T_{ref} (heating), or rises above T_{ref} (cooling). The T_{ref} at which the weather-dependent curve equals the base load, $E = a + b \cdot T_{ref} = B_L$, is calculated from

$$T_{ref} = (B_L - a)/b.$$

NAC is then obtained from

$$NAC = 365 \cdot B_L + (E_m - N_m \cdot B_L)C/C_1 \cdot N_{T,15}/N_T \cdot N_m/365; \text{ if } T_{DB} < T_{ref}(\text{heating})$$
$$\text{or } T_{DB} > T_{ref}(\text{cooling})$$
$$NAC = 365\, B_L; \text{ if } T_{DB} \geq T_{ref}(\text{heating})$$
$$\text{or } T_{DB} \leq T_{ref}(\text{cooling})$$

where E_m is the total energy consumption from the utility bills (kWh/m^2.y), N_m is the total number of days covered by the utility bills, B_L is the base load energy consumption (kWh/m^2. d), and $N_{T,15}$ the number of hours when $T_{DB} < T_{ref}$ for gas heating or $T_{DB} > T_{ref}$ for electrically cooled buildings; T_{DB} is the external dry bulb temperature (°C). The following relations define the parameters C and C_1:

if $T_{DB} < T_{ref}$ (heating) or $T_{DB} > T_{ref}$ (cooling):

$$C = 1/15 \sum_{j=1}^{15} \sum_{i-1}^{n} (a + b\, T_{DB,i,j} - B_L)BIN(T_{DB,i,j})$$

$$C_1 = \sum_{i-1}^{n} (a + b\, T_{DB,i,j} - B_L)BIN(T_{DB,i,j})$$

if $T_{DB} \geq T_{ref}$ (heating):

$C = 0$

if $T_{DB} \geq T_{ref}$ (heating) or $T_{DB} \leq T_{ref}$ (cooling):

$C_1 = 0$

where $BIN(T_{DB}, j)$ is the number of hours of occurrence of the dry bulb temperature bin having T_{DB} as the centre during the operation of the HVAC system. For the heating model, the value of 'b' is negative whereas, for the cooling model, it is positive.

The Zmeureanu rating system consists of a combination of the following features:

- the index of energy performance assessed based on the previous history of the building;
- comparison of the energy performance results with those of reference buildings;
- information to the owner on the potential energy savings.
- for a new building evaluation of the energy based on drawings and specifications at the design stage; and
- for an existing building assessment of energy consumption and parameters such as the thermal performance of the exterior envelope and occupant behaviour.

Two basic approaches exist:

- the absolute approach, which is important for the building owner; and
- a relative approach where potential savings are emphasised.

Two steps are generally followed:

- normalisation of utility bills for weather conditions and size of the building (conversion of all energy units to kWh); and
- evaluation of the energy used for heating purposes.

The analysis of utility bills is made using AHEM software (Zipperer *et al.* 2013) in which inputs are year of construction, total heated floor area, location, energy source used for heating, energy consumption, and cost for at least 12 months. Outputs are energy consumption and cost for each energy source, normalised energy consumption and cost, the contribution of each energy source to NAC, and a comparison between the NAC for test and reference houses.

In the *analytical* approach, the energy consumption is calculated using a building's design data, characteristics of the HVAC systems, and standard meteorological conditions. The following brief descriptions are extracted from surveys that evaluated and classified existing methods (Santamouris 2001, Nikolaou *et al.* 2011.

Vermont-Hers (Home Energy Rating Systems) method[5]: This characterisation refers to a set of energy rating schemes applied in the USA (also referred to as Home Energy Saver). The home energy rating is a standard measure of a home's energy efficiency. The implemented HERS may be one of the following:

- a points system, which evaluates the energy performance of a house by awarding performance scores to each subsystem;
- a performance system, which assigns an index of performance in terms of annual heating performance or cost; and
- an awareness system, which recommends the annual total and heating energy consumption, and the corresponding cost, in terms of the year of construction, the climate zone, and the energy source.

Initially, an on-site inspection of a dwelling is realised by an energy efficiency professional, the home energy rater, aiming to measure energy characteristics such as insulation level, wall-to-window ratios, heating and cooling system efficiency, solar orientation, and water heating system type. These data are then input into an application program and translated into points. The home receives a score between 1 and 100 according to its relative efficiency. The home's energy performance is then star-rated, ranging from very inefficient to highly efficient (5-star). Finally, the homeowner receives a report listing cost-effective options for improving the home's rating. The point score is defined as follows:

$$\text{Pointscore} = 100 - 20(E_R/E_C)$$

where E_R is the estimated purchased energy consumption (kWh) for heating, cooling, and hot water for the rated home, and E_C the estimated purchased energy for the same three consumptions for a reference home. For $E_R = E_C$, the points score is 80.

Key Number method: The key number trade name was issued by Norway (Roulet *et al.* 2002), although such numbers are used in several countries to compare energy performance between buildings and existing references. Several levels are considered. The lowest level is simply to compare the total yearly energy use of a building to reference cases. The energy use must first be normalised. At a second level, the energy use can be graded in various categories: heating, cooling, ventilation, hot water, lighting, appliances, and so on, and then compared to corresponding references. At a third level, building characteristics can be compared with conventional values.

Energy Barometer for Sweden: This method (Westergren *et al.* 1999) addresses building performance through the continuous monitoring of energy and weather variables in a random sample of houses with reporting of changes in energy use at the individual building level. According to the measurement protocol, three variables are measured:

- electrical energy use for household appliances;
- energy use for heating, including domestic hot water; and
- indoor temperature.

---

[5]https://publicservice.vermont.gov/efficiency/building-energy-standards/residential-building-energy-standards/hers-provider

Weather data are collected from the nearest meteorological station, and building technical data are collected either by interviews or by onsite inspections. The required weather data includes outdoor temperature and solar irradiation. The required technical data includes:

- household size and composition;
- year of construction, house type, and heated area;
- construction data;
- service systems information; and
- appliance data.

The measured energy use is standardised with respect to time and climate. This is done by statistically regressing energy data against weather data for a normal or average year, with the obtained regression coefficients used to determine the average annual energy use. The model used for the regression analysis, called energy-signature, is a three-parameter model: the first parameter is the base line energy use when there is no heating; the two others are the intercept and the slope coefficient for the heating regime. This model can be automated and applied to a range of buildings. The energy used for heating, E_h (kWh), during an average year, is calculated from

$$W = c \cdot H + b \cdot Q + f \cdot I + d \cdot P$$

where,

H is the average hourly energy use, independent of outdoor temperature and solar irradiation (kW),
Q the heat loss factor expressing how well the building copes with the outdoor climate (kW),
I the solar aperture or window factor (m^2),
P the average hourly energy use during the unheated period (kW), and
c, b, f, and d have specific values for each house.

3.1.5 Environmental impact

Impacts connected with the construction industry, whether local or remote, are many and derive from the materials used, pollutants emitted, and/or unhelpful physical interactions. The materials used in construction and equipment manufacture have an embodied energy in addition to unintended consequences during extraction, supply, use, and disposal. Minimising such impacts is a principal driver of design innovation and the circular economy.

Although not a pollutant, the CO_2 associated with the burning of fossil fuels is widely legislated against because of its perceived global warming potential. Other compounds associated with fossil fuels, such as SO_2 and ethylene, can cause acidification and photogenic smog, respectively.

To quantify impact, emissions of resource groups are quantified by an appropriate method such as the Tally tool (Zhou *et al.* 2023), which analyses the operational energy demand of an input BIM model and combines its material

Table 3.8   Environmental impact categories and indicators (credit: choosetally. com)

Category	Indicator
Climate change	kg CO_2 equivalent
Ozone depletion	kg CFC-11 equivalent
Ecotoxicity for aquatic fresh water	CTUe (comparative toxic unit for eco-systems)
Human toxicity, cancer effects	CTUh (comparative toxic unit for humans)
Human toxicity, non-cancer effects	CTUh (comparative toxic unit for humans)
Particulate matter/respiratory inorganics	kg PM2.5 equivalent
Ionising radiation, human health effects	kg U^{235} equivalent (to air)
Photochemical ozone formation	kg NMVOC equivalent
Acidification	mol H+ equivalent
Eutrophication, terrestrial	mol N equivalent
Eutrophication, aquatic	Fresh water, kg P equiv.; marine, kg N equiv.
Resource depletion, water	m^3 water use related to local scarcity of water
Resource depletion, mineral, fossil	kg antimony (Sb) equivalent
Land transformation	kg (deficit)

attributes and assembly details with the tool's database containing similar environmental impact categories and indicators as shown in Table 3.8.

Quantifying such impacts allows investigators to weigh each category by its significance and determine mitigating actions. However, it may not be possible to capture this level of detail in every project due to myriad complexities. For example, cement, a common production source, not only consumes resources in its manufacture but also exudes pollutants (CO_2, SO_2, NO_X, PM), with studies showing that high-grade cement results in higher emission concentrations (Guo *et al.* 2023).

A number of tools and methods have been developed that address the different environmental impacts during a project's life cycle. A common method is Life Cycle Assessment, which evaluates environmental impacts at project stages such as procurement of materials, construction, operation, and decommissioning (Casini 2022). Section 3.3 elaborates on this prospect in the context of BPS+.

3.1.6   Unmet hours

'Unmet hours' is a term with regional connotations. In North America, where there is an expectation that buildings always deliver environmental conditions that are within specified lower and upper set points, an 'unmet hour' is a serious breach of protocol. In Europe and elsewhere, expectations are often more relaxed, and occasional discomfort is an accepted trade-off to save on capital and operating costs. In such dominions, zero 'unmet hours' would indicate a system that was overdesigned. In either event, it is straightforward to use a BPS + tool to track unmet hours as an indicator of unacceptable departures from an ideal state.

3.2   Performance rating

Buildings are thermodynamically complex, and most energy transfer processes are non-linear and three-dimensional. Yet, rather than take advantage of the extraordinary capabilities of building simulation technology, most regulatory authorities base performance assessments on obsolete methods. The objective is not to understand the performance of a particular design in any realistic sense but rather to come up with a few numbers that can be used as part of a subjective 'A to G' rating based on some notional measure of acceptability as depicted in Figure 3.1.

Building control authorities are usually unfamiliar with and even distrustful of the output from modern computational methods and resort to a lowest common denominator approach. This limits compliance to a few indefensible rating numbers as evidence of a building meeting current performance regulations. These rating systems take little or no account of how the subject building is to be used by the owner, or how it will perform over its lifetime. Instead, an assessor chooses occupancy patterns and operational data from predetermined lists, and weather data from a few national locations. Modifications are made to the building model to allow for deficiencies in the simplified modelling method (e.g., the design uses a night purge to address overheating, but this is omitted because the method does not support it).

In Europe the schemes have become complex, and the calculations are done using government-sponsored, simplified methods applied by individuals trained in their use. Dynamic simulation software is allowed but is considered optional rather than mandatory (as is the use of the compliance tool). Many practitioners are under the mistaken belief that compliance equates with actual energy performance or perceptions of comfort. This can result in the misapplication of the compliance tool

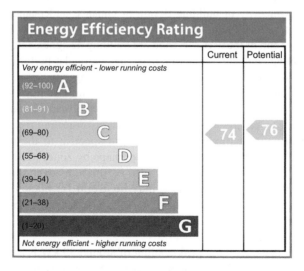

Figure 3.1   A performance-rating example

as a design tool. Unfortunately, the sparse palette of descriptors of the building and its use means that compliance tools are able to support only a few of the what-if questions typically posed in the design process. This has impeded the adoption of building simulation tools by a wide range of users and discouraged the incorporation of advanced technologies in building design that the simplified calculation method cannot process.

Schemes require the calculation of an asset rating (Energy Performance Certificate, or Building Energy Rating), which means that they are independent of the actual occupancy and usage of the building. The rating is usually for the entire building, including fixed services such as heating and lighting, but not plug loads (the equipment and appliances installed in the building by the occupant). Ratings are reduced from a kWh/year calculation to a letter A through G and are required for new buildings and existing buildings when they change owner. Ratings relating to the calculation of CO_2 emissions are expressed relative to a so-called notional building. The calculations, therefore, have to be carried out twice. The notional building has the same geometry as the actual building but uses prescribed values of heat transmittance for walls and glazing, whilst windows are adjusted where necessary to maximise their size. Factors relating to plant are predefined, and an activity database specifies hours of occupancy for both the actual and notional buildings.

Much debate has circulated around the topic of the 'performance gap'. In the current context, this refers to the difference between a rating tool outcome and actual performance in use. It would be surprising if there was not a performance gap to which poor plant commissioning, contractor design changes, and basic modelling errors contribute. How much does this matter? If the intention is to provide a single number to an undiscerning customer, is the full capability of complex dynamic simulation needed? Probably not, but such attitudes do nothing to improve the integrity of building design or enhance understanding of the performance of discrete building fabric or plant elements.

At the other end of the user spectrum are building designers (architects, service engineers, lighting engineers, etc.) who may have particular performance targets to meet as specified by a client or local government agency; or need to evaluate alternative design approaches (e.g., to find a compromise between solar gains and winter heat loss). As building designers incorporate substantial quantities of thermal insulation, new criteria arise, such as summer overheating risk.

The performance parameters required are specific to the purpose of the investigation. Summations of energy use, perhaps by zone, would be needed to estimate annual or longer term energy use. Estimations over shorter periods – monthly, weekly, daily, and hourly – may be useful for plant sizing purposes and plant efficiency assessment. In that case, a statistical summary showing energy demand categorised into 'bins' is sometimes helpful, as shown in Figure 3.2.

This shows that a reported peak cooling of around 39 kW is a rare occurrence during the week, whilst a peak heating demand of 60 kW is an outlier. A control engineer could easily mitigate these rare peaks. Simulation over shorter periods,

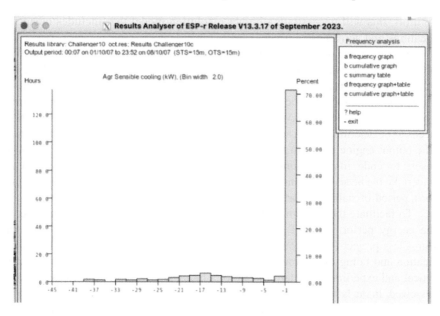

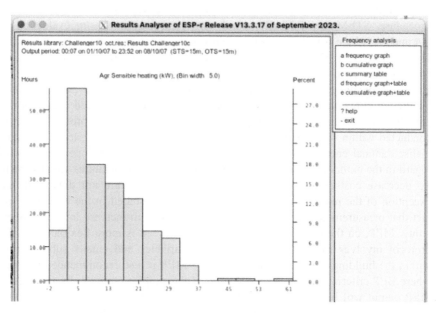

Figure 3.2 Frequency of occurrence of cooling (top) and heating demand

and perhaps with shorter time steps, is needed to assess modelled plant performance. A time series output of temperatures in critical zones is needed, and more attention should be paid to the modelling of the thermal coupling between air, fabric, and solar inputs. If the investigation is focused on thermal or visual comfort, this will necessitate a higher resolution model that includes occupant locations and activities.

In general, the metrics to be generated will be time series outputs of specified properties of air, water, or fabric (temperature, humidity, velocity), and these are relatively straightforward to produce. An averaging capability will be useful in most applications. Extrinsic properties need to be expressed per unit of area, volume, or flow rate (e.g., kWh/y or kWh/m^2.y in the case of energy), which in turn needs to be determined as a summation of selected parts of zones or multiple zones. The output engine needs to be sophisticated in these regards if the user is not to resort to code modifications or post-processing (e.g., using a spreadsheet to undertake the necessary manipulations of simulation results to produce a PPD value for a period of stable indoor conditions)[6].

To facilitate the implementation of the (then) proposed European directive on the energy performance of buildings in the residential and tertiary sectors, the EuroClass project (Santamouris 2001) set out to develop a method for the classi-fication and rating of existing buildings. Within the project, methods for the ana-lytical and experimental determination of a building's thermal characteristics were reviewed. In the latter case, the emphasis was on the fabric thermal (U) and window solar (g) transmittance values. The reviewed methods covered a range of approa-ches: various university projects, Energy Barometer, House Energy Labeling Procedure, STEM and PSTAR, neural networks, and the *in situ* evaluation of UA and gA, where A is the area of the fabric and window components as necessary. The focal point of the project was the development of experimental protocols and normalisation procedures.

Experimental protocol: The aim here was to determine actual energy use and identify its constituents. Two procedures were developed, the billed energy proto-col (BEP) and the monitored energy protocol (MEP). The simpler protocol, BEP, is conducted within 4 person-hours per residence. The output is the actual and nor-malised annual energy use. Normalisation of energy use is only performed with regard to the outdoor climate for the reason that indoor temperatures are unknown. To decrease costs for the rating process, measurements are not done with the exception of the reading of meters that are already installed in the building. The fact that measurements are not conducted will lead to an increase in result uncer-tainty. MEP, on the other hand, makes use of bills but is more flexible since the protocol involves measurements of weather variables and entails sub-metering within the building to enhance data accuracy. MEP is also recommended in cases where BEP criteria prevent the application of the simpler protocol. In most cases, MEP output will be more reliable than that of BEP. Moreover, normalisation of energy use, taking into consideration references to outdoor and indoor climate, is performed. MEP may involve a monitoring scheme lasting for over 10 weeks: the duration depends on which service, aside from rating, is being purchased. Technical information from MEP is important to document, as this information can

---

[6]The need to frequently post-process results is a clear sign that a tool needs urgent refinement.

be used to establish default values for BEP. An example of this is boiler efficiency as a function of fabrication, type, and age.

Normalisation: Normalisation takes into account the size of the building, the external climate, and the internal climate. Typically, heated floor areas are used to represent building size. This relates energy use to the parts of the residence that are utilised and conditioned. Normalisation by external climate takes into consideration annual variations. By placing the building in a reference climate, the performance of different buildings may be directly compared. Normalisation for heating is done using heating degree-days, whereas normalisation for cooling must be done using a more sophisticated approach such as the climate severity index proposed by Clarke *et al.* (1984). Normalisation of the internal climate can be done under a set of predefined conditions. This type of normalisation standardises end-user behaviour and allows a comparison between different occupant types in the same building.

A rating procedure is a comparison scheme that associates a score with a specific building. It is based on three aspects:

- the performance variable or set of variables to be compared;
- the comparison scenario, i.e., the group of buildings that will provide a distribution of the performance variable, thus creating the framework for comparison; and
- the rating score, i.e., the criteria and limits that give the score when the performance variable is compared within the comparison scenario.

The performance variables are the total supplied energy or the total delivered energy, both expressed in kWh/m^2. Both quantities are normalised and attributed to the different energy streams, namely heating, cooling, hot water, and lighting. Space heating and cooling require weather normalisation. The score of the rated building allows comparison with other buildings for a given similarity level or normalisation scenario. Several possibilities for expressing the rate exist:

- points on a scale of 0 to 100 based on the percentile that the energy consumption of the building has when compared to the frequency distribution of the energy consumption of the existing building stock;
- a star rating whereby the cumulative distribution is divided into a number of sectors and the building is assigned to one sector in terms of its energy consumption; and
- a distance to a reference value obtained by dividing the area-related energy by a reference value.

When it is not possible to employ statistics about energy consumption for buildings similar to the rated one, the frequency distribution of the consumption follows an analytical expression derived from existing distributions in other countries. In all cases, allowance is made for the intermittent operation of HVAC systems by means of an intermittency factor, which is used to normalise the actual operation of the building to a standard scenario. The calculation of this factor is based on a representative day.

3.2.1   *Appraising design options*

BPS+ enables building performance to be tracked in terms of principal parameters such as the spatial/temporal variation of temperature, air movement, and heat flux; and derived parameters relating to aspects such as required heating/cooling system capacity, operational energy consumption, and indoor comfort conditions. This allows the user to consider trade-offs by comparing relevant performance profiles.

Although such data are readily available from a simulation, they need to be interpreted in some way. A starting point is to collate the principal and derived data that underpin the answer to a particular design question. Rules of interpretation can then be applied to rank problem severity and help identify issues requiring remedial action. To demonstrate such a procedure, consider a BPS+ session aimed at ensuring adequate thermal comfort conditions throughout a building over time. The required principal data will include intra-zone air and mean radiant temperature, relative humidity, and air movement. This will allow the rank ordering of occupied zones in terms of their overheating (or underheating) based on, say, the magnitude and duration of a simple index such as resultant temperature or a higher-level index such as TPV as elaborated in Section 3.1.1.

For problematic zones, the convective heat flux relating to surface heat transfer, natural/mechanical ventilation, solar penetration, and casual gains can be presented in a manner (such as that elaborated in Section 4.6) that indicates changes to the design that will improve performance. Figure 3.3 shows this procedure as encapsulated in an

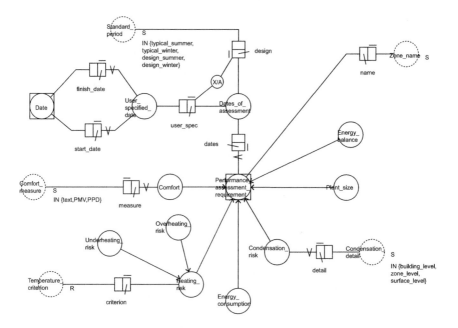

Figure 3.3   Performance assessment encapsulation

ATLIAM diagram produced within the EC Combine project[7] (Augenbroe 1992, Hand and Strachan 1998) – see Section 2.4 for details on the approach.

Similar encapsulations of relevant data can be established for other performance evaluations. Lighting, for example, would rank order zones based on the distribution of illuminance and glare.

The next stage is to apply weightings to the different performance evaluations. These weightings are likely to change significantly between designs, so it must be possible for the user to influence them. A future possibility is the existence of a 'performance evaluation function' (currently the user) that will automate decisions based on rules relating to several performance assessments.

3.2.2   Rating carbon action

Other performance presentation formats are possible, depending on the assessment focus. For example, in the case of an assessment of the carbon reduction potential of energy efficiency actions, financial indicators might include capital cost, payback period, and internal rate of return (IRR), whilst a carbon reduction indicator might be tonnes of CO_2 mitigated. The decision-making process becomes more challenging when there is more than one decision-maker, with one party preferring to use financial indicators and the other preferring a carbon reduction indicator. Issues such as carbon taxes and emission trading schemes put pressure on program vendors to extend the range of indicators.

Presentations such as the Marginal Abatement Cost Curve (MACC; Somar 2010) and the Emissions Reduction Investment Curve (ERIC; Lavery 2011) can be used. MACC is a method to present and compare available carbon abatement opportunities in a graphical manner. The output curve provides the carbon abatement potential of each opportunity (tonnes of CO_2 on the x-axis) versus the cost of abatement (£ per tonne of CO_2 on the y-axis). Lavery (2011) proposed ERIC as an alternative to MACC, arguing that MACC is unhelpful since it does not display IRR and cumulative IRR, which are generally more acceptable metrics for CEO- and CFO-level officers within a company.

Ali (2013) used these methods to assess 10 carbon reduction opportunities within the (then) Barr Construction Company[8] as follows.

- Transport energy reporting system. A telematics system linked to a monthly reporting tool provides key performance indicators on transport energy use.
- Plug-in timers. Applied to space and water heating units to reduce electricity usage during non-working hours.
- Coating plant burner. The replacement of the fuel burner in an asphalt coating plant with a more efficient model.
- Drying room dehumidifier. The replacement of electric heating elements in drying rooms used to dry operatives' clothes with low-energy dehumidifiers.

---

[7]This project established an integrated data model to support the input needs of different building design tools.
[8]https://web.archive.org/web/20130724104821/http://www.barr-construction.co.uk/

- Solar PV. The installation of a 50 kW array to generate electricity for local use and export.
- Sheds for aggregates. The installation of sheds in a quarry to keep the aggregates dry – within the asphalt production process, moisture in aggregates results in an increase in the fuel used to dry the aggregate before it can be coated with bitumen.
- Vertical bitumen tanks. Bitumen storage tanks require a high temperature (160 °C–220 °C) to maintain the viscosity of the thick fluid. This option proposes the replacement of existing storage tanks with highly insulated tanks to reduce heat loss.
- IT server cooling. The use of passive cooling techniques eliminates the need for a mechanical cooling plant.
- LED lighting. The replacement of fluorescent, metal-halide, and sodium lamps with LEDs.
- Wind Turbine. The installation of a 75 kW wind turbine to generate electricity for local use and export.

ESP-r and other methods were used to assess these opportunities, and a MACC was constructed using a spreadsheet tool developed by Somar (2010). Figure 3.4 shows the outcome relating to the above carbon reduction opportunities. The width of each opportunity on the x-axis represents its carbon abatement potential in tonnes of CO_2, whilst the y-axis represents the marginal abatement cost (MAC) of that opportunity.

The MAC is based on an opportunity's net present value (NPV), lifetime (LT), and annual carbon emissions abatement potential (CEA):

$$MAC = (NPV \cdot LT)/CEA.$$

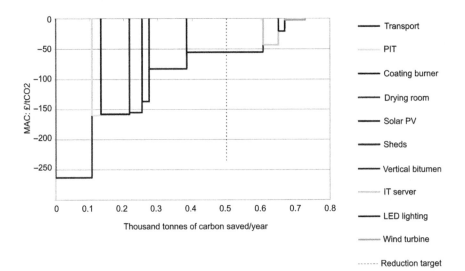

Figure 3.4   MAC curve for company opportunities (credit: Ali 2013)

Using this method, decision-makers can compare the cost of abatement to the Carbon Reduction Commitment (CRC) scheme allowance price[9], which was £12 per tonne of equivalent CO_2 emissions at the time. A discount rate of 6% was assumed, corresponding to company policy. The dashed vertical line shows the company's carbon reduction target. Here, all opportunities have negative abatement costs, implying that they would be viable even if the company did not participate in a carbon-trading scheme. The transport energy reporting system had the lowest carbon abatement cost ($-£236$ per tonne of CO_2), which means that this action would reduce CO_2 emissions by 105 tonnes and save £263 for each tonne of CO_2 reduced. The plot shows the wind turbine opportunity as the least attractive, although even this was expected to annually reduce CO_2 emissions by 57 tonnes and save £2 for each tonne of CO_2 reduced. The dashed vertical line on the graph represents the carbon reduction target of 500 tonnes, indicating that the company will exceed its target by implementing the first seven opportunities (from left to right).

In ERIC, as with MACC, the *x*-axis displays the carbon reduction potential of the opportunities. Here the *y*-axis is a logarithmic scale showing the IRR of each opportunity. The IRR is the rate of return that makes the NPV of all cash flows (both positive and negative) from a particular investment equal to zero. In ERIC, the emphasis is on IRR rather than the money spent per tonne on emissions reduction. ERIC gives the IRR of individual options as well as the cumulative IRR of multiple options. Since there is no assumption of a discount rate, decision-makers are free to choose their preferred risk level by merely examining the IRR. Figure 3.5, shows an ERI curve plotted for the same opportunities at Barr with the opportunities sorted from highest IRR on the left to lowest on the right.

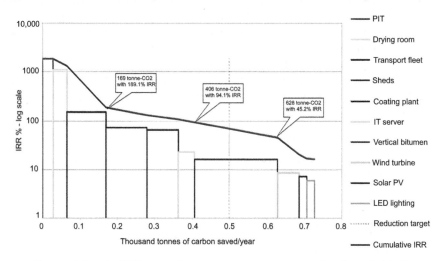

Figure 3.5   An ERI curve for company opportunities (credit: Ali 2013)

[9]The CRC Energy Efficiency scheme ended on 31 March 2019; gov.uk/guidance/crc-energy-efficiency-scheme-allowances.

Each opportunity is now either less or more viable in comparison to the other opportunities based on the IRR it offers. Plug-in timers had the highest IRR (at 1,849%), whilst this option was the second best in the MACC analysis, though it had little impact on the company's overall carbon reduction. The transport reporting system, which was the most attractive option suggested by MACC, had the third-best IRR of 151% in ERIC; LED lighting, with a 6% IRR, was the least attractive.

From an investor's point of view, all carbon reduction opportunities considered had an IRR above 6%. However, if there was an opportunity with an IRR less than the interest rate offered by a bank, the opportunity is unlikely to be implemented, and the investor would be financially better off putting available funds in a bank account. On the other hand, if the company invests in an opportunity by taking a loan, then the IRR of the opportunity must be reasonably higher than the interest rate on the loan.

As with the MACC analysis, the dashed vertical line on the graph shows the carbon reduction target of 500 tonnes, indicating that the company will cross its carbon reduction target by implementing the first seven opportunities (from left to right).

The best opportunities identified from MACC and ERIC were compared, as shown in Table 3.9. The positions of these opportunities vary between the two techniques; with the largest change in position being for Solar PV because it offers a poorer rate of return than most of the other opportunities.

Ali (2013) explored the usefulness of the two methods and concluded the following.

- ERIC can deal with MACC's issues of negative scale, whilst inflation may also be considered when calculating IRR and cumulative IRR.
- ERIC does not give an indication of an opportunity's value (abatement cost in £ per tonne of CO_2) against the carbon allowance price. It is therefore not possible to identify the allowance price that would make an opportunity viable when participating in an emissions trading scheme such as the CRC.
- To display the range of IRR observed in the ERIC data (6%–1,849%), a y-axis logarithmic scale is required. This makes it difficult to detect the difference between two opportunities: the wind turbine (IRR 8.4%) and IT server room

Table 3.9   Comparison of scoring outcomes

Opportunity	ERIC	MACC
PIT	1	2
Drying room dehumidifier	2	4
Transport fleet management	3	1
Aggregate sheds	4	6
Coating plant burner	5	3
IT server room	6	8
Vertical bitumen tanks	7	7
75 kW wind turbine	8	10
50 kW solar PV	9	5
LED lighting	10	9

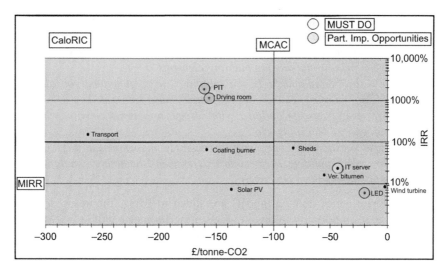

Figure 3.6   A CALoRIC output (credit: Ali et al. 1999)

(IRR 23.2%) options do not appear significantly different, despite the latter being around three times the former.

To obtain the benefits of both approaches, Ali constructed a blended model, CALoRIC, as shown in Figure 3.6.

This provides decision-makers with both MAC and IRR and is particularly useful when comparing a large number of carbon reduction opportunities.

3.2.3   Illuminating trade-offs

All design solutions are sub-optimum given the high number of oft-incomparable objective functions. This dilemma is well summarised by Kesik (2015):

> *"Every advanced industry relies on key metrics and indicators to convey the performance of its products. Building performance assessment, including simulation, has to deliver meaningful information that is consequential. Here are some questions that need to be answered before design of new or retrofit buildings goes beyond concept.*
>
> *During prolonged energy outages, how long before the indoor temperature becomes too high or low for inhabitants?*
>
> *Over what fraction of the building's floor area can acceptable indoor air quality be maintained through natural ventilation alone?*
>
> *What fraction of the building floor area can enjoy adequate daylighting during typical periods of daytime use?*
>
> *What is the building's base metabolism independent of occupancy? (e.g., peak and annual energy demands of the unoccupied enclosure)*
>
> *How durable is the building enclosure assuming recommended maintenance?*

(e.g., service life of cladding and control layers, recommended inspection, cleaning, maintenance intervals, etc.)

What is the initial and recurring embodied energy and associated carbon footprint of the passive building elements?

How flexible/adaptable is the building? (e.g., functional obsolescence, adaptive reuse, change of occupancy, churn rates, etc.)

How resilient is the building with respect to seismic activity, wind, flooding, energy blackouts, etc.?

These questions are becoming increasingly important and there is a need to ensure building performance assessment, including simulation, begins to address them effectively."

As more performance domain models are added to BPS+, the need for output constructs that support cross-domain views of performance will grow (Mahdavi 2003). Only then can designs be adequately assessed in terms of diverse considerations. A useful concept for design engineers is the use of an 'integrated performance view' (IPV) whereby key performance metrics are displayed together. They may be somewhat unrelated, e.g., maximum heat demand, visual comfort, and CO_2 emissions. They may represent processed time series, e.g., an averaged comfort index over the occupied period, or an annual energy consumption normalised by floor area. The idea is that as work progresses on one aspect of performance, the user can track the impact on several other performance parameters. IPVs are usually set up for a particular project, and sometimes post-processing may be required.

Figure 3.7 illustrates a typical IPV (Prazeres and Clarke 2005, Prazeres 2006) whereby performance criteria related to energy use, occupant comfort, and environmental emissions are brought together and represented in a manner that is deemed suitable for the related project and user type.

Different IPV 'flavours' are possible with alternative levels of detail made available. At the time of writing, such images are produced via ESP-r data exports to Excel. The IPV instances shown here are technical and include the following entities.

- *Annual energy demand* for heating, lighting, and equipment when normalised by floor area.
- *Maximum heating demand* related to the installed capacity of the building's environmental control system.
- *Energy demand profiles* during typical seasonal days throughout the year.
- *Thermal comfort* expressed as predicted percentage of dissatisfaction as defined in the ISO 7730 standard.
- *Daylighting,* the profile of daylight factors (ratio between internal and external horizontal illuminance) at the level of the work plane, perpendicular to the window and in the middle of the room.
- *Visual comfort* represented here by a Visual Comfort Probability (Guth 1966), which defines the percentage of persons satisfied by the visual environment when looking in a given direction from a particular location.

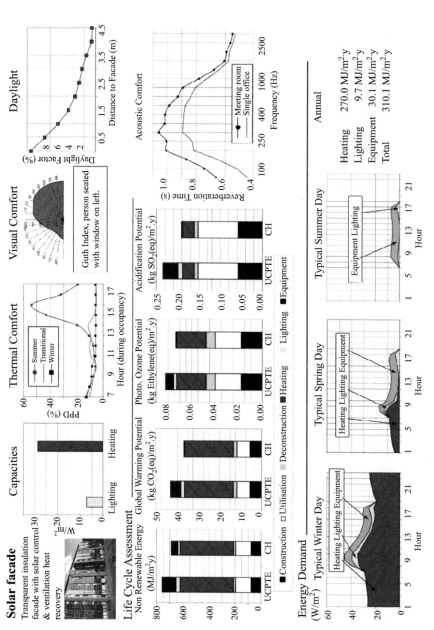

Figure 3.7 Example of an Integrated Performance View (IPV) as output by ESP-r

- *Room acoustics* expressed in terms of the reverberation time based on the Sabine method.
- *Environmental impact*[10] expressed in terms of emissions related to atmospheric warming, photochemical ozone production, and acidification potential.

Such entities can be quantified using alternative indices and displayed in a variety of styles, depending on the target user. One prospect is to hyperlink each entity to enable 'drill-down' performance explorations post-simulation. Within ESP-r, the entities to be displayed are defined as part of the input model. Notably, at the time of writing, IBPSA-USA has issued a 'Request for Proposals' to investigate the standardisation of BPS+ outputs (IBPSA-USA 2023).

3.2.4   Stakeholder types

Regardless of the presentation format, BPS+ tools must support the needs of individual stakeholders if the benefits of the approach are to be fully realised. There is a role here for professional bodies to define these needs in consultation with members.

Building owners/occupants

Where a post-occupancy evaluation is possible, a calibrated model can be used to identify performance deficiencies and areas for improvement. Regular feedback on progress can incentivise individuals to act differently, especially where this feedback is referenced to equivalent facilities to encourage competition. As time passes, opportunities to correct inaccuracies in the model should be taken. These can arise for many reasons, e.g., a change in use, structural modifications, and system control changes. Ideally, the model is created early in the design process and is continually updated and improved as more information becomes available. It can then become a tool for facilities managers to understand the design intent and ensure that their building is performing as envisaged. Many parameters can be generated for the evaluation of performance. Whilst energy data will be of interest to the building owner, environmental parameters will likely engage occupants. BPS+ tools should also output performance summaries that can be readily understood by laypersons.

Government agencies

Government agencies would benefit from the application of performance simulation to explore better rating schemes or to confirm the benefit and replicability of alternative energy actions. At present, ratings are principally concerned with CO_2 emissions, but other parameters are just as important, especially those related to wellbeing, such as indoor and outdoor air quality. The computational approach can also be used to confirm the operational robustness of renewable energy technologies and, in collaboration with geospatial assessment tools (see Chapter 10), identify suitable deployment sites. Simulation-derived outputs may be the result of a one-off exercise, or repetitively generated for online displays for widespread dissemination. BPS+ tools should offer automated performance appraisals for use by policymakers.

---

[10]Here the emission estimates were made using two life cycle inventory databases to characterise the level of agreement.

Experimentalist

An important aspect of all BPS+ development is to ensure that the predicted performance parameters match reality to an acceptable degree. Many detailed comparison studies have been carried out between the various simulation programs, where the computed metrics are compared with careful measurement of the equivalent parameters in a real case, which may be anything from a simple test chamber to a large building. Deviations may arise from simplifying assumptions in the simulation tool or errors in the input model. The properties of materials may differ substantially, and differing assumptions about surface properties may have an impact. Many studies have used the measured data to calibrate the computer model, by judiciously adjusting model parameters until satisfactory correspondence is achieved between simulations and observations. BPS+ tools should offer support for such activities.

Embedded simulation engines

Energy-related product developers often want to assess the performance of their products when deployed in representative situations under typical operating conditions. This requires a BPS+ tool with a simple, bespoke front end, with simulations carried out to produce precisely the performance parameters required. In this case, the usual features of the tool are disabled, and outputs are fixed to comply with the developer's specifications. Examples of such applications are given in Section 4.9.

Simulation-assisted control

The industry is moving towards smart control, whereby actions can be better informed by events. BPS+ can be embedded into the code of a controller to carry out real-time simulation of a building and its systems with multiple live metric inputs, and output metrics to enable the controller to position system actuators associated with valves, dampers, window blinds, etc. (see Section 6.9). This enables the simulation of complex, interacting physical processes and supports the deployment of smart control. A simulation engine can also be interfaced with a real controller device as a means to field test the controller behaviour under a wide variety of virtual building types, occupancy regimes, and weather influences. The alternative of testing in real buildings is time-consuming and difficult to organise.

Simulation community

The simulation community is primarily interested in exploring new applications and adding new capabilities to BPS+ programs. They will have access to the code and tend to work with open-source modelling and simulation systems. They are free to determine the metrics and performance parameters to use. Initially, only thermal energy modelling was carried out, but today the emphasis has shifted to airflow, humidity, lighting, acoustics, and electrical power flow at user-specified spatial and temporal resolution. In addition, traditional plant modelling capabilities are being extended to all types of energy conversion and storage systems, especially those related to renewables. All BPS+ developers should recognise this trend and seek to align their developments appropriately.

Future users

The use of BPS+ to assess overall building resilience would represent a major step forward from the present situation of compliance based on a single set of emission parameters, calculated for one predetermined average year. An intelligent agent would carry out resilience studies (see Section 11.3) by running long-term simulations and assessing the response to extremes of weather and randomly imposed equipment failures. Virtual-reality-type interactions would allow designers to assess visual appeal under different lighting schemes and internal surface finishes.

Ultimately, prospective customers and end users want to know how a product will perform, and the parameters that define the performance are usually easy to verify at the prototyping stage or after the product is put into service. Buildings are no exception, but the true energy performance of a building cannot be known until some years after commissioning. Before the advent of computers, assessments were carried out by constructing scale models, testing the models (in a wind tunnel, for example), and scaling the results. These studies were almost exclusively applied to large and expensive structures, where the consequences of failure could be profound. As computers evolved, simulation migrated to the computational realm, and now dynamic simulation can be used at the early design stage to establish the performance parameters that a building should meet.

Even quite basic computer hardware can bear the computational load of contemporary performance simulation (e.g., see Section 9.3.1). Furthermore, as building standards tighten, construction heat loss becomes a lesser consideration, and ventilation performance, solar utilisation, and novel approaches to energy supply become increasingly dominant. Two issues are impeding the development and use of dynamic simulation, as follows.

- Perceived complexity and the mistaken belief that only simulation experts can use such tools. There is also resistance because many commercial tools are effectively black boxes. Whilst the principles of the calculations may be described, it is impossible to check what actual calculation process led to a particular result. Open source programs (where the user can scrutinise the computer code and make changes to solve a particular problem) do exist but are oriented towards the academic or specialist user.
- There is no plan for what the use of the simulation tool is to achieve. Blindly building models, running them through some representative weather data, and generating pages of output, provides little understanding of why the building performs as it does and how performance might be improved. Worse still, is the endemic belief that generating the ratings required by standards authorities and achieving an arbitrary metric means that the building is performing well.

The first issue is the responsibility of developers. Building modelling, often seen as a time-consuming and fault-prone process, could be revolutionised with automated tools. These might, for example, take the initial architect design (as a BIM model), identify the construction types and materials, model the intersections (heat bridges), construct a suitable airflow model, calculate plant component sizes, propose a lighting scheme, and deliver a ready-to-simulate model to the user. Some initial automated simulation tasks could be conducted, e.g., assessing building orientation issues, identifying overheating risk, minimising the need for heating and cooling, assessing the visual quality of internal spaces, and quantifying the acoustic environment.

The second issue is the focus of the rest of this section. In general, the results generated by a BPS+ tool should be comparable with real-world equivalent measures; otherwise, there is no way to validate the computational outputs. This applies even if there is no intention to carry out a post-occupancy evaluation. For example, the temperature in a zone as measured by a sensor will be a complex synthesis of air and surface temperatures (of walls, windows, and heat emitters) and their respective areas and view factors. Calculating the performance parameter, which synthesises metrics, is necessary for comparing the measured zone temperature with the simulated temperature. The performance parameter is the value of most interest to the user or commissioning client. The metrics are what the computer simulation program must calculate to generate the performance parameters, to whatever degree of accuracy is deemed acceptable. For example, an operative temperature consisting of the average of the air and the mean radiant temperature may be acceptable as the performance parameter of interest. Likewise, any such parameters that represent the controlled variable for plant items must be correctly calculated from computed metrics. Otherwise, particularly with short time steps, the modelled plant will not behave realistically.

Many of the calculations to produce the required performance parameters may be carried out by post-processing the core simulation metrics. For example, rating schemes usually require the calculation of CO_2 emissions. The computational metric used is the energy required per time step. This is converted to energy per unit of time (e.g., kWh/y), which is then multiplied by the emission content per unit of fuel used (kg CO_2/kWh) to produce an annual emission rate. This may be further adapted depending on transmission losses, primary energy conversion efficiency, and so on. The accuracy of this parameter is highly dependent on the accuracy of the various external factors in the calculation and is impossible to verify by experimental means.

Building owners and potential tenants/purchasers need these ratings to determine that a building meets the required energy efficiency standards and that its energy performance is acceptable to potential purchasers or lessees.

3.3 Environmental impact assessment

An Environmental Impact Assessment (EIA) is a decision-making tool that spans the lifecycle of a development (Brown and Thérivel 2000). After predicting various environmental impacts that are likely to occur due to the initiation of the project, the EIA can identify alternative solutions to minimise or avoid problems (Brady *et al.* 2011). The EIA can be applied in the design and feasibility stage, or sometimes the conceptual stage, of a project, as its main aim is to ensure that potential environmental consequences are foreseen and avoided as early as possible, and preferably within the planning stage of the project lifecycle (Colombo 1992). An

EIA, similar to an economic analysis or an engineering feasibility study, is a management tool for engineers and decision-makers. For example, a designer developing a project that suits the local environmental conditions is more likely to complete the project within the available budget and time constraints. It should be noted that an EIA is a mandatory procedure and, as such, must be performed with due diligence to ensure maximum efficacy (Morrison-Saunders and Bailey 1999).

Enabling life cycle assessments of buildings and their technical systems requires a data model that covers the construction, operation, and demolition phases. Product modelling research has been underway since the mid-70s (Eastman 1991, Eastman *et al.* 2008) with efforts to develop a standard model for the construction industry (Hannus *et al.* 1994, Tolman 1999) although no universal model yet exists. As the data model for BPS+ becomes more extensive, it will tend to become assessment method independent.

All phases of a building's lifecycle generate environmental impacts. The annual UK carbon emissions from domestic buildings for 2021 was 408 $MtCO_2e$ (BEIS 2021), the embodied carbon emissions produced by the construction sector was estimated at 9.4 $MtCO_2e$ (2.3%) and are responsible for 6.7% of the UK total CO_2 emissions (Drewniok *et al.* 2023, BEIS 2021, Giesekam *et al.* 2014). At the end of its life, a building generates material waste flows that must be included in the assessment of its environmental impact. Construction waste also causes environmental impacts during transport and processing, and although such waste is increasingly recycled, around 8% of the approximately 68 million tonnes of construction and demolition waste produced each year in the UK ends up in landfills (CPA 2022).

A number of projects are executed in a manner that imparts, at best, a modest level of attention to environmental issues, especially during the concept and planning stages (Tukker *et al.* 2008, Scrase and Sheate 2002). This can easily result in issues being identified at a much later stage of the project lifecycle, such as during the detailed design and construction stages, with very little opportunity to find or implement a viable solution. Therefore, the late focus on such environmental issues could severely hinder the entire project (Therivel 2004, Curran 2008).

Due to these possible inadequacies, it is recommended that the level of environmental attention be as high as possible, particularly during the initial stages of a project. Integrating the EIA mechanism early in the project planning cycle allows for more solutions and options to be made available whilst incorporating required changes with minimum difficulty. Figure 3.8 depicts the integration of environmental considerations in the lifecycle of a project and shows how and when an EIA can contribute positively to the project's progress.

To enable a comprehensive assessment of the environmental impact of a building, an EIA method and supporting data are required. Such a method, comprising the following steps, is defined in ISO standards 14040 to 14043.

- *Goal definition,* defining the purpose of the study, its scope, and the time and spatial resolution. This step also includes a description of the system boundaries, the level of detail required, and the origin of the collated information.

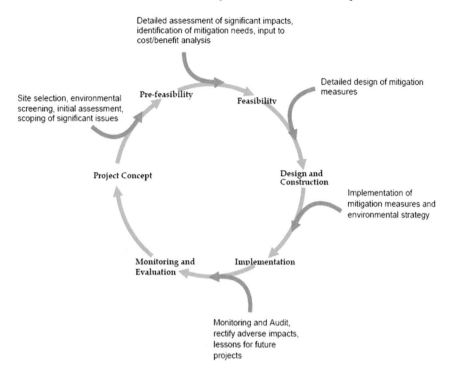

Detailed assessment of significant impacts,
identification of mitigation needs, input to
cost/benefit analysis

Detailed design of mitigation
measures

Site selection, environmental
screening, initial assessment,
scoping of significant issues

Pre-feasibility

Feasibility

Project Concept

Design and
Construction

Implementation of
mitigation measures and
environmental strategy

Monitoring and
Evaluation

Implementation

Monitoring and Audit,
rectify adverse impacts,
lessons for future
projects

*Figure 3.8   Integration of an EIA mechanism with a project's lifecycle (credit:
National Environment Commission 2012)*

- *Inventory analysis,* quantifying the environmental loads associated with the building, such as resource depletion (material and energy) or pollutant emissions.
- *Life cycle impact assessment,* quantifying the adverse effects of the loads identified from the inventory analysis, including their classification into impact categories.
- *Results interpretation, leading to* the identification of options for redesign where required.

The process must accommodate all the materials and processes that occur during the different phases of the building's life cycle and have a potential effect on the environment (Edwards and Hobbs 1998). Further, the system boundaries must encompass all energy and mass flows related to the analysed product. For instance, material delivery/disposal requires transportation energy. The result is an assessment of the environmental impacts generated by the manufacturing, transport, assembly, operation, maintenance, replacement, decommissioning, and elimination of the construction materials used in the building.

Citherlet (2001) established an EIA procedure for BPS+ as summarised in Figure 3.9 and tested the approach through implementation within ESP-r (Citherlet *et al.* 2000).

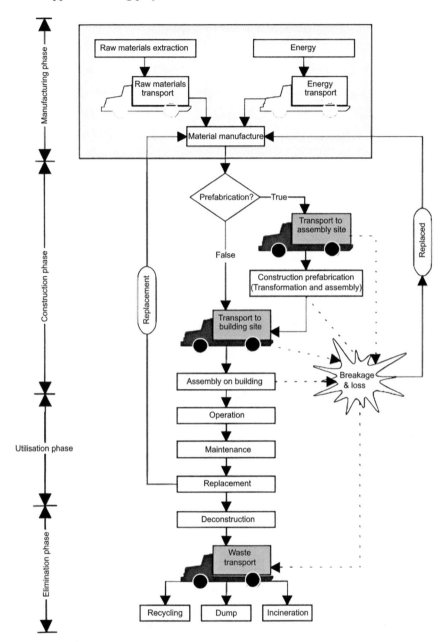

Figure 3.9   Components of an EIA for BPS+ (credit: Citherlet 2001)

The EIA requires the calculation of extrinsic (energy consumed for services) and intrinsic (building entity) contributions. The existing capabilities of ESP-r allowed the calculation of annual energy consumption, which was used to calculate

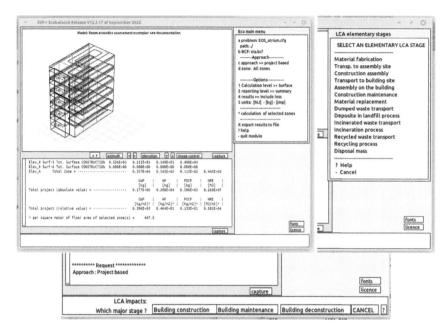

Figure 3.10   ESP-r ecobalance tool focused on an office atrium

the extrinsic impacts. For the intrinsic contribution (e.g., materials), an ecobalance life cycle analysis (LCA) module was developed, the interface of which is shown in Figure 3.10. This supports the integration of LCA and BPS+.

A project is loaded, and the corresponding geometry is displayed in the graphic feedback area. The sub-menu on the right gives access to elementary stages, allowing the user to focus on aspects of the building life cycle. Based on the loaded LCA data for the region and the level of detail specified, the EIA calculation proceeds, and the result is displayed or exported to file for processing elsewhere (e.g., to produce the emissions summaries of an IPV as shown below). To demonstrate the approach, ESP-r was used to assess the Energie Ouest Suisse SA (EOS) headquarters building located in Lausanne[11], which has several innovative features relating to natural light utilisation.

The design process usually involves multi-criteria decision-making, and ideally, the selection of the building fabric should take into account thermal, acoustic, and embodied energy performance. In relation to LCA, a particular mix of construction materials may emerge that satisfy several design team goals but adversely affect acoustic performance (Strachan 1997). The atrium in the EOS building is a large volume intended as a social space. To carry out an acoustic assessment, acoustic properties were added to the ESP-r model (Citherlet and Macdonald 2003). These properties define material absorption at 1/3 octave middle

---

[11]https://rdrarchitectes.com/en/project/eos-headquarters

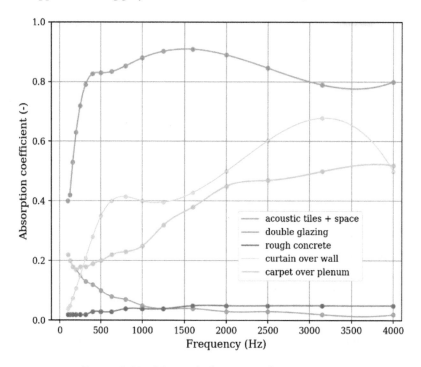

Figure 3.11   Material absorption characteristics

frequencies. Different combinations of materials and thicknesses can have sub-stantially different acoustic signatures, as shown in Figure 3.11.

Rough concrete and glazing have little absorption at any frequency, whilst carpet over a floor plenum absorbs well at higher frequencies. Acoustic tiles over batt insulation or air spaces offer considerable scope for improving space acoustics. Specialist acoustic simulation tools such as ODEAN[12] offer an extensive list of material properties, and absorption coefficient tables are available online[13].

Figure 3.12 shows ESP-r being used to control the acoustic assessment process and presents an analysis outcome for the EOS atrium.

An initial prediction of a high reverberation time was deemed to be challenging for conversation and led to the introduction of acoustic treatments that reduced the reverberation time to within an acceptable range for the intended use but were neutral in terms of an LCA assessment.

Figure 3.13 summarises the overall assessment outcome for the EOS building in the form of an IPV (see Section 3.2.3) comprising the performance indicators listed in Table 3.10.

[12]https://odeon.dk/downloads/materials/acoustic-absorption-data/
[13]For example, https://acoustic.ua/st/web_absorption_data_eng.pdf

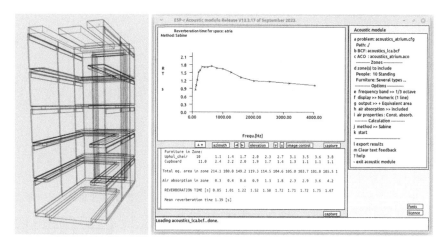

Figure 3.12   Acoustic assessment of the EOS building atrium

The LCA analysed all materials' phases, focusing on intrinsic and extrinsic environmental impacts. The intrinsic impacts were categorised by construction, operation, and deconstruction, whilst the extrinsic contribution was based on energy sources used for building services. Results were normalised per gross floor area and year, with a fixed 80-year building life span. The impact of different energy sources was illustrated using two regional sets, one obtained with an UCPTE[14] electricity mix (Euro) and the other with a Swiss-mix (CH). The Swiss-mix showed lower environmental impacts due to the predominance of hydropower in the region.

The results demonstrated that when all building phases are included in the impact assessment, the extrinsic and intrinsic contributions are of the same order of magnitude. The largest intrinsic contribution was 'operation' as it encompasses maintenance and replacement (i.e., the sum of all downstream impacts generated by the replacement of a building element, including its manufacture, transport, and assembly). During the building's life span, only reinforced concrete is persistent. The facade cladding is replaced once, whilst the fitted carpet might be replaced several times. The sum of all contributions generated for these material replacements is therefore significant. As can be seen in Figure 3.13, extrinsic aspects generate the major contribution to the Non-Renewable Energy outcome (73% Euro, 70% CH) and the Global Warming Potential outcome (69% Euro, 64% CH). Conversely, the intrinsic aspects generate the major contribution to Photochemical Ozone Creation Potential (58% Euro, 65% CH) and Acidification Potential (68% Euro, 84% CH).

Depending on the region, information gathering can be problematic – factors for materials, transport, and disposal may be available, whilst material loss rates are difficult to obtain. Nevertheless, the study concluded that the EIA approach could

[14]https://docstore.entsoe.eu/news-events/former-associations/ucte/Pages/default.aspx

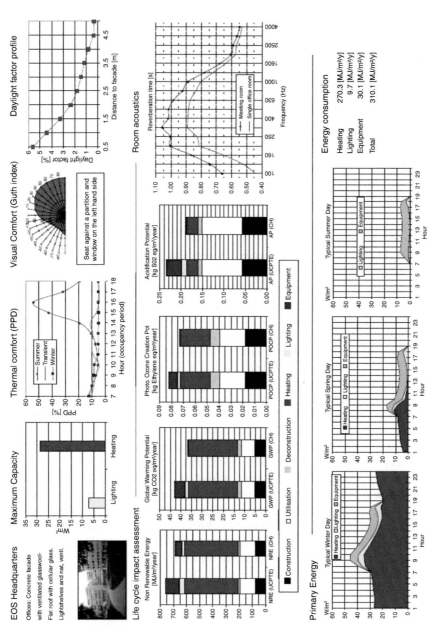

Figure 3.13   An IPV for the EOS office building

Table 3.10   Performance indicators included in the IPV

Aspect	Indicator
Energy consumption	Maximum heating capacity required. Annual energy consumption.
Thermal comfort	Percentage Persons Dissatisfied.
Daylight availability	Daylight factor profile.
Visual comfort	Guth index.
Room acoustic	Reverberation time.
Environmental impact	Emissions of pollutants, energy consumption. Emission of pollutants, construction materials.

deliver a comprehensive design stage assessment of the environmental impacts over the building lifecycle.

3.4   Weather boundary conditions

In most BPS+ applications, the aim is to test alternative design possibilities against relatively short-period data that characterise typical or extreme weather conditions for the location in question. Undesirable options may then be disregarded before the near-final scheme is subjected to long-term, annual, or lifetime simulations to determine energy consumption trends. The selection of an annual weather collection should be done in a manner that ensures that:

- portions of the collection correspond to the different levels of severity under which the building will operate, such as extreme and typical conditions in the winter, summer, and transition seasons; and
- the collection overall will support an assessment of cost-in-use.

The former condition is easier to satisfy since it is usually straightforward to locate representative short-term sequences within weather collections, even where these do not correspond to the location in question. This task is assisted by reference to exceedance data as depicted in Figure 3.14, here related to external air temperature at three UK locations.

The latter condition will require data corresponding to the building location and a means to rate its overall severity. Degelman (1997) helpfully demonstrated that the results from typical week simulations could be made to correlate well with results from annual simulations, thus allowing short-term simulations to be scaled up.

Table 3.11 lists the (usually hourly) weather parameters required for simulation. Annual collections of these data are widely available online and often come packaged within a specific BPS+ program or can be acquired separately in the required format.

There are many repositories of weather data suitable for use with BPS+ (Crawley and Barnaby 2019). Practitioners may have their own preferences as to

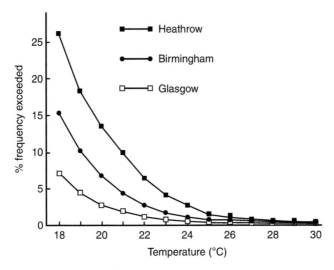

Figure 3.14   Air temperature exceedance curves (credit: CIBSE Applications Manual 10, Ventilation in Buildings)

Table 3.11   Weather parameters required by BPS+

Dry bulb temperature (°C)
Wet bulb temperature (°C)
Wind speed (m/s)
Wind direction (° clockwise from north)
Atmospheric pressure (bar)
Net longwave radiation (W/m²)
Precipitation (mm)
Global horizontal (or direct normal) solar radiation (W/m²)
Diffuse horizontal solar radiation (W/m²)
and, where solar radiation data is not available:
Cloud cover and type (%, −)
Sunshine hours (hr)

the source organisation or the specific regime of data – e.g., a typical meteorological year or a test reference year – and each of these can be packaged into a range of file formats that can be understood by a specific tool or for which a translation facility exists. In most cases, vendors will provide pre-packaged weather data in the required format.

One widely accessed repository is climate.onebuilding.org. This hosts Typical Meteorological Year (TMY) data in multiple file formats for thousands of sites around the world. The site is arranged by the seven World Meteorological Organisation regions. The distribution of sites is, of course, variable: Antarctica has 114 sites, whilst North America has more than 4000.

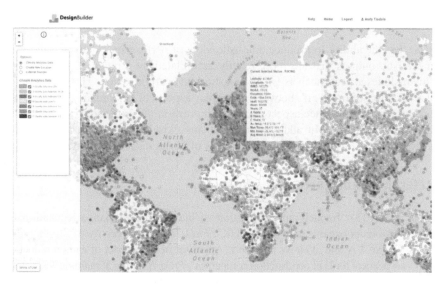

Figure 3.15   DesignBuilder Climate Analytics' weather coverage (credit: DesignBuilder)

In addition to the weather data collected by formal organisations, individuals have also established weather stations. An example of this is Weather Underground[15], a site comprising more than 250,000 weather stations. Although the quality and continuity of what is shared vary, the density of sites in some regions allows for the possibility of 'hyperlocal' data. For example, in one retrofit study involving housing across a 30 km coastal region that was known to have considerable variability, there was one official weather station but 12 sites on Weather Underground.

Another popular repository is DesignBuilder Climate Analytics[16], an extensively populated database of global, hourly weather data, and analysis tools (Figure 3.15) to help select the most appropriate collection (e.g., design, typical, or actual years).

To standardise the format of weather files, the EPW[17] file format was introduced by Crawley *et al.* (1999) and this has been widely adopted within BPS+ programs. Further, Crawley and Lawrie (2019, 2021) have offered advice on the selection of weather files for building performance assessments in terms of the corresponding level of severity required for different tasks. A typical meteorological year (TMY) should be used to represent median or typical conditions, whilst an eXtreme Meteorological Year (XMY; Crawley and Lawrie 2015) can be used to represent extreme conditions. Using both file types will capture the wide range of conditions likely to be experienced over a building's lifetime.

[15]https://sunderground.com/
[16]https://designbuilder.co.uk/cahelp/Content/ClimateAnalyticsData.htm/
[17]https://designbuilder.co.uk/cahelp/Content/EnergyPlusWeatherFileFormat.htm

3.5   Chapter summary

BPS+ has the ability to produce a holistic view of performance delivered in terms of realistic outputs that are more readily assimilated by a range of stakeholders. This includes aspects such as LCA, acoustics, and carbon action. This supports the identification of trade-offs between dissimilar aspects of performance, thus ensuring that one performance aspect does not improve at the expense of another to an unacceptable degree. In addition, mechanisms exist to collate such outputs for presentation in forms that support the needs of disparate stakeholders.

References and further reading

Ali S M (2013) 'Impacts and mitigation of latest climate change legislation on participant organisations in the UK', *PhD Thesis*, University of Strathclyde.

Ali S, Clarke J A and Weir W S (1999) 'Appraisal of the impact of carbon reduction assessment tools in construction companies', *Proc. Building Simulation'99*, Kyoto.

Augenbroe G (1992) 'Integrated building performance evaluation in the early design stages', *Building and Environment*, 27(2), pp. 149–161.

BEIS (2021) *Final UK Greenhouse Gas Emissions National Statistics: Government Document*, Department for Business, Energy and Industrial Strategy, Stationary Office, London.

Bodmann H W and Söllner G (1965) 'Glare evaluation by luminance control', *Light and Lighting*, 58, p. 195.

Brady J, Ebbage A and Lunn R (2011) *Environmental Management in Organizations – The IEMA Handbook* (2nd Edn.)', Routledge, Milton Park, ISBN 9781849710626.

Brager G S and de Dear R J (1998) 'Thermal adaptation in the built environment: a literature review', *Energy and Buildings*, 27(1), pp. 83–96.

Brown A L and Thérivel R (2000) 'Principles to guide the development of strategic environmental assessment methodology', *Impact Assessment and Project Appraisal*, 18(3), pp. 183–189.

Casini M (2022) *Advanced Technology, Tools and Materials for the Digital Transformation of the Construction Industry*, Woodhead Publishing Series in Civil and Structural Engineering, ISBN 978-0-12-821797-9.

Citherlet S (2001) 'Towards the holistic assessment of building performance based on an integrated simulation approach', *PhD Thesis*, LESO-PB, EPFL, Lausanne.

Citherlet S and Macdonald I A (2003) 'Integrated assessment of thermal performance and room acoustics', *Energy and Buildings*, 35, pp. 249–255.

Citherlet S, Clarke J A and Hand J (2000) 'A life cycle based data model for multiple-domain assessment of building performances', *Automation in Construction*.

Clarke J A, Markus T A and Morris E (1984) 'The influence of climate on housing: a simple technique for the assessment of dynamic energy behaviour', *Energy and Buildings*, 7, pp. 243–259.

Colombo A G (Ed.) (1992) *Environmental Impact Assessment, Springer*, Springer, Berlin, ISBN 9789401051163.

CPA (2022) 'How much waste is produced by the construction sector?', Construction Products Association.

Crawley D B and Barnaby C (2019) 'Weather and climate in building performance simulation', In Hensen J L M and Lamberts R (eds.), *Building Performance Simulation for Design and Operation* (2nd Edn.), Chapter 6, pp. 384–395, Routledge, Milton Park, ISBN 9781138392199.

Crawley D B and Lawrie L K (2019) 'Should we be using just 'Typical' weather data in building performance simulation?', *Proc. Building Simulation'19*, Rome, pp. 4801–4808.

Crawley D B and Lawrie L K (2021) 'Our climate conditions are already changing – Should we care?', *Building Services Engineering Research and Technology*, 42(5), pp. 507–516.

Crawley D B, Hand J W and Lawrie L K (1999) 'Improving the weather information available to simulation programs', *Proc. Building Simulation'99*, II, pp. 529–536.

Crawley D B and Lawrie L K (2015) 'Rethinking the TMY: is the 'typical' meteorological year best for building performance simulation?', *Building Simulation '15*, Hyderabad.

Curran M A (2008) 'Development of life cycle assessment methodology: a focus on co-product allocation', *PhD Thesis*, Erasmus University, Rotterdam.

Degelman L O (1997) 'Examination of the concept of using "typical-week" weather data for simulation of annualized energy use in buildings', *Proc. Building Simulation '97*, Prague.

De Wilde P (2018) *Building Performance Analysis*, John Wiley & Sons Ltd., New York, ISBN 9781119341925.

Drewniok M P, Dunant C F, Allwood J M, Ibell T and Hawkins W (2023) 'Modelling the embodied carbon cost of UK domestic building construction: today to 2050', *Ecological Economics*, 205.

Eastman C M (1991) 'The evolution of CAD: integrating multiple representation', *Building and Environment*, 26 (1), pp. 17–23.

Eastman C, Teicholz P, Sacks R and Liston K (2008) *BIM Handbook: A Guide to Building Information Modeling for Owners, Managers, Engineers and Contractors*. Hoboken, NJ: Wiley.

Edwards S and Hobbs S (1998) 'Data collection and handling for environmental assessment of building materials by architects and specifiers', *Proc. Conf. on materials and Technologies for Sustainable Construction*, Gävle, Sweden, pp. 569–576.

Fanger P O (1972) *Thermal Comfort. Analysis and Applications in Environmental Engineering*, McGraw-Hill, New York.

Gagge A P, Fobelets A P and Berglund L G (1986) 'A standard predictive index of human response to the thermal environment', *ASHRAE Transactions*, 92(2B), pp. 709–731, Atlanta.

Giesekam J, Barrett J, Taylor P and Owen A (2014) 'The greenhouse gas emissions and mitigation options for materials used in UK construction', *Energy and Buildings*, 78, pp. 202–214.

Guo X, Li Y, Shi, H, *et al.* (2023) "Carbon reduction in cement industry – an indigenized questionnaire on environmental impacts and key parameters of life cycle assessment (LCA) in China", *Journal of Cleaner Production*, 426, 139022.

Guth S K (1963) 'A method for the evaluation of discomfort glare', *Illuminating Engineering*, 58(5), pp. 351–364.

Guth S K (1966) 'Computing visual comfort ratings', *Illuminating Engineering*, October, pp. 634–642.

Hand J W and Strachan P A (1998) 'ESP-r data model decomposition', *Project Report TR 98/9*, ESRU, University of Strathclyde.

Hannus M, Karstila K and Tarandi V (1994) 'Requirements on standardised building product data model', *Proc. Conf. on Product and Process Modelling in the Building Industry Dresden*, Germany, pp. 43–50.

Humphreys M and Nicol J F (1998) 'Understanding the adaptive approach to thermal comfort', *ASHRAE Transactions*, 104(1), pp. 991–1004.

IBPSA-USA (2023) 'Request for Proposals: Review of BPM Protocols and BEM Reporting Output', https://mailchi.mp/ibpsa.us/rfp-bpm-protocols?e=e376797a0d/ (submission deadline July 13, 2023).

ICI (2019) Discomfort Caused by Glare from Luminaires with a Non-Uniform Source Luminance International Commission on Illumination, ISBN: 978-3-902842-15-2.

Kesik T (2015) 'Vital signs: towards meaningful building performance indicators', *Green Paper,* Faculty of Architecture, University of Toronto.

Lavery G (2011) 'ERIC replacing marginal abatement cost curves', https://laver-ypennell.com/replacing-marginal-abatement-cost-curves-maccswith-erics/.

Mahdavi A (2003) 'Computational building models: theme and four variations', *Proc. Building Simulation'03*, 1, pp. 3–17, Eindhoven.

Mahdavi A and Ries R (1998) 'Towards computational eco-analysis of building designs', *Computers and Structures*, 67(5), pp. 375–387.

Markus T, Whyman P, Morgan J, *et al.* (1972) *Building Performance*, Applied Science, London.

Morrison-Saunders A and Bailey J (1999) 'Exploring the EIA/environmental management relationship', *Environmental Management*, 24, pp. 281–295.

National Environment Commission (2012) *Environmental Assessment Guideline for Mines and Quarries*, Royal Government of Bhutan.

Nikolaou T, Kolokotsa D and Stavrakakis G (2011) 'Review on methodologies for energy and thermal comfort benchmarking, rating and classification of office buildings', *Advances in Building Energy Research*, 5(10), pp. 53–70.

Prazeres L (2006) 'An exploratory study about the benefits of targeted data perceptualisation techniques and rules in building simulation', *PhD Thesis*, ESRU, University of Strathclyde.

Prazeres L and Clarke J A (2005) 'Qualitative analysis of the usefulness of perceptualisation techniques in communicating building simulation outputs', *Proc. Building Simulation'05*, Montreal, 15–18 August.

Roulet C A, Flourentzou F, Labben H H, *et al.* (2002) 'ORME: a multicriteria rating methodology for buildings', *Building and Environment*, 37(6), pp. 579–586.

Santamouris M (2001) *Final Report of the EUROCLASS Project, SAVE Program*, Directorate General for Transports and Energy, European Commission, Brussels.

Scrase J I and Sheate W R (2002) 'Integration and integrated approaches to assessment: what do they mean for the environment?', *Environmental Policy and Planning*, 4(4), pp. 275–294.

Somar (2010) 'MAC curve chart', https://somar.co.uk/tools/marginal-abatement-cost-curve.php?Pageid=121/.

Strachan P A (1997) 'Noise and vibration', *Sensor Systems for Environmental Monitoring*, Campbell M (Ed), 2, pp. 283–310, Blackie Academic and Professional.

Therivel R (2004) *Strategic Environmental Assessment in Action*, Routledge, ISBN 9781849772655.

Tolman F P (1999) 'Product modeling standards for the building and construction industry: past, present and future', *Automation in Construction*, 8, pp. 227–235.

Tukker A, Eder P and Suh S (2008) 'Environmental impacts of products: policy relevant information and data challenges', *Industrial Ecology*, 10(3), pp. 183–198.

van Treeck C and Wölki D (2019) 'Indoor thermal quality performance prediction', In Hensen J L M and Lamberts R (Eds.), *Building Performance Simulation for Design and Operation* (2nd Edn.), Chapter 5, pp. 146–190, Routledge, Milton Park, ISBN 9781138392199.

Wagner A (2007) 'Post occupancy evaluation and thermal comfort: state-of-the-art and new approaches', *Advances in Building Energy Research*, 1, pp. 151–175.

Westergren K-E, Högberg H and Norlén U (1999) 'Monitoring energy consumption in single-family houses', *Energy and Buildings*, 29(3), pp. 247–257.

Zhou Y, Tam V W & Le K N (2023) 'Sensitivity analysis of design variables in life-cycle environmental impacts of buildings', *Journal of Building Engineering*, 65, 105749.

Zipperer A, Aloise-Young P A, Suryanarayanan S, *et al.* (2013) 'Electric Energy Management in the Smart Home: Perspectives on Enabling Technologies and Consumer Behavior', *Technical Report*, National Renewable Energy Laboratory, USA.

Zmeureanu R (1992) 'A new method for evaluating the normalized energy consumption in office buildings', *Energy*, 17(3), pp. 235–246.

Chapter 4

Application in practice

This chapter describes the issues to be confronted when applying BPS+ in new design or retrofit situations and identifies opportunities for simplifying the process in contrast to the high-resolution modelling examples of the next chapter. The contention is that such simplifications can introduce the use of a simulation approach in a non-demanding manner that is helpful for the novice.

Despite the advanced capabilities of contemporary BPS+ tools, users continue to face challenges in their applications. The first is the inherent conflict in tool interfaces between user friendliness and the complexity of the input model. This issue is exacerbated by the divergence of conceptual frameworks between design-oriented program users and technical-oriented program developers. Next, there is little agreement on the data model used to define a building and its energy systems. Program-specific data models frustrate the validation process and present a barrier to collaborative design. Third, the absence of agreed performance assessment methods has forced users to develop personalised appraisal strategies that require coordination expertise that is difficult to replicate. The establishment of standard methods would harmonise program use and make the application experience less fraught for the novice. Lastly, user expectations are evolving and to service these new technical features will need to be added.

BPS+ application requires teamwork due to its diverse demands. Individuals with appropriate domain knowledge[1] and modelling skills are needed to translate client requirements into modelling and simulation specifications, collaborate on model development, service activities like model calibration and automation, commission simulations, analyse outcomes, and prepare design team and client reports. Design team members can operate intra- or inter-company, with analyses based on a common program or multiple specialist programs. A BPS+ team typically includes a process manager, model creators, simulation coordinators, and results analysts, with defined procedures for model quality assurance and outcome archiving in support of later reuse and audit. The stages involved in undertaking a computational approach to design are as follows.

---

[1]Building physics, thermodynamics, fluid mechanics, heat and mass transfer, control theory, etc.

Creating an input model

Modelling is the process of expressing a design hypothesis to a sufficient level of detail to address the performance issues of interest. There are often lively discussions as to what constitutes a model that is fit for purpose. It depends on the context, the scale of the project, the design stage, the availability of data, and the specific performance appraisal objectives. Unlike visual assessments, energy modelling is often abstract, involving judicious simplification of complex geometry or the grouping of rooms of similar temperatures. It is possible to mix levels of detail within a model, with critical spaces or facades represented at high resolution and surrounding rooms and facades having simplified representations. Such abstract modelling is in contrast to the high-resolution approach as proffered in Chapter 5.

Although the aim is usually to preserve surface areas (the principal influence on convective/conductive/radiative heat transfer), material volumes (the principal influence on thermal capacity), and important spatial relationships (the principal influence of radiation exchange), determining a workable approach that delivers useful information may require exploratory work. An initial simplified model might be systematically improved to determine the impact on predictions. It is time to stop refining the model when the predicted impact becomes negligible. A reasonable strategy is to select a portion of the building and use this to explore different design options. Lessons learned can then be applied to the full-scale model.

Each BPS+ tool will have use patterns that reduce resource requirements. This can be as simple as noticing repeating elements and fully attributing initial definitions before replicating them. For a building type that has not been assessed in some time, a review of similar projects may provide helpful clues. It may even be possible to use a past project as a starting point in a new project. The risk, of course, is that corporate memory may not be long enough to recall the downsides of that past model or the archetype may include assumptions that are no longer relevant.

Base case and reference models

Many projects involve maintaining a suite of models. An initial base case model might correspond to an existing building scheduled for refurbishment or to a proposed new-build. Several reference models are then spawned to represent different design options. The base case model should support the range of performance metrics that will allow comparison with the results for the reference cases.

For example, a daylight utilisation study might modify the base case model to create reference models encapsulating combinatorial variations of advanced glazing, daylight capture, and daylight-responsive luminaire control. The required outputs would cover energy saving due to artificial lighting displacement as well as aspects of visual comfort such as illuminance level and glare distribution. In a refurbishment context, the reference models would be clones of the existing building with upgrade options introduced separately and together.

Managing a suite of models is often outwith the capability of the BPS+ tool, requiring a clear model naming convention and robust documentation to track and manage changes implemented over time and make these clear to others.

Usage and environmental systems

HVAC systems can be modelled in various ways spanning a spectrum. At one pole, only the impact on the building is considered via time- and space-dependent flux inputs (often termed idealised control). At the other pole, a network of connected components is defined allowing the energy exchanges within and between individual components to be simulated. As with building modelling, a reasonable rule of thumb is to represent systems as simply as possible in relation to the required outputs but in a manner that can be extended and refined as the project evolves.

Another way to express this incremental approach is to delay 'investment in detail' until there is evidence of need. What is the point of including a variable air volume air conditioning system with advanced control before heating/cooling load studies have demonstrated that air conditioning is the only option? The sunk cost leaves little time to identify options that offer better value to the client.

Boundary conditions

An essential pre-simulation task is to determine the weather influences to which the building should be subjected. This can be done by obtaining statistics for the site and matching them with local weather data or 'reference year' data. This helps isolate design and typical sequences for different seasons, supports performance exploration, and allows design improvement strategies to evolve. Some simulation programs will have embedded rules for selecting weather patterns of interest. Long-term simulations based on historic or projected weather data can also be conducted to encapsulate the conditions likely to occur during the building's lifetime.

This task can be reframed as identifying boundary conditions that could cause the building to fail, for example, a sequence of warm sunny days may stress the design more than a single extreme day. Other conditions of interest might include cold and clear/windless or overcast/windless periods to determine operational resilience in the absence of solar and/or wind inputs. Such patterns can be readily located from a weather file search or otherwise synthesised.

Model calibration

Before simulation, the base case model should be calibrated against measured or benchmark data (see Section 4.4). Where comparisons are poor, remedial action is necessary. This might include a judicious adjustment to the model or the imposition of monitored data. The purpose of calibration is to gain confidence in the simulation results, not to validate a tool's mathematical model, which is a separate activity applied to a program's theoretical constructs (see Section 2.6).

Simulations and results

It is normal to undertake multiple simulations as changes are made in response to lessons learned. This often necessitates design parameter sensitivity studies that require simulation automation (see Section 4.5). In any event, the results need to be collated, interpreted, and presented in a client-friendly manner (perhaps involving multimedia) to deliver an engaging message.

4.1   Performance assessment method

The application of BPS+ in design and policy-making contexts requires the coordination of the above actions. Successful simulation teams tend to have made early and substantial investments in codifying methods and work practices that have been proven to work. Table 4.1 presents a generic performance assessment method (PAM) that can be tailored to specific cases as required. Here, the italicised text indicates the context-specific knowledge required at each stage.

The stage 1 base case model may be constrained or unconstrained. Constrained models are common in new design contexts where innovative features are initially omitted so that they may be added at stage 9 to determine their impact separately or together. Unconstrained models are suitable in refurbishment contexts where existing systems are included and systematically adjusted or replaced thereafter.

The stage 2 model calibration process depends on the target building's characteristics. If the building exists, an empirical approach may be used, comparing simulation outputs with metered or monitored data. Where agreement is unacceptable, adjustments may be made to the model with a focus on uncertain parameters. If the building is a proposed design, calibration may involve inter-program comparison or adjustments to align it with standard test cases for which benchmark data are available (e.g., a BESTEST test case, see Section 2.6). At its simplest, calibration may involve ensuring that the results comply with common sense when the model is prodded in creative ways.

The stage 3 activity involves setting boundary conditions as mentioned above and discussed in Section 3.4. As an aside, initial no-plant simulations will provide early feedback on building performance to the design team and client.

Table 4.1   A generic performance assessment method for BPS+

1. Establish initial model for a *constrained or unconstrained base case design.*
2. Calibrate model using *reliable techniques.*
3. Assign boundary conditions of *appropriate severity.*
4. Undertake integrated simulations using *suitable applications.*
5. Express multi-domain performance in terms of *suitable criteria.*
6. Identify problem areas as a function of *criteria acceptability.*
7. Analyse results to identify *cause of problems.*
8. Postulate remedies by *relating parameters to problem causes.*
9. Establish reference model to *required resolution* for each postulate.
10. Iterate from step 4 until overall *performance is satisfactory.*
11. Repeat from step 3 to establish *design resilience or replicability.*

The stage 4 activity may involve the use of a single tool or multiple tools operating cooperatively. In the latter case, it is important to ensure that the information being transferred between tools is not constrained by the input/output foibles of the tools involved in an exchange.

The stage 5 activity requires that careful consideration be given to the organisation and presentation of the simulation results in line with the assessment criteria and ratings of Section 3.1 and Section 3.2, respectively.

BPS+ tools will normally support the stage 6, 7, and 8 activities by providing data analysis facilities. For example, the components of overall energy use and their variation over time might be investigated or the variation in the mean age of air throughout a zone examined. Well-structured interrogations will lead to the identification of design parameters that could be adjusted to improve outcomes.

At stages 9 and 10, the initial model is adapted as required by adding or removing features and initiating further stage 4 simulations.

Finally, at stage 11 the weather boundary condition can be changed to assess resilience under extremes or the replicability of solutions should the intention be to 'export' the design to other locations.

Establishing a high-resolution model of a building at stages 1 and 9 requires extensive information, including 3D geometry, hygro-thermal properties of construction materials, thermal bridge details, air leakage distribution, internal thermal mass, operational data, HVAC descriptions, domestic hot water, and renewable energy components. These data are often collected from various sources such as construction drawings, site surveys, manufacturers' product data, building standards, city cadastres, and existing computer models in a variety of formats (SAP, PHPP, BIM, etc.). In the case of estates with many buildings (planned or existing), information collation is typically undertaken for a representative subset, making the investment in model construction less onerous and modest compared to the overall upgrade cost and effort. As the BIM standard evolves and new buildings replace old, digital models will likely become widely available.

Table 4.2 summarises a systematic approach to model evolution that provides significant design support against the maxim that 'behaviour follows description' or, alternatively stated, 'reward follows effort'.

The approach allows detail to be progressively added to a model as the design hypothesis evolves and domain interactions grow. It is a paradox that simulation is most powerful when used with uncertain data because it allows exploration of the impact of likely ranges in design and operational parameters to inform decision-making. Unfortunately, most BPS+ users do not follow the stages of Table 4.2, instead prioritising legislation compliance over occupant wellbeing.

Consider the following illustrative example of the incremental behaviour-follows-description approach based on the ESP-r system when its underlying data model is cumulatively refined.

A project manager module (Hand 1998) provides access to databases, BIM imports/exports, an integrated simulation engine, performance assessment tools, and third-party applications for CAD, visualisation, and report generation. It coordinates problem definition and data model exchange between support

Table 4.2   Progressive problem description mapped to assessment outcome

Cumulative model data	Typical assessment enabled
Pre-existing databases	Simple performance indicators (e.g. U-values, consumption benchmarks, etc.).
+ Geometry	Visualisation, photomontage, shading, insolation, etc.
+ Constructional attribution	Material quantities, embodied energy, etc.
+ Operational attribution	Casual gains, electricity demands, etc.
+ Boundary conditions	Photo-realistic imaging, illuminance distribution, no-systems thermal and visual comfort levels, etc.
+ Special materials	Photovoltaics and switchable glazing evaluation, etc.
+ Control system	Daylight utilisation, energy use, system response, etc.
+ Flow network	Ventilation and heat recovery evaluation, etc.
+ HVAC network	Psychometric analysis, component sizing, etc.
+ CFD domain	Indoor air quality, thermal comfort, etc.
+ Electrical power network	Renewable energy integration, load control, etc.
+ Enhanced geometry	Thermal bridging, etc.
+ Moisture network	Local condensation, mould growth, and health.

applications, allowing incremental evolution of the model description process and providing access to corresponding simulation engine functionality at each stage.

A new project typically begins with the review and updating of support databases. These include hygro-thermal, embodied energy, and optical properties for construction materials and composites, typical occupancy profiles, external surface pressure coefficient sets, plant components, mould species data, and weather collections. These databases may already include entities that match the project's needs, whereas a bespoke database may be required to explore the impact of innovative materials or new renewable components.

Embedded within such databases is knowledge that aids design conceptualisation. For example, the construction materials database contains hygro-thermal properties for a range of materials. These properties can be used to deduce behaviour, such as thermal diffusivity for temperature change response and thermal transmittance for heat loss behaviour allowing selections that match requirements.

Problem definition involves specifying building geometry via an in-built facility, a tightly linked CAD tool included in a distribution, or an industry application such as AutoCAD (Autodesk Ltd 1989). This allows for the creation of complex building geometry, which can then be attributed within the project manager. The route selected is largely a matter of personal preference although it is advisable to avoid unnecessary details that will make the attribution process time consuming and results interpretation fraught.

Wireline or false-coloured images can be generated at this stage as an aid to the communication of design intent or the study of solar/daylight access. The project manager provides wireframe perspective views and coloured, textured images via Radiance[2] (Larson and Shakespeare 1998) or Blender[3], automatically generating the required input models, driving these third-party applications and receiving back their outputs. It is possible to specify eye and focus point locations, angle of view, and scaling factors so that outputs can be directly merged with other media (say site photographs) to create a photomontage.

The construction and operational attribution process involves selecting products and entities from support databases and associating these with previously defined geometrical entities. It is at this stage that a simulation novice will appreciate the importance of a problem abstraction that minimises the number of entities requiring attribution, simulation processing, and performance data processing. The skills needed to create models that are fit for purpose are acquired over time and through practice. It is likely that this will benefit from corporate knowledge in addition to skills acquired via vendor training and in formal workshops.

Time-varying temperature, wind velocity, solar radiation, air pressure, and luminance boundary conditions of the required severity are now associated with the model to enable an appraisal of environmental performance – e.g., indoor thermal and visual comfort levels throughout the year – and to gain insight into the extent of required remedial action. As appropriate, these boundary conditions can be modified to represent extreme weather events or microclimate phenomena such as wind hollows, vegetative shading, and evaporative cooling.

As required, geometrical, constructional, or operational changes can be applied to the model to determine the impact on performance. This could involve exploring alternative constructional systems, imposing different occupancy loadings, or assessing approaches to daylight utilisation. The possibilities are limited only by the user's imagination.

To improve environmental performance and reduce energy use, special facade features might be introduced, such as PV components, transparent insulation, and adaptable glass (electro-, photo- or thermo-chromic) to manipulate parts of the solar power spectrum in desirable ways. Such systems may also have unintended consequences, and the study should include steps to identify conflicts, e.g., where a PV facade reduces daylight penetration to interior spaces and thereby increases electricity use for artificial lighting.

A luminaire control system, consisting of photocells connected to a circuit switch or dimming device, might be introduced to maximise energy displacement potential from daylight penetration. Simulations can be conducted to optimise the control system's parameters, examining the conflict between daylight capture benefits and the negative effects of reduced heat gain on heating load.

The integration of environmental control can now be explored through a control system consisting of open or closed loops that act to dictate heating, cooling, ventilation, and lighting availability, and resolve conflicts when required. The results might suggest revisiting the architectural aspects of the model at this point

[2]https://www.radiance-online.org/download-install
[3]https://blender.org

to fine-tune the building's dynamic response to better accommodate the intended control action.

Although some practitioners will assume a building requires an HVAC system, in many cases there will be periods when alternative approaches can play a role. An airflow network can be linked to a building model to determine the feasibility of natural or mixed-mode ventilation. Such networks can represent window opening and flow damper control, and thereby support the assessment of ventilation schemes including user interaction.

Where mechanical intervention is deemed necessary, a component network may be defined to represent the HVAC system for association with both the building model and any active flow network. The control definition previously established may be extended to provide internal component control and link the room states to the supply condition. Such a model can be used to study the operational characteristics of individual plant components as well as the behaviour of the system overall.

To examine indoor air quality, one or more spaces within the building model can be further discretised to enable the application of computational fluid dynamic (CFD) procedures to evaluate the intra-space air movement and the distribution of temperature, humidity, and species concentration. These data may then be combined to determine comfort levels and air quality at different points within the space. A useful indication of indoor air quality is the distribution of the mean age of air.

One of the defining features of BPS+ is support for multiple assessment domains. Each addition to the model may bring into play additional solvers and deeper analysis facilities. Whilst the building, fluid flow, HVAC networks, and CFD domain may be processed independently (Hensen 1999), it is helpful to subject them to an integrated assessment whereby the dynamic interactions are explicitly represented. For example, a building model might be assigned a flow network to represent natural ventilation, an HVAC network to represent ventilation heat recovery, a CFD domain to enable the analysis of air quality, and a moisture flow network to assess indoor humidity levels.

A further network might be added to represent the building's electrical power circuits. This can be used in conjunction with the previously established models for facade-integrated photovoltaics, luminaire control, HVAC, and flow networks to study scenarios for the utilisation of the outputs from building-integrated renewable energy components, cooperative switching with the public electricity supply, and the shedding of electrical load as a contribution to grid stability. Other technologies, such as heat pumps, combined heat and power plants, and fuel cells, can also be assessed.

For specialist applications, the resolution of parts of the model can be selectively enhanced to allow a focused study of particular issues. For example, a portion of a multi-layered construction might be finely discretised to enable the study of the behaviour of an innovative building component incorporating a potential thermal bridge. This can be connected to a moisture flow network to support an assessment of the potential for interstitial or surface condensation.

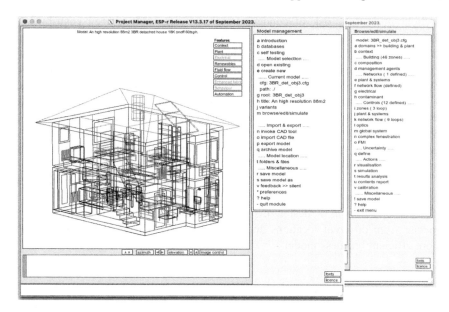

Figure 4.1 An ESP-r Project Manager session focused on a low-energy dwelling

By associating the time series pairs of near-surface temperature and relative humidity (to emerge from the integrated building, CFD and network air/moisture flow models) with the growth limit data as held in a mould species database, it is possible to determine the risk of mould growth. In this way, the different possible remedial actions can be explored – from eliminating moisture at the source, to modifying the constructional material or arrangement to prevent optimum growth conditions from occurring in the first place.

Within ESP-r, the project manager (Figure 4.1) is the agent that manages the creation and evolution of the problem composition. Here a model has been loaded and its active features highlighted. This model activates a subset of the available features as indicated. The menu on the right gives access to high-level facilities such as managing databases, import/export facilities, and exporting the model for use in other tools. Command menus are arranged hierarchically. Selecting the 'browse/edit/simulate' option, for example, leads to the menu shown on the right, which gives access to further options for managing and editing model features. Such menus reflect the model composition; here there are 46 thermal zones in a dwelling with nine physical zones plus 12 control loops.

In the 'actions' section are options to commission assessments and support data mining activities. Given sufficient directives, the project manager is able to commission sensitivity analyses to explore the impact of uncertainties or test building performance under a range of operating conditions. It is possible to bring together the results for different performance aspects to explore trade-offs between issues such as seasonal fuel use, environmental emissions, thermal/visual comfort, daylight utilisation, risk of condensation, and renewable energy contributions. Outputs might be requested directly by the user or recovered via automated agents. In the latter case, the project manager operates in silent mode with no user display.

The core message is that any problem – from a single space with simple control and prescribed ventilation, to a group of buildings with systems, distributed control, and enhanced resolutions – can be passed to the simulator where its multivariate performance is assessed and made available to inform the process of design evolution. By integrating the different technical domains, the approach supports the identification of trade-offs. This, in turn, nurtures sustainable approaches to building design and operation.

Integrated modelling fosters teamwork by allowing diverse professional perspectives to contribute equally to the outcome. Its electronic form allows for efficient updating as a design hypothesis evolves, allowing team members to operate from different locations and time zones, leading to collaboration and a more efficient business model.

4.2   Resilience testing

The preceding generic example application is predicated on user process control. To realise the true potential of building performance simulation, it is appropriate to change the application intent from performance prediction to automated operational resilience testing. BPS+ then assumes the role as an emulator of building lifetime operations. This is regarded as a more effective way of reducing the gap between design intent and the operational reality (Menezes *et al.* 2012, de Wilde 2014).

Instead of requiring tool users to define the assessment conditions and subsequently analyse assessment outcomes, they would instead submit their proposal as an appropriately configured input model to a 'resilience testing environment' (RTE) resident on an in-house or cloud server. The RTE, with an approved BPS+ 'engine', would automatically perform multiple annual simulations, with standardised perturbations imposed to represent events such as severe weather, equipment breakdown, energy tariff change, and the like. Outcomes over appropriate intervals (short, medium, and long term) would be evaluated against expectations relating to performance aspects such as indoor environmental conditions, running cost, and emissions. Whilst the capabilities and features of an RTE would be the same for all users, it could be powered by any BPS+ tool that is able to demonstrate an ability to represent the causal relationships underlying the particular resilience test to be applied. For example, to test the resilience of a community energy scheme, changing sky conditions might cause greater daylight penetration to indoor spaces. This, in turn, might result in artificial light dimming, giving rise to a reduced electrical load, requiring the power flow from a local photovoltaic array to be exported, resulting in a power quality impact on the low voltage network, requiring an export refusal or tariff adjustment, and so on. Such a simulation would result in an entirely different outcome than one that merely determined the output power from a photovoltaic array for a given solar irradiance, and would lead naturally to a robust practical scheme. This implies the need for a high-resolution input model that has relinquished reliance on simplifying prescriptions relating to weather, occupancy schedules, occupant behaviour, equipment efficiency, infiltration rates, and the like in favour of an explicit representation of these phenomena that encapsulate realistic impacts. It also implies that the simulation engine embedded in the RTE (perhaps formed from several coordinating tools) be approved in terms of algorithm validity as is done via procedures such as BESTEST (Judkoff and Neymark 1995). This may well result in a situation where specific simulation tools are approved for use with only a subset of the resilience tests supported by an RTE corresponding to different archetypes and standards of performance.

A prototype RTE, named Marathon[4], has been established within a project funded by Built Environment – Smarter Transformation[5] (Clarke and Cowie 2020) and as elaborated in Section 11.3. The operational procedure is envisaged as follows. A high-definition model (HDM) is prepared at the required abstraction level (early/detailed design stage) and scale (building, estate, district, etc.) using suitable model definition tools. The HDM is introduced to the RTE via an open standard dashboard and the required resilience test(s) are selected from an approved list. It is expected that the HDM will be fully BIM compliant once the BIM standard evolves to level 3 and beyond and that the list of resilience tests will grow as this descriptive capability matures. In any event, the resilience tests would:

- be a priori established for different target types (new low-energy housing, housing refurbishment, high-performance commercial buildings, community energy schemes, demand management/response, etc.), and increasing performance stringencies;
- impose target-specific perturbations corresponding to events such as adverse weather, equipment malfunction, and tariff change, and applied randomly throughout a simulation;
- assess test-specific performance metrics relating to occupant satisfaction, facility management, legislative compliance, etc.;
- be approved by professional bodies representing built environment professions.

After an HDM is introduced to the RTE, an automated calibration is undertaken to ensure that the outcomes are aligned with observations in the case of a refurbishment project, or with benchmarks corresponding to standard validation tests for new-build cases. The calibrated model is then simulated over the multi-year lifetime of the proposal, the test-specific perturbations applied, and relevant performance metrics repeatedly extracted corresponding to the evolving system state and test strictness: occurrences of thermal or visual discomfort, high CO_2 levels, excessive utility cost, problematic power quality, excessive local emissions, etc. Such outputs would initially be treated by the RTE in a passive manner, merely flagging issues, until the related problem reaches a threshold where the RTE would

[4]https://www.esru.strath.ac.uk/applications
[5]https://be-st.build

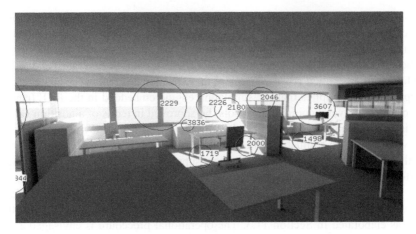

Figure 4.2   Luminance (glare) sources in an office

suspend (but not terminate) the simulation process and request remedial action. On delivery of a revised input model that embodies a proposed remedy to the reported issues, the simulation process would recommence and continue until resilience is assured (i.e., the corn kernel stops popping). The outcome would be a compliance certificate corresponding to the selected resilience test level. It is suggested that professional bodies be the approvers of the content – perturbations and performance metrics/criteria – of the resilience tests corresponding to different problem types and increasing stringencies.

The significant point is that the RTE-embedded simulator(s) would enact standard performance assessments for issues related to overall performance. Consider, for example, a test that assesses the risk of glare for standard viewing directions relative to the external facades within a commercial office. Here, the model perturbations might relate to changing sky conditions or the impact of loss of shading device control, whilst the performance metrics and target values might relate to the BS EN 12464-1 standard (BS 2011). Figure 4.2 shows one possible outcome indicating unacceptable luminance values (cd/m^2) corresponding to a critical viewing direction at a particular time.

The further (automated) processing of such outcomes might calculate the spatial variation in the Unified Glare Rating for comparison with the thresholds defined in the standard. Where deemed acceptable (over time), this aspect of performance would pass the resilience test. Otherwise, the RTE user would be invited to accept a certificate at a lower test level or suggest a remedial action before recommencing the suspended test.

Similar automated procedures can be introduced for other wellbeing aspects such as thermal comfort and indoor air quality as shown in Figure 4.3 for the latter case, as well as for energy utilisation, system control, embedded renewables, equipment operational efficiency, integrated energy storage, demand management, and the like.

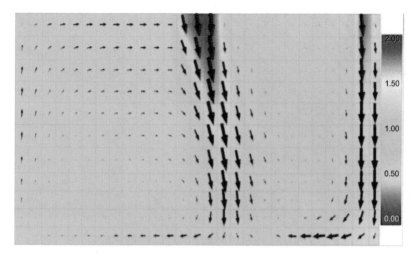

Figure 4.3   Mean age of air (seconds) distribution for an office cross section

An emulation of reality has significant benefits. To state the obvious, it relates to reality and not to an inappropriately simplified, partial representation. It does not require the simulation tool user to define performance assessments and interpret outcomes, thus standardising the assessment process and facilitating the inter-comparison of alternative proposals in terms of whether or not they passed the test. Such attributes leave the design team free to innovate, supported by a computational environment that assures acceptable performance under a range of conditions that are likely to be encountered in practice, and in terms of a range of standard performance criteria that can be readily understood. It also provides an approach to building regulation compliance that overcomes the constraints of the present rudimentary methods that fail to respect the thermodynamic integrity of the building.

It is stressed that models can still be configured against standard prescriptions to obtain outputs from simulation tools for legislative compliance purposes, to size system components for peak demand, or to obtain deeper performance insight, all as is done at present.

4.3   Application example

The following example is taken from the EC Daylight-Europe project (Kristensen 1996), which aimed to generate daylighting design guidelines for architects and engineers. The project comprised two research activities: the monitoring of 70 existing buildings throughout Europe and the in-depth simulation of a subset of 6. The simulation work adhered to the PAM elaborated in Section 4.1 (Clarke *et al.* 1996), with BPS+ tools used to compare the thermal and visual performance of a base case model and reference models arrived at by subtracting the as-built

Figure 4.4   Principal daylighting features in the Brundtland Centre

daylighting features. ESP-r was used to simulate the models and environmental controls, whilst Radiance was used to assess light distributions and the visual impacts of different blind systems, with ESP-r invoking Radiance on a time-step basis to assess light levels at specific locations for input to lighting control systems. In the context of 1996, such a cooperative use of tools represented a serious computational burden and so an approach of 'targeted detail' was followed by which Radiance was only invoked when the loaded ESP-r model changed (e.g., due to blind actuation or radical sky condition change).

The case considered corresponds to the Brundtland Centre located at Toftlund in Denmark. This is an 18,000 m² conference and exhibition centre designed to demonstrate new energy strategies for office buildings. The base case model was created from information extracted from architectural drawings, design team discussions, and site visits. Figure 4.4 shows the building and its principal daylight capture features. An atrium provides borrowed light and acts as a preheating buffer for the environmental control system as well as a platform for an array of PV modules positioned to diffuse the light entering the atrium. Internal daylight is enhanced by the use of redirecting blinds, automatic lighting control, and reflective ceilings. The southeast facade relies on blind control to limit glare and overheating.

To support the assessments, an explicit representation of the atrium, adjacent offices, and southeast-facing offices was constructed as summarised in Figure 4.5. The project focus was insensitive to furniture and fittings, so these were not included. Similarly, the ventilation towers were abstracted, as were the displacement ventilation details.

Other portions of the building were included at sufficient resolution to support shading and light distribution determination and to define dynamic boundary conditions for the main zones. To respect the daylighting focus of the project, the reflectivity of all internal surfaces was measured and the glazing layers were

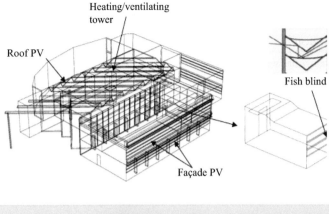

Heating/ventilating tower

Roof PV

Fish blind

Façade PV

Figure 4.5   Brundtland Centre, base case model

represented explicitly so that the intra-pane temperature and radiation processes could be examined. Table 4.3 lists the thermal and optical characteristics of the glazing in terms of the total visible transmittance (TVT) and direct solar transmittance (DST), both at normal incidence angle, and the overall thermal transmittance (U, $W/m^2.K$). The total diffuse shortwave reflectivity (R) of opaque surfaces is also given.

The centre is multipurpose, and occupancy varies considerably. The exhibition spaces cycle between full occupancy, none, and half occupancy during weekdays and Saturdays, and half occupancy on Sundays. Small offices were assigned one occupant and a computer and are not occupied on Sundays. Small power loads were set at 20 W/m^2 during working hours and 8 W/m^2 otherwise. The service core has a constant small power load for the energy management system. Domestic hot water energy use was based on average CIBSE recommended consumption for offices (0.513 kWh/person.day): assuming an average occupancy of 110 persons, the demand was estimated at 56.4 kWh between 6 a.m. and 6 p.m. corresponding to an average power draw of 4.7 kW. The hydraulic lift consumed 1.66 kWh during occupied periods (CIBSE Guide Table B18.19).

Table 4.3   Glazing and surface properties[6]

Element	TVT	DST	U
Low-e glazing	0.79	0.54	1.6
Curved blinds	0.63	0.43	1.6
Fish blinds	0.34	0.23	1.7
PV modules	0.20	0.20	1.7
Surface	**R**	**Surface**	**R**
Atrium floor	0.20	Office floor	0.42
Ceiling (metal)	0.95	Ceiling paint	0.89
Atrium wall	0.30	Office wall	0.89
Metal grills	0.45	External wall	0.35
Window frame	0.45		

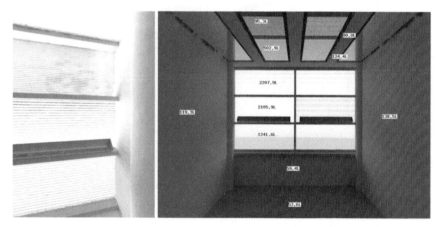

Figure 4.6   Prediction of light levels for different blind types

All spaces adjacent to the external facade and atrium have upward facing, parabolic luminaires affixed to the upper facade and conventional blinds in the lower facade. These operate in conjunction with the reflective ceiling to provide indirect light. Figure 4.6 shows the illumination levels (lux) as recorded in a model calibration study.

In spaces remote from facades, high-efficiency luminaires are used with an overall casual gain of 4 W/m^2. Luminaire control is on a room-by-room basis and is linked with the control of facade blinds in the building management system but with manual override. This was modelled via time-step invocation of Radiance in two offices. In a modern context with greater computation power, this approach might be readily extended to all offices.

---

[6]Note that simulations require such data at various angles of incidence of the solar radiation.

Table 4.4   Monthly max and min temperatures (°C), Copenhagen

	Jan	Feb	Mar	Apr	May	Jun	Jul	Aug	Sep	Oct	Nov	Dec
Min	−12.4	−13.5	−7.5	−0.7	3.1	5.6	7.4	5.2	6.6	−2.5	−3.0	−7.8
Max	5.0	6.5	9.5	20.8	24.8	25.2	27.4	28.8	21.4	19.0	10.2	7.5

When the indoor daylight level is high, the luminaires are off and the central blind is closed. As the daylight level falls, the blind is rotated and the luminaires are switched on (at 1–100% of maximum output). Occupants can override this action and increment or decrement the room controls. The system reverts to automatic operation after 2 hours. Given the computational load associated with stochastic processes, occupant behaviour modelling was not included. Instead, lighting control was enacted by establishing the internal illuminance at each time step and depending on the level opening/closing the blinds and deactivating/activating the luminaires. Lighting control in the lecture rooms, classrooms, and cafe was prescribed based on observations.

A gas-fired condensing boiler supplies radiators in the offices and classrooms and underfloor heating in the exhibition areas. The heating set point was 20 °C when the rooms were occupied, otherwise 15 °C. The centre is mechanically ventilated, with two heat recovery towers in the atrium. There are separate heat recovery units for the offices. Flow rates are varied depending on conditions at several points in the building. The atrium acts as a central buffer for the ventilation air and the atrium doors, and roof vents are opened automatically during warm periods to limit over-heating. These systems were represented explicitly by an airflow network comprising suitable components. The fan energy demand was 20 kWh/m^2.y (Johnsen *et al.* 1993), and the fan capacity for the two ventilation towers was set at 5.6 kW.

Weather data were not available for the site, so TRY data for Copenhagen was used as summarised in Table 4.4 from which the severe winter and low minimum temperatures during most months are apparent. Maximum temperatures during the summer can result in overheating.

In addition to annual simulations, three periods were simulated corresponding to typical winter, spring/autumn, and summer weeks[7]:

- 4–11 January, which included a design cold condition;
- 6–13 April, with average values for spring/autumn; and
- 19–26 July, which included a design warm condition.

Two (of many) reference models (here tagged 1.1 and 3.0) were established by removing the PV modules from the atrium roof and the southeast facade and downgrading the daylight features – white painted rather than reflective ceilings and standard blinds rather than reflective. The first reference model included office luminaire control based on indoor illuminance, whilst the second reference model had conventional lighting operation. The removal of the atrium PV modules was expected to increase glare in atrium-facing offices.

---

[7]ESP-r has a facility to scan weather files to help identify such periods.

Before commencing the simulation study, judicious adjustments to the models were made until reasonable agreement was obtained with daylight measurements; the models were then declared fit for purpose. (See Section 4.4 for a description of the formal approach to model calibration.) Figure 4.7 shows the simulation results for the base, reference 1.1, and reference 3.0 cases. It draws on the IPV concept described in Section 3.2.3 where the goals of the project were codified within the model and relevant metrics automatically extracted.

A comparison of the performance entities contained within these IPVs gave rise to the following conclusions.

On maximum capacity: The diversified total heating peak demand (W/m^2) represents required plant sizes and hence capital costs. The results show little difference between the two reference cases although both are $\sim$10 W/m^2 below the base case. This capacity drop was attributed to the greater solar gains in the atrium. There is no difference in the maximum electrical load for lighting, fans, lifts, and domestic hot water.

On annual energy performance: The normalised annual energy requirement for the base case was 159.1 kWh/m^2.y, whilst the reference 1.1 value was 142.0 kWh/m^2.y and the reference 3.0 value was 136.4 kWh/m^2.y. Compared to normalised performance indicator (NPI) ranges for similar centres in the United Kingdom as shown in Table 4.5, these results lie in the transition between 'good' and 'fair' categories. Given the severity of the Copenhagen weather, the performance was rated 'good'.

The primary difference between the various cases was in heating demand, with the base case requiring 1.3 kWh/m^2.y more for heating and 3.4 kWh/m^2.y more for lighting than reference 1.1. The removal of the PV modules from the atrium roof increased the passive solar gain to the atrium although this had a minor impact on the heating demand. In reference 3.0, the heating demand is 0.1 kWh/m^2.y more than reference 1.1 and 1.05 kWh/m^2.y less than the base case.

On seasonal energy demand profiles: These data are expressed as cumulative daily profiles for each season. In terms of heating demand, there is little difference during the winter, whereas the higher transmission of solar energy in both reference cases results in minor reductions in heating during occupied periods in the transition and summer seasons. In line with the contribution of lighting to the NPI figures above, in all cases, lighting is one of the smaller bands in the energy demand profiles. Rooms with advanced daylight control use little artificial lighting; such control is not applied throughout the building and the effect is muted. The inclusion of PV modules on the atrium roof slightly increases the demand for artificial lighting for atrium-facing rooms. There is no change in the fans, lifts, and small power demands.

Environmental emissions: The annual energy data were converted to primary energy units based on the EU average generating inefficiency conversion factors of Table 4.6.

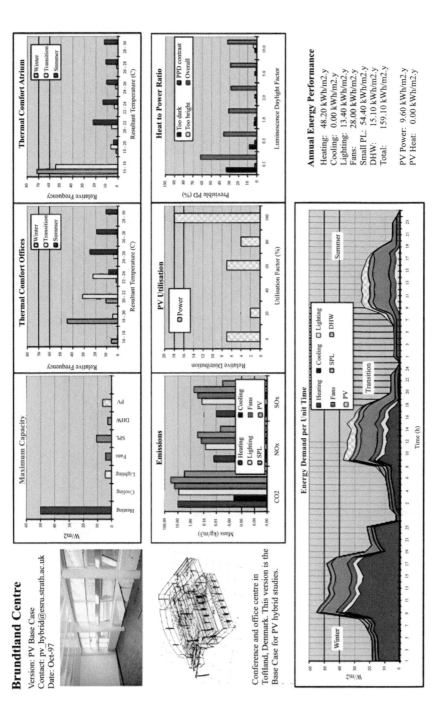

Figure 4.7 IPVs for the base case and two reference models

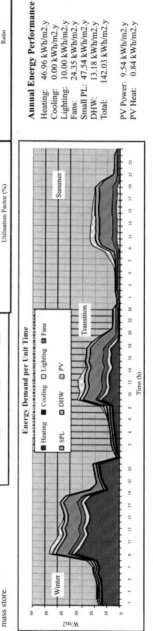

Brundtland Centre

Version: PV Reference 1.1
Contact: pv_hybrid@esru.strath.ac.uk
Date: Dec-97

Conference and office centre in Totfland, Denmark. As Base Case with warm air in upper atrium directed to mass store.

Maximum Capacity

Thermal Comfort Offices

Thermal Comfort Atrium

Emissions

PV Utilisation

Heat to Power Ratio

Energy Demand per Unit Time

Annual Energy Performance

Heating: 46.96 kWh/m2.y
Cooling: 0.00 kWh/m2.y
Lighting: 10.00 kWh/m2.y
Fans: 24.35 kWh/m2.y
Small PL: 47.54 kWh/m2.y
DHW: 13.18 kWh/m2.y
Total: 142.03 kWh/m2.y

PV Power: 9.54 kWh/m2.y
PV Heat: 0.84 kWh/m2.y

Figure 4.7 (Continued)

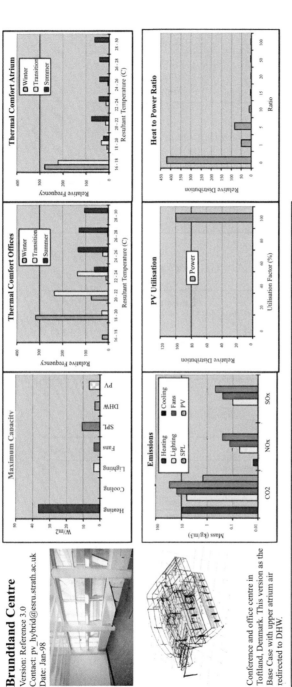

Brundtland Centre

Version: Reference 3.0
Contact: pv_hybrid@esru.strath.ac.uk
Date: Jan-98

Conference and office centre in Toftland, Denmark. This version as the Base Case with upper atrium air redirected to DHW.

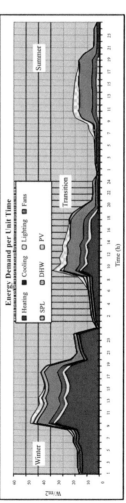

Annual Energy Performance

Heating:	47.15 kWh/m2.y
Cooling:	0.00 kWh/m2.y
Lighting:	10.00 kWh/m2.y
Fans:	24.35 kWh/m2.y
Small PL:	47.54 kWh/m2.y
DHW:	7.37 kWh/m2.y
Total:	136.41 kWh/m2.y

PV Power:	9.53 kWh/m2.y
PV Heat:	7.37 kWh/m2.y

Figure 4.7 (Continued)

Table 4.5   *NPI ranges*

Category	Range (kWh/m^2.y)
Good	<190
Fair	190–260
Poor	Poor > 260

Table 4.6   *EU average generating factors*

Fuel	Use	Delivered-to-primary energy modulus
Gas	Heating	1.05
Electricity	All	1.73
Gas	DHW	1.05

Table 4.7   *EU emission conversion factor*

Fuel	Emission (g/kWh)		
	CO$_2$	NO$_2$	SO$_2$
Gas	190.	0.3	0.2
Electricity	360.	3.0	5.7

Table 4.8   *Impact on gaseous emissions*

Model	CO$_2$	NO$_2$	SO$_2$
Base case	+2.2	+0.5	0.0
Reference 1.1	0.0	0.0	0.0
Reference 3.0	+1.9	+3.0	+3.1

The primary energy data were then converted to equivalent gaseous emissions using EU-averaged conversion factors as given in Table 4.7.

Table 4.8 summarises the changes in environmental emissions between the three cases relative to reference 1.1. The additional heating required in the base case results in a slight increase in CO$_2$ emissions when compared with reference 1.1, whilst the additional luminaire use in reference 3.0 results in increased NO$_2$ and SO$_2$ emissions.

Daylight availability: The daylight factor is a common metric that is well understood by the design community, the magnitude and distribution of these factors being a reasonable indicator of daylight availability and, consequently, artificial lighting requirements. In the southeast-facing offices, the base case design enhances natural light near the external facade without noticeably reducing light at the opposite end. In the atrium-facing offices, the base case design exhibits lower daylight factors due to the relative opacity of the atrium roof although levels are still sufficient to limit the use of artificial lighting. The daylight factor distributions in both reference cases were deemed to meet most visual requirements.

Thermal comfort: To determine if the improved daylight performance adversely affects thermal comfort, the frequency of occurrence of the spatially averaged resultant temperatures in offices and classrooms was examined during the occupied period in each season. As expected, the winter performance was essentially the same. During the spring, the most common temperature range was 18 °C–20 °C, with the reference cases being slightly cooler. During summer, both reference cases were around 4 °C warmer due to the clear atrium roof and removal of shading on the southeast facade. Clearly, the increased solar gain to the atrium and offices, which is beneficial during cooler summer periods and in the transition seasons, results in discomfort during warm periods.

Glare sources: This output in the IPV highlights potential glare sources within a 3D colour picture, with luminance values given in cd/m^2. For offices facing the atrium within the base case, glare sources are clustered where PV modules are not included, that is, at the vertical south facade and at the ends of each 'saw tooth' roof section. Where an occupant looks directly towards the atrium, the PV modules eliminate most glare sources. In both reference cases, glare sources are widely distributed for views parallel and perpendicular to the atrium wall. Note that these observations correspond to an overcast sky condition. In the worst case, a clear sky and a clear atrium roof will result in additional glare sources.

Visual comfort: The 'J index of Previsible Percentage Dissatisfied' or JPPD index (Meyer 1993, Carlucci *et al.* 2015) relates visual discomfort to any excess/ lack of light or inadequate contrasts in the field of view of a person performing a reading task. For atrium-facing offices, the base case is less comfortable under a dull sky but similar to the reference case under brighter conditions. As with glare sources, JPPD is sensitive to viewing position. For views towards the atrium facade, the base case has the advantage, whilst the reference cases are better for views parallel to the facade. In southeast-facing offices, the JPPD indices are similar for both facade treatments. This does not entirely diminish the benefits of the advanced blind system because, for the specific viewing direction, the light falling on the target is from the lower portion of the facade, which employs the same standard blinds in all cases.

System optimisation: The base case model was used to establish the specific contribution of each daylighting feature (the fish blinds and borrowed light) and to improve the system overall through better lighting control. Initial assessments indicated that there was little benefit from the use of fish blinds in the atrium-facing offices. Indeed, the diffusing properties of the PV modules resulted in acceptable daylighting in the adjacent offices and minimal glare (except for brief periods in the late afternoon). As a result, the design team was able to forego the expense of the advanced blind control on the atrium-facing offices.

Table 4.9   Monthly max and min temperatures (°C), Bolzano

	Jan	Feb	Mar	Apr	May	Jun	Jul	Aug	Sep	Oct	Nov	Dec
Min	−11.5	−9.1	−4.0	−1.7	2.4	6.8	9.0	7.9	5.4	1.5	−7.1	−9.1
Max	12.2	18.1	20.5	25.4	30.2	32.4	34.6	33.6	27.4	26.0	18.0	17.6

Table 4.10   Change in heating capacities and energy demand (%)

	Copenhagen		Bolzano	
Item	Base case	Reference 1	Base case	Reference 1
Heating capacity	0	−13	−2.2	−4.2
Cooling capacity	0	0	0	0
Lighting capacity	0	0	0	0
Heating energy	0	−2.7	+0.2	−2.1
Cooling energy	0	0	0	0
Lighting energy	0	−1.4	−5.9	−4.1
Lighting energy (vs. no control)	−25.5	−26.5	−29.8	−28.5

An alternative computational approach was now invoked to test whether luminaire switching based on predetermined daylight factors is acceptable. Now, the indoor illuminance was determined by passing the (changing) problem description from ESP-r to Radiance at each computational time step. The results indicated that the use of static daylight factors under-predicted luminaire use by 13% during many winter periods although the overall difference in lighting energy demand was small at ∼3%. For this problem case at least, assessing the energy impact of light redirecting systems did not appear to require an appraisal of indoor illuminance at the time-step level.

Replication: To test the sensitivity of the design to location, a replication study was undertaken in which the model was relocated from Copenhagen (latitude 55.6761 °N) to Bolzano, Italy (latitude 46.4983 °N). The monthly maximum and minimum temperatures for Bolzano are listed in Table 4.9.

As before, three periods were selected corresponding to typical winter, spring/ autumn, and summer weeks:

- 4–11 January, which includes a design cold condition;
- 19–26 April, with average values for spring/autumn; and
- 9–16 August, which includes a design warm condition.

In comparison with the Copenhagen weather data, this reduced the winter season heating degree-days from 1,996 to 1,880 and the spring heating degree-days from 1,217 to 692, while increasing the summer cooling degree-days from 174 to 490. Table 4.10 summarises the simulation results for the replicated base and reference 1.1 cases. These results show that moving to a southern climate has a minor impact on heating capacity and heating energy demand. The mild spring at Bolzano is balanced by harsher periods during the winter and cooler summer mornings.

In terms of solar radiation, the 10° shift south increases the sun azimuth in the winter and transition seasons sufficiently to increase the aperture for passive solar gains in the atrium. The atrium-facing offices, therefore, require fewer hours of artificial lighting in the base case. In reference case 1.1, the high daylight factors result in little reduction in lighting use as luminaires are seldom on. For offices facing southeast, the additional radiation in Bolzano has an effect early in the day. Overall, the base case requires less lighting than reference 1.1 in the southern location. Indeed, it would be possible to reduce the area of glazing without incurring additional luminaire use.

In terms of comfort, the warmer spring shifts resultant temperatures 2 °C–3 °C upwards in the base case and 3 °C–4 °C in reference 1.1. In the summer, the degree of discomfort is considerable, especially in reference 1.1. The number of hours over 28 °C during a typical summer week in five typical rooms in Copenhagen is 34 for the base case and 58 for reference 1.1. In Bolzano, the corresponding values were 94 and 207. This indicates that the building requires adjustments to its design to improve warm weather performance. The clear atrium roof allows passive solar gains that are excessive during warm periods. Here, the PV modules of the base case are beneficial; without them, there would be a requirement for seasonal shading of the atrium.

Shading on the southeast facade is helpful in Bolzano. However, given the orientation, it is difficult for fixed shading to be effective at low sun angles. Instead, a reduction in the glazing area could be implemented without serious degradation of daylight factors. Note also that the ventilation strategy was the same in both locations. Moderation of unacceptable temperatures could be affected by increasing the venting of the atrium and by implementing a night purge.

Based on the IPV performance indicators for all cases, it was concluded that the building's special features improved performance by a significant margin.

4.4   Calibrating a model

High-integrity model calibration requires the use of measured data. Whilst it is relatively easy to collect data from a building, monitoring regimes are often poorly matched to the needs of simulation tools. There is a long history of using monitored data to extract an overall heat transfer coefficient (HTC) for a building. For example, Subbarao *et al.* (1988) described the PSTAR method, and Mangematin *et al.* (2012) proposed the QUB method. The co-heating method, based on initial work by Sonderegger *et al.* (1980), is also widely used.

Whilst each method provides an indication of building performance based on observed conditions over a few days to several weeks, they have limitations in relation to the calibration of simulation models. An HTC is a steady-state metric

Figure 4.8   A monitored dwelling and its digital twin

that is incompatible with the dynamics represented by a BPS+ program. Where a test corresponds to static indoor temperatures, it is unable to represent the response characteristics of the actual building. Such tests are also insensitive to the impact of thermal mass, air movement, and radiation exchange. To be useful for BPS+ model calibration, tests should assist in judging the response of buildings to step changes in indoor temperatures as well as free-floating conditions (Clarke *et al.* 1993).

4.4.1   Domestic scale

A Pulse Test (Hand *et al.* 2021), which applies consideration of dynamic behaviour to a traditional co-heating regime, entails the following phases.

Design of experiment

A BPS+ model of moderate resolution is used to design the test regime by confirming the required capacity and placement of heaters. Leakage characteristics to represent infiltration and natural ventilation can be introduced later following pressurisation tests. Figure 4.8 shows a typical model with thermal zones for each room, attic, and crawl space. Control is imposed to mimic the phases of the experiment.

Dwelling preparation

Figure 4.9 (left) shows a monitoring setup: doors are open, reflective foil is installed on the windows, and all appliances including the heating system. Portable heaters and fans are then installed along with sensors in each room. Interventions normally associated with a blower door test are carried out (e.g., sealing windows and doors).

Pre-conditioning

Monitoring begins by allowing the building to free-float towards ambient conditions for 24–48 hours before commencing the heating phase for a further 24–48 hours.

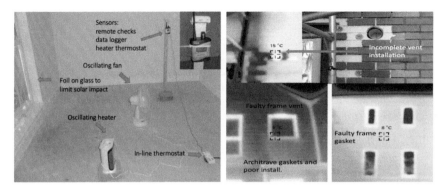

Figure 4.9   Monitoring setup and fault detection

Heating step change

Heating is applied to each room until indoor conditions are approximately 20 °C above ambient. Fans ensure air mixing and good heat transfer to the room surfaces. The elevated temperatures help in the identification of facade faults and ensure sufficient heat flows for the heat transfer paths to be tested within the digital twin. The heat output from each heater is recorded for later comparison with predictions.

Setpoint maintenance'

Heating is applied to maintain a specific set point temperature for 24–36 hours. Fans continue to ensure mixing and high surface heat transfer. The goal is to 'saturate' the structure. At the end of the maintenance period, an infrared survey is carried out to identify facade faults (Figure 4.9 right) to help update the model.

Cool down

The heaters and fans are switched off and the indoor temperatures are left to decay until they reach approximately 1/3 of their starting value.

Model calibration

The differences between the monitored and predicted indoor conditions can now be investigated, adjustments made to the input model and, if necessary, the pulse test repeated. Figure 4.10 shows the difference between observations and predictions after a number of pragmatic tweaks had been made to the starting model. Such a model is now considered fit for purpose.

4.4.2   Commercial scale

The Hit2Gap project (2023) established a calibration platform as part of a data-centred approach to the facilities management of large estates. Within the project, Calibro[8] was used to calibrate key input parameters of an ESP-r model (Monari 2016, Monari and Strachan 2017). The input required for the calibration comprises

[8]https://www.esru.strath.ac.uk/Applications

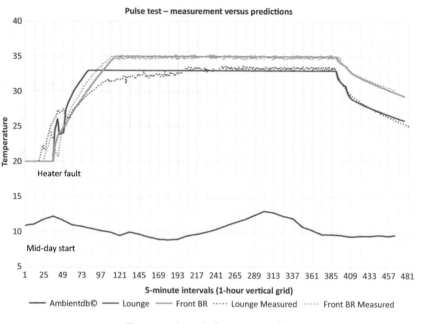

Figure 4.10   Pulse test results

a dataset of input–output data pairs for multiple simulation cases corresponding to judicious input parameter perturbation. Whilst these input–output parameter pairs will typically depend on the intended application of the BPS+ tool, they have no particular meaning within Calibro. The calibration also requires measured values of the targeted output parameters (e.g., indoor conditions, metered energy use, etc.) and time-matched weather conditions. Using the input–output pairs, Calibro constructs an emulation (meta-model) of the BPS+ tool and uses this to determine the input parameter values that will cause the tool to reproduce the measured performance. The approach utilises four statistical techniques:

- principal components analysis (Jolliffe 2002) to reduce the dimensionality of the input data sets;
- sensitivity analysis (Ratto 2001) to select the parameters for emulator inclusion;
- training of a Gaussian process emulator with optimisation techniques (Nelder and Mead 1965); and
- a Markov Chain Monte Carlo method (Rosenthal 2009) to infer the parameter values and related uncertainties.

The best-fit input parameter values are then incorporated in the BPS+ input model before use in performance assessment mode. The procedure was demonstrated within the Hit2Gap project by the application of ESP-r to a portion of the Challenger building situated in the headquarters complex of Bouygues Construction in Saint-Quentin-en-Yvelines as shown in Figure 4.11 (upper left).

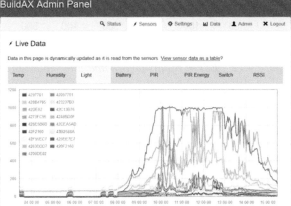

Figure 4.11   Monitoring indoor conditions using BuildAX sensors

Also depicted in this figure are BuildAx multi-sensors as deployed to collect indoor environmental conditions data at 5-minute intervals (Clarke and Hand 2015), with corresponding weather data acquired from local sources. These wireless sensors transmit high-resolution spatial and temporal data (one sensor per six occupants at 10-minute resolution) and cover several significant parameters (air temperature, humidity, illuminance, and movement).

The inputs required by Calibro were generated by invoking ESP-r's Monte Carlo sensitivity analysis feature (Macdonald and Clarke 2007). Figure 4.12 shows typical outcomes before and after calibration.

The final goodness-of-fit achieved was satisfactory so that ESP-r was deemed suitable for application in a facilities management context as targeted in the project. Some lessons emerged from the calibration exercise as follows.

Input–output pair generation

The production of the input–output simulation sets required by Calibro for calibration poses challenges as follows.

• The number of required Monte Carlo runs in the ESP-r model calibration was set to 10 times the number of parameters being considered as suggested in the literature (Jones and Schonlau 1998). Whilst reducing this value is

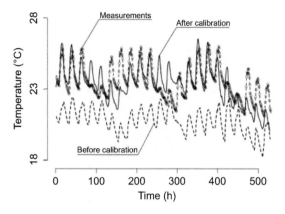

Figure 4.12   ESP-r results before and after calibration

computationally desirable, investigations indicated increasing problems as the value is lowered.

- In models with few parameters for calibration, Monte Carlo simulations will comprise, at worst, a few hundred runs. In a whole-building, domain-integrated simulation, the amount of input parameters can be substantial: with multiple zones and modelling domains as in the Challenger building, the number of runs would be of order 10^4 if, unusually, all input parameters were considered uncertain. As each simulation requires several CPU minutes, the computational burden is high. One way to reduce this burden is to use a proxy parameter as described below.
- ESP-r has an in-built parametric sensitivity analysis capability that automates the production of the required input–output pairs. Where such a capability does not exist, it will be necessary to externally script the BPS+ tool. In this case, the quality assurance of outcomes will be problematic as it is difficult to detect errors in simulations when dealing with aggregated data from (potentially) thousands of runs.
- Experience to date indicates that the calibration of complex models is best undertaken incrementally as the model is constructed, as opposed to attempting to calibrate a final, complex model.

Calibration objectives

Model calibration must be purposeful, as a model calibrated for one purpose will generally perform badly when applied elsewhere.

- Another calibration target in the project was SBEM (2023) to demonstrate the process when applied to low- and high-resolution models (e.g., annual energy consumption in SBEM and unmet comfort hours in ESP-r). Although a successful calibration was possible in both cases, applications of the former model showed poor agreement on an hourly basis, whilst models calibrated against high-resolution considerations can perform badly when used for annual energy consumption prediction.

- For cases where time resolution is important, 5-minutely monitored data gave good results when associated with disaggregated energy meter data and the spatial distribution of indoor conditions. At this resolution, Calibro was capable of assessing the model response to rapid changes in internal and boundary conditions although this comes at considerable cost in computational time requiring access to a high-performance computer.

Proxy parameters

Many input parameters will vary throughout a simulation, e.g., those with a strong stochastic component, or related to airflow or internal gains. The calibration of such parameters is challenging.

- Where the target for calibration is a prescribed profile (e.g., time-varying casual gains or air change rates), it is possible to calibrate the individual values comprising the profile and then assume the calibrated profile thereafter (although this would still result in a large number of parameters for calibration (e.g., 24 hourly values per day-type and zone).
- An alternative approach, successfully exploited in the Hit2Gap project, is to associate a scaling factor with such a profile in a manner that preserves the profile shape. The input–output pairs required by Calibro are then generated based on the perturbation of this factor. The suggested value emanating from the calibration effectively nudges the profile to fit better the observation. This approach results in a significant reduction in the number of parameters for calibration. Because Calibro is tool agnostic it makes no distinction between individual input parameters with physical meaning, or proxy parameters representing multiple parameters grouped as a profile, as long as the variation in the proxy parameter is indicative of the profile it represents.
- Where a domain model exists, for airflow or stochastic occupant behaviour for example, it is possible to transform the simulation result to a profile and then apply the scaling factor technique as above. This allows the use of a calibrated profile in place of the domain model but without the need to generate a prescriptive profile in the first place. Furthermore, complex profiles can be decomposed using techniques such as Fourier analysis or B-splines, and the parameters corresponding to these harmonics are used as the calibration target.
- An intriguing future prospect is to apply the above approach but where alternative domain model configurations are the inputs (e.g., CFD models using different grid resolutions, turbulence models, wall treatments, numerical schemes, etc.) with outputs nominated as before. The results of the proxy parameter calibration can then be used to determine which domain model provides the best representation.

Quality assurance

Deficiencies in models presented for calibration are provided as an interim output from a Calibro analysis to aid model quality assurance.

- Data on the sensitivity between output and input parameters are provided. These data can be used by the modeller to adjust the input–output sets for delivery for final calibration.
- Data on the correlation between calibration residuals and boundary conditions is provided. Such data can help to identify modelling errors, e.g., a strong correlation between solar radiation and residuals led in one case to the identification of problems with the modelling of shading control. The challenge here is to automate such detections.

After calibration, models may be used to undertake performance assessments as long as the calibration target relates to the assessment intention. The frequency at which a model is calibrated depends on the nature of the assessment. In the Hit2Gap project, the intention was to recalibrate the ESP-r model daily to enable the identification of areas of concern to be rectified before use within the online facilities management system the following day.

There are many targets for model calibration. For example, Borg and Kelly (2012) applied the technique to a dynamic absorption chiller model whilst Murphy *et al.* (2013) calibrated a model for controller design.

4.5 Automating performance assessment

The normal mode of operation of BPS+ requires the user to control the application process. Whilst this is required in relation to input model creation, is it not helpful to expect each user to develop their own performance assessment procedure – essentially plotting a path through a program to obtain customised outputs. A more rational approach is to embed automated procedures within BPS+ tools that encapsulate standard assessments based on criteria agreed with practitioners.

The possibilities for the design of such an intelligent BPS+ are vast. Expert systems are typically restricted to problems that are themselves rules based, such as regulations advising, process control, and data interpretation. In the approach, a knowledge base is established to contain domain-specific facts and relations. The front end then acts on this stored knowledge to make logical deductions in response to a posed problem. The expert system will then reproduce the expert view, which may or may not be correct. For example, in the case of building regulations or HVAC component selection the system's advice will be 'correct' in that the problem domain is totally rules based: obey the rules and the answer emerges. But given another application – say building comfort and energy performance assessment at the design stage – problem solution requires more than just a set of prescriptions expressed as predicates. Now an important new element is needed: performance assessment at the design stage to quantify the performance criteria identified in the knowledge base. In other words, even an expert will have need to refer to data, monitored or computed, that characterise building performance.

One approach to the automation of performance assessment is to create a wrapper that utilises one or more tools against a predefined appraisal procedure. Typically, this wrapper will be implemented via a high-level programming language such as Python[9], which combines the functionality of a high-level language, command shell programming, and Web page design. To elaborate on the technique, consider the automated procedure for overheating assessment as implemented within ESP-r (Clarke 1986) based on the Unix Bourne shell (Bourne 1982) and several software tools within the Linux environment. This 'shell script' is designed to coordinate the operation of ESP-r program modules when on a mission to identify the location and severity of overheating throughout a building.

Shell script rules, although hardwired, can be replaced by the user at script invocation. Within the script, ESP-r and shell commands cooperate to perform set operations as a function of the in-built rules that correspond to each performance appraisal. The computational path to be followed at any stage in the script will depend on the performance data to emerge at run time and on the embodied rules. Each shell script can be viewed as a design assistant: the performance assessment and program operation knowledge is known to the assistant; the designer is free to use the results to influence their design decision-making.

A thorough appreciation of shell script programming can only be obtained from the study of the syntax involved[10]. However, to demonstrate the possibilities and the technique, one script is described here. Its purpose is to undertake a comfort analysis to:

- determine an appropriate simulation boundary condition by selecting a weather year with a severity rating matched to the building's geographical location and function;
- initiate and control simulation processing over a period of time determined as a function of severity criteria;
- identify building zones that are uncomfortable according to relevant comfort criteria;
- recover and present statistics on conditions in uncomfortable areas;
- determine the cause of the problem;
- initiate a sensitivity study, focusing on the causal energy flow paths, and to so rank order the options for design intervention; and
- provide a comprehensive report on comfort, including problem causes and potential cures.

This script is constructed as six sub-scripts as listed in Appendix C. The first runs the ESP-r climate module (*clm*) to determine the weather context then runs the building/plant simulator (*bps*). The second script recovers the state variables that quantify comfort by running the results recovery module (*res*) and then grouping

[9]https://python.org
[10]https://open.win.ox.ac.uk/pages/fslcourse/lectures/scripting/all.htm

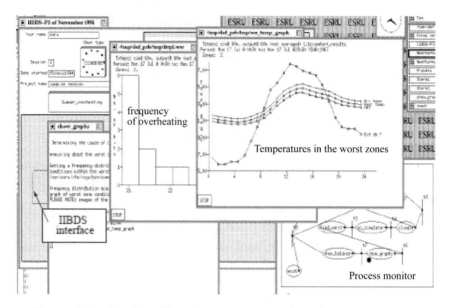

Figure 4.13   Results collage from an automated overheating assessment

zones according to whether or not they violate the comfort criteria. Summary statistics on the worst offenders are then output. The third script investigates the cause of any discomfort, whilst the fourth commissions a sensitivity analysis. The last two scripts are special in that they exist to complement the Linux data manipulation tool (*awk*): in essence, they control the extraction of information – such as the cause of discomfort or the location of the worst zone – from the data sets made available to *awk* from *bps*. Because of the complexity of the syntax, these two scripts are not reproduced in the appendix. Also, since script 4 is effectively the iterative operation of the first three, it is only cursorily explained.

When invoked, and after a computational effort that depends on the complexity of the building model and the extent of the required simulation, the comfort script will produce outputs as shown in the screen image of Figure 4.13.

Only the more salient features of the scripts are discussed here since much space would be required to explain the subtleties of script syntax. The interested reader should consult the appropriate Linux manual entry (*man sh*) to fully grasp the power of the data redirections used throughout (the '>', '>>', '<', '<<', and '|' symbols).

With reference to the listing of Appendix C, the environment variable, ESPdir, is used everywhere. This is set (in .profile or .login) to define the location of the main shell script directory. This allows ESP-r files to exist in, and the shell scripts to run from, any directory. The 'echo' process is used throughout to inform the user of progress or to write to a trace file (comfort_trace) to provide a record of the performance assessment.

The script is executed by typing the command *comfort*, perhaps followed by one or more of the command line options -h, -i, -o, -f, -c, -b, -p, -t or -d to modify the script's action. At the start of script 1, variables are assigned default values. The command line is then scanned to determine if the user wishes to invoke one or more of the options. For example, the command *comfort -i set -o 25* executes the script with standard effective temperature defined as the comfort index and the cut-off temperature for overheating set at 25 °C. There are many possible permutations. *Clm* is run to determine a weather collection of appropriate severity. *Bps* is then run to determine the building's behaviour against this weather boundary condition. Selection of the -c option allows the user to force the use of a specified weather collection.

At a certain point in its operation, script 1 suspends until a special file is created by script 2 to indicate that the latter has finished. Script 2 will have filled this file with data that identifies those zones that are uncomfortable in terms of the selected criteria. If the file is empty, all zones are within the comfort zone. Otherwise, script 3 is run to continue the appraisal. The *newwin* process opens a new window according to the specified arguments. The *graph* process is used to display any ESP-r graphics in a display window.

In script 2, the back quotes (') cause whatever is between them to be run, with the output replacing the quoted string. To generate the input for *awk*, the two commands *echo* and *cat* are placed in brackets. This tells the shell to execute them sequentially, but to treat them as one command as far as the rest of the line is concerned. First, the echo output is piped to *awk*, then the *cat* output is sent down the same pipe. This means that the variable PZ will contain the results of the *awk* process: an ordered list of the problem zones in this case. The variable WZ is then set to the worst zone – that is, the first one in the PZ list. This zone is then tested for occupants. Process *impb* does this. It returns 0 if a zone is unoccupied; it provides no output. The worst zone is then written to the lock file to restart script 1 and feed script 3.

Script 3 again uses the *awk* process to determine the cause of any thermal discomfort. OHC contains the rank-ordered causal list – for example, *Infilt Solair Surfconv* for infiltration, glazing solar absorption convected to the zone air point, and internal surface convection, respectively.

Script 4 is then run. For each causal flowpath, *impc* will modify the building description as a function of a relational table linking flowpaths to possible design changes. Processes *bps* and *res* are then rerun to establish the effect of each design change before the final performance data are output. Finally, a script is run in a new window to produce a perspective view of the building under analysis for zone identification purposes.

Such scripts might typically exist for a range of standard tasks such as heating/cooling plant sizing, equipment control strategy appraisal, condensation assessment, summer overheating, annual and seasonal energy use, air quality assessment, glare potential, and so on. The important point is that the request to run such appraisals can be included in the model so that the process can commence automatically on model submission.

4.6   Determining cause and effect

A unique attribute of simulation is the interconnectedness of all energy flows. This enables system performance to be assessed by interrogating simulation results to obtain an insight into the underlying causal relationships and so identify beneficial changes to the design parameters. Following this seemingly simple procedure can be problematic due to the size and complexity of the temporally varying data to result from a BPS+ simulation.

Whilst it is a relatively simple task to formulate PAMs, interpreting outputs can be problematic where a simulation has yielded copious data on a system's changing state over time rendering the analysis task complex. For example, a data set of the order of 15,000 items might be generated in the case of a single zone, one-day simulation. Little wonder that decisions on acceptability and design intervention can become a process of random chance.

One solution to this problem is to develop a mechanism that ensures that the user's attention is focused on the issue(s) of greatest impact. Causal energy breakdown tracking is such a mechanism. In the approach, an energy balance is formulated for the overall building, identifying the energy gains/losses of zones and supply/consumption of systems. From this, the points of greatest demand and consumption can be determined and new energy breakdowns formulated for each in turn. This process is repeated until the required resolution is attained and an acceptable design modification is identified. To illustrate the process, and with reference to Figure 4.14, consider the following example.

- Assuming that the simulation results indicate overheating in an occupied space, an energy balance breakdown at the zone air point would yield information on convective plant flux (heating or cooling), shortwave flux absorption, zone-coupled advection flux, surface convection flux and convective casual gain.
- Assuming that the surface convection flux was the dominant causal flowpath, an energy breakdown for the corresponding surface(s) would yield the relative

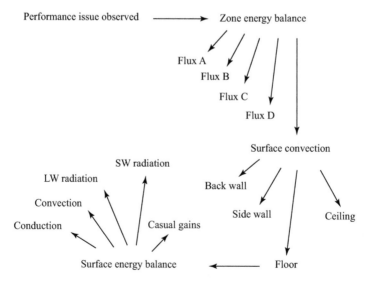

Figure 4.14   Causal chaining of energy balances

contributions of constructional conduction, surface convection, radiant casual gain, shortwave flux, longwave flux, and radiant plant.

- Assuming that the shortwave flux was now the dominant flow path, an energy breakdown for the room windows would identify the worst offenders.
- A window energy breakdown (not shown here) would then identify the relative contributions of the incident direct and diffuse shortwave flux (at both external surfaces and at internal surfaces due to internal inter-reflections), the transmitted/reflected/absorbed components, the conductive and convective processes, the longwave radiation exchanges at internal and external surfaces, the radiant casual gains, and any radiant plant interaction.
- At this level of resolution, informed decisions can be made on the most appropriate design change – for example, the incorporation of a shading device or changing the glazing type – to lessen the impact of the offending energy flow path(s). Significantly, this change can be made with the full knowledge that the impact will be beneficial. If necessary, further tracking of other components of the window energy breakdown will allow any other impacts of the proposed change to be assessed.

In applying the technique, it is important that the initial design hypothesis be not overly constrained by the incorporation of potentially conflicting energy-saving devices. If this is achieved and if the user has a good appreciation of the spectrum of design and operational options, then the approach will ensure that performance will proceed in the direction of improvement. It is left to the designer's judgement as to when improvements move out of the region of cost-effectiveness. A strength of this approach is that it can be automated so that the BPS+ tool is able to suggest design changes automatically based on simulation outcomes.

4.7 Embedding simulation in practice

Built environment simulation is a technology of considerable potential because of its ability to quantify the likely performance of a proposed design in a realistic manner and at a relatively low cost and effort. Indeed, a growing number of practitioners will recognise the maxim 'if you can't simulate it, don't build it'. This is not to imply that all projects should be simulated. Indeed, as summarised by Hensen and Lamberts (2019) based on a paper by Banks and Gibson (1997), there are compelling reasons not to simulate:

- the problem can be solved using common sense;
- the problem can be solved analytically;
- it is easier to change or perform experiments on the real thing;
- the cost of the simulation exceeds possible savings;
- the required resources are not available;
- there is no time for the model results to be useful;
- there are no data – not even estimates;
- the model cannot be calibrated or even pragmatically verified;

- project expectations cannot be met; or
- system behaviour is too complex, or cannot be defined.

Furthermore, simulation tool application needs to be considered in the context of traditional work practices such as embodied in the RIBA Plan of Work (RIBA 2024):

- strategic definition;
- preparation and brief;
- concept design;
- development design;
- technical design;
- construction;
- handover and close-out; and
- in use.

Since its emergence in the late 1970s as part of the personal computing revolution, BPS+ has slowly matured driven by academic enquiry initially and then by commercial pressures in a regulations-driven industry. Whilst the present generation of programs has impressive capability, they remain a work in progress when viewed against the future prospect of a fully embedded computational approach to design whereby arbitrarily complex models are evolved on a task-sharing basis, readily exchanged, subjected to industry-standard assessments, and easily integrated within the design process. The tendency however is for BPS+ to cover gradually all aspects of design, construction, and operation.

The goal eventually is to enable a high-integrity representation of the dynamic, connected, and non-linear physical processes that govern the disparate performance aspects that affect the overall acceptability of buildings. The approach offers a means to conflate the different performance domains to represent explicitly the interactions and conflicts that occur, requiring practitioners to make performance trade-offs. The effective application of the technology requires that it be integrated within the design process. This, in turn, requires adapting work practices to embrace the exploratory manipulation of a design hypothesis with design changes introduced in response to unacceptable performance feedback. This ability to prototype and test solutions at the design stage defines a new best practice that will allow practitioners to

- integrate all technical domains;
- respect temporal interactions;
- attain appropriate levels of comfort and indoor air quality;
- embody new and renewable energy supply technologies;
- incorporate innovative energy efficiency and demand management solutions;
- link life cycle performance to health and environmental impact; and
- conform to legislative requirements.

BPS+ enables experiential appraisals that can convey performance issues to non-specialists and thereby foster the engagement of stakeholders in the design

	Concept design —			Schematic design		Development design	Construction document
	Step 1	Step 2	Step 3	Step 4	Step 5	Step 6	Step 7
Design solution	· Identify energy goals and strategies · Site and climate analysis · Available energy sources · Functional requirements	· Compare with reference buildings	· Building form design · Energy system studies	· Envelope design · Solar shading design · Natural ventilation system definition	· HVAC system design · DHW production system design · PV system design	· Design solutions and control schedule · Prove concept design's energy goal compliance	· Prove regional energy code compliance · Prove rating system compliance
BPS+ model	· Establish weather data	· Collect model input data	Simulations to establish: · Building orientation · Building form · Zone heating and cooling requirements	Simulations to define: · window-to-wall ratio · opaque envelope arrangement · glazing arrangement · solar shading · natural ventilation	Simulations to define: · HVAC system component sizing · HVAC system configuration and control · DHW production system · PV system	Simulations to: · optimise use and control of air-conditioning, lighting, ventilation, solar shading, DHW and PV systems	· Baseline building creation and comparison with hypothesis

Figure 4.15   Applying BPS+ in practice (adapted from Ferrero et al. 2015)

process. As well as application in building design, simulation can be applied at the district or city scale in support of energy action planning. This requires the modelling of potential deployments and the identification of best-value approaches.

An important feature of BPS+ is its ability to support cooperative working on new design or refurbishment proposals. As the problem description evolves, simulation is able to deliver ever more detailed performance information that fosters further design evolution. In this way, a virtuous cycle is established and the design team is able to take an integrated view of performance whereby different aspects may be held in balance.

When deployed in practice, the use of a simulation tool requires adherence to a formal PAM by which issues such as process reliability, repeatability, and applicability can be cost-effectively assured. Such a method will depend on the stage of the design process being addressed. Ferrero *et al.* (2015) have divided the application of building simulation into seven steps as shown in Figure 4.15.

Several knowledge exchange mechanisms exist to support the uptake of BPS+ in practice. These include training courses and CPD offerings, simulation-assisted design frameworks, modeller accreditation schemes, and the services and publications of professional bodies. Nevertheless, many barriers remain that stymie the collaborative application of multiple tools in a design context (Hand 1998).

There are also different BPS+ user types with differentiated needs. An architectural practice might employ the technology at the early stage of a project to incorporate passive features that deliver acceptable performance with minimal need for services. An engineering practice might employ the technology to explore environmental control system configurations that meet user expectations at low-energy costs whilst complying with regulations. Facilities managers might employ the technology as an aid to operational performance tuning. In future, such application boundaries will blur and new user types might emerge, such as client bodies or occupants accessing the technology as a means to participate in the design and/or facilities management process. In a world accepting of AI agents, BPS+ might even provide a mechanism to train these agents in support of an automatic assessment of options.

Despite this diversity of users and project goals, specific BPS+ tools are implemented with a largely fixed user interface and set of facilities, with tool workflows evolved for ease of use. Typically, training materials and example models will steer users through a narrow band of options and perhaps suggest that there are standard approaches to common project types. Users then habituate these approaches[11] resulting in less-used features being deprecated or passing beyond the memory of the user community. Such feature obscurity can have downsides. Many of the projects described in Chapter 6 are predicated on the user driving the tool 'off-piste'. Whilst it is undoubtedly the case that tool developers have an advantage in this situation, a user who expands their application comfort zone will often be amply rewarded by the discovery of a powerful but hitherto unknown tool feature.

Whilst accepting the potential benefits of BPS+, many businesses still perceive barriers to uptake, not least in relation to the cost/complexity of software, the required changes to work practices, and the degree of preparedness of employees. For this reason, innovative approaches to technology transfer are required whereby program application is supported within live projects. Although most practitioners will be aware of BPS+, few are able to claim expertise in its application. This situation is slowly changing driven by

- performance-based standards;
- societies dedicated to the effective deployment of simulation, such as the International Building Performance Simulation Association (IBPSA) and its many regional affiliates[12];
- technology transfer initiatives supporting deployments; and
- growth in small-to-medium-sized practices pioneering simulation-based services.

For example, the Scottish Energy Systems Group (SESG, an IBPSA affiliate) operated from 2002 to 2008 with support from a Scottish Knowledge and Innovation Transfer Programme (SEEKIT) and the European Regional Development Fund (ERDF), with matched funding provided by the (then) Strathclyde European Partnership. The intention was to assist companies to adopt BPS+ for the design of low-energy buildings. By allowing design teams to gain risk-free access to modelling and simulation in the context of live projects and normal work practices, the industry was able to identify the financial and human resource barriers to routine tool deployment. In addition, by placing specialists within the design team, a two-way flow of information was supported: simulation expertise was passed directly to practitioners, whilst simulation specialists were exposed to real design issues.

This proactive approach to technology transfer gave rise to several independently verified benefits (ERC 2008). Companies were able to examine the impact

---

[11]Code developers are guilty of habituating short-circuit application routes, resulting in program parts being rarely exercised, and giving rise to program bugs confronted by normal users.
[12]https://ibpsa.org

of BPS+ technology on their current work practice and so develop an appreciation of the required organisational development path. They were able to refine their design process by providing rapid and accurate assessments of a variety of design options, hitherto impossible within the time scales available to practitioners. By being equipped to address the dynamic interactions inherent in energy systems, companies were able to enhance the quality of their process and products. Several distinct messages emerged from the project:

- contemporary modelling systems can be cost-effectively deployed where appropriate support is available;
- the largest portion of the cost relates to staff training, not to the acquisition of hardware and software;
- a change in work practice is needed if the profession is to move to a new best practice based on a computational approach to design;
- companies reported that they did not anticipate an impact on their professional indemnity insurance due to the uptake of simulation; and
- project fees were likely to remain the same despite the added service value (modelling and simulation engendered the confidence to implement innovative solutions that were not possible by conventional methods).

To guide future applications, quality assurance procedures are required covering the spectrum of modelling tasks. These include identification of objectives, mapping of objectives to simulation tasks, identifying uncertainties and risks, commissioning simulations, maintaining an audit trail, translating simulation outcomes to design evolution, client reporting, and model archiving. In addition, to guide results analysis, benchmark data are required to provide a means to judge the integrated performance of a building against others in the same class.

In the longer term, the creation of a modelling-oriented construction industry will consolidate and strengthen the market position and growth potential of companies due to the increased depth of service they can provide. It will also give rise to a better balance between energy supply and demand, reflecting the fact that demand-side measures impact supply-side decision-making and that opportunities exist for embedding new and renewable energy generation within society at all levels. Finally, BPS+ deployments will assist design process decentralisation by enabling cooperative working between distributed partners. Whilst the benefits are palpable, so too are the barriers derived from the industry's vested interest in maintaining the status quo.

What follows is a summary of the lessons learned from supporting 50 companies of various types and scales in their attempts to deploy BPS+ technology (McElroy 2009). Such outcomes were underpinning factors in CIBSE's Applications Manual for Building Energy and Environmental Modelling (CIBSE 2015).

Getting started

A first step can be daunting due to the many pertinent questions that are far from straightforward to answer.

- What are the costs and benefits of simulation?
- How do I identify the best-fit program(s) for my needs?
- Who provides independent program validation and accreditation?
- How do I embed modelling tools in my business?
- What are the different roles in a simulation team?
- What training will my staff require and who can provide this?
- In what ways will I need to change my business work practices?
- Are there recognised procedures for making models and undertaking assessments?
- How is a model quality-assured?
- How is a model documented and archived?
- How is a model calibrated before use?
- What performance metrics and criteria should I adopt when appraising performance?
- What are the risks to my business?
- Where will I find approved databases of material properties, weather data, equipment types, and the like?

Such questions are best addressed through participation in the activities of a local IBPSA affiliate[13] where the experience of others at various deployment stages will be available. From an individual user's viewpoint, it is probably best to adopt a 'minimum regret' approach to program application that starts simple and builds gradually. In that way, answers to the above questions will emerge naturally in the context of their specific application needs.

Start simple. Apply your chosen programs to familiar tasks such as the assessment of summer overheating or winter heating requirements under design weather conditions. There is no point in turning on modelling features just because they are there and getting hopelessly lost in the process. Rely on simplifying assumptions as you have previously done. For example, use design air change rates rather than succumb to the temptation to use that fashionable airflow prediction feature. By initially staying on familiar ground, your competence and confidence will grow apace, forming a solid foundation upon which to build.

Learn as much as possible via a constrained model before increasing the descriptive complexity. For example, where the issue is overheating potential in a highly glazed multi-storey office, it is best to define initially a single office with a minimal description of its bounding spaces. Then identify and simulate a week of weather associated with overheating rather than commissioning annual simulations. Commence with no environmental control and observe the office's response characteristics, incrementally adapting the design until a robust solution is achieved. Such an approach will deliver useful information to another member of the design team much faster than you or they expected.

Get support. The simulation of building and community systems is part art, part science. No one person is an expert in all aspects, so teamwork is required.

---

[13]See ibpsa.org for a list of regional affiliates and local contact details.

In some cases, it may be possible to form an in-house team comprising individuals with complementary skills in modelling and simulation. Where this is not possible, strategic alliances with the vendor organisation or a local university group might be appropriate. Remember that whilst the rewards from the successful adoption of the computational approach can be great, the pitfalls are many and varied. Take care, be cautious and designate someone as the in-house paranoid checker (better that they identify a model or reporting error than the client).

Pace your progress. After you have acquired basic skills, you will be ready to delve deeper into the exciting world of integrated performance simulation. This step is best taken in a focused and systematic manner rather than through a scattergun approach. Select a topic – air movement, control action or renewable energy system deployment for example – and explore the program's related data requirements and analysis capabilities. If possible, attend a training course to ensure that you understand the technical details (a little knowledge is a dangerous thing). Within a short period, you will be ready to try out your new skill on a real project. As long as you resist the temptation to over-elaborate and ensure that specialist help is to hand, the outcome will be rewarding. In this way, you will be able to develop your expertise in one topic before moving to another.

Reflect on your learning. Against the maxim that reflection helps consolidation, you should take every opportunity to discuss with colleagues your approach to modelling and the usefulness of the results to be generated. The aim is to evolve a performance assessment procedure that aligns appropriately with existing design processes, improves productivity, and adds value to the product delivered to the client. Remember that different people will progress at different rates. For example, some individuals will be good at managing modelling and simulation processes, some will be good at coordinating simulation tasks, and others will be effective at explaining outcomes to clients.

Now, jump forward a few years. You have mastered the capabilities of a few programs. You are now able to provide others with quantitative performance insights into issues such as energy use, environmental impact, component performance, comfort levels, and indoor air quality, and so help to improve the overall performance of the related product, process, or system. Significantly, you are able to do this within the real-time, real-scale, resource-constrained context of design practice. Can you think of a more apt contribution to sustainable development?

In the longer term, as company in-house or cloud-based computing power permits, the simulation approach might give rise to an exciting new prospect: the design of energy systems in real time through the on-screen manipulation of component parts. Such a virtual design environment would enable the rapid prototyping of solutions that strike the correct balance between product performance and sustainable development; Section 12.9 considers this prospect further.

Hand (2018) has provided examples of the translation of design goals to issues such as tool selection, workflow planning, and resource allocation. Table 4.11 gives some examples.

Being pedantic during the planning stage often speeds up subsequent tasks. It is essential to be clear about the assessment goals – first to drive understanding within the

Table 4.11   Translating design goals to simulation-related tasks

Design goal	Modelling and simulation issue
What performance issues are related to the design of big ideas?	Ascertain what needs to be quantified, and what tools and staffing can deliver assessments.
Under what conditions does the building fabric deliver comfort without mechanical assistance?	Establish how the building performs without environmental controls. Explore architectural variants such as shading or thermal mass. Establish if there are patterns associated with remaining discomfort.
Is natural ventilation an option for overheating control?	Investigate weather patterns during overheating and implement a natural ventilation scheme to determine if this reduces unmet hours.
Does the building require air conditioning?	Test whether mechanical ventilation can limit unmet hours. If not, determine the demand patterns for air conditioning.
If so, which HVAC system type is a good fit?	Establish candidate system models and, for each, explore performance under design and part load operation.
A facilities manager would like to know if a building could accommodate changes in use.	Establish a matrix of possible uses and establish a model for each one. Simulate the models and review the results to see if a general rule set might be provided.
Are there conflicts between daylighting and glare?	Consider both general illumination and occupant position-specific visual comfort. Consider focusing on a few principal zones at high resolution.
What is the design team debating now and what are they likely to consider in future?	Consider creating (temporary) model variants to deal with what-if questions. Is the current model fit for answering likely follow-up questions?

simulation team and then to translate that knowledge into a digestible form for the client. The nature of planning decisions that follow from the client brief is illustrated below.

- Detail: decide locations in the model where measurements should be taken and the level of modelling detail required.
- Scope: establish the extent of the model in terms of what needs to be included and what can be excluded.
- Robustness: confirm that the planned model composition and simulation procedure make sense to other team members.
- Credibility: consider what level of model calibration will increase confidence in the process.
- Resilience: consider how the design might fail and determine boundary conditions and operating regimes that will test failure modes. Take time to notice the unintended consequences of design changes.

- Opportunities: notice patterns beyond the usual and communicate opportunities for better performance.

The concept of delayed complexity is recommended. Asking, 'What is the simplest model that can deliver useful results to the client?' limits risk. More detail can be added gradually.

Common simulation procedures

Simulation-based projects are team efforts. It is critical that someone considers the broad sweep of issues relating to the client's goals. The *team manager* tasks tend to be undertaken by an individual who is not also tasked with creating models. They are responsible for ensuring that models are fit for purpose and that staff are working within their limits and the limits of the tool. They review models to anticipate possibilities as well as notice when deadlines slip.

Although it is everybody's job to notice opportunities as well as errors and omissions, there is a role for a *quality manager*. Those who work on models can find it difficult to recognise whether the model is appropriately detailed and whether the latest predictions are consistent.

With few exceptions, the simulation projects described in this book were undertaken with such separation of tasks. Working procedures tended to allow time for pedantic planning as well as time to 'live-with-the-model' and thus identify opportunities as well as faults. See Section 4.8 for further discussion on simulation project roles.

Interpretation of results and client reporting

Essentially, time and energy are invested in the creation of models to enable simulations that illuminate performance over time. User skills in identifying patterns and recognising the chain of thermodynamic dependencies can deliver value to the client.

Recognition of patterns in time series data is, for some users, an existing skill. Others require such data to be presented in a more intuitive form: e.g., the frequency binning of space temperatures may give a better insight into comfort than a table of numbers.

For other users, patterns emerge in the relationships between different aspects of building performance. For example, where a surface temperature elevation is evident and a causal relationship with solar radiation and the radiant component of casual gains is suspected as the cause, then a mix of data needs to be reviewed.

Patterns may only emerge when data is viewed from different perspectives. Elevated surface temperatures might have been noticed from a graphical presentation and a coincidence with solar radiation by overlaying these data on the same graph, but the magnitude of risk will only be made clear when all surface heat gains that constitute an energy balance are considered together.

Influence of simulation on the final design

Where BPS+ is relegated to compliance tasks after the design process is largely complete, there is little scope for influencing the design. Otherwise, various approaches can be taken to bring influence to bear at any design stage.

- The first is a 'coffee break' model: Isolate a portion of a large model and adapt it for specific what-if questions. For example, take an office model, copy it, keep one version as a base case, alter the second, undertake short-period simulations, and review the differences. If the altered version performs better, consider imposing its features on the overall model.
- The second is to test a base case model against reference cases arrived at by introducing alternative design features, systems, and control to the base case. Use the differences to guide actions.
- The third is not to assume that there is only one BPS+ tool and input model suitable for the project. Just as the design process is iterative and changes focus over time, so models and tools may need to evolve continually.

Sources of training

Most vendor organisations provide training courses in the application of their specific BPS+ tool. Where an academic department makes tools available to their students, it is also possible to contract training support, sometimes delivered in-house.

For example, IBPSA hosts an education page[14] that offers Webinar recordings, some relating to special issues of the *Building Performance Simulation* journal and member publications. ASHRAE maintains an archive of courses[15], most recently focused on indoor air quality and the mitigation of infection agents, otherwise addressing topics such as airflow patterns, managing HVAC systems, and highlights of recent conferences. The ASHRAE Learning Centre covers several topics, including modelling requirements for complying with Standard 90.1.

Program vendors, such as IESVE[16] and EQUA Simulation[17], offer webinars and online training for the products they sell. Accreditation is often awarded to individuals who demonstrate specific competencies. Some training is ad hoc, some free, and some delivered via multi-day courses tailored to practitioner needs, e.g., EDSL[18] offers free online training and fee-based, formal workshops.

Support organisations

Many organisations provide support to the simulation novice. Most notable is IBPSA[19] and its many affiliates throughout the world. IBPSA hosts a biennial

---

[14]https://ibpsa.org/education
[15]https://ashrae.org/technical-resources/training
[16]https://iesve.com/training
[17]https://equa.se/en
[18]https://edsl.net/training
[19]https://ibpsa.org

international conference, whilst its affiliates host regional conferences of various flavours, e.g., ASIM in Asia, eSim in Canada, Building Simulation and Optimization in England, BauSIM in Germany, Simbuild in the USA, and uSIM in Scotland.

4.8   Simulation-based, consultancy services

Some simulation-based studies are undertaken within well-resourced projects, whilst other applications are part of relatively low-budget, rapid turnaround consultancies. In relation to the latter, a standard procedure should be put in place. What follows is a description of the procedure implemented by ESRU[20], which is summarised here for possible replication within SMEs and other academic groups wishing to support the construction industry.

An initial client enquiry is handled by a designated project initiator (PI). In collaboration with senior management, a project manager (PM) is appointed to handle the consultancy. A meeting is then arranged between the client and PM to understand the client requirements and agree the scope of the project. The PM prepares a quote using a standard template and passes this to the PI for vetting and transmission to the client after sign-off by senior management. After the quote is accepted, a project team is formed comprising the PM and one or more technical assistants (TAs). The PI initiates a project log and transmits this to an archive along with a record of the client's instruction to proceed.

The project team then plans and executes the project against the agreed plan. The PM liaises with the client, coordinates the team's activities, and keeps the PI informed of progress. All interim and final reports as prepared by the team using a standard template. These are approved by the PM and passed to the PI for sign-off and transmission to the client. All models are quality-assured by an individual not directly involved in the consultancy.

It is the PM's job to conclude the consultancy and ensure that the client accepts the outcome. This being the case, the PM then informs the PI and provides the project final report and related models to the PI for archiving.

Whilst it is desirable that a project team comprises individuals with complementary skills – to service model definition, simulation coordination, results analysis, and client interaction – it is likely that staff will have different BPS+ use aptitudes. The following user types are typical.

Previous experience

Having absorbed the syntax and the philosophy of one simulation program, these individuals typically find it easy to adapt to the idiosyncrasies of another program. Previous exposure to simulation will have made them receptive to ideas of design abstraction and model planning. Such users will appreciate the potential of the

---

[20]A note to fellow academics: it is important to deploy BPS+ to support businesses, not to compete with them.

simulation approach and be willing to accept the increased demand for information that is necessary to unleash its potential.

There is an unhelpful variant of this user type: a person who prefers to retain previously acquired work patterns even when confronted with a new program that is demonstrably more powerful.

Background in building physics

These individuals have a conceptual grasp of building energy flows and an insight into the design parameters that influence them. Typically, they take a methodological approach to problem solving and are comfortable with technical detail. They make effective PMs.

Technicians

These individuals have the skills required to manage the modelling and simulation process but are less able to translate client requirements into strategies for simulation. They are effective TAs. That said they tend to be 'tool-led' (i.e., follow a presented path) and inclined to evolve complex problem descriptions and employ brute force simulation approaches. Having little understanding that simulation is best when it targets the essence of a problem, they are determined to enter details until the descriptive limits of the system are reached. With no appreciation that a subset of a program's functionality might be sufficient to answer a particular question, they strive to make sure that every possible program toggle is set to 'on'. The PM must ensure that they adhere to the project plan.

Single-issue users

Some users have a tendency to fixate in one of two ways: either judging the worth of a program by the manner in which it handles a particular issue or demanding that one particular facet of the simulation be enhanced to a level that is disproportionate to the other issues. The PM's task is to constrain such a user to adhere to a balanced approach.

Traditionalist

This individual does not cope well with the BPS+ approach. Design abstraction and simulation planning are foreign concepts, and they prefer to represent building performance in terms of a limited set of numbers rather than the spatial and temporal distribution of state variables. Having been conditioned by the constraints of traditional calculation methods, they tend not to consider incremental problem evolution with simulations focused on specific issues along the way. If well managed, they can contribute significantly to project plan formation and client interaction.

User training

The learning curve associated with BPS+, which is capable of modelling a diverse set of problem types, is non-trivial. As with most advanced engineering applications, there is the problem of syntax in relation to the procedures to 'drive' a program and the data underlying its input model. The power and, at the same time, the weakness of simulation is the ability to offer users multiple ways to represent and analyse problems in an attempt to accommodate aspects at different levels of abstraction. Whilst an experienced modeller may derive power from this situation, a novice will typically feel unnerved.

To be effective, training will require more than access to a computer, some documentation, and sufficient time to sort things out by trial and error. This is because most new users are content to be led by the tool rather than to develop a view on what is required and then pursue this by creative tool use. The present state-of-the-art in energy modelling requires that problems be realistically constrained in terms of focus and scale. Learning by reflective practice is a good place to start. Being part of an experienced team is the best way to acquire tacit knowledge.

4.9   Simulation-based, simplified design tools

BPS+ tools are normally applied directly by experienced users to accomplish a range of performance assessments based on models of arbitrary complexity. This section presents an alternative, indirect approach that reduces the burden on inexperienced users. The principal attraction of the approach is that it can be tailored to the needs of particular user types and applied in ways that are familiar to them.

Most BPS+ tools can be driven by non-human agents and can therefore be easily embedded behind bespoke interfaces or employed to generate performance maps from massive parametric excursions. Such agents can take various forms: automation scripts allowing coordination for a given purpose (see Section 4.5), GUI or Web-based interfaces that communicate with modules via file transfer and remote invocation, or simplified interfaces linked to a fully embedded simulation engine.

The following list categorises alternative flavours of the embedded approach, with examples based on ESP-r given in the following sections as indicated in parentheses.

- The simulator acts as a building emulator to advise a building's energy management system (Section 4.9.1).
- Real-time simulation is carried out by linking the simulator to physical sensors/actuators that enable predictions based on currently available measured data to influence an actuator signal or raise an operator alert (Section 4.9.1).
- The simulator is run over a range of input conditions to develop a performance map that can be embedded in look-up tables, with interpolation where necessary, or to develop a set of correlations by curve fitting (Section 4.9.5, Section 4.9.4).
- The simulator is placed behind a constrained interface whereby the user can alter only a relatively few parameters to address a particular design issue (Section 4.9.2, Section 4.9.3, Section 4.9.7).

- The simulator automatically imposes constraints on a model to align it with the assumptions required to demonstrate regulatory compliance (Section 4.9.6)
- The simulator provides pre-processed results for a large estate to enable rapid appraisals of retrofit options (Section 4.9.5).
- The simulator is run for a variety of cases to provide performance data that can be utilised by an existing simplified method to improve its veracity (Section 4.9.8).
- The simulator is used to produce benchmark performance data for use in post-occupancy studies (Section 4.9.9).

4.9.1   Intelligent BEMS

Simulation has not traditionally been used for operational building control although the possibility has been long recognised. For example, simulation programs have been used in place of a building and its HVAC system to aid in commissioning and the training of operators, e.g., within the SIMBAD emulator[21] (Husaunndee *et al.* 1997). More commonly, simulation is used to evaluate alternative control strategies (e.g., Chua *et al.* 2007) and this has evolved to the Building Controls Virtual Test Bed[22] (Wetter 2011) that supports co-simulation between BPS+ and control algorithms implemented in Simulink[TM23].

Another possibility is to embed BPS+ within BEMS for use in semi-real-time mode to evaluate control options and make a selection based on criteria such as minimum energy use and acceptable thermal comfort. Conventional control functions in BEMS have limited or no model of the building and therefore cannot take into account relationships between design and performance parameters. Simulation-assisted control is most suited to control schemes with the following characteristics:

- significant look-ahead times (e.g., night ventilation);
- where there are complex but known system interactions (e.g., glare, requiring blind repositioning, causing luminaire actuation, and leading to increased cooling loads);
- for supervisory control where several alternatives need to be evaluated (e.g., load shedding); and
- where the building use varies and this is known in advance (e.g., large occupancy variations).

The concept is summarised in Figure 4.16, which depicts the usual BEMS control structure: inputs obtained from weather and building state sensors, with an internal control algorithm deciding appropriate control actions. The new elements are the simulator, which models the building and HVAC system using sensed data as boundary conditions; and an evaluator, which scans the simulation

---

[21]https://scribd.com/document/68016530/SIMBAD-Building-and-HVAC-Toolbox

[22]https://simulationresearch.lbl.gov/projects/building-controls-virtual-test-bed

[23]https://mathworks.com/products/simulink.html

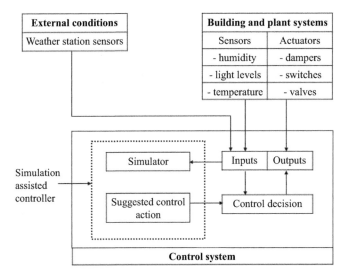

Figure 4.16 Simulation-assisted control in BEMS

results to suggest an appropriate control action to the main simulator-assisted controller.

A prototype control structure of this type was developed and tested in an environmental test room operated by Honeywell at Newhouse in Scotland (Clarke *et al.* 2002). Experiments tested the developed controller with realistic time constants. The environmental test facility consisted of two realistically dimensioned rooms surrounded by temperature-controlled voids. The constructions used in the test rooms were similar to those in a real dwelling (insulated cavity walls with double-glazed windows) and each room was heated by low-temperature radiators supplied from a central boiler. Within the study, the controller was in optimum start mode.

The graphical user interface of the simulation-embedded tool allows the user to select from five house types, five plant types, and five control regimes (giving a total of 125 variants). For each component in the selected system, individual attributes may be altered. For example, a radiator can have its supply rate, inlet and return temperatures, and thermal mass set by the user.

In subsequent studies, involving a comparison of controllers from two manufacturers, a broader comparison of WCH systems was included and coordinated with the manufacturers' test facilities. See Section 6.10 for further details.

4.9.2 Control strategy

For rating purposes in the United Kingdom, the energy performance of domestic buildings is evaluated using the Standard Assessment Procedure (SAP), which takes a rudimentary approach to assessing the control of heating plant. The impact of the method of control in a house can be significant, and the availability of electronic products with embedded control algorithms creates new opportunities to reduce energy demand. The quantitative energy use and potential savings may be influenced by the type of building construction and heating system, as well as the control logic. Control improvements can be readily applied to existing housing stocks: Palmer *et al.* (2006) recognised that because 70% of the 2050 housing stock already existed, a modest reduction in energy use due to better control could make a greater cumulative impact on energy consumption than improvements due to new building standards.

A project was carried out to establish a modelling tool, Advanced Domestic Energy Prediction Tool (ADEPT; Cockroft *et al.* 2009), for typical domestic house types with wet central heating systems and incorporating a variety of conventional and advanced control approaches. The short timescale dynamics of the control model were integrated into an ESP-r dynamic construction model with relatively long time constants so that accurate estimates of seasonal and annual energy consumption could be made. The ADEPT user interface is shown in Figure 4.17.

This interface facilitates the selection of combinations of house type, heating system and control scheme, with access to a range of system and control parameters

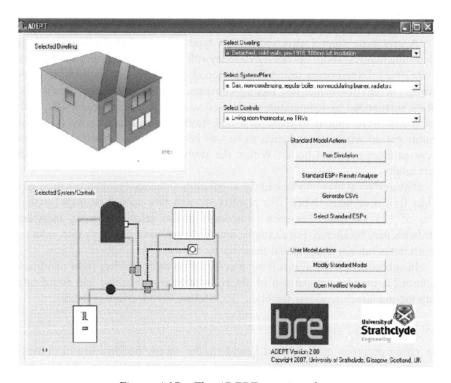

Figure 4.17   The ADEPT user interface

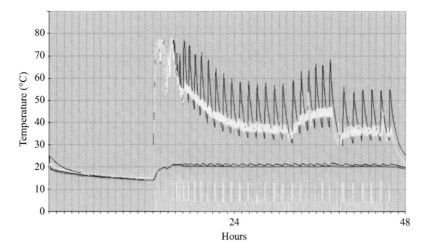

Figure 4.18   Comparison of on/off thermostat and PI control

including set points, proportional-integral control characteristics, and boiler and construction thermal response characteristics. Standardised outputs relating to control system behaviour and energy use are produced (Cockroft *et al.* 2007), allowing users to evaluate control options in a variety of circumstances, thus taking advantage of the power of simulation whilst avoiding the learning curve associated with navigating a BPS+ tool directly.

As an example of tool use, a thermostat-controlled radiator system is compared with a system using a PI controller directly modulating the burner. A house meeting the then-current building regulations fitted with a combination boiler was assumed for the comparison. Figure 4.18 shows the living room temperature control and the water temperatures for the two cases. The room thermostat cycles approximately twice per hour and is set so that the temperature does not fall below 21 °C. The PI controller is mostly cycling on and off at the bottom of the boiler modulating range (30%) and is able to maintain the room temperature using a lower average water temperature and run the boiler at a lower firing rate, resulting here in an annual saving of 6.2%.

After selecting a building type, the user is able to alter a subset of the model description related to control set points and parameters. This allows users to concentrate only on aspects of interest and avoid distractions from other aspects of performance that are out of scope.

4.9.3   Natural ventilation

In the previous example, a focused user interface called upon the simulation tool in a manner that was invisible to the user. Sometimes, just a subset of the simulation

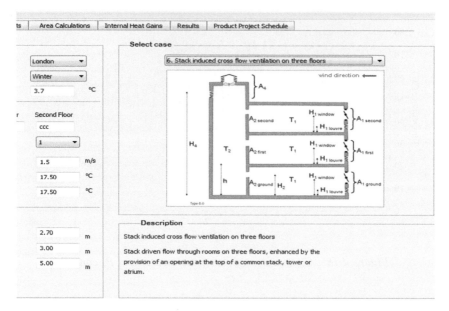

Figure 4.19   A portion of the ventilation selection tool

capability is required, thus giving direct access to simulation but without the overheads of run-time delays.

A louvre system manufacturer required a sales-oriented selection tool that could assess company products using predefined airflow networks corresponding to natural ventilation processed by a constrained version of ESP-r's mass flow solver. A bespoke interface was developed and airflow network topologies were embedded. For example, Figure 4.19 shows a stack-induced cross-flow ventilation scheme for a three-floor building. A range of possible airflow topologies were covered, including natural and fan-assisted schemes.

Users can alter temperatures, wind velocity, and level of internal gains as well as the height of the louvre. After the iterative solution, the output flow is overlaid on the building cross section. Rapid adjustments can be made to any scheme, with output in the form of a product schedule that may be incorporated into a client offer.

4.9.4   Biomass boiler sizing

Biomass boiler performance is notably sensitive to oversizing and short cycling. With support from the Carbon Trust, this project involved the creation of a tool for sizing commercial biomass boilers and their associated thermal storage systems (ESRU 2013). Figure 4.20 shows the tool interface.

A typical design strategy is to use thermal storage to minimise the required biomass boiler size as shown in Figure 4.21. To model the charge/discharge process between the boiler and thermal store, the tool needs the hourly heat demand profile

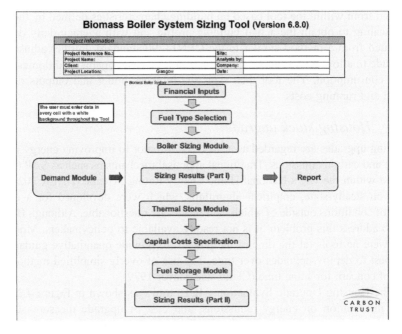

Figure 4.20   The biomass boiler sizing tool

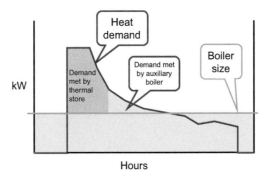

Figure 4.21   Biomass boiler size

for a design day to carry out plant sizing, and the hourly heat demands on other days to calculate the relative contributions of the biomass and auxiliary boilers to meeting that demand.

In this case, it was not feasible to run dynamic simulations in real time and a series of pre-simulations were carried out on a typical building, modelling variations in fabric insulation, fraction of glazed wall area, thermal mass of fabric, and duration of occupancy. Some 81 hourly demand profiles for average outside temperatures ranging from −3 °C to 14 °C were created. These profiles are then selected from within the tool based on building characteristics defined by the user, with scaling to obtain the actual building profile and with average daily demand estimated from a method similar to the CEN13790 standard. Further adjustments are made to allow for variations in ventilation loss, casual gains, and domestic hot water consumption. The tool then sizes the biomass boiler and outputs data on capital and running costs.

4.9.5   Housing stock upgrade

Dwelling upgrades are regarded as a major contributor to improving energy performance and carbon emissions. Traditional appraisal mechanisms such as SAP (2023), as used within the UK's National Home Energy Rating Scheme (NHER 2023), are based on steady-state, empirical algorithms, which were developed for a narrow range of conditions outside of which their results are questionable. Although BPS+ is able to address this problem, it is not readily available to policymakers. Moreover, there were no tools (at the time of the project) to provide quantitative guidance on how best to deploy upgrades over time. The use of overly simplified methods has been of concern for some time (Clarke and Maver 1979).

The Housing Upgrade Evaluation (HUE) tool[24] as shown in Figure 4.22 provides information on energy, emissions, and cost of upgrade measures for the domestic stock (see Section 8.1 for details). It is based on an extensive set of

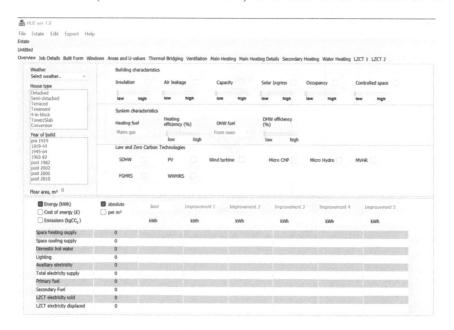

Figure 4.22   The HUE tool interface

---

[24]https://www.esru.strath.ac.uk/Applications

simulation models that are automatically generated and pre-simulated using the technique elaborated in Section 7.7, with results captured as performance maps for installation in HUE.

In total, 18,750 prototype models were generated by considering combinations of governing design parameters at quantised levels – exposure, insulation level, air tightness, position of thermal capacity, solar ingress, occupancy level, floor area fractions, and different set point temperatures – with parameter values inferred from National House Condition Surveys (e.g., SHCS 2019) and the building regulations prevailing at different times. These prototypes were then pre-simulated against weather conditions representative of the United Kingdom (75 locations spanning 30 years), and the time series results were subjected to regression analysis to produce equations that express energy demand as a function of weather parameters. These equations are used as a proxy for the actual simulation results.

Users can determine dwelling energy performance, cost, and emissions by answering questions relating to location, built form, and year of build. These choices determine a unique prototype based on equivalence rules for the governing design parameters. Additional model detail is inferred from building regulations for the age band and house conditions survey data. The input model can be refined by adding more detail. Once a dwelling has been selected, it can be improved by choosing upgrades relating to form and fabric and low carbon technologies. This automatically upgrades the dwelling by the selection of a replacement prototype. Results for both prototypes are immediately available to the user, who can accept or reject the upgrade based on the impact. The approach may be applied at different scales from a single dwelling to a national housing stock.

HUE was applied to develop an upgrade plan to 2020 for the Scottish Housing Stock (Clarke *et al.* 2004). It was demonstrated that heating energy demand could be reduced by up to 60% by the phased deployment of insulation improvements and air leakage reduction. In another project, HUE was applied to assess upgrade strategies for Scottish Local Authorities (Tuohy *et al.* 2006). In this manner, advice based on explicit simulation is provided rapidly but without recourse to model making.

4.9.6 *Regulatory application*

SBEM[25], the official UK non-domestic calculation tool for the generation of Energy Performance Certificates (EPCs) and demonstrating regulation compliance, is restricted in that it can only be applied to non-complex building forms. To assess buildings possessing form, fabric, and system complexity, official guidance is to use BPS+ tools. This project undertook developments of ESP-r to enable the program to be accredited for demonstrating compliance with the National Calculation Methodology (NCM; BRE 2010). The NCM imposes constraints on the users of BPS+ by prescribing libraries of activities and operational patterns, standard

---

[25]https://www.uk-ncm.org.uk

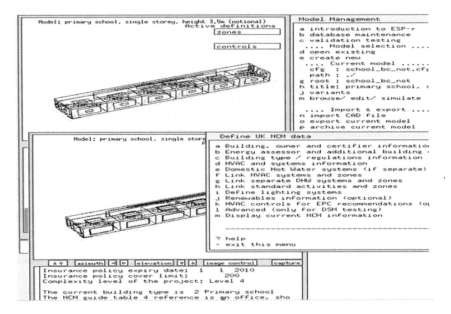

Figure 4.23   ESP-r's NCM interface showing initial and morphed models

treatments of thermal bridges, standard weather conditions, etc., and requiring accreditation of programs and users.

Accreditation tasks involved demonstrating that results were within specified tolerances (CIBSE 2006) and the automatic generation and simulation of prescribed models. Data input for the models was similar to standard iSBEM (the interface to SBEM) but instead of using the SBEM calculation engine ESP-r was employed. Any data that were not provided within the NCM are inferred from building regulations or default values are assumed. This ensures simulation models for buildings of any complexity can be created and assessed for NCM purposes. ESP-r follows NCM guidelines and generates the simulation models automatically. As depicted in Figure 4.23, these models are generated according to rules and represent various minimum regulation compliance permutations.

Once generated, the models are subjected to annual simulation and the results are post-processed to provide inputs that are relevant to the creation of energy reports and EPCs.

Another example of this type of application is HOT2000[26] (2023) for regulation compliance and energy performance assessment (Purdy *et al.* 2005). This tool comprises a Web interface that elicits simplified inputs and uses these to generate ESP-r models that are then simulated to generate the required assessment outcomes.

---

[26]https://natural-resources.canada.ca/energy-efficiency/homes/professional-opportunities/tools-industry-professionals/20596

4.9.7 Advanced glazing selection

The European Commission's IMplementation of Advanced Glazing in Europe (IMAGE) project (Kristensen 1996) involved the application of BPS+ to existing and proposed building designs incorporating advanced glazing systems. To facilitate the wider dissemination of the project's outcomes, a Glazing Design Support Tool (GDST; Citherlet *et al.* 1999) based on ESP-r was established to allow the glass industry to assess the multi-variate impact of applying a given advanced glazing component to a given building located in a given climate. Figure 4.24 shows examples of the tool interface.

GDST stores ESP-r models of exemplar and project-specific buildings, weather data relating to typical and project-specific climates, and data defining the optical and thermal properties of advanced glazing components. In use, it allows the association of glazing components with buildings, and buildings with climates, and the retrieval of pre-formed, multi-variate views of performance (encapsulating energy use, gaseous emission, and thermal/visual comfort aspects) for pre-selected combinations of building, climate, and glazing. It is possible to invoke ESP-r for combinations not previously processed or for the analysis of new glazing components.

4.9.8 Regulatory compliance tool refinement

National compliance tools are reviewed regularly and existing/new capabilities are modified/added. For example, a review of SAP in the United Kingdom identified that the underlying monthly method was a poor fit for the needs of variable utility pricing, load balancing, smart metering, hot water distribution and storage, and the transition from gas boilers to heat pumps. The SAP algorithm would also require refinement to accommodate emerging construction standards for new builds and retrofits. To address these requirements, the compliance tool would need to undertake higher frequency calculations based on a higher resolution input model. The time-sensitive nature of the issues suggested the replacement of the monthly method with a half-hour time-stepping method supported by typical weather patterns and a more detailed building description. It was also expected that users would require the ability to undertake 'what-if' studies concerning the impacts of different occupancy patterns and control regimes.

A re-imagining of the underlying solution method suggested a development process that involved robust checks against a well-validated existing BPS+ tool. ESP-r was deployed in this role because of its ability to accept input models at different levels of abstraction and also assess multi-variate performance impacts. Such an approach is akin to inter-program validation as often applied to determine the degree of alignment of new programs with other, more scrutinised equivalents. A suite of test cases was developed, including dwelling models that capture the essential characteristics of existing and new-build configurations (Figure 4.25). These models were used to test the new SAP outputs for a 'matrix' of dwelling types when located in a variety of climates.

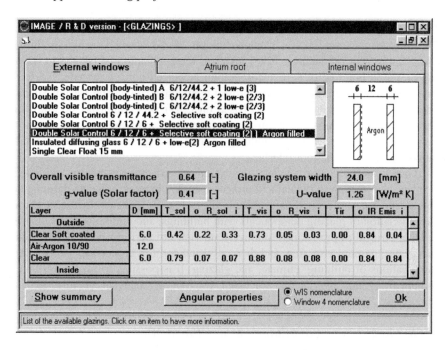

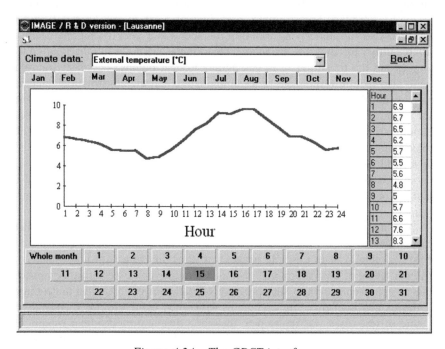

Figure 4.24   The GDST interface

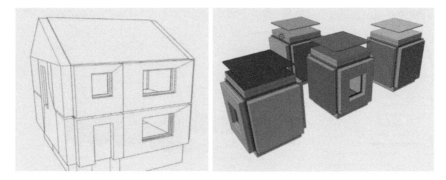

Figure 4.25   ESP-r dwelling and part models used in SAP refinement

The simulation results were then used in inter-program comparison mode to confirm the performance of new SAP models as they were developed.

4.9.9   Upgrade quality assurance

A Post Operations Evaluation Tool (POET[27]) was developed to provide a method for the automatic confirmation of the quality of dwelling upgrades. POET implements a procedure to assess the energy and environmental efficacy of the upgrade based on meter readings and indoor environmental conditions monitoring pre- and post-upgrade. This provides an evidence-based quality assurance test of the upgrade that supports project sign-off. The user interface is shown in Figure 4.26.

Based on the known property details, a computer model is constructed and calibrated against metered data. The model is simulated over the heating season to determine the theoretical heating energy-saving potential associated with the proposed upgrade scheme. The result is displayed in the 'Benchmark saving (%)' field. (A further role for this model is to support investigations into the causes of post-upgrade problems). In a project for Glasgow City Council, social housing in several neighbourhoods was assessed; Table 4.12 lists the predicted energy savings for several cases. All dwellings received external wall insulation except where marked ([#]underfloor insulation, [*]internal wall insulation).

The pre- and post-upgrade energy consumption of the target dwelling, as determined from meter readings, is normalised using outdoor temperature data from a local weather station. This determines the savings achieved by the property's upgrade and is displayed in the 'Savings achieved (%)' field. The benchmark saving is compared to the actual saving achieved, and a rating is assigned that shows how close the property came to achieving its theoretical target. After monitored environmental conditions are imported, a preview image is generated as depicted in Figure 4.27 along with indicators marking the pre- and post-upgrade meter reading periods. Clicking on the image allows the user to

---

[27]https://www.esru.strath.ac.uk/Applications

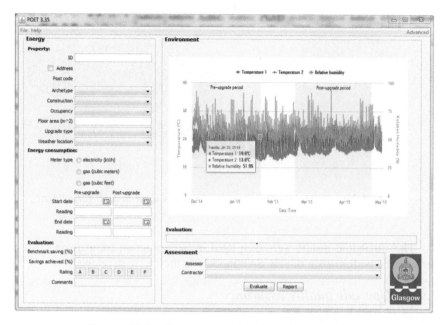

Figure 4.26   Constructs within the POET interface

*Table 4.12   Energy-saving benchmarks for a subset of
properties in Glasgow*

Location	Dwelling ID	Benchmark saving (%)
Shettleston	SH1	24.7
	SH2	18.6
	SH3	24.7
	SH4	13.1
	SH5	24.7
	SH6[#]	14.8
	SH7[#]	14.8
	SH8[#]	14.8
Mosspark	MO1[*]	16.4
	MO2	16.4
	MO3	12.9
	MO4	12.9
Shawlands	SL3	24.6
	SL4	15.9
	SL5	22.9
	SL6	10.7
	SL7	14.2
	SL8	10.7
	SL9	14.2
	SL10	22.9
Anniesland	AN1[*]	13.5
	AN2[*]	13.5
	AN3[*]	13.5
	AN4[*]	13.5

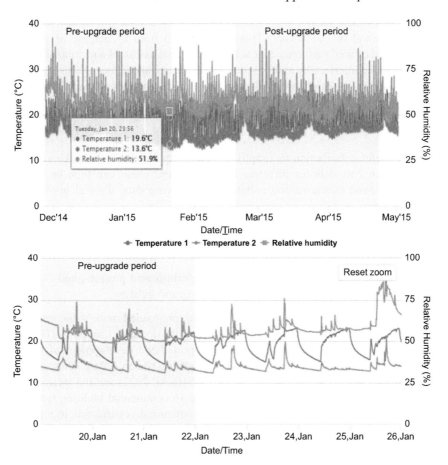

Figure 4.27   Interactive POET graph – upper, all monitored data, lower, a one-week period

remove specific data profiles or zoom in to examine the monitored data over different periods.

A field is provided under the preview image that gives textual feedback from an underlying environmental conditions algorithm that assesses the monitored, post-upgrade indoor temperature and humidity profiles to determine if these are within acceptable ranges. The tool also provides an interface for changing the acceptable performance ranges.

The interactive viewing facility, when used in conjunction with the written feedback from the environmental assessment algorithm and survey information, provides the assessor with an analysis tool for identifying acceptable indoor environmental performance post-upgrade and isolating possible causal factors where this performance is deemed unacceptable.

The assessment section of the tool allows the user to input information regarding the assessor and the contractor. These are selected from a database of previous assessors and official partnered contractors. Additional assessors and contractors can be added to the interface by authorised users via the advanced menu. This section of the tool launches the evaluation of the algorithms, and generates a compliance report.

4.10   Archetype models

Another useful contribution to simplifying BPS+ application is to make available pre-constructed models for different building types. These can then be utilised without the need to source and collate the underlying data. Typical applications might include:

- student and practitioner training in modelling and simulation;
- rapid options appraisal in support of the refinement of building energy codes and standards;
- use as starter models in new design and refurbishment projects; and
- testing and quality assurance following program updates.

Whilst some BPS+ programs will come with pre-loaded models, the need is to establish representative models that have wide industry support. The US Department of Energy, for example, has supported the development of models covering the US commercial and residential estates (Hirsch 2016). These models are compliant with ASHRAE standard 90.1 (ASHRAE 2022) and the International Energy Conservation Code (IECC 2023). Some 16 commercial building types and two residential types have been subjected to parametric diversification to represent specific configurations as observed in the field (different heating systems, foundation types, and climate zones for example). This resulted in 3,344 commercial models and 3,552 residential models that are available for download. (There is also a set of 152 'manufactured housing' models available.) See Section 8.1 for a description of an approach to parameter diversification.

To illustrate the potential, consider archetype models for domestic and commercial buildings in Rotterdam and London as established in ESRU consultancy and research projects (Clarke and Hand 2016, INDU-ZERO 2021). Figure 4.28 summarises this set of models, which can be accessed as exemplar models within ESP-r and interrogated and/or simulated.

Typical assumptions underpinning these exemplar models include the following.

Commercial

- Solid brick construction ($U = 2.1$ W/m^2.K) in older premises, and cavity brick walls ($U = 1.6$ W/m^2.K) and insulated cavity brick walls ($U = 0.35$ W/m^2.K) in modern buildings.
- Retail park buildings are assumed to have facades following 2008 standards. Ceilings and floors in offices and older retail are assumed concrete with dropped ceilings.

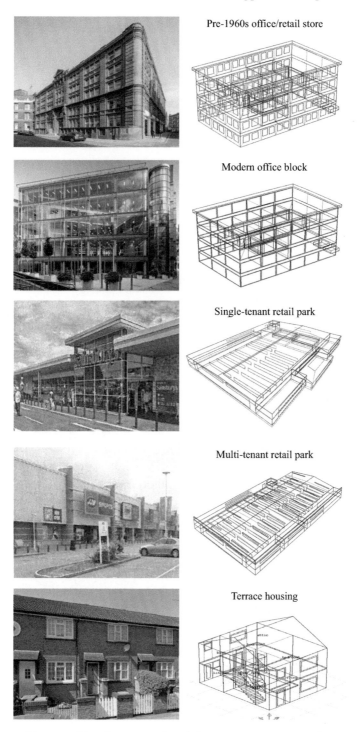

Pre-1960s office/retail store

Modern office block

Single-tenant retail park

Multi-tenant retail park

Terrace housing

Figure 4.28   Commercial and domestic archetype models

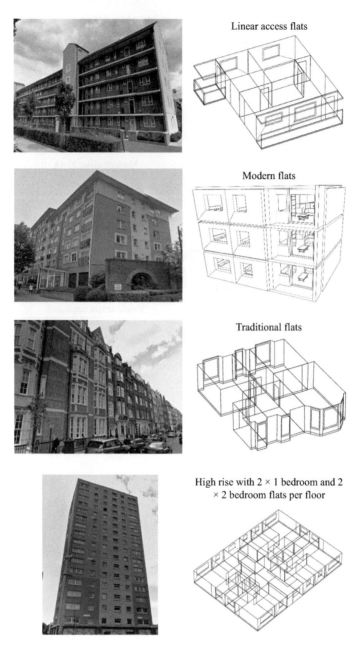

Linear access flats

Modern flats

Traditional flats

High rise with 2 × 1 bedroom and 2 × 2 bedroom flats per floor

Figure 4.28   (Continued)

- The windows in pre-1960s buildings are single glazed with newer buildings having low-quality double glazing. Higher-performance double glazing is assigned to modern office blocks.
- Older properties are assigned an infiltration rate of 1.0 ACH, medium-efficiency properties of 0.9 ACH, and newer properties of 0.7 ACH.
- Offices maintain traditional hours with a few people in on Saturdays and closed on Sundays. Retail premises are assumed open all day but with reduced trading hours on Sunday.
- The models assume wet central heating systems with radiators sized for peak demands. Heating temperature set points are 20 °C during operating hours with night and holiday setbacks to 15 °C. It is assumed that thermostatic radiator valves are used to control zone temperatures.

Domestic

- Floor areas are derived from the English Housing Survey.
- Solid brick construction ($U = 2.1$ W/m^2.K), cavity brick walls ($U = 1.6$ W/m^2.K), and insulated cavity brick walls ($U = 0.35$ W/m^2.K).
- Ceilings and floors in terrace houses are of wood structure with carpet and plaster ceilings, carpet over slab-on-grade for the ground floors, and a cold roof attic with 100 mm of insulation.
- Windows are single, low-quality double glazing or higher-performance double glazing.
- Low-rise flats have solid brick construction with wood intermediate floors, single glazing, and wooden window frames.
- Older common access flats have a solid brick construction on concrete floor plates with single glazing and wooden frames. Medium-efficiency common access flats have cavity brick and low-performance double glazing. High-efficiency common access flats have filled cavity brick and high-performance double glazing.
- System-built, high-rise flats have a typical facade construction and a concrete floor plate.
- Internal walls are brick or blockwork with a plaster finish before 1960 and plasterboard and stud in buildings constructed post-1960.
- Occupancy patterns include diversity – people move between zones and timings on weekdays and weekends are different.
- Properties with lower energy efficiency are assigned an infiltration rate of 1.0 ACH, medium-efficiency properties have 0.8 ACH, and newer properties 0.7 ACH.
- A minimum of two separately controlled zones are assumed with temperature set points aligned with SAP (18 °C in bedrooms, 20 °C in living areas, and night setback to 15 °C).
- Zone temperature control is via radiators with TRVs.
- The models assume wet central heating systems with radiators sized for peak demands.

Of course, such assumptions can be updated before model use.

4.11   Supporting energy rating systems

BPS+ can be used to provide performance data as required by national energy rating schemes[28]. Such schemes have common attributes as follows.

- A governing body defines the specifics of the schemes, how the scheme is administered, project entry requirements, and responsibilities for participants as well as those who support compliance with the scheme.
- Each scheme has a target market of building types and performance metrics that can be rated. Ratings are often subdivided into categories based on levels of performance as assessed or measured.
- Each scheme has specific rules for its trademarks and intellectual property in terms of use in marketing and communications.
- Each levy fees for administration, training, and review that often, but not always, relate to the building type, size, and the specific rating(s) on offer.
- Each provides a mechanism for ensuring the quality of submissions, often via independent assessors, who comply with specific competency and training rules. Some governing bodies undertake additional reviews. Most establish specific examinations for agents who support a scheme.
- There is a considerable variation in the requirements for numerical assessments. Some are measurement only, some use spreadsheets, some are prescriptive, and some accept performance simulation results as an option.
- Rating schemes tend to form an ecosystem including resources for participants, incentives for acquiring skills, recurring licensing and training requirements. Governing bodies tend to evolve and expand their ecosystem and seek to strengthen their market. A considerable investment in time and resources appears to be required to stay within these ecosystems. A degree of rent seeking is apparent.

Consider the following examples of contemporary rating systems.

4.11.1   NABERS

This scheme[29] was established by the New South Wales government in 1998. It is based on measured energy, water, waste, and indoor environmental conditions of buildings in use. It targets a range of building types, including apartments and residential care, hospitals, offices, and retail shopping. There are separate fees for energy, water, indoor environment, and waste ratings as well as for setting up the initial performance agreement. Fees are also charged for the training, examination, and licensing of assessors. Training is separate for each building type and specific ratings. Assessors are independent parties, who set their own fees.

---

[28]A possible future alternative to following such schemes is elaborated in Section 11.3 whereby a rating depends on extended BPS+ simulations that test operational resilience.
[29]https://nabers.gov.au/about/

The scheme is based on building energy use monitoring, and the governing body distributes spreadsheets to convert monitored data to a rating. There are also reverse calculators that convert building attributes into desired levels of fuel demands.

NABERS UK[30] was adapted from the Australian programme. It is owned and overseen by the New South Wales government and managed by BRE. It is focused on commercial properties with ratings based on a mix of numerical simulations, design reviews, and monitoring of energy use. Through a 'Design for Performance' element[31], a developer/owner enters into an agreement to design and then commission and operate a building to achieve a specific target performance. The project design, operational regime, and use are numerically assessed, with the design and simulation models reviewed by an independent assessor. After one year of monitoring, predictions and measurements are compared to judge whether an operational rating can be issued.

Assessments must use software that meets the requirements of ASHRAE 140. This may be supplemented by other tools for minor energy demands. Some wording suggests that plant assessments require more than look-up-table performance estimates. The model includes the whole building, its contents and environment controls, with the building zoned to correspond to the control capability. General guidance is given on the treatment of shading devices, glazing systems, thermal bridges, plant, and control. Operational patterns and occupancy are meant to be realistic and provision must be in place to mitigate for instances of tenant fit-out posing a risk to the performance goals. The choice of the simulation tool, the methodology used, as well as any simplifications of form, composition, use, and environmental control are subject to review. To establish risks, at least four variants should be assessed – including alternative occupancy, environmental control regimes, and maintenance regimes, plus a variant that combines aspects that have been seen to degrade performance. Guidance is provided as to what predicted performance metrics are expected, their frequency, and their level of detail. Guidance is also provided as to what is to be monitored and at what level of detail.

As with other rating schemes, NABERS UK defines specific relationships between the parties taking part in a Design for Performance, a scheme to help developers ensure a project's intended energy consumption is verified through post-occupancy performance monitoring (Roumi *et al.* 2021). There is a Code of Practice for those acting as independent assessors who are licensed by BRE and who have undergone BRE-approved training.

Figure 4.29 shows a NABERS-rated building located in Southwark, England. This comprises 3.4 ha of predominantly office space and accommodation for retail and leisure (Better Buildings Partnership 2022). The design utilised hybrid steel and cross-laminated timber, a combination that resulted in a carbon intensity around half of a typical London office. It is expected that the building will receive a 5-star NABERS UK Energy Rating once in operation.

---

[30]https://bregroup.com/products/nabers-uk/nabers-uk-about/
[31]https://bregroup.com/products/nabers-uk/nabers-uk-products/nabers-design-for-performance/

Figure 4.29   Timber Square office (credit: betterbuildingspartnership.co.uk/
nabers-uk-names-landsec%E2%80%99s-timber-square-first-
certified-design-performance-project/)

Figure 4.30   Angel Square office (credit: placenorthwest.co.uk/nomas-4-angel-
square-breaks-nabers-rating-ceiling-in-manchester)

A second example is a building located in Manchester, England as shown in
Figure 4.30. This comprises 1.9 ha and was awarded five stars for Design for
Performance due to its use of air source heat pumps and onsite energy generation
from photovoltaic panels (North West Place 2023).

4.11.2   LEED

This rating scheme[32] is administered by the US Green Building Council. It addresses
energy efficiency, water conservation, material selection, daylighting, and waste

---

[32]https://usgbc.org/store

reduction. LEED comprises a suite of ratings covering building design and construction (BD+C), operations and maintenance (O+M), and interior design and construction (ID+C). These overall ratings are adapted for new construction projects, core and shell (partially fitted out) commercial buildings, and data centres. In general, a rating is applied to the whole building. As with other schemes, a suite of administrative fees applies, separated into phases of a project, with additional fees related to the floor area of the building.

LEED Associates work with the design team to build up a description of the project attributes and design intents relating to mandatory and optional credits. Associates undergo training to support specific credit evaluations and set fees as independent licensed agents in addition to the governing body fees.

Compliance is based on a suite of credits[33]. Some require a review of project documentation; others involve some degree of numerical assessment. For example, an 'Integrative Process' credit can be supported with a rectangular box model of the building to establish broad-brush impacts of site constraints, massing and orientation, facade attributes, lighting levels, comfort, and operational parameters. As another example, the 'Minimum Energy Performance' credit is based on ANSI Standard 90.1, with a mix of prescriptive provisions, normative exceptions via a comparison between a baseline and proposed building, and exceptional calculations in accordance with guidelines. There is a 'minimum energy performance calculator' spreadsheet for one level of credit. Alternatives are available such as the 'Optimize energy performance via building simulation' (OEPBS) as summarised in Table 4.13.

The governing body takes submissions from associates and has additional processes to determine if the requirements of the rating have been achieved. The example shown in Figure 4.31 is the Herman Miller office building located in Cheltenham, England, which achieved a high LEED rating (BSRIA 2009). The first two floors have natural, cross-ventilation controlled by sensors monitoring internal conditions.

Another example is Bloomberg's European headquarters in London, England (Figure 4.32), which achieved LEED Platinum (Bloomberg 2023). The building consumes 70% less water and 35% less energy than a typical office due to conservation measures, LED lighting, and natural ventilation that can be augmented by mechanical means in specific areas.

4.11.3   PassivHaus standard

This standard[34] combines superior thermal comfort with minimum energy consumption achieved predominantly by installing high-quality insulation material, triple-glazed windows with insulated frames, increased airtightness, and the fitting of an MVHR system. The aim is to provide optimum levels of comfort and indoor air quality, heating by natural means to remove the dependence on fossil fuels, and

---

[33]https://usgbc.org/credits
[34]https://passivehouse-international.org

Table 4.13   Summary of the OEPBS credit

EA Credit: Optimize Energy Performance

Demonstrate increasing levels of energy performance going beyond prerequisite standards to lower economic and environmental impacts from excessive energy use. Points are allocated to specific building purposes: schools (1–16 points), health care (1–20 points), multi-family mid-rise (1–30 points), and all other purposes (1–18 points).

Energy performance targets (measured in $kWh/m^2.y$) must be established before the schematic design phase. During the design process, efficiency measures must be analysed for comparison with similar building projects. Furthermore, efficiency measures related to load reduction and HVAC strategies must be analysed to determine potential energy savings and holistic project cost implications.

EA Prerequisite: Minimum Energy Performance

Demonstrate a minimum reduction of 5% for new construction, 3% for major renovations, or 2% for core and shell projects in the proposed building Performance Cost Index (PCI) below the Performance Cost Index Target (PCIt) calculated in accordance with Section 4.2.1.1 of ANSI/ASHRAE/IESNA Standard 90.1–2016, Appendix G (with some exceptions). For mixed-use buildings, the required PCI is calculated using an area-weighted average of the building type. Projects must meet the minimum PCI before taking credit for renewable energy systems.

Unregulated loads should be modelled accurately to reflect the actual expected energy consumption of the building. If unregulated loads are not identical for both the baseline and the proposed building performance rating, and the simulation program cannot accurately model the savings, follow the exceptional calculation method (ANSI/ASHRAE/IESNA Standard 90.1–2016, G2.5).

Once implemented, the building's energy use must be monitored for a full 12 months of continuous operation and achieve the required levels of efficiency. All building projects are required to establish an energy portfolio containing energy consumption data, building space types, and usage patterns.

Figure 4.31   Herman Miller HQ office (credit: www.bsria.com/uk/news/article/ model-project-herman-miller-uk-hq)

Figure 4.32 Bloomberg's European HQ office (credit: https://bloomberg.com/ company/stories/eco-friendly-features-bloombergs-new-european- headquarters/)

low-energy consumption resulting in low emissions. Various factors must be considered when seeking compliance: orientation, compactness, fabric heat loss, thermal bridging, airtightness, glazing, shading, heat recovery, heating system, and appliances.

The orientation of a dwelling determines whether it can maximise the benefit from solar gains, whilst its compactness (the ratio of external surface area to internal volume) will influence its overall energy performance: two dwellings of similar orientation and construction but different compactness will have substantially different energy use. In addition, larger buildings may be more prone to thermal bridges particularly where geometries are complex.

The standard recommends that walls, floors, and roofs should have U-values of ≤ 0.15 W/m^2.K and ≤ 0.85 W/m^2.K for windows. Other modern building standards have similar goals; however, exceptional performance in PassivHaus is predicated on minimal faults in design, construction, and commissioning. Another way to express this is 'no snagging', and the use of BPS+ beyond the standard compliance tools can help attain this goal. Thermal bridges are a major factor in a passive house where complex construction junctions result in a bypass route for heat loss. It is common for a certified dwelling to employ external insulation to avoid this possibility. Linear thermal bridges must have a Ψ-value ≤ 0.01 W/m.K to conform to the standard. Obtaining a negative Ψ-value is possible, indicating that the level of insulation is more than satisfactory.

One approach to attaining good indoor air quality and reducing the chances of mould growth is to prevent warm moist air from entering the dwelling. This can be achieved by creating an airtight barrier around all building elements. For openings such as windows or doors, this can be achieved by the application of proprietary tape. Airtightness is measured by a 50 Pa pressurisation test, with the resulting air leakage to be ≤ 0.6 ACH. It is advised that this test be performed during the construction phase to provide the opportunity to resolve any issues.

Normally, habitable rooms such as bedrooms and living rooms are located on the southerly facade (in northern latitudes), with large windows to capture solar gain. Conversely, non-habitable rooms such as kitchens and bathrooms are located on the northerly facade where large windows are unnecessary. In cold climates, triple-glazed windows are used to reduce heat loss and increase the surface temperature of the inner pane to minimise cold draughts. As the controlled solar gain is the fundamental aim, window glazing must have high solar energy transmittance (G-value): for triple glazing, a G-value ≥ 0.5 would be expected.

To avoid overheating during summer, the dwelling is equipped with shading devices to control solar gain. A fixed shading device can be critically positioned to ensure that solar gain is maximised in winter and minimised in summer. Alternatively, shading devices can be operated mechanically in response to indoor temperature conditions and/or external solar radiation levels.

The standard specifies that indoor temperatures should not exceed 25 °C for more than 10% of the year. If this is not possible by passive means, an MVHR system can be installed. Such a system will also control ventilation losses that can otherwise contribute greatly to heat loss. To minimise losses without incurring a detrimental impact on air quality, the MVHR system recovers heat from exhaust air. Opening windows to provide natural ventilation is not then required. The warm indoor air passes through a heat exchanger where the heat is transferred to incoming outdoor air in a separate circuit. The standard mandates a heat recovery efficiency of $\geq 75\%$ in conjunction with a fan power ≤ 0.45 Wh/m^3.

PassivHaus buildings are inherently insensitive to changes in ambient conditions, and it would be expected that solar and internal gains would offset facade losses. Only in the harshest conditions would supplementary heating be required, and the MVHR system be operated in heating mode. Primary energy demand for heating, hot water, ventilation, and electrical appliance use is limited to 120 kWh/m^2.y, including losses via process conversion and transmission. This is important in the PassivHaus standard as minimising the energy demand also minimises CO_2 emissions. To further lower energy consumption, it is recommended that electrical appliances be highly rated.

Two PassivHaus examples follow. The first, shown in Figure 4.33, is located in Hertfordshire, England, and is constructed with an insulated timber frame and an airtight membrane erected on a concrete slab. Other key features include the installation of LED lighting throughout the house, solar-derived underfloor heating, and recycled paper insulation. Natural timber cladding is used on the exterior and interior, wrapping around the front and rear facades.

The second example is a zero-carbon PassivHaus dwelling in the Scottish Borders, which sits on a south-facing slope as shown in Figure 4.34. All main rooms are south facing to capture daylight and give access to the countryside view. Focus is on achieving a balance between thermal insulation and solar gain,

Figure 4.33   A PassivHaus dwelling (credit: homebuilding.co.uk/ideas/discover-this-passivhaus-a-masterclass-in-timber-design/)

Figure 4.34   A PassivHaus dwelling (credit: Venner Lucas, vennerlucas.co.uk/projects_zero_carbon.html)

allowing for minimal heating and electrical demand both of which are serviced by photovoltaic-thermal micro-generation. The dwelling has an MVHR system.

In summary, the aim of a dwelling built to PassivHaus standard is to deliver good indoor conditions at all times of the year and with reduced energy use and emissions. Given the price of energy and the pressures of net zero legislation, it is likely that such dwellings will increasingly be seen as a good investment. Given the increased thermodynamic complexity inherent in such designs, BPS+ is an effective tool to ensure good performance post-construction.

The question remains, why employ simulation when there is a compliance process that can be serviced via a simplified model. Consider the small dwelling of

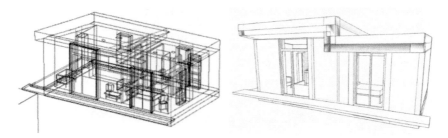

Figure 4.35   ESP-r High-resolution dwelling model based on PassivHaus techniques

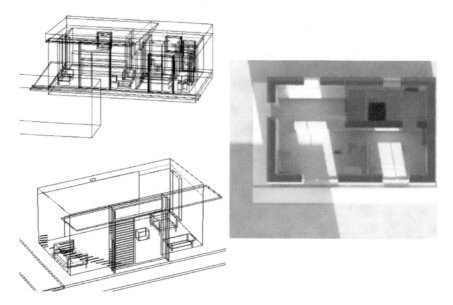

Figure 4.36   Exploring sun patches via CAD views, calculated patches and ray tracing

Figure 4.35, which has been attributed in keeping with a PassivHaus regime but with the goal of supporting multiple assessments.

Whereas compliance focuses on zone-averaged comfort, such a model tracks position-dependent thermal and visual comfort. Whereas compliance focuses on the outer facade, the model includes both the inner and outer faces of the facade, partitions, and the mass of furniture and fittings. Hidden aspects of the design, such as voids above the kitchen and bathroom are treated with the same level of rigour. Whereas compliance looks at seasonal solar radiation ingress, BPS+ supports a range of enquiries. With reference to Figure 4.36, views from the sun at specific dates and times (left, upper) can be supplemented by calculated sun patches (left, lower) and ray tracing can take into account architectural elements and furnishings (right).

Whereas compliance is silent on thermostats, the model includes explicit sensor placement and control logic representation. Whereas compliance airflows are abstract, the model solves for air and contaminate movement between and within rooms. Whereas compliance imposes simple patterns of infiltration and natural ventilation, the model represents leakage distribution and the impact of window control strategies in response to changing weather patterns. Whereas compliance has (at best) a monthly resolution, the model reports the state of all elements at high temporal resolution (hourly or less). Whereas compliance deploys low-resolution methods to assess overheating, the model can identify the specific conditions associated with internal temperature excursions as well as take into account the specific distribution of thermal mass within the building. Whereas compliance treats the building as one thermal zone with one air temperature, the model can resolve conditions in each room including all containing constructions and the impact of sensor positions.

4.12   Chapter summary

Performance assessment methods are required to guide the application of BPS+, enable automated appraisal, and assist with inter-program results comparison. The technology can be applied by individuals to understand the impact of design changes on performance via short-period simulations focused on extreme events and system failures or via life cycle simulations that test resilience overall. Before using a model, it must be calibrated to ensure that it is fit for purpose.

Whilst performance simulation is potentially better, quicker, and cheaper than traditional methods, its effective application will require the development of industry standards for model making, performance assessment, and outcome reporting. This, in turn, will require a change in business work practices. To reduce the burden on users, it is possible to establish pre-constructed exemplar models and provide simulation-based tools that are endowed with a simplified interface.

References and further reading

ASHRAE (2022) 'Standard 90.1', https://ashrae.org/technical-resources/bookstore/standard-90-1/.

Autodesk Ltd (1989) *AutoCAD Release 10 Reference Manual*, Autodesk Ltd.

Banks J and Gibson R R (1997) 'Don't simulate when: ten rules for determining when simulation is not appropriate', *IIE Solutions*, Institute of Industrial and Systems Engineers.

Better Buildings Partnership (2022) 'NABERS UK names Landsec's Timber Square as the first certified design for performance project', https://

betterbuildingspartnership.co.uk/nabers-uk-names-landsec%E2%80%99s-timber-square-first-certified-design-performance-project/.

Bloomberg (2023) '5 Eco-Friendly Features at Bloomberg's Sustainable New European Headquarters', https://bloomberg.com/company/stories/eco-friendly-features-bloombergs-new-european-headquarters/.

Borg S P and Kelly N J (2012) 'The development and calibration of a generic dynamic absorption chiller model', *Energy and Buildings*, 55, 533–544.

Bourne S R (1982) *The UNIX System*, Addison-Wesley.

BRE (2010) 'NCM modelling guide for buildings other than dwellings', *Technical Report*, Building Research Establishment, Garston.

BS (2011) *BS EN 12464-1 Light and Lighting. Lighting of Work Places*, British Standard Institute, London.

BSRIA (2009) 'Model project - Herman Miller UK HQ', https://bsria.com/uk/news/article/model-project-herman-miller-uk-hq/.

Carlucci S, Causone F, de Rosa F and Pagliano L (2015) 'A review of indices for assessing visual comfort with a view to their use in optimization processes to support building integrated design', *Renewable and Sustainable Energy Reviews*, 47, 1016–1033.

Chua K J, Ho J C and Chou S K (2007) 'A comparative study of different control strategies for indoor air humidity', *Energy and Buildings*, 39(5), 537–545.

CIBSE (2006) *TM33: Tests for Software Accreditation and Verification*, CIBSE, London.

CIBSE (2015) 'Applications Manual 11: building performance modelling', cibse.org/knowledge-research/knowledge-portal/applications-manual-11-building-performance-modelling-2015/.

Citherlet S, Clarke J A, Hand J, *et al.* (1999) 'IMAGE: A simulation-based tool for the appraisal of advanced glazing', *Proc. Building Simulation '99*, Kyoto.

Clarke J A (1986) 'An intelligent approach to building energy simulation', Proc. Seminar on Expert Systems for Construction and Services, BSRIA, Bracknell.

Clarke J A and Cowie A (2020) 'A simulation-based procedure for building operational resilience testing', *Proc. CIBSE/ ASHRAE Technical Symposium*, 16–17 April, Glasgow, United Kingdom.

Clarke J A and Hand J W (2015) 'An overview of the EnTrak/BuildAX eService delivery platform', *ESRU Occasional* Paper O1–2015.

Clarke J A and Hand J W (2016) 'Archetype models for London district heating feasibility study', *Consultancy Report to BuroHappold*, ESRU, University of Strathclyde.

Clarke J A and Maver T W (1979) 'Deemed to satisfy?', *Architects Journal*, 872–874.

Clarke J A, Cockroft J, Conner S, *et al.* (2002) 'Simulation-assisted control in building energy management systems', *Energy and Buildings*, 34(9), 933–40.

Clarke J A, Hand J W, Hensen J L H, *et al.* (1996) 'Integrated performance appraisal of daylight-Europe case study buildings' *Proc. Solar Energy in Architecture and Urban Planning*, Berlin, March.

Clarke J A, Johnstone C M, Kondratenko I, *et al.* (2004) 'Using simulation to formulate domestic sector upgrading strategies for Scotland', *Energy and Buildings*, 36, 759–770.

Clarke J A, Strachan P A and Pernot C (1993) 'An approach to the calibration of building energy simulation models', *ASHRAE Transactions*, 99(2), 917–927.

Cockroft J, Ghauri S, Samuel A and Tuohy P (2009) 'Complex energy imulation using simplified user interaction mechanisms', *Proc. Building Simulation '09*, University of Strathclyde, Glasgow, July 27–30.

Cockroft J, Samuel A and Tuohy P (2007) 'Development of a methodology for the evaluation of domestic heating controls', *Market Transformation Programme Report RPDH19-2*, Department for Environment, Food and Rural Affairs, UK.

Darby S, Hinnells M, Killip G, Layberry R and Lovell H (2006) 'Reducing the environmental impact of housing', Technical Report, Environmental Change Institute, University of Oxford.

de Wilde P (2014) 'The gap between predicted and measured energy performance of buildings: a framework for investigation', *Automation in Design*, 41, 40–49.

ERC (2008) 'Evaluation of the Scottish energy systems group activities', *Report to Strathclyde European Partnership and Scottish Enterprise*, Eclipse Research Consultants, Cambridge.

ESRU (2013) *Biomass Boiler System Sizing Tool: User Manual*, ESRU, University of Strathclyde.

Ferrero A, Lenta E, Monetti V, Fabrizio E and Filippi M (2015) 'How to apply building energy performance simulation at the various design stages: a recipes approach', *Proc. Building Simulation '15*, Hyderabad, India.

Hand J W (1998) 'Removing barriers to the use of simulation in the building design professions', *PhD Thesis*, ESRU, University of Strathclyde.

Hand J W (2018) 'Strategies for Deploying Virtual Representations of the Built Environment', https://esru.strath.ac.uk/Courseware/ESP-r/tour/Downloads/strategies_may_2018.pdf/.

Hand J, McElroy L and Sinclair C (2021) 'Evolving building testing regimes to deliver performance metrics and support model calibration', *Proc. Building Simulation '21*, Bruges.

Hensen J L M (1999) 'A comparison of coupled and de-coupled solutions for temperature and air flow in a building', *ASHRAE Transactions*, 105(2).

Hensen J L M and Lamberts R (2019) 'Building performance simulation – challenges and opportunities', In: Hensen J L M and Lamberts R (eds.), *Building Performance Simulation for Design and Operation* (2nd ed.), Chapter 1 pp. 1–10, Routledge, ISBN 9781138392199.

Hirsch J J (2016) *'DOE-2 Related Programs', https://doe2.com/*.

Hit2Gap (2023) https://hit2gap.eu/.

HOT2000 (2023) 'HOT2000 software suite', https://natural-resources.canada.ca/energy-efficiency/homes/professional-opportunities/tools-industry-professionals/20596)/.

Husaunndee A, Lahrech R, Vaezi-Nejad H and Visier J C (1997) 'SIMBAD: a simulation toolbox for the design and test of HVAC control systems', *Proc. Building Simulation '97*, 2, pp. 269–276, Prague.

IECC (2023) 'International Energy Conservation Code', https://codes.iccsafe.org/content/IECCComm2021P1/.

INDU-ZERO (2021) 'Smart Renovation Factory', https://northsearegion.eu/indu-zero/.

Johnsen K, Grau K and Christensen J (1993) *TSBi3 User Manual*, Danish Building Research Institute, Copenhagen.

Jolliffe I T (2002) *Principal Component Analysis*, Springer.

Jones D R and Schonlau M W W J (1998) 'Efficient global optimization of expensive black-box functions', *Journal of Global Optimization*, 13, 455–492.

Judkoff R and Neymark J (1995) 'Building energy simulation test and diagnostic method', *Final Report for International Energy Agency cooperation between Solar Heating and Cooling (Task 12B) and Energy Conservation in Buildings and Community Systems (Annex 21C)*, https://nrel.gov/docs/legosti/old/6231.pdf.

Kristensen P E (1996) 'Daylight Europe', *Proc. 4th European Conference on Solar Energy in Architecture and Urban Planning*, Berlin.

Larson G W and Shakespeare R (1998) *Rendering with Radiance - The Art and Science of Lighting Visualization*, Morgan Kaufmann, San Francisco.

Macdonald I A and Clarke J A (2007) 'Applying uncertainty considerations to energy conservation equations', *Energy and Buildings*, 39(9), 1019–1026.

Mangematin E, Pandraud D and Roux D (2012) 'Quick measurements of energy efficiency of buildings', *Comptes Rendus Physique*, 13, 383–390.

McElroy L B (2009) 'Embedding integrated building performance assessment in design practice', *PhD Thesis*, ESRU, University of Strathclyde.

Menezes C, Cripps A, Bouchlaghem D and Buswell R (2012) 'Predicted vs. actual energy performance of non-domestic buildings: using post-occupancy evaluation data to reduce the performance gap', *Applied Energy*, 97, 355–364.

Meyer J J (1993) 'Visual discomfort: evaluation after introducing modulated light equipment', *Proc. Energy-Efficient Lighting*, Arnhem, The Netherlands, pp. 348–357.

Monari F (2016) 'Sensitivity analysis and Bayesian calibration of building energy Models', *PhD Thesis*, University of Strathclyde.

Monari F and Strachan P (2017) 'CALIBRO: an R package for the automatic calibration of building energy simulation models', *Building Simulation '17*, San Francisco.

Murphy G B, Counsell J, Allison J and Brindley J (2013) 'Calibrating a combined energy system analysis and controller design method with empirical data', *Energy*, 57, pp. 484–494.

Nelder J A and Mead R (1965) 'A simplex method for function minimization', *Computer Journal*, 7(4), 308–313.

NHER (2023) 'UK National Home Energy Rating Scheme', https://ratioseven.co.uk/what-is-national-home-energy-rating/.

North West Place (2023) 'NOMA's 4 Angel Square breaks NABERS rating ceiling in Manchester', https://placenorthwest.co.uk/nomas-4-angel-square-breaks-nabers-rating-ceiling-in-manchester/.

Purdy J, Lopez P, Haddad K, *et al.* (2005) 'The development of a test protocol for an on-line whole building energy analysis tool for homeowners', *Building Simulation '05*, Montreal, Canada.

Ratto M (2001) 'Sensitivity analysis in model calibration: gsa-glue approach', *Computer Physics Communications*, 136(3), 212–224.

RIBA (2024) RIBA Plan of Work, Royal Institute of British Architects, https://ribaplanofwork.com.

Rosenthal J S (2009) 'Markov Chain Monte Carlo Algorithms: theory and practice', In: Ecuyer P and Owen A (eds.), *Monte Carlo and Quasi-Monte Carlo Methods*, Springer, doi.org/10.1007/978-3-642-04107-5_9/.

Roumi S, Stewart R A, Zhang F and Santamouris M (2021) 'Unravelling the relationship between energy and indoor environmental quality in Australian office buildings', *Solar Energy*, 227, 190–202.

SAP (2023) 'UK Standard Assessment Procedure for Energy Rating of Dwellings', https://gov.uk/guidance/standard-assessment-procedure/.

SBEM (2023) 'UK Simplified Building Energy Model', https://bregroup.com/a-z/sbem-calculator/.

SHCS (2019) 'Scottish House Conditions Survey', https://gov.scot/collections/scottish-house-conditions-survey/.

Sonderegger R, Condon P and Modera M (1980) 'In situ measurement of residential energy performance using electric coheating', *ASHRAE Transactions*, 86(1), 394.

Subbarao K, Burch J, Hancock C, Lekov A and Balcomb J (1988) *Short-Term Energy Monitoring: Application of the PSTAR Method to a Residence in Fredericksburg Virginia*. SERI, Golden, Colorado.

Tuohy P G, Strachan P A and Marnie A (2006) 'Carbon and energy performance of housing: a model and toolset for policy development applied to a local authority housing stock', *Proc. Eurosun*, Glasgow.

Wetter M (2011) 'Co-Simulation of building energy and control systems with the building controls virtual test bed', *Building Performance Simulation*, 4(3), 185–203.

Chapter 5

High-resolution modelling and simulation

This chapter makes the case for a computational approach to design based on high-resolution models that support holistic, multi-variate appraisals of performance over extended periods and under a range of conditions likely to be experienced in practice. Although such an approach is not routinely practiced at present, it could become widely accepted if it was demonstrated to be better, cheaper, and quicker than the present fragmented approach to design/upgrade stage options appraisal. The chapter that follows then summarises a wide range of applications, in some cases following judicious simplifications applied to an otherwise high-resolution input model.

Many factors can cause a building to underperform. These relate to the poor performance of construction materials, the occurrence of thermal bridging and condensation, the incorrect sizing and operation of heating, cooling, ventilation, lighting, and power supply systems, and the failure of control systems. Whilst there is wide consensus that building performance needs to be substantially improved, conflicting factors relating to finance, occupant requirements, and technology selection often confound effective action. Consider Table 5.1, which lists some oft-touted options for energy efficiency and clean supply.

Such options are typically unable to provide an effective solution on their own and, in many cases, may be mutually exclusive. The need is to identify fit-for-purpose, context-dependent blends as a function of the requirements and constraints of a particular target building, new or existing. This is an obvious role for BPS+ when applied to buildings and their related energy supply systems within the urban environment, at whatever scale. The technology provides an effective mechanism to identify appropriate solutions within the time and resource constraints of design practice (Hong *et al.* 2000, Clarke and Hensen 2015).

Less obvious but no less powerful is the use of BPS+ to ensure the best temporal match between energy demand and supply, in the latter case when derived from blends of conventional and renewable energy sources (at the national and urban scale) assisted by energy storage and/or smart utility grid interaction. The conflicts that underlie policy and business agendas that seek to electrify building heating and transport, encourage net zero emissions within the building stock, establish community energy schemes with smart control, or encourage urban renewables can best be resolved by utilising this dynamic matching capability to bring forward hybrid schemes.

Table 5.1   Options for energy demand reduction

Passive features	Heat-related	Electricity-related
Daylight utilisation	Heat pumps	Integrated photovoltaics
Adaptive facades	Solar thermal	Smart meters
Solar ventilation pre-heat	Electricity-to-heat	Smart grids
Switchable glazing	Condensing combi-boilers	Urban wind power
Selective films	Smart heating control	Fuel cells and hydrogen
Advanced insulation	Heat recovery	Electric vehicles
Desiccant cooling	Biomass/biofuel heating	Demand management
Evaporative cooling	Culvert heating/cooling	Demand response
Moveable devices	District heating/cooling	Smart water heating
Breathable walls	Energy storage	Combined heat/power
Phase change materials	Tri-generation	Ducted wind turbines

Table 5.2   Merits of the holistic BPS+ approach*

Relevant issues addressed	Relevant processes encapsulated
Technical feasibility	Building physics
Adaptive thermal comfort	Thermofluids
Indoor and outdoor air quality	Heat and mass transfer
Life cycle operation	Radiant heat exchange
Energy and carbon economics	Plant and systems processes
Environmental emissions	Electrical power flows
Controllability assurance	Micro-climate effects
Operational resilience	Renewables stochasticity
Wellbeing and productivity	Control system characteristics
Uncertainty and risk	Innovative materials and systems
Increasingly stringent legislation	Stochastic weather and events

*Lists not matched.

The merits of an approach that enables whole system, multi-variate performance appraisal against realistic operational scenarios are summarised in Table 5.2. The contention is that only holistic performance simulation can accommodate the underlying physics, address all relevant performance aspects, and satisfy the time and cost constraints of business.

Built environment resilience typically refers to the ability of communities to recover from adverse disturbances associated with disasters such as flooding (Cere *et al.* 2019). In the present context, resilience refers to the ability of a building to perform as intended over its lifetime despite changing use and weather patterns. The absence of performance resilience in buildings has been extensively documented through post-occupancy evaluations (POE) (Bordass *et al.* 2001), and POE procedures have been incorporated into many standards and rating systems (Bordass and Leaman 2005, Doan *et al.* 2017). On the other hand there is no present method to ensure operational resilience at the design stage. This is an obvious role for BPS+ in the future.

Although the power of simulation is widely acknowledged, users do not generally appreciate that the approach does not generate design solutions, optimum or otherwise. Instead, it supports user understanding of complex systems by providing relatively rapid feedback on the performance implications of proffered designs. In this way, improvement follows for both the design and the user.

In addition, because tool users are required to create input models that are a mix of physical objects (walls, windows, equipment, etc.) and virtual objects (methods for system discretisation, heat transfer, turbulence, etc.), to define the required simulation conditions, and to interpret complex time-varying results, simulation outcomes from different tools/users are generally incomparable and, in the hands of a novice, often suspect. In any event, it is hubris to suggest that the future performance of complex entities as found in the built environment can be predicted in any meaningful way via the application of a 'user-friendly' program.

In order to realise the true potential of building performance simulation – as an emulator, not a predictor of future reality – it is helpful to change the application intent from performance prediction to operational resilience testing. Such a shift implies the need for a high-resolution input model that has relinquished reliance on simplifying prescriptions relating to weather, occupant presence/behaviour, equipment efficiency, infiltration rates, and the like in favour of explicit representations of these phenomena in order to encapsulate realistic life cycle operation.

This chapter considers the creation and use of high-resolution models as a means to test the resilient operation of buildings during the design, refurbishment, and in-use phases. This, of course, requires that the BPS+ tools being used are high-resolution compatible.

5.1   High-resolution models

Given the growth in BPS+ capability, how might this be applied in practice, especially in the context of future buildings with embedded supply technologies and advanced control? Consider the following generalised application example.

It is possible to determine the optimum combination of zone layout and constructional schemes that will best complement a high degree of operational automation or provide a weather-responsive design of high efficacy. Several simulations are conducted to determine a zoning strategy that not only satisfies the functional criteria but will also accommodate sophisticated multi-zone control and the re-distribution of excess energy. Some simulations might focus on the choice of construction materials and their relative positioning within constructions so that load and temperature levelling are maximised. In addition, alternative facade fenestration, ventilation, and shading control strategies might be investigated in terms of thermal/visual comfort and cost criteria.

Once a fundamentally sound design has emerged, well tested in terms of its performance under a range of anticipated operating conditions, a number of alternative control scenarios can be simulated. For example, basic control studies will lead to decisions on the potential of multi-zone control, optimum start/stop times, weather anticipation, sensor location, and nighttime temperature setback.

Further analysis might focus on 'smart' control, whereby the system is designed to respond to occupancy levels, fuel and power prices, or prevailing levels of illuminance and solar irradiance. As the underlying relationships emerge, the designer is able to assess the benefits and problems of any given course of action before it is implemented.

The appraisal permutations are essentially without limit, with BPS+ used to provide answers to pertinent questions such as the following examples.

- What are the maximum demands for heating and cooling, where and when do they occur, and what are the causal factors?
- What will be the effect of a particular design strategy, such as adopting heat recovery, advanced glazing systems, or sophisticated control regimes?
- When is the optimum start time for the heating plant or an effective algorithm for weather anticipation?
- How is energy consumption affected by daylight capture schemes?
- How will air movement or temperature distribution be affected by a particular management strategy, and will condensation become a problem?
- What is the contribution to energy savings and comfort levels of particular passive solar features?
- How can renewable energy systems be best deployed to match building loads or cooperate with the electricity grid?

The BPS+ approach allows the designer to understand the interrelation between design and performance parameters, identify potential problem areas, and implement appropriate design modifications. The design to result is more energy efficient, with better comfort levels attained throughout.

5.1.1   Building models

To benefit from such a wide-ranging capability, it is necessary to operate with a high-resolution input model. Figure 5.1 summarises the constituent parts of such a model for the case of a dwelling when serviced by a wet central heating system for space and water heating, a solar thermal collector for supplementary water heating, and/or a photovoltaic (PV) array for local electricity generation. The aim is to include all relevant phenomena: boiler dynamics, contents thermal inertia, indoor air quality, thermal bridge effects, occupant behaviour, and low voltage grid connection. The contention is that BPS+ will realise its true potential only when it is used to emulate reality in a manner that allows design proposals to be subjected to the types of influences and stressors experienced in practice.

Highlights of this model include:

- each apartment in the dwelling is represented in a literal 3D manner to support high-fidelity heat flow tracking and enable assessments of occupant comfort, air quality, and heating/cooling system control;
- all constructions (walls, ceilings, floors, windows, etc.) are modelled as multi-layered representations, with enhanced resolution around material junctions where thermal bridging might occur;

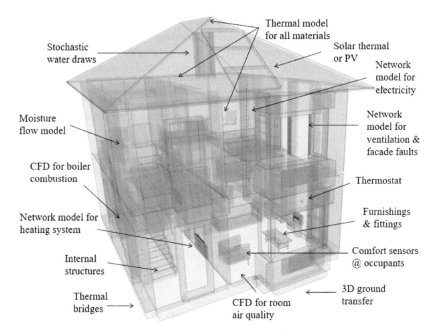

Figure 5.1   A high-resolution dwelling model

- the air leakage characteristics associated with windows, doors, construction details, extract fans, etc. are explicitly included to enable heat and mass transfer considerations at simulation time;
- plant and control system parts are represented as objects explicitly located within the dwelling to support building/plant thermodynamic interactions and the simultaneous solution of the combined model;
- inclusion of the underfloor crawl space, attic, and thermal bridges at exterior junctions;
- inclusion of internal features such as furnishings and stairs that add thermal mass and affect longwave radiation exchange and shortwave radiation distribution;
- an explicit representation of the condensing boiler, solar thermal collector, and PV array that do not require efficiency factors as input;
- electrical equipment (renewable energy components, lights, small power, etc.) defined to a level of detail that permits modelling of pertinent power flows and environmental heat transfers;
- inclusion of the connection of the PV array to the low-voltage network to support grid interaction evaluation;
- imposition of computational fluid dynamics domain models on a dwelling zone and the boiler combustion chamber to support comfort, air quality, and emissions studies;
- an explicit representation of the hot water store to evaluate approaches to dual heat source control and unintentional heat leakage avoidance;

- control action comprising multi-zone temperature sensing and solar water heating priority with the explicit placement of heating system control thermostats;
- imposition of air, water, and electricity flow networks with parameter adjustments that are responsive to dynamic boundary conditions and control system signals; and
- the imposition of a stochastic model of occupant presence and behaviour in relation to small power usage and heating system control adjustments.

The information required to establish a high-resolution model is extensive: 3D geometry; the hygro-thermal properties of construction materials; details of thermal bridges, air leakage distribution, and internal thermal mass (e.g., furnishings); operational data relating to occupancy, lighting, and small power; definitions of heating, ventilation, domestic hot water, and embedded renewable energy components; and weather time series. Although such data are often readily available, they must usually be collated from disparate sources such as construction drawings, site surveys, manufacturers' published data, building standards, city cadasters, and existing computer models in a variety of formats (e.g., SAP, PHPP, and BIM). A point to note in the case of retrofits applied to estates comprising multiple buildings is that the information collation exercise is required only for representative cases. The model creation effort is therefore modest compared with the cost and effort directed towards the overall upgrade. Further, it is likely that pre-constructed models will be more widely available in the future as the BIM standard evolves and new buildings replace old ones. The following data types are typically required to construct a high-resolution model.

Context

- Site details: location, building orientation, ground conditions, shading objects.
- Weather data: dry bulb temperature, relative humidity, wind speed and direction, direct and diffuse solar irradiance, and atmospheric pressure.

Building

- Elevations and plans conveying 3D information.
- Construction abutment details (to enable thermal bridge representation).
- Construction materials and their hygro-thermal properties (thermal conductivity, specific heat capacity, density, vapour diffusivity, surface longwave emissivity, and surface solar absorptivity).
- Internal fixtures and their positions: furniture, electrical appliances, lights, and small power outlets.
- Special materials (phase change, switchable glass).

Plant and systems

- Component and network layout for heating and cooling systems:
  - supply type (gas, electricity, heat);
  - heat source (boiler, heat pump, district heating);
  - thermal storage (water tank, phase change material, ice);
  - pipes and ductwork (material, dimensions, location, insulation);

- delivery mechanism (radiators, chilled beams, underfloor, diffusers); and
- design flow rates.

- Component and network layout specifications for mechanical ventilation systems:
  - air handling units;
  - air inlet and extract locations;
  - fresh air intake rate;
  - fan types; and
  - ductwork and mixing boxes (material, dimensions, location, insulation).

- Component specification for passive ventilation systems:
  - design air tightness;
  - openable windows;
  - trickle vents; and
  - other deliberate openings.

- Component specification for building-integrated renewables:
  - photovoltaic panels;
  - solar thermal panels;
  - heat pumps; and
  - micro wind turbines.

Control

- Component specification for thermal and air quality-related components:
  - controller type and logic (thermostatic, PID response, bespoke);
  - control schedules and temperature set points;
  - location and type of user controls; and
  - automated passive measures (night ventilation, solar ingress control).

- Component specification for lighting-related components:
  - control type and logic (daylight modulation, on/off, movement detection);
  - control schedules and illuminance set points;
  - location and type of user controls;
  - photocell location and orientation;
  - facade shading devices; and
  - internal blinds (manual, automated).

Occupant behaviour

- building operating hours;
- occupant types, activities, and clothing types;
- workstation locations; and
- stochastic behavioural model.

Because collating such information in the form required by a BPS+ tool can require significant effort, most tools assist by providing CAD-like interfaces and pre-constructed entities corresponding to constructions, plant components, and even pre-formed exemplar models. Figure 5.2 (upper) shows an office model that can be

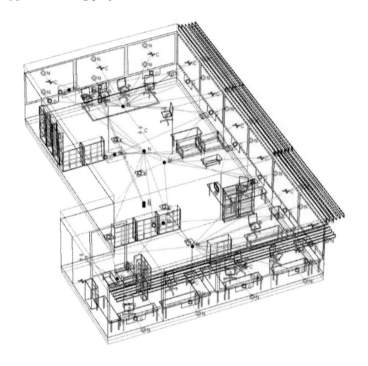

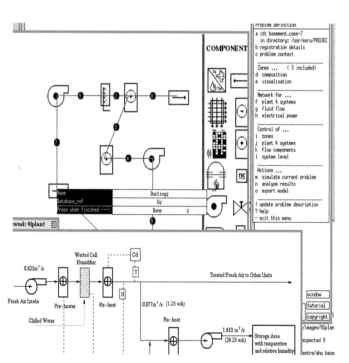

Figure 5.2   An office model and plant definition session

assigned alternative air leakage paths depending on the ventilation system, whilst the lower image depicts a session underway to define ventilation and heating systems via the manipulation of pre-formed component models. Whilst defining a building's leakage distribution is not straightforward, BPS+ tools will offer support in the form of databases of ventilation components (cracks, doors, windows, fans etc.) and sets of external surface pressure coefficients relating to different exposures.

In order to lessen the burden on BPS+ tool users, it is helpful to pursue the creation of a high-resolution input model as a two-stage process. The first stage encapsulates the physical entities comprising a proposal – construction layout and material attribution, internal fittings and furnishings, heating system and domestic hot water components, low voltage electrical network, solar thermal/PV panels, and control system components – all of which are in the domain of the designer. In the second stage, this information can be used to guide the selection of computational entities relating to heat transfer, network air/moisture/electricity flow, room air movement, combustion-related emissions, occupant behaviour, numerical gridding, etc. – all in the domain of the program creators. In this way, the burden of simulation moves partially from the user to the computational environment.

Although a high-resolution input model has enhanced utility, it is sometimes appropriate to apply simplifications to model parts. For example, where the intention is to evaluate the impact of alternative constructions, it may be appropriate to prescribe infiltration rates rather than model air leakage distribution explicitly; or, where occupancy patterns are definite, these may be imposed in place of a stochastic occupant behaviour model.

In the other direction, the resolution of a model may need to be increased to include corner effects or represent a possible thermal bridge, as depicted in Figure 5.3.

Here, a refined 3D grid is being applied to a corner element, and this new nodal network will be linked to the uni-directional conduction scheme applied to the joining walls (Nakhi 1995, Nakhi and Aasem 1999). This may significantly increase the model preparation and computing effort, although most programs providing the feature will offer substantial support. A solution to this problem is to apply a 1D model generally with grid refinements only at critical locations. Alternatively, it is possible to impose Ψ-values (pronounced psi) on the 1D model

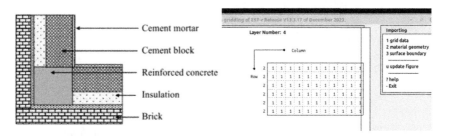

Figure 5.3   Increasing model resolution to include a corner effect

to enhance the conduction heat loss:

$$U' = U A \left(1 + \frac{1}{A} \sum_{i=1}^{n} \Psi i \, Li \right)$$

where U' is an enhanced construction U-value (W/m^2.K), U the U-value with no thermal bridge, A the construction area (m^2), and Ψ_i and L_i the Ψ-value (W/m.K) and length of the ith thermal bridge element, respectively (e.g., the column height in this example). Because a simulation program requires the individual thermo-physical properties of the materials comprising a construction (i.e., it does not use a U-value), a Ψ-value must be imposed via a suitable adjustment to the conductivity of the affected construction layer.

Ψ-values can be determined analytically (Tang and Saluja 1998, BS 2017) or on the basis of a separate high-resolution simulation using a numerical model such as AnTherm (2023).

Similarly, high-resolution models can be created for commercial buildings, where data defining form, fabric, contents, control systems, HVAC equipment, renewable energy components, artificial lighting, and electrical networks are augmented by virtual entities describing occupant presence/behaviour, domain gridding, etc. Figure 5.4 shows part of a model of the Challenger Building situated within the Bouygues Construction Company's headquarters complex in Saint-Quentin-en-Yvelines, France.

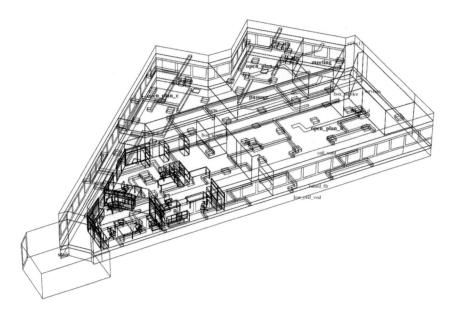

Figure 5.4 High-resolution model of the Challenger building.

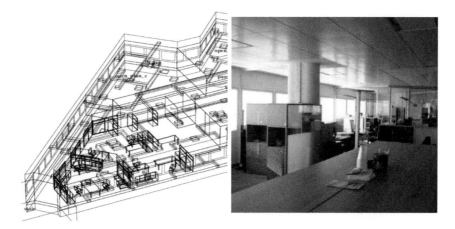

Figure 5.5   High-resolution representation of furnishings and service void ducting

This model encapsulates the following features of the office:

- furnishings and fittings are included as literal representations, including their level of thermal mass;
- service voids are explicitly represented; and
- occupant representations are positioned throughout so that visual (glare) and thermal comfort can be individually assessed.

Two features of the high-resolution modelling approach are emphasised here. First, the approach is holistic and therefore enables performance to be represented in terms of a range of relevant criteria. Second, it is possible to semi-automate the generation of models for large building stocks based on design parameter diversification applied to 'seed' models corresponding to particular archetypes (Clarke *et al.* 2008). The former feature provides a means to make trade-offs explicit (e.g., between energy reduction and occupant wellbeing); the latter feature reduces the model creation burden and helps harmonise the simulation process because the same model can be utilised by different users and for different applications.

The effort expended in creating a model is rewarded by the depth and range of the appraisals that are then enabled. Figure 5.6, for example, shows typical outputs from an indoor air quality study: air velocity distribution within a selected plane, and local surface temperature/humidity levels superimposed on mould growth isopleths.

Further, control action influences the whole-system model by driving parameter adaptation in response to changing system states and linking problem parts that have typically been decoupled in earlier tool generations. For example, the power output from a PV array is delivered to the electrical system, while any generated heat is absorbed by the building fabric or transferred to a heat recovery device.

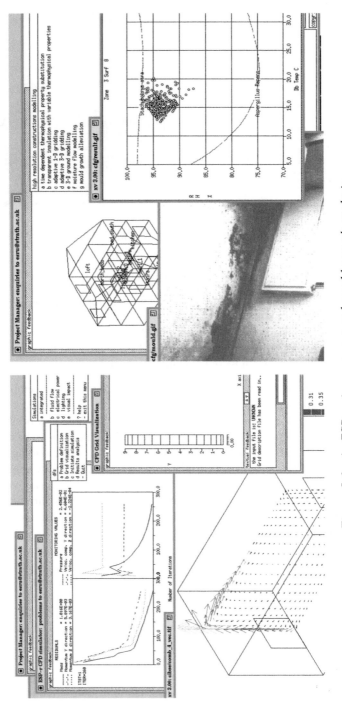

Figure 5.6   An ESP-r air movement and mould growth analysis

Figure 5.7 Deep reveal facade details

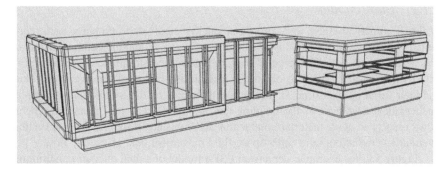

Figure 5.8 Constrained complexity in model facade representations

Design teams can produce facade details that pose modelling challenges. Commercial facades, such as those shown in Figure 5.7 may feature deep framing and offset glazing as well as spandrel panels that act as low-efficiency solar collectors. The exposed area at both the outer and inner faces is significantly greater (∼23%) than an elevation view would suggest. Such framing sections may be replete with thermal bridges. In this case, the framing is sufficiently dense to intercept a considerable portion of the incoming solar radiation. Such details may be replicated over large facade areas and are thus worth investigating.

Enabling realistic heat transfer computation in such a context, whilst constraining the complexity of the simulation input, is demonstrated in the model shown in Figure 5.8.

Two portions of a typical office level are included to capture the horizontal and vertical facade patterns at moderate resolution and are zoned to account for space use and perimeter exposure. The facade below the glazing includes the intermediate ceiling void, structure, and raised floor platform. The vertical framing is literal, but the density of framing in the horizontal regime is abstracted to preserve the orientation and offsets whilst aggregating the framing into fewer surfaces. This approach marginally underestimates the shading of the glass by the adjacent framing.

Another aspect of a high-rise office is the considerable facade area associated with the building structure and ceiling voids. Whether such voids are active air plenums or passive elements, ignoring heat transfer at their perimeter adds unnecessary assessment doubt. To constrain the required simulation resource, this model does not include ducting and piping in the voids.

Spandrel panels can also be a challenge. A dark inner void with high-quality glazing surrounded by framing is essentially a solar collector. Predictions indicated a panel temperature roughly 30 °C greater than ambient, and the recommendation here was to adapt the detail. The approach taken was to model a dozen panels of different orientations in detail to support an investigation of insulation up-rating, alternative glazing, and spandrel surface treatments that would not alter the appearance.

5.1.2   Plant models

Within a numerical BPS+ tool, an input model is subdivided into parts, each represented by conservation equations relating to energy, mass, and momentum balance as and where appropriate. These equations are then solved simultaneously and repeatedly by numerical methods to obtain the spatial distribution of state variable values and their variation over time. In the approach, there is no differentiation between the mathematical treatment of the building and the plant. Each has interacting energy flows associated with material conduction and moisture flow, surface convection, inter-surface radiation, intra-/inter-space fluid movement, and electricity flow.

Figure 5.9, for example, depicts an ESP-r model of a dual duct air conditioning system coupled to a two-storey building as established by Aasem (1993), whilst

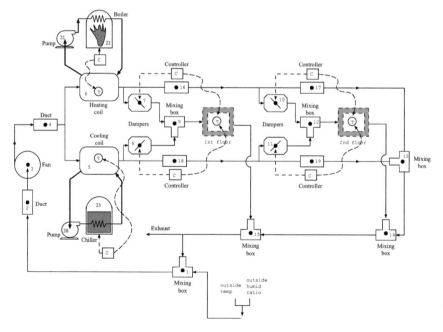

Figure 5.9   ESP-r model of a dual duct air conditioning system (credit: Aasem 1993)

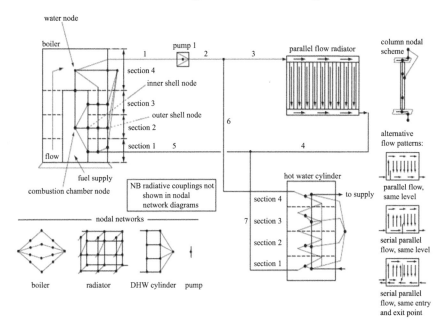

Figure 5.10   Nodal network representation of a wet central heating system

Figure 5.10 depicts a wet central heating system with the discretised nodal network superimposed.

Dynamic plant modelling can thus be carried out by interconnecting components to form a network. The associated pipework and ductwork may be subdivided into zones, each zone treated in the same way as any other building-related zone – a node with properties, and 'walls' associated with duct/pipe surfaces. Of course, the outside surfaces of some components (e.g., a radiator) are located in other thermal zones and thermally interact with them. An account then has to be taken of the transit times of fluids passing through such plant components, and shorter simulation time steps (15 minutes or less) will usually be required. Care needs to be taken if it is likely that flow regimes could switch between laminar and turbulent behaviour, as this will in turn affect flow/pressure relationships. Motorised valves and pumps that modulate flow in response to demand can be included in the network model.

By degrading this high-resolution model (for example, by imposing published boiler efficiency factors or disregarding content capacity), it is possible to emulate the lower-order modelling approaches as embodied in some tools at present and thereby characterise the inherent loss of appraisal capability and realism. Since this loss is demonstrated to be significant (see Section 5.2), the implication is that tool users require better support for plant component representation with equipment efficiency delivered as an output rather than required as an input corresponding to performance data produced under test conditions that will not prevail in practice.

Table 5.3   ESP-r plant components available as multi-node representations

WCH focused:	CHP:
non-condensing boiler	internal combustion engine
condensing boilers	fuel cell for cogeneration
water heater with storage	Stirling engine
variable speed pump	residential scale fuel cell
radiator	PEM fuel cell
flow control valve	electrolyser
insulated piping	hydrogen compressor
calorifier	hydrogen-to-electricity conversion
stratified tank (with or without coils)	metal-hydride storage unit
electrically heated calorifier	adsorption storage unit
gas-water heat exchanger	compressed gas storage unit
thermosyphon heat exchanger	hydrogen stove
oil-filled electric radiator	Heat pump:
thermostatic radiator valve	water loop
HVAC:	air source
mixing box	ground source
humidifier	hydronic heating or cooling
centrifugal fan	in slab
heating and cooling coil	Miscellaneous:
conduit	mechanical thermostat
three-way damper/diverter	fixed temperature source
converging/diverging junction	load generator
heat exchanger	sub-surface water pipe
in-line duct heater	mains water temperature source
cold water storage tank	water draw (profile or stochastic)
cooling tower	flat-plate solar collector
	TRNSYS component coupler

A downside of this dynamic approach to plant system modelling is the low availability of component models – Table 5.3 lists the models available in the 2024 release of ESP-r.

To increase the number of component models that could participate in an integrated simulation of the building and plant, Beausoleil-Morrison *et al.* (2012) developed a co-simulation capability between ESP-r and TRNSYS[1]. Figure 5.11 shows the co-simulator interface with a cooperating simulation in progress.

Chow (1995) tackled the scarcity of plant component models in a different way by developing a set of primitive part (PP) equations that can be combined to form a new plant component at the resolution required. Table 5.4 lists the PPs relating to the components of air conditioning systems.

Consider the formation of a chilled water, cooling coil. The coil model can be constructed in several stages. A straight tube section is established as a combination of PPs 4.3 and 4.4 (assuming here that the thermal resistance of the tube wall is negligible). Then the cooling coil can be visualised as a network of these tube

[1]https://trnsys.com

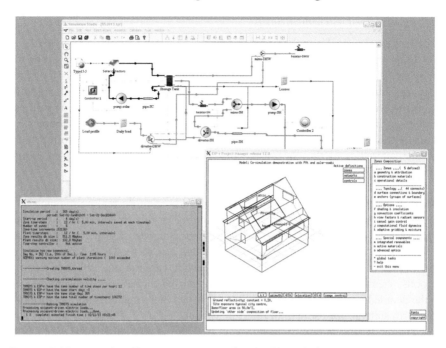

*Figure 5.11   A co-simulation session involving ESP-r and TRNSYS
(credit: Beausoleil-Morrison et al. 2012)*

Table 5.4   PPs for rapid AC component creation

Thermal conduction	Flow divider and inducer
1.1 solid to solid	5.1 flow diverger (for all fluid)
1.2 with ambient solid	5.2 flow multiplier (for all fluid)
Surface convection	5.3 flow inducer (for all fluid)
2.1 with moist air	Flow converger
2.2 with 2-phase fluid	6.1 for moist air
2.3 with 1-phase fluid	6.2 for 2-phase fluid
2.4 with ambient	6.3 for 1-phase fluid
Surface radiation	6.4 for leak-in moist air from outside
3.1 with local surface	Flow upon water spray
3.2 with ambient surface	7.1 for moist air
Flow upon surface	Fluid injection
4.1 for moist air; 3 nodes	8.1 water/steam to moist air
4.2 for 2-phase fluid; 3 nodes	Heat injection
4.3 for 1-phase fluid; 3 nodes	9.1 to solid
4.4 for moist air; 2 nodes	9.2 to vapour-generating fluid
4.5 for 1-phase fluid; 2 nodes	9.3 to moist air

sections connected in the same configuration as the actual coil – single pass, serpentine, or double serpentine. If a counter-flow coil is to be modelled, PP 5.2 is added to adjust the fluid flow rates accordingly. The procedure is explained in detail elsewhere (Chow 1995, Chow and Clarke 1998).

5.1.3   Control system models

Within the numerical simulation approach, a control system is represented as a collection of control loops, each acting to regulate a condition (e.g., zone air temperature, boiler supply temperature) by adjusting a related model parameter (e.g., radiator flow via TRV adjustment, boiler fuel supply rate). A collection of loops then regulates the system overall. As shown in Figure 5.12, a control loop comprises three elements: a sensor to detect one or more model state variables, an actuator to deliver the required control action, and a regulation law to represent the characteristics of the control response. Also shown are some typical target parameters for each element.

Any model parameter may be regulated in this way as a function of intrinsic or extrinsic events and influences, whether it corresponds to a real entity such as a dimmer switch, or an abstract entity such as room heat flux input/extract. The latter possibility is a useful feature in that ideal control regimes can be imposed at an early design stage (e.g., the room temperature will be exactly 21 °C between 9 a.m. and 5 p.m. on weekdays) in support of design exploration. Realistic control response characteristics can then be imposed later when the design has progressed to the detailed stage. MacQueen (1997) established hierarchical control algorithms in a form suitable for use within the ESP system.

The modelling of control at this level of detail is often neglected in practice, with dry bulb temperature being the universal parameter that is passed between model components and output analysis routines. This may be acceptable when the temperatures being recorded are for flowing air in ductwork. However, to model a thermostat in a boiler, the air temperature in a zone, the comfort temperature of an occupant, or the temperature 'seen' by a thermostatic radiator valve, three conditions must be satisfied. First, the spatial location of the sensed condition must be respected (within the boiler jacket or a location in a zone where the radiator is located). Second, several parameters may need to be

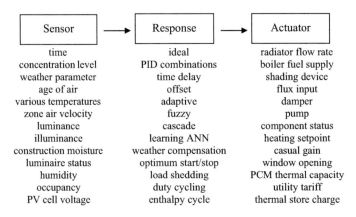

Sensor	Response	Actuator
time	ideal	radiator flow rate
concentration level	PID combinations	boiler fuel supply
weather parameter	time delay	shading device
age of air	offset	flux input
various temperatures	adaptive	damper
zone air velocity	fuzzy	pump
luminance	cascade	component status
illuminance	learning ANN	heating setpoint
construction moisture	weather compensation	casual gain
luminaire status	optimum start/stop	window opening
humidity	load shedding	PCM thermal capacity
occupancy	duty cycling	utility tariff
PV cell voltage	enthalpy cycle	thermal store charge

Figure 5.12   Elements of a control loop and typical parameters

Table 5.5   Examples* of simulation-assisted control

Control focus	Optimised parameter
HVAC operation	Start/stop time
District heating	Match to load
Night cooling	Hours of operation
Underfloor heating	Period of operation
Night set-back	Set-back temperature
Mixed ventilation	Avoid overheating
Boiler sequencing	Heating system efficiency
Ice store charging	Hours of operation
Load shedding	Energy consumption
Ground heat pump	Thermal storage
Combined heat and power	Hours of operation

*Lists not matched.

combined (e.g., local air and radiant temperatures in the case of a TRV). Third, the characteristics of the sensor must be included to capture any intrinsic time delay. One way to do this is to explicitly represent the sensor as a small thermal zone in its own right so that thermal capacity and other effects are included (see Section 5.1.4). Cockroft *et al.* (2009) demonstrated that this could have a significant impact on control performance, with simulation results showing better agreement with test results from environmental test chambers when the sensor model is literal.

Clarke *et al.* (2001) investigated the possibility of placing simulation at the core of the supervisory element of an energy management system in order to compare the efficacy of alternative control actions before enacting the best one (see Section 6.10 for details). Of the many possible control strategies (Martin and Banyard 1998), Table 5.5 lists examples that might be enabled by the simulation-assisted control approach.

5.1.4   Pre-constructed models

In addition to pre-constructing archetypical building models as described in Section 4.10, it is helpful to pre-define models of frequently used internal entities to improve modelling resolution and minimise the data input burden (Hand 2016). Consider the following examples as included in ESP-r's 'predefined entities' database.

5.1.4.1   Furniture and fittings

Lighting and visual assessments, as well as an accurate representation of thermal response, require knowledge of internal objects with non-negligible thermal mass. The approach taken in ESP-r is to define and place such objects in a support database for use at model definition time. Two examples are shown in Figure 5.13.

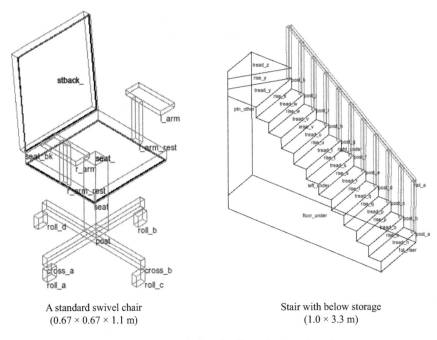

A standard swivel chair
(0.67 × 0.67 × 1.1 m)

Stair with below storage
(1.0 × 3.3 m)

Figure 5.13   A predefined office chair and staircase

Other office furniture (desks, filing cabinets, computer monitors), residential furniture (chairs, tables, beds, wardrobes), facade elements, lighting fixtures, kitchen fittings, and environmental control components are similarly available for embedding in models. Such entities can be scaled and located in a model as required and will participate in subsequent thermal/visual simulations. New entities can be added to the database depending on the needs of a project.

5.1.4.2   Control thermostat

The temperature sensed by a thermostat will depend on its local thermal environment, with a typical response time of the order of minutes or less. Where the mounting location falls within a temporal sun patch, the delivered temperature will be unrepresentative of conditions elsewhere in a controlled zone. Such behaviour can be accommodated by including the materials of the device. In essence, a thermostat is no different from an occupied space in terms of the heat flow paths and thus can be treated in a similar manner where a corresponding input model is available. Figure 5.14 shows a predefined thermostat and its placement on the right wall of a room.

This model comprises simple representations of the entities found in a real thermostat (plastic case, circuit board, sensor in the upper right corner, battery, LCD display) and will participate fully in subsequent thermal/visual simulations. Does such a representation make a difference? Consider Figure 5.15, which shows

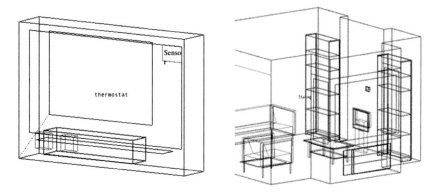

Figure 5.14 A high-resolution thermostat model

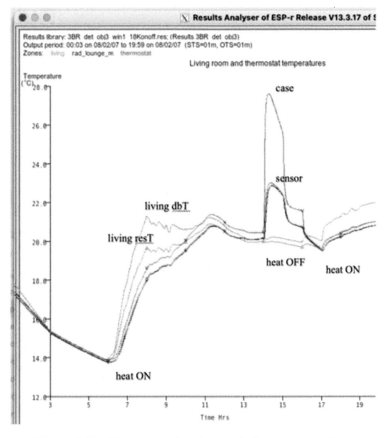

Figure 5.15 Thermostat time lags and placement sensitivity

the conditions prevailing in a room. Heating is enabled just before 7 a.m., and it takes approximately 1 hour for the room's dry bulb and resultant temperatures to settle. As also shown, both the circuit board and sensor lag room conditions. At 2 p.m., the thermostat is within a solar insolation patch and the circuit board and sensor rise by about 3 °C, signalling the heating to stop. The case temperature rises by about 8 °C. The room temperature drops until the case cools, and then heating resumes. Such time lags are endemic to physical devices but are rarely considered in building simulations.

5.1.4.3   Smart storage heaters

These heating devices comprise an internal thermal mass that can be pre-heated during off-peak, low tariff times and then discharged when required. The key issue is discharge control. Figure 5.16 (left) shows a pre-defined component comprising an inner core of thermal mass and an outer casing with venting capability to allow the unit to interact with the building airflow network and be charged under the control of an external agent. A control system injects heat into the core and separately actuates fan-driven heat extraction in the room. The image to the right shows the addition of explicit heater models to individual rooms within a floor in a high-rise apartment block.

Contemporary storage heaters support discrete level charging under the control of agents concerned with extraneous issues such as excess wind power storage and network frequency balancing, e.g., Quantum heaters from Glen Dimplex[2]. Section 9.2 describes such a smart grid and an ESP-r model established to address it.

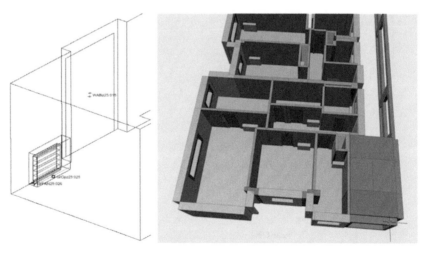

Figure 5.16   Using a predefined, high-resolution storage heater model

[2]https://dimplex.co.uk

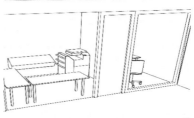

Figure 5.17 A high-resolution photocopier model

5.1.4.4   Office equipment

Equipment such as photocopiers, printers, phone chargers, and fans, generate heat when in use or in standby mode. The infrared image in Figure 5.17 shows the substantially increased surface temperatures of a photocopier. Also shown is a pre-defined photocopier object, the use of which can be scheduled throughout a simulation.

5.1.5   Simulation directives

Once a model has been prepared and calibrated, it can be offered up for simulation. This stage will require additional user input:

- single or multi-year weather collections to define boundary conditions;
- simulation periods corresponding to design, typical, and long-term assessments;
- the spatial and temporal resolution necessary to deliver the required performance information; and
- coordination directives where tool application is automated.

It is usual to include sets of simulation directives as part of the input model so that other users can repeat standard simulations with ease. These sets will typically correspond to periods known to yield results that are useful for plant sizing, overheating assessment, extreme glare cases, and the like. It is also a useful feature where models must be tuned to work well when selecting appropriate computational fluid dynamics (CFD) solver parameters. The directives ensure that the model will run and save other users from having to relearn the parameters.

5.2   Advantages of the high-resolution approach

To demonstrate the benefits of the high-resolution modelling approach, consider the results obtained for four modelling scenarios corresponding to the high-resolution modelling case of Figure 5.1 and three model reductions applied to represent simplifications that are often made in practice.

- S0 – the high-resolution case, giving performance outputs that are absent or required as input in the degraded scenarios;

Table 5.6   Model degradation impact

	Boiler energy consumption (kWh)	Gross boiler efficiency (%)
S0	545	90
S1	487	95
S2	–	–
S3	468	–

- S1 – as S0 but with explicit models of plant replaced by simplified models that rely on empirical data defining plant efficiency;
- S2 – as S0 but with pseudo-steady-state plant behaviour imposed and the removal of internal dwelling features that add significant thermal mass; and
- S3 – as S0 but with the removal of explicit occupant behaviour modelling relating to plant control.

Simulation of the S0 model provides the spatial and temporal distribution of all relevant performance parameters, as depicted in the following snapshots. The other modelling scenarios give insight into the functional loss when a model is degraded.

Table 5.6 lists the boiler energy consumption for each scenario over a 1-week period corresponding to a typically cold condition within the UK heating season and as predicted by ESP-r.

Outputs in each case include boiler energy consumption and, where appropriate, gross efficiency. Note that the intention here is not to pass judgement on the performance, good or bad, of the case studied but to illuminate the functionality loss and impact of imposing modelling assumptions that are not aligned with reality. The message is that only the high-resolution model is able to deliver relevant information on the spatial and temporal performance of the design.

The S0 model is able to deliver information on boiler emissions and operational efficiency since the combustion chamber has an active CFD domain, including heat recovery from the flue gas. As building regulations in the UK mandate the use of condensing boilers, and as low-temperature heat distribution systems (e.g., underfloor) integrated with low-temperature heat sources (e.g., heat pumps) become the norm, the importance of carefully considering the performance of boilers in the condensing range is paramount. In the S0 simulation, the flame is modelled so temperature-dependent chemical reactions that typify various modes of pollutant production (NO_x, CO_2, etc.) can be studied and flue gas recirculation assessed. This is particularly applicable to the study of emissions reduction from gas, oil, and wood burners. In the present case, a methane-fuelled burner is considered with a condensation-enabled heat exchanger.

Modelling of plant components, at these high-resolution levels, depends on the availability of detailed data describing the primary space (e.g., boiler combustion chamber) and the full characterisation of ancillary devices such as valves, fans, and pumps that will affect the primary space. A CFD model enables the calculation of the chemical/combustion processes that occur at high temperatures. Manufacturers typically do not publish these properties, and product development does not routinely deploy simulation tools. The functioning of digital control processes also has to be replicated, as these can have a significant effect on plant performance.

A practical approach is to move to an S1-level model and use data derived from experimental testing. For a boiler, a matrix of the thermal output from the heat exchanger and the associated flue/fabric heat losses over a range of gas, air, and water flow rates enables simulations to be carried out using interpolated values from the performance matrix representing a black box boiler model. Some allowance for thermal mass could be incorporated where there are rapid fluctuations, e.g., if the boiler controls were being separately modelled. The consequence would be some deterioration of model validity, but a relatively simple modelling process and a quasi-dynamic simulation.

Further simplifications are possible, e.g., constant gas/air ratios, combustion efficiencies, etc. Ultimately, a boiler model can be simplified to a single node with an efficiency attached to the output at a given gas-firing rate. This level of abstraction decreases the accuracy of the model but may be appropriate for investigations that are unaffected by detailed plant performance.

Whilst BPS+ tool developers should aspire to the highest level of modelling integrity, it is usual to adopt a lower level of fidelity. Heat pump modelling is often treated this way, with system outputs dependent on outside air and water temperatures, the main efficiency determining factors. Alternatively, the heat pump can be modelled taking the compression process into consideration, and with detailed modelling of the various heat exchangers.

Returning to the S0 case, Figure 5.18 shows the predicted variation of boiler efficiency, combustion chamber temperature, boiler water temperature, and return water temperature over a period of 5 hours corresponding to a typical boiler start-up event, with the boiler water temperature rising from around 20 °C to its set point of

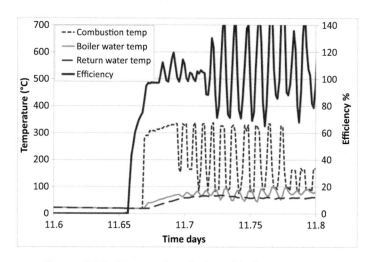

Figure 5.18   Temporal variation of boiler parameters

80 °C, followed by regular on/off cycling of the burner. The average temperature of the flue gases in the combustion chamber fluctuates as the burner operates, reaching a peak of around 300 °C. Note that the boiler efficiency continuously varies depending on the firing rate, water return temperature, and heat exchanger temperature.

Here, boiler efficiency is determined as the instantaneous ratio of heat output to fuel input. Because of the dynamic thermal interactions occurring within the boiler, these time-step level values are actually meaningless. Only the average value over some period (typically days) can convey a true indication of the fuel conversion effectiveness. This 'so-called' efficiency drops at times when the return temperature rises above the dew point temperature of the flue gas; such behaviour often occurs during the morning heat-up period. Occasionally, the instantaneous efficiency rises above 100%, indicative of the situation where, over a short duration, the burner may be off but heat is still being extracted from the waterside of the heat exchanger (again a manifestation of system dynamics).

On other occasions, such as when starting from cold, the burner is firing but little heat is being extracted from the cold water system, so efficiency is low. Such behaviour will typically result in significantly different overall performance compared to that predicted by using fixed values embedded in regulatory codes such as SAP. This is a graphic illustration of the gross simplification assumed in the S1 scenario. The difference in predicted energy consumption here is 11% (Table 5.6) but may be significantly greater with biomass boilers, where there are complex issues to be considered. Further, the results shown here correspond to a typical winter week when average boiler efficiency is closer to published data. Such deviations will increase in the case of annual simulations or if the boiler capacity is incorrectly sized.

Figure 5.19 shows two combustion chamber temperature distribution snapshots corresponding to different levels of stoichiometric excess air.

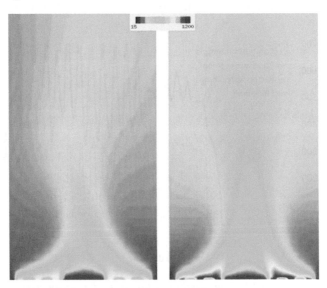

Figure 5.19 Boiler combustion chamber temperatures with close (left) and poor excess air control

These temperature distributions are significantly different. With close control of excess air, a maximum flame temperature of 900 °C is predicted, indicating that with a pre-mix burner, NO_x emissions should be well controlled. In the uncontrolled excess air case, the flame temperature rises to 2000 °C, raising concern about increased rates of NO_x emissions and possibly CO and particulates. These are current issues for biomass boiler design in particular. The high-resolution model supports such emission-related studies, including the production of animations showing the change in emissions over time or under specific operating conditions.

This appraisal functionality is absent in the S1 scenario, where the boiler CFD domain is removed and efficiency is treated as a model input corresponding to a standard test condition, a typical modelling assumption. Many questions arise from this simplification. Who ensures that the prescribed efficiencies are an adequate representation of what will happen in practice? Moreover, how can such a degraded model support user understanding of how the system is likely to operate in practice as a prerequisite for effective design decision-making? The comparative results for the S0 and S1 entries in Table 5.6 highlight the dilemma in quantitative terms: the results for the S1 model correspond to optimistic performance expectations that may not be realised in practice.

A CFD domain has also been imposed on the lounge zone, and Figure 5.20 shows the predicted mean age of air, which further highlights the enhanced appraisal functionality of the S0 model. This, along with other environmental

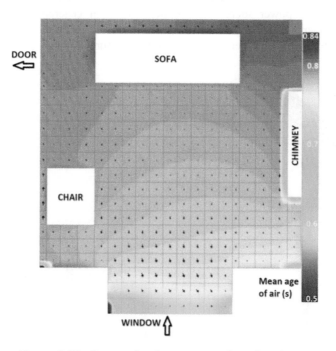

Figure 5.20   Lounge local mean age of air distribution

parameters such as air and radiant temperature, local airspeed, relative humidity and CO_2 concentration, supports an assessment of thermal comfort and air quality. Here, it is evident that poor air quality is not an issue since the space is effectively vented (i.e., it has a low mean age of air).

The next issue addressed is a local one: even in new-build situations, construction details at junctions often create thermal bridges that degrade performance. As indicated in Figure 5.1, the high-resolution model includes enhanced spatial resolution around a constructional element to support the elimination of any thermal bridging that might cause surface condensation and mould. Analyses outcomes such as the one depicted in Figure 5.21, along with corresponding near-surface temperature and relative humidity outputs from an adjacent space CFD domain model, support the mitigation of fabric deterioration and poor air quality alongside consideration of occupant thermal/visual comfort, energy use, emissions, and approaches to control. This ability to switch the appraisal focus between issues at different scales to establish acceptable performance trade-offs is a unique feature of holistic, high-resolution BPS+ applications.

The S0 model also incorporates a PV array and a representation of the connection to the low-voltage (LV) electricity network to enable consideration of building-integrated micro-generation. In the UK, over 2.5 GW of PV rooftop panels had been installed in the 5 years following the introduction of the Feed-in Tariff in 2010 (FIT 2015). Where electrical demand is supplied by local PV generation, the load visible to the network diminishes significantly during periods of high solar insolation. Where sufficient renewable capacity exists, a building or community may become a power source for the network when generation exceeds local demand. Given the dynamic, fluctuating nature of solar radiation, a building may then switch rapidly from being a load to a power source.

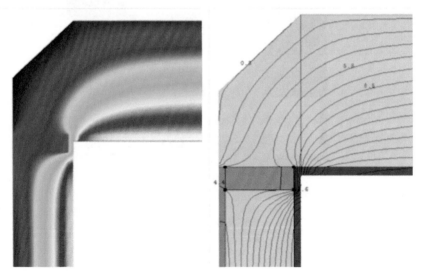

Figure 5.21   Thermal bridge assessment

The expansion of renewables at small and large scales and the changing nature of building energy demand pose challenges for utilities as they attempt to maintain a reliable and compliant power supply to consumers. Potential issues include voltage rises due to excessive power export, increased losses and thermal failures in the LV network due to high power flows, and less predictable domestic and community demand profiles leading to problems in power system stabilisation.

Utilities are deploying demand management as a means to maximise the utilisation of local renewable generation[3], help stabilise local network operations, and improve power quality. Demand management mechanisms include:

- direct intervention (Strbac 2008) in which the utility takes control of flexible loads such as electrical space and water heating;
- demand response (Palensky and Dietrich 2011) in which consumers are expected to adapt their behaviour in response to price signals; and
- domestic energy storage, in which capacity-enhanced space and water heating appliances are charged in a manner that absorbs excess renewable energy generation (Clarke *et al.* 2014, Allison *et al.* 2018).

The problem is intrinsically dynamic, with causal relationships existing at a range of scales from the small (power quality regulation) to the large (thermal storage fluctuation). The inclusion of an explicit electrical network model allows consideration of such issues alongside the other factors that will affect the overall acceptability of a proposed building design, most notably occupant satisfaction.

The S0 model incorporates a 4 kW PV array, with the generated power supplied to an electrical network model that is also connected to the dwelling's electrical loads. This is the average size of domestic PV installations under the Feed-in Tariff. In ESP-r, PV panels are modelled explicitly, with the physical construction of the panel treated in the same manner as constructional elements within the building model. The PV model uses the calculated solar insolation for the surface, the associated construction node temperature, and an electrical model of the solar cells to calculate the time-varying voltage and current.

The S0 model was simulated over a week in summer at a 1-minute time resolution, with the irradiance data being modified using the approach outlined by Hand *et al.* (2014) to reflect the variability in solar radiation over short time scales. Figure 5.22 shows the predicted power flows, highlighting the fluctuation of power between the consumer and LV network and significant power export (shown as negative power); such a result indicates the need for load control, and various approaches could then be studied using the S0 model.

Figure 5.23 illustrates the effect of electricity import/export on the supply voltage to the dwelling; here the voltage drops as the load increases and rises with the export of power to the LV network.

The electrical model can be applied at the individual building or community scale. To illustrate the latter, similar PV arrays on 200 houses were simulated (i.e., a portion of the S0 model is scaled) and Figure 5.24 shows the simulated voltage

[3]The raison d'être of smart meters perhaps.

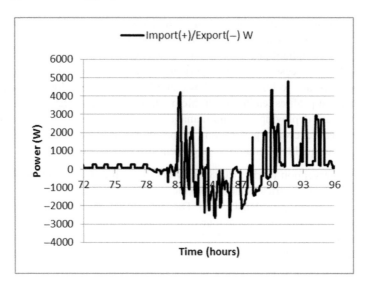

*Figure 5.22   Electricity import/export on a summer day for a dwelling equipped
with PV*

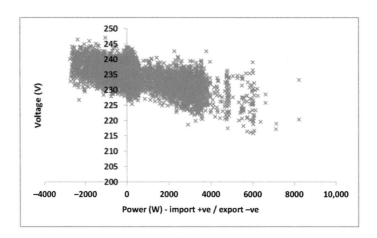

Figure 5.23   Voltage excursions with power import/export

levels with and without PV generation. Such an output highlights a potential problem: the tendency for general voltage levels to rise in PV-rich sections of the
network. PV in a small geographical area will tend to be synchronised with solar
radiation, whilst the demand will be subject to load diversity. Consequently, significant surplus power can be available at times of relatively low demand, leading
to voltage rises in the LV network, with an increased risk of voltage levels
exceeding tolerances (230 V +10%/−6% in the UK).

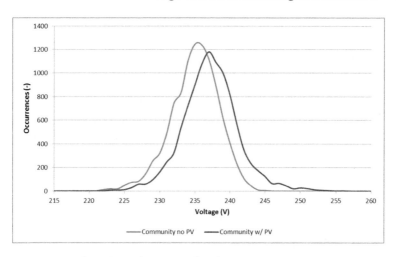

Figure 5.24   Supply voltage for a 200-dwelling community with and without PV arrays

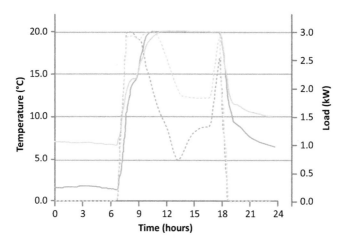

Figure 5.25   Impact on heating load of internal thermal mass

The next implication considered arises from the S2 results. In this case, the additional thermal mass associated with dwelling contents and plant components is absent resulting in significantly different heating loads and response times as shown in Figure 5.25.

Here the different room responses corresponding to the S0 and S2 simulations result in markedly different heating requirements (dashed lines). This situation typifies the usual application context for BPS+, with users offered no convenient way to populate models with content objects and forced to employ component models that apply the logic of the plant system but do not represent their composition as is done on the building side. Such a model is destined to deliver incorrect response times and temperature excursions that might lead to the inappropriate introduction of mechanical plant or a false impression of control robustness leading to post-occupancy disappointment.

Occupant behaviour also has a major impact on the energy performance of buildings as indicated by the S3 results. Individuals behave in ways that reflect preferences that depend on circumstances. To capture this important aspect of reality, it is necessary to include an explicit model of occupant behaviour: such models are available in some BPS+ tools and enhanced models have been compiled for use with BPS+ generally (IEA66 2015). Figure 5.26 shows a possible impact on the room temperature of occupants turning off radiators in response to an initial temperature rise caused by extraneous factors (the more variable curve). The difference between the simulation results is marked, corresponding to a 15% increase in energy consumption (see Table 5.6). This underlines the importance of respecting the stochastic aspect of occupant behaviour.

Finally, as depicted in Figure 5.27, all models except S2 (no content representation) are able to provide photorealistic images, including indoor illumination and glare.

Figure 5.28 shows some possible outputs for the case of a commercial office when addressing indoor environmental quality.

Clarke *et al.* (2012) have reported the data requirements of the performance domains incorporated in a high-resolution model but which are missing at present from BIM schemas. Rectifying this situation is an essential prerequisite of a computational approach to design in which simulation provides the means to facilitate the understanding of thermodynamically complex buildings.

The enhanced fidelity of a high-resolution model will have insignificant resource implications for model creation where future tools have access to

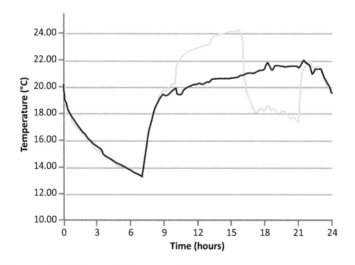

Figure 5.26   Impact of occupant behaviour on room temperature

Figure 5.27   Daylight and glare

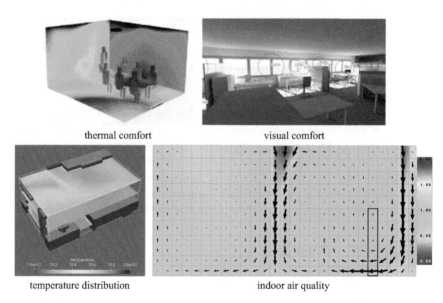

Figure 5.28   Spatial/temporal wellness data from an office simulation

pre-constructed, parameterised models relating to the additional physical and mathematical aspects needed by BPS+. Some examples include:

- detailed space content and plant component models to further attribute the building model defined via a CAD interface;

- network flow models corresponding to typical cases (e.g., single-sided natural ventilation, rooms surrounding an atrium, displacement ventilation, hydronic heating, and low voltage electrical networks) to define the intra- and inter-connections between building and plant parts;
- CFD models for association with problem parts that require high-resolution representation (e.g., boiler combustion chamber, thermal storage tanks, building zones, and urban canyons); and
- control networks connecting sensor and actuator points throughout the unified model via linkages representing the required control actions.

In the absence of such parameterised models, it is unlikely that building performance simulation will ever be routinely used by practitioners to pursue robust design solutions in the face of thermodynamic complexity and operational uncertainty. In this case, performance assessment will become the domain of specialist groups.

The dynamically evolving performance data of Figures 5.18–5.28 and the profiles underlying the integrations of Table 5.6 can be repeatedly probed (spatially and temporally) and the outcomes compared with performance benchmarks to confirm overall compliance using agreed criteria. As required, 'gremlins' can be imposed randomly on a simulation to enact typical contingencies – extreme weather events, equipment failure, tenant change, control system malfunction, etc. – as a way to ensure resilience in practice. This issue is discussed further in Section 11.3.

Whilst high-resolution modelling increases the computational burden, the real question is whether this will be compatible with the time-constrained, resource-limited design process. There are at least two positive responses to this question. First, the fact that quantum computers are gaining computational power at a doubly exponential rate (Hartnett 2019) will accommodate the increased CPU requirement. Second, it is possible in the integrated simulation approach to ensure that CPU-intensive domain models are only invoked when required. For example, it may not be necessary to process the CFD domain model associated with building zones at the computational time step required by other model parts. Instead, a CFD domain solution might be invoked only when specific events occur, such as heating system start-up or when room environmental or occupancy conditions change significantly.

5.3   Chapter summary

BPS+ tools can be applied in a low-cost, focused manner, where this can be justified in relation to the aspects then excluded. Alternatively, the technology offers a more exciting prospect: a computational approach to design based on high-resolution models that enable holistic, multi-parameter appraisals across the building life cycle. Realising this potential will require adherence to industry standard performance assessments and a change in business work practices. The model definition process can be greatly enhanced by the existence of pre-defined objects for incorporation as required and archetypal building designs that can be tailored to specific cases.

References and further reading

Aasem E (1993) Practical simulation of buildings and air-conditioning systems in the transient domain, *PhD Thesis*, ESRU, University of Strathclyde.

Allison J, Bell K, Clarke J A, *et al.* (2018) 'Assessing domestic heat storage requirements for energy flexibility over varying timescales', *Applied Thermal Engineering*, 136, pp. 602–616.

AnTherm (2023) 'A program for the analysis of thermal behaviour of building constructions with heat bridges', https://antherm.at/antherm/EN/.

Beausoleil-Morrison I, Kummert M, Macdonald F, Jost R, McDowell T and Ferguson A (2012) 'Demonstration of the new ESP-r co-simulator for modelling solar buildings', *Energy Procedia*, 20, pp. 505–514.

Bordass B, Cohen R, Standeven M and Leaman A (2001) 'Assessing building performance in use 2: technical performance of the Probe buildings', *Building Research & Information*, *29*, pp. 103–113.

Bordass B and Leaman A (2005) 'Making feedback and post-occupancy evaluation routine 1: a portfolio of feedback techniques', *Building Research and Information*, 33(4), pp. 347–352.

BS (2017) 'BS EN ISO 14683:2017 Thermal bridges in building construction – linear thermal transmittance, simplified methods and default values', https://thenbs.com/PublicationIndex/documents/details?Pub=BSI&DocID=319619/.

Cere G, Rezgui Y and Zhao W (2019) 'Urban-scale framework for assessing the resilience of buildings informed by a delphi expert consultation', *Disaster Risk Reduction*, 36.

Chow T T (1995) 'Atomic modelling in air-conditioning simulation', *PhD Thesis*, University of Strathclyde, Glasgow.

Chow T T and Clarke J A (1998) 'Theoretical basis of primitive part modelling', *ASHRAE Transactions*, 104, pp. 299–312.

Clarke J A and Hensen J L M (2015) 'Integrated building performance simulation: progress, prospects and requirements', *Building and Environment*, 91, pp. 294–306.

Clarke J A, Cockroft J, Conner S, *et al.* (2001) 'Control in building energy management systems: The role of simulation', *Proc. Building Simulation '01*, Rio de Janeiro.

Clarke J A, Ghauri S, Johnstone C, Kim J and Tuohy P G (2008) 'The EDEM methodology for housing upgrade analysis', *Proc. eSim 2008*, Quebec, Canada.

Clarke J A, Hand J W, Kelly N, *et al.* (2012) 'A data model for integrated building performance simulation', *Proc. Building Simulation and Optimisation*, University of Loughborough.

Clarke J A, Hand J, Kim J, Samuel A and Svehla K (2014) 'Performance of actively controlled domestic heat storage devices in a smart grid', *Power and Energy*, 229(1), pp. 99–110.

Cockroft J, Kennedy D, O'Hara M, Samuel A, Strachan P and Tuohy P (2009), 'Development and validation of detailed building, plant and controller modelling to demonstrate interactive behaviour of system components', *Proc. Building Simulation'09*, July, Glasgow.

Doan D, Ghaffarianhoseini A, Naismith N and Zhang T (2017) 'A critical comparison of green building rating systems', *Building and Environment*, 123(1), DOI:10.1016/j.buildenv.2017.07.007.

FIT (2015) https://www.ofgem.gov.uk/environmental-programmes/feed-tariff-fit-scheme/feed-tariff-reports/feed-tariff-quarterly-report.

Hand J (2016) 'Performance implications of fully participating furniture and fittings in simulation models', *Proc. Building Simulation & Optimisation*, Newcastle, IBPSA England.

Hand J W, Kelly N and Samuel A (2014) 'High resolution modelling for performance assessment of future dwellings', *Proc. Building Simulation and Optimisation*, University of Loughborough.

Hartnett K (2019) 'A new law to describe quantum computing's rise?', https://quantamagazine.org/does-nevens-law-describe-quantum-computings-rise-20190618/.

Hong T, Chou S K and Bong T Y (2000) 'Building simulation: an overview of developments and information sources', *Building and Environment*, 35(1), pp. 347–361.

IEA66 (2015) 'IEA-EBC Annex 66: definition and simulation of occupant behaviour in buildings', https://annex66.org/.

MacQueen J (1997) 'The modelling and simulation of energy management control systems', *PhD Thesis*, University of Strathclyde.

Martin A J and Banyard C P (1998) *Library of System Control Strategies*, Building Services Research & Information Association, ISBN 9780860224976.

Nakhi A E (1995) 'Adaptive construction modelling within whole building dynamic simulation', *PhD Thesis*, ESRU, University of Strathclyde.

Nakhi A E and Aasem E O (1999) 'Conflation of thermal bridging assessment and building thermal simulation', *Proc. Building Simulation '99*, Kyoto.

Palensky P and Dietrich D (2011) 'Demand side management: demand response, intelligent energy systems, and smart loads', *IEEE Transactions*, 7(3), pp. 231–388.

Strbac G (2008) 'Demand side management: benefits and challenges', *Energy Policy*, 36(12), pp. 4419–4426.

Tang D and Saluja G S (1998) 'Analytic analysis of heat loss from corners of buildings', *International Journal of Heat and Mass Transfer*, 41(4–5), pp. 681–689.

Chapter 6

Example applications

High-resolution modelling provides a way to consider simultaneously the many factors that influence overall performance by enabling holistic simulation (whether delivered by a single program or several cooperating programs). That said not all performance assessments require a fully developed high-resolution model. Where a problem boundary, weather or otherwise, can be reasonably prescribed, it is sensible to restrict the simulation focus and thereby reduce the model definition and computational burden. For example, where different glazing systems are being compared, a model may be restricted to a simple multi-zone representation with detail concentrated only at the facades. Alternative glazing solutions can then be applied in each zone and compared in terms of their energy use and thermal/visual impact based on a single simulation. It is important to record simplifying assumptions alongside a model to avoid the use of results outside their range of applicability. This is rarely, if ever, done at present and should be mandated in future.

The following 'quick-fire' descriptions of BPS+ applications are intended to demonstrate the flexibility and relevance of the modelling and simulation approach when applied to the myriad issues encountered in practice. Many of the examples relate to housing because these are easier to convey. In most cases, the procedure highlighted is equally applicable to commercial buildings. The examples are drawn from research and consultancy projects undertaken by staff of the Energy Systems Research Unit (ESRU) over a 45-year period. These projects were variously resourced and mostly had aims defined by a client. This means that the input models were judiciously simplified to exclude those performance aspects that were not of immediate interest. The underlying message is that modelling and simulation are part science and part art and that skills are best developed through practice.

Although no specific details are given on the input models because of space constraints, it may be assumed that these are generally aligned with the principles of Chapter 5 but restricted in scale or scope as stated or implied[1]. In many cases, corresponding models are available as part of an ESP-r distribution so that they may be explored, edited, and simulated as required. It is stressed that the presented outputs are intended to illustrate the useful information available from simulation

---

[1]The importance of documenting and justifying assumptions alongside a model cannot be over-emphasised. How else can simulation outcomes be properly understood and applied.

rather than be an attempt to convey recommendations for the design types studied: such outputs are always a function of the context and do not translate readily to design paradigms. The aim is to enthuse users new to the field, challenge practitioners to deepen their offerings, and encourage vendors to extend the application scope of their software tools.

In all cases, the software tools used are referenced, and where these relate to applications developed by ESRU, the source code is available for download (even where developments have ceased, in the hope that the reader might be motivated to build on what has gone before).

A cautionary point: the projects that follow were serviced by an experienced team of modellers. Whilst it is possible that they could be replicated by an individual with aptitude, they should be regarded as an inducement to form or become part of a dedicated team. Do not expect to download a BPS+ tool and reproduce the featured results shortly thereafter. As stated or implied throughout the book, the requirement is for well-resourced groups that possess the complementary skills necessary to service the many aspects of performance simulation – not least, in the case of ESP-r, acquiring the source code[2], installing the system under Linux[3], and then learning to drive.

6.1   ESP-r application features

Because most of the example applications that follow involve the open-source ESP-r program, this section summarises this system's application features, whilst Appendix A presents complementary information on ESP-r's theoretical basis – knowledge that helps in planning a model's scope and depth. The ability to replicate or extend the example applications using other programs will depend on the capabilities of those programs as reported by their vendors and users. It should be noted that the aim of the ESP-r development project has always been to explore the boundaries of design-oriented simulation in terms of creating high-fidelity models of real systems and simulating these in support of design decision-making. The system is therefore not commercial although it has underpinned other systems that are. It is emphasised that the intention is not to predict a future state but to emulate reality in a manner that subjects a model to realistic stressors to ensure resilience.

ESP-r application is essentially a three-stage process as follows.

- A design hypothesis is specified by offering information on building geometry, construction, plant, and control. This is supplemented by patterns of building use as well as weather boundary conditions and directives for simulation. The model is calibrated.
- An initial simulation is now performed and this can take from a few minutes to many hours depending on the complexity of the problem and the period

[2]https://www.esru.strath.ac.uk/applications
[3]https://www.esru.strath.ac.uk/Documents/ESP-r/Configuring_Ubuntu%2020.04_for_ESP-r.pdf

covered. This produces time-series evolutions of the variables of state throughout the building.

- System performance is assessed by interrogating simulation outputs to obtain an insight into the underlying causal relationships and so identify cost-effective changes to the design hypothesis. This process may take from a few minutes to many hours depending on the skills and experience of the analyst.
- The initial design hypothesis can now be modified in light of insights gained and further simulations commissioned. The analysis, synthesis, and appraisal process then iterates.

As summarised in the collage of Figure 6.1, ESP-r's native mode of operation[4] presents a 'project manager' (PM) interface that provides user control via a hierarchy of menus, with graphical views of the model accompanied by performance simulation outputs in graphical and textual format.

The PM can send models, in whole or part, to ESP-r support modules or third-party programs as required. In the former case, this includes an integrated energy simulator, partial simulators (e.g., for network fluid flow, air movement, acoustics, external shading, and embodied energy), results analysis tools, and report generators. ESP-r also makes use of third-party productivity aids such as Gtool (CAD for geometry definition), Radiance (visualisation), Blender (model manipulation), and ParaView (multi-platform data analysis and visualisation).

In use, BPS+ programs like ESP-r offer a means to answer design questions such as the following.

- What and when are the peak building/plant loads and what are the rank-ordered causal influences?
- What is the effect of design changes such as increasing wall insulation, changing glazing type/distribution, re-zoning the building, re-configuring the plant, or changing the control regime?
- What is the optimum heating system start time or the most effective algorithm for weather anticipation to activate storage system charging?
- How will comfort levels vary throughout the building in response to sustained warm weather conditions?
- How will temperature distribution, in terms of zone sensor and terminal unit location, affect energy consumption, and comfort control?
- What contribution does building infiltration and zone-coupled airflow make to the total boiler or chiller load, and how can this be minimised?
- What is the contribution to energy saving of a range of passive solar features?
- What is the optimum arrangement of constructional elements to encourage good load levelling and hence efficient plant operation?

---

[4]ESP-r has evolved on a Unix/Linux platform and makes use of features such as environment variables, input/output redirection, shell scripting, and pre-configured software tools. The system can also operate under Windows either directly or using Cygwin. The source code is provided to enable multiple platform installation.

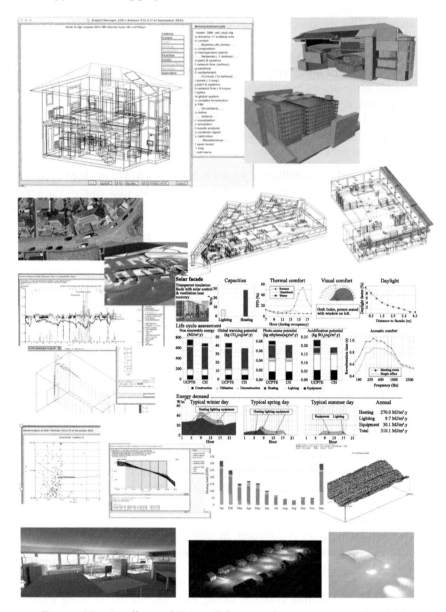

Figure 6.1   A collage of PM model manipulation aspects within ESP-r

- What are the energy consequences of non-compliance with prescriptive energy regulations or, conversely, how should a design be modified to converge on a deemed-to-satisfy performance target?
- Which heat recovery system performs best under a range of typical operating conditions?

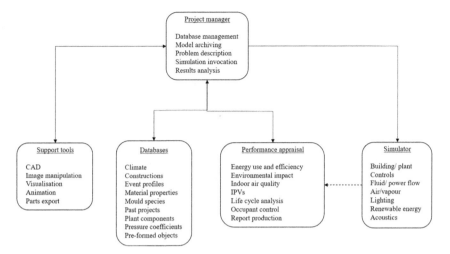

Figure 6.2 Modules of the ESP-r system

- How can the output from building-integrated photovoltaics (PV) be matched to a building's electrical load?

ESP-r is able to process plant networks by which the heat transfer processes within individual components are modelled explicitly and coupled to the building and control systems[5]. It is then possible to examine the relationship between component design parameters (e.g. boiler firing rate, fan capacity) and overall system performance, and determine component sizes under realistic operating conditions. Alternatively, it is possible to model plant in a conceptual manner by which characteristics (e.g. the time availability of cooling/heating flux and application points) are imposed on the building model so that the processing of actual plant components is obviated.

As shown in Figure 6.2, the system comprises a central project manager (PM; Hand 1998, 2020) around which is arranged support databases, a simulator, performance assessment tools, and a variety of third-party applications for geometry definition, model visualisation, and report generation. The PM's function is to enable incremental problem definition and give/receive the data model to/from the support applications.

This mode of operation has proven to be powerful in the hands of an experienced operator but challenging for the infrequent user as often found in design practice. An alternative application approach is to automate design appraisals using computerised assessment agents in place of user mouse clicks and keystrokes. These agents take the form of shell scripts[6] encapsulating rules corresponding to

---

[5]A drawback of this approach is the necessity to formulate detailed models of plant components in advance and acquire detailed component descriptions.
[6]https://www.shellscript.sh

particular assessment tasks. These rules can be inspected and if required adapted by the user before invocation of the script. The computational path to be followed depends on the requirement of the assessment goal as well as the performance data emerging at run time. Such scripts can be viewed as a design assistant: the performance assessment and program operation knowledge are known to the assistant; the user is free to focus on design decision-making. Section 4.5 in conjunction with Appendix C describes a script for building overheating assessment.

The typical starting point for a project is to make ready the support databases in relation to the needs of the project, perhaps adding hypothetical materials, new plant components, or unusual weather patterns to explore innovative design features and lifetime resilience. Database management facilities are provided. In undertaking this task, it may be necessary to source new information and establish the provenance of data sources. Regional variations in products may come into play and it may be necessary to consider some parameters as uncertain and assign likely ranges. Other questions might need to be answered, such as is it sufficient to impose standard occupancy patterns or must a stochastic occupant behaviour model be used?

The project site will require weather collections that support tests under different levels of severity. It may be necessary to source local weather data if microclimate effects are important. If natural ventilation is to be modelled then a review of pressure coefficient sets and available flow components will be necessary, perhaps requiring wind tunnel tests in extreme cases or the use of a model such as CpCalc[7] to generate surface wind pressure data including the impact of surrounding structures (Chiesa and Grosso 2019).

Planning a simulation project requires consideration of how the building will be used. It is usually appropriate to delay investing effort in defining environmental control systems until indoor conditions and energy demand patterns have been studied. Projects will often consider model variants to future-proof a design – e.g., different constructions and control regimes, or system failures and site development affecting shading patterns and sky access.

Although the procedure for creating and evolving a model is a matter of personal preference, it is usual to commence the process with the specification of building geometry using a CAD tool[8]. ESP-r has an intrinsic geometry manipulation capability and can import models from third-party tools supporting gbXML file exchange[9]. Figure 6.3 depicts an ESP-r model for a terrace of houses following import from an external CAD program.

Typically, there will be minor glitches and omissions requiring attention. It will be necessary to use the CAD tool in a compatible manner. For example, ESP-r requires a list of zones, each comprising bounding surfaces. There is no point in providing a list of disassociated lines that display well but do not convey the

---

[7]https://www.iris.polito.it/handle/11583/2579969

[8]Alternatively, the process can skip to technical systems with the building represented by load profiles generated elsewhere or guesstimated.

[9]https://www.gbxml.org/Schema_Current_GreenBuildingXML_gbXML/

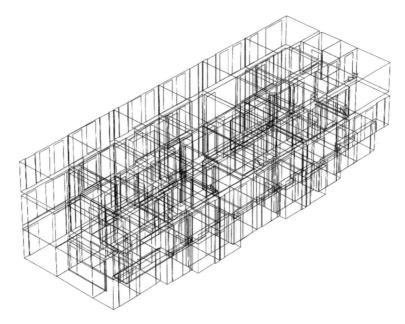

Figure 6.3   ESP-r Geometry imported from a CAD program

required semantic. It is important to constrain geometric complexity to minimise the subsequent effort required to attribute the model and the time to simulate it.

Geometry attribution is achieved by selecting building entities (e.g., wall constructions, usage profiles) and plant components (e.g., heat pump, solar panel) from the support databases and associating these with the surfaces and spaces comprising the problem. It is at this stage that the simulation novice will appreciate the importance of a well-conceived problem abstraction, which achieves adequate thermodynamic resolution whilst minimising the number of entities requiring attribution.

As ESP-r holds a superset data model, the PM is able to export project information to third-party tools such as Radiance[10] (Ward 1993) or Blender[11] to support visual comfort assessments, model visualisation/animation, part extraction, and file format mapping.

Where required, network models are defined to represent HVAC systems (Aasem 1993), building airflow paths (Clarke and Hensen 1991), plant working fluids (Chow 1995), and electrical power circuits (Kelly 1998). These networks are then associated with the building/plant model so that essential dynamic interactions are preserved. Figure 6.4 shows the conflation of a building and airflow network in a simple model.

[10]https://www.radsite.lbl.gov/radiance/download.html
[11]https://blender.org

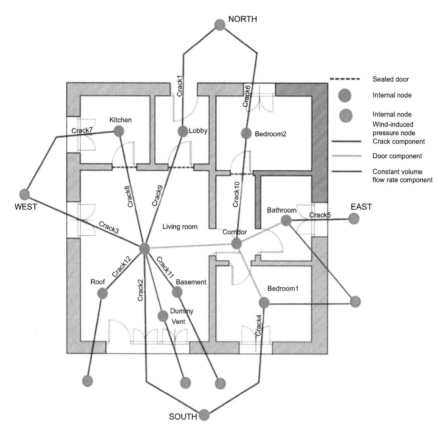

*Figure 6.4   Conflation of building and airflow models (credit: Monari and
Strachan 2016)*

Control system definitions can now proceed depending on the appraisal
objectives. This involves the establishment of multiple closed or open loops, each
one comprising a sensor (to measure one or more simulation parameters at each
time step), an actuator (to deliver the control signal), and a regulation law (to relate
the sensed condition to the actuated state). These loops are used to regulate plant
components, associate components with building zones, manage building-side
entities such as luminaires and blinds, and coordinate flow components (window
opening, fan/pump operation, etc.) in response to changing environmental conditions. Control loops can also be used to change portions of a problem over time
(e.g., to substitute alternative constructions to emulate a movable wall) or impose
replacement parameters (e.g., the thermal properties of a material undergoing a
change in phase or moisture content).

For specialist applications, the resolution of model parts can be selectively
increased, for example:

- the default one-dimensional construction gridding scheme applied to represent transient conduction can be selectively enhanced to a two- or three-dimensional scheme to represent a complex geometrical feature or thermal bridge (Nakhi 1995);
- a three-dimensional grid can be imposed on selected zones to enable a thermally-coupled computational fluid dynamics simulation (Negrao 1995, Clarke *et al.* 1995, Beausoleil-Morrison 2000);
- special behaviour can be associated with a material, e.g., electrical power production in the case of photovoltaic cells (Clarke *et al.* 1996); and
- models can be associated with material hygro-thermal properties to define their moisture and/or temperature dependence in support of moisture flow simulation and mould growth studies (Anderson *et al.* 1996).

A model also includes text and images to document the project's intent and any special features, and it is possible to embed directives for the performance metrics to be recorded during a simulation – e.g., the entities of an 'integrated performance view' (IPV) as shown in the example of Figure 6.5, which summarises energy demand, environmental emissions, and thermal/visual comfort.

Table 6.1 shows a portion of a model file defining the content of an IPV. This drives both the commissioning of simulations and the subsequent data mining.

In this example, a typical week in each season is being assessed and scaled for inclusion in IPV reporting. The requested metric is resultant temperature as a proxy for thermal comfort with two groups of thermal zones identified for reporting; the first group corresponds to an open plan office, the other a conference room. The seasonal assessment periods and scaling factors for heating and cooling are determined from a scan of the weather file, with the latter set as the ratio of seasonal to assessment period heating and cooling degree-days. This enables short-period simulations to be translated into seasonal estimates. The factors for fans, small power, etc. are scaled by time. Of course, the assessment period can be set to encompass the whole season in which case the scaling factors are unity. That said it has been observed that the scaling approach is often acceptably close to the result for a whole season simulation. Whereas simulating all days at short time steps might take 11 minutes, generate 13 GB of data, and require considerable effort for data mining, short period assessments will typically be an order of magnitude less demanding. Table 6.2 shows a resulting report for post-simulation processing.

The report includes tokens with titles, subtitles, and values from which a third-party tool can produce the requisite entities of an IPV montage.

The problem – from a single space with simple control and prescribed ventilation to a building with systems, distributed control, and enhanced resolutions – is passed to the simulation engine where, in discretised form, the underlying conservation equations are numerically integrated at successive time intervals over a period of interest (day, week, month, year, lifecycle, etc.).

For problems involving daylight utilisation, ESP-r 'drives' Radiance (Figure 6.6) and probes the final resolved image to quantify the internal

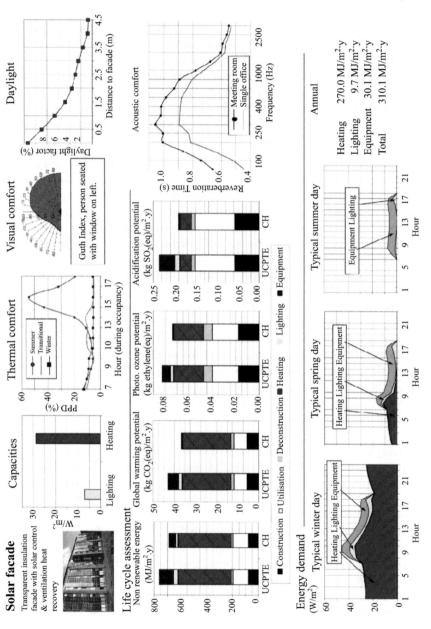

Figure 6.5   An ESP-r integrated performance view

Table 6.1   IPV directives

Title, office upgrade						
images,3						
plan.jpg, section.jpg,system_sketch.jpg						
# assessments to run						
simulations,3						
# tag	start	finish	focus	description		
*period	36	42	37	winter run		
*period	107	113	108	spring run		
*period	184	190	185	summer run		
# season scaling factors						
# name	heating	cooling	lights	fans	power	DWH
*win	9.7	1.	11.	11.	11.	11.
*spr	13.4	1.	9.	9.	9.	9.
*sum	15.	1.	14.5	16.	16.	16.
# metric			floor	scaling		Zone
# name			area	factor	Topic	list
*comfort_office			43.7	1.0	Res.T	1,2,3,4
*comfort_conference			41.9	1.0	Res.T	6,7,8,9
*end						

Table 6.2   IPV results for post-processing

Heating/cooling capacity (kW)
All,13.4,0.0
Lighting load (W)
All,890.0
Fan and pump load (W)
All,350.0,0.0
Small power load (W)
All,760.0
Domestic hot water load (kW)
All,4.4
Heating/cooling energy (kWh/a)
8223.6,0.0
Lighting energy (kWh/a)
1978.2
Fan and pump energy (kWh/a)
338.5
Small power energy (kWhr/a)
2348.9
Domestic hot water energy (kWhr/a)
6451.6
Resultant temperature ($°C$)
Zone, Maximum, Minimum, Mean, Std-dev
All,30.9,17.6,22.1,2.6

(Continues)

Table 6.2   (Continued)

Bin	Range	hours (per week)		
		Winter	Spring	Summer
1	0.00–16.00	22	14	0
2	16.00–18.00	55	61	3
3	18.00–20.00	76	56	12
4	20.00–22.00	14	21	53
5	22.00–24.00	0	11	62
6	24.00–26.00	0	3	32
7	26.00–28.00	0	0	6

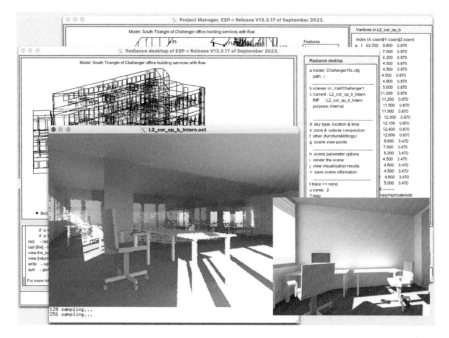

Figure 6.6   Obtaining positional illuminance from Radiance

illuminance distribution for input via a control sensor (photocell) to an artificial lighting control loop

Simulations result in a time series of 'state information' (temperature, pressure, voltage, etc.) for interrogation via the results analysis module: changes to the model parameters then follow depending on appraisal outcomes. Whilst the range of analyses is essentially unlimited, interrelating the different performance indicators and translating these indicators to design changes is problematic. How is a table of numbers corresponding to conduction heat loss, solar/casual gains, and infiltration load to be interpreted when these heat flows interact with different parts of a zone and have different time-shifts? Fortunately, the time-series simulation

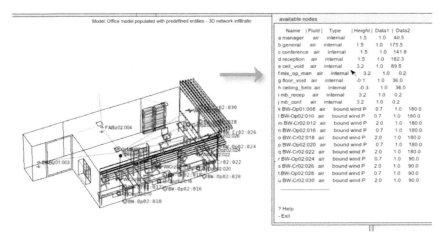

Figure 6.7   Defining an airflow network

results support the 'causal chaining' of energy balances whereby the underlying cause of poor performance can be traced (Section 4.6 describes the approach).

The PM also offers model management whereby past designs are stored as fully attributed 3D models. Exemplar models are included with each software distribution to assist with application training (Hand and Hensen 1995). These range from simple problems demonstrating basic model construction, through real-scale designs, to systems involving special components such as photovoltaic cells, advanced glazing, and adaptive materials. Some typical modelling and simulation activities follow.

An office block with natural ventilation represented as a multi-zone system with an airflow network superimposed is shown in Figure 6.7. Typical application: summertime temperature estimation against postulated occupant interactions and weather severities.

An air-handling plant with temperature and humidity modulation serving spaces requiring critical environmental control is shown in Figure 6.8. Typical application: component sizing, alternative layout appraisal, and control system tuning.

A constant air volume air-conditioner connected to a building is shown in Figure 6.9. Typical application: study of zone environmental control and plant psychrometric states.

A house with enhanced resolution around a thermal bridge and explicit construction moisture flow is shown in Figure 6.10. Typical application: estimation of condensation and mould growth risk, with appraisals of the potential of retrofits to alleviate problems.

A factory space with radiant heating and a gridded computational fluid dynamics (CFD) domain is shown in Figure 6.11. Typical application: assessment of spatial temperature distribution to achieve workplace comfort at minimum energy consumption.

An office with a hybrid photovoltaic facade and electrical network is shown in Figure 6.12. Typical application: appraisal of power generation and heat recovery potential, and a comparison of autonomous utilisation versus grid connection.

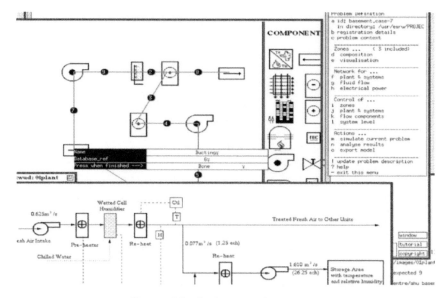

Figure 6.8   Defining a plant system

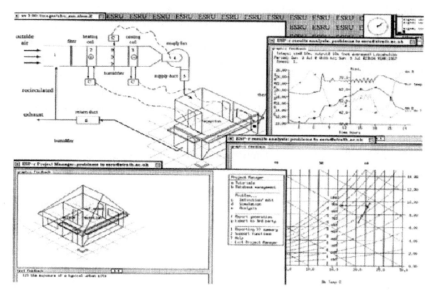

Figure 6.9   Tracking plant performance

An office employing daylight utilisation and artificial lighting control is shown in Figure 6.13. Typical application: assessment of the electrical power reduction potential and checking that this is not achieved at the expense of other performance parameters such as thermal/visual comfort.

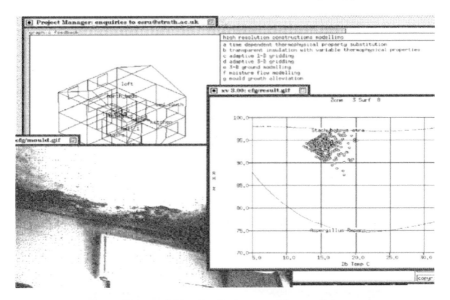

Figure 6.10   Predicting mould growth

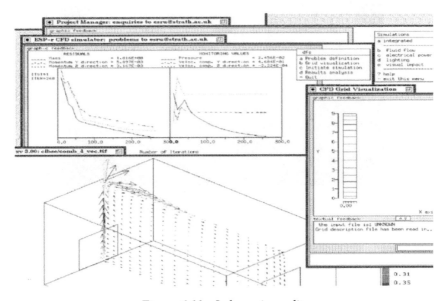

Figure 6.11   Indoor air quality

An assessment of the reverberation time in a space is shown in Figure 6.14. Typical application: an appraisal of speech interference in a performance space as judged by the timing difference between direct and reflected sound waves arriving at a user's location.

Figure 6.12   Photovoltaic facade

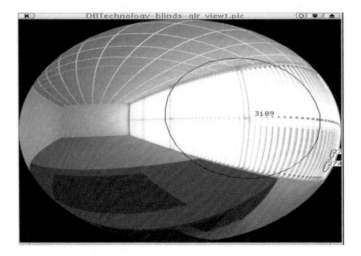

Figure 6.13   Glare assessment

An assessment of embodied energy is shown in Figure 6.15. Typical application: contrasting embodied and operational energy/carbon to ensure that the latter is not reduced at the expense of the former.

Figure 6.16 shows a subset of the pre-constructed models folder as included in an ESP-r distribution.

A model folder can include several related models corresponding to design variants. Some models are multi-zone with alternative design options and/or technical systems applied to each zone. Other models focus on specific problem types such as a smart facade, building-integrated PV, heat pumps, district heating,

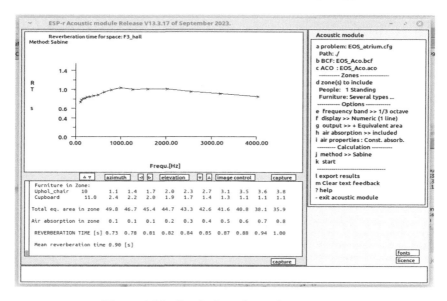

Figure 6.14   Prediction of reverberation time

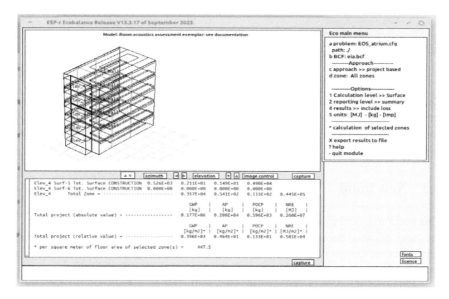

Figure 6.15   Estimating embodied energy

combined heat and power, and indoor air quality. Also included are archetype buildings, to demonstrate the use of a past project as the starting point for a new one. The following sections elaborate on the modelling and simulation approach by describing a variety of projects undertaken by ESRU staff over an extended period.

Figure 6.16   Example exemplar model folders as distributed with ESP-r

The intention is to illuminate application strategies as opposed to reporting performance outcomes for specific cases. Most of the projects were undertaken as consultancies serviced by nominated staff operating under the procedures elaborated in Section 4.8[12].

6.2   New build

This section describes the use of BPS+ to compare and contrast design variants in the case of new housing. The challenge was to manage multiple models and accommodate design issues raised by the client intermittently and over an extended period.

Although there is a degree of conformity in new housing in the United Kingdom, some developments commissioned by Housing Associations seek to test alternative construction methods and environmental control systems. The example presented here was a 2019 project in North East England to develop a site comprising 50 two- and three-bedroom, semi-detached dwellings (Hand 2020). The mix

---

[12]The aim of ESRU's consultancy activities is to empower the industry by offering computational support, never to compete for project work.

would include houses that were building code compliant and some with an enhanced facade *U*-value. Some dwellings were supplied with electric storage heaters, some with air-source heat pumps, some with ground source heat pumps, and some with direct electric heaters.

The Housing Association commissioned the project to guide future developments in terms of ease of deployment, occupant satisfaction, and contribution to their net zero targets. In relation to the construction process, one system was ready for occupancy after minimum site work whilst others involved considerable post-delivery site activity. In support of project goals, a monitoring programme and numerical assessments were carried out by the Building Research Establishment with the support of an ESRU staff member.

ESP-r dwelling models, each comprising 10 thermal zones, were created for each construction type. The models were of moderate resolution because information on the distribution of thermal and visual comfort was not required given uncertainties about occupancy at the time. Each dwelling was initially assigned the same control set points and occupancy patterns to support intercomparison. At a later stage, control variants and alternative occupancy were introduced to reflect anticipated diversity in use. Over time, the matrix of model variations grew large: whilst an individual dwelling model required hours of effort, the creation and management of multiple models required days. It was therefore critical that a robust model management process was implemented.

The approach taken was to adopt abbreviations related to the client's naming scheme for dwellings, construction, and environmental control systems. For example, 'P14MXFStdHHpD3V2.1' corresponded to plot 14, manufacturer X, facade standard *U*-values, heating via heat pump, design stage 3, and model version 2.1. The separately created dwelling models were concatenated into a street-scale community model as shown in Figure 6.17.

A feature of the project was the in situ testing capability, which was used to implement a pulse test regime for model calibration purposes (see Section 4.4.1). A variant of each model was created to act as a digital twin of the pulse test. This included control system adjustment to follow the phases of the test – that is, free-floating to converge with ambient, heating to raise the indoor temperature to 20 °C

Figure 6.17 ESP-r model of ten diversified semi-detached dwellings

Table 6.3 Space heating demand (kWh/m²)

Variant	Jan to Feb		Apr to May		Annual	
	GF	**HP**	**GF**	**HP**	**GF**	**HP**
Traditional (S)	4.7	4.8	1.4	1.0	11.1	11.2
Traditional (E)	3.4	3.3	1.1	0.8	9.3	9.1
Supplier X (S)	4.9	4.9	1.3	1.0	10.2	10.1
Supplier X (E)	7.3	7.4	0.6	0.5	17.9	17.8
Supplier I (S)	7.5	7.4	1.6	1.3	18.1	18.0
Supplier I (E)	5.5	5.5	1.5	1.2	12.8	12.7
Supplier P (S)	5.9	5.8	1.8	1.5	12.3	15.3
Supplier P (E)	3.4	3.4	1.2	1.0	9.1	8.9
Supplier M (S)	3.7	3.7	0.7	0.6	9.4	9.2
Supplier K (S)	2.9	2.8	0.6	0.5	8.4	8.3

S = standard construction; E = enhanced construction; GF = gas boiler; HP = heat pump.

above ambient, indoor condition held for 24–36 hours, and heating off to record the temperature decay. Before a pulse test, the models were used to determine the required heater capacity and when one pulse test failed due to an electrical fault, the associated model was used to determine the increased heat input rate that would rectify the situation without affecting other site work.

The data from the pulse tests (and ongoing monitoring) were used to adjust the models. In one case, it was discovered that insulation had not been correctly installed in a roof space; in another, a door draught excluder was missing. A thermographic survey was undertaken to detect facade weaknesses for inclusion in the models. Adapting the model with lower specification framing at some windows and doors indicated possible contributions to performance degradation.

Simulations were undertaken to determine which construction types were the better fit for specific heating systems. The assessments examined the response characteristics of the dwelling and the temporal distribution of room temperatures and heat demand. Tables 6.3 and 6.4 give a flavour of the results obtained for gas boiler and heat pump cases.

There were minimal differences in overall energy demand between GF and HP systems, although some enhanced facades performed less well than the standard facades of other manufacturers. Some of the dwellings were comparable with results for a dwelling built to PassivHaus standard.

For the January to February period in all construction types, the number of hours of required heating was greater with the heat pump although the energy use was essentially the same for both systems due to the higher heat delivery rate of the gas boiler during the start-up period. The main heat loss was associated with infiltration, whilst facade losses were 15%–25% of the infiltration loss, the latter being the dominant flow path. The temperatures were essentially the same for the different configurations with few occurrences of resultant temperatures over 25 °C.

Table 6.4   Overheating potential (hours resultant temperature >25°C)

Variant	Jan to Feb		Apr to May		Jul to Aug
	GF	**HP**	**GF**	**HP**	**GF**
Traditional (S)	30 (0.3%)	30 (0.3%)	488 (4.2%)	467 (4.1%)	1115 (9.5%)
Traditional (E)	96 (0.8%)	95 (0.8%)	808 (7%)	775 (6.7%)	640 (6%)
Supplier X (S)	16 (0.1%)	16 (0.1%)	220 (1.9%)	212 (1.8%)	2648 (22.6%)
Supplier X (E)	11 (0.1%)	11 (0.1%)	507 (4.4%)	492 (4.3%)	4409 (37.6%)
Supplier I (S)	0 (0%)	0 (0%)	74 (0.6%)	71 (0.6%)	837 (7.1%)
Supplier I (E)	4 (0%)	4 (0%)	–	125 (1.1%)	935 (8%)
Supplier P (S)	20 (0.2%)	19 (02%)	414 (3.6%)	384 (3.4%)	6065 (51.8%)
Supplier P (E)	76 (0.7%)	75 (0.7%)	784. (6.8%)	734 (6.4%)	5638 (48%)
Supplier M (S)	54 (0.5%)	54 (0.5%)	536 (4.7%)	527 (4.6%)	3215 (27.5%)
Supplier K (S)	123 (1.1%)	123 (1.1%)	748 (6.5%)	741 (6.4%)	380 (54%)

S = standard construction; E = enhanced construction; GF = gas boiler; HP = heat pump.

For the April to May period in all construction types, the number of hours of required heating was again greater with the heat pump, whilst heating energy use was always greater with the gas boiler. There were small differences between temperatures within the standard and enhanced models although the latter resulted in a higher percentage of time above 25 °C. As window openings were required to limit overheating, the infiltration load was around double the winter value.

At this stage, storage heaters were added to the mix to assess their operation within each house type. Assessments showed that uncontrolled heat emission would result in approximately 1 °C higher temperatures in rooms with no or low heat demand. It was found that the heaters aligned best with houses with standard construction where the demand-supply match was more favourable. Other findings were that the heaters would require variable charging (schemes were explored) and that even on days with no charge there would sometimes be sufficient residual heat stored to meet the next-day demand. The initial intention to deploy storage heaters only in dwellings with enhanced facades was reconsidered.

The issue of indoor temperature elevation above ambient was exacerbated by the enhanced construction of all heating systems. This is to be expected where developers position dwellings on site with scant regard to orientation, increasing facade heat retention and providing no shading. Some construction techniques tend to resist overheating: a consequence of the off-site manufacture of housing intended for bulk transport is that there is little thermal inertia to absorb excess heat. In relation to natural ventilation, the window opening logic in the assessments was conservative. Only when overheating was detected were the windows opened as a delayed action. A night purge and/or advice to occupants on how to enact appropriate ventilation strategies was demonstrated to reduce substantially the overheating tendency.

In another project (Samuel and Cockroft 2008) the focus was narrowed to the condensation risk associated with a proposed window unit. In this case,

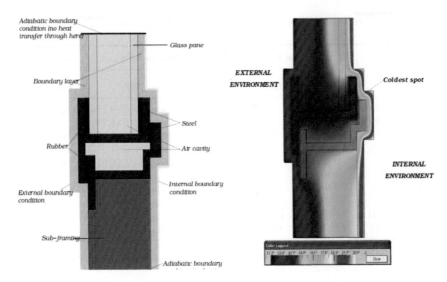

Figure 6.18   Model and simulation outcome

Table 6.5   Condensation analysis outcomes

RH (%)	T_d (°C)	T_e (°C)	Hours $< T_e$ (%)
35	5.2	−1.5	154 (1.8)
50	10.3	6.1	2540 (29.1)
65	14.3	11.7	6086 (69.7)
80	17.5	16.1	8346 (95.3)

the THERM[13] program was used to undertake a 2D heat transfer analysis, as summarised in Figure 6.18.

With an internal temperature of 21 °C imposed, the external temperature, T_e, at which internal surface condensation will occur, was determined for four levels of internal relative humidity. Table 6.5 presents the outcome. Also shown are dew point temperature, T_d, and the number of hours in the year that the external temperatures will be lower than this critical value.

At low relative humidity (RH), the condensation risk is minimal but becomes problematic as the RH rises to typical levels and beyond. The outcome was a change of product.

6.3   Retrofit

This section describes the retrofitting of a historic building with the intention of retaining the existing facade by creating a double facade. The aim of the project

---

[13]https://windows.lbl.gov/therm-software-downloads

Figure 6.19   Retrofit proposed behind this historic facade

was to provide evidence in support of regulatory exemption. The project used multiple BPS+ tools to this end.

A retrofit can be focused, as with the deployment of energy efficiency measures, or involve major construction works. BPS+ is applicable in both cases. The project summarised here was in the latter category: the creation of office accommodation behind a historic facade in Edinburgh as shown in Figure 6.19 (McDougall and Hand 2003). It was anticipated that, after refurbishment, the building would require cooling for much of the year.

At the time, a new set of building standards was coming into effect and a concern was to ensure that a proposed perimeter atrium would satisfy the anticipated compliance regime. The regulations offered two routes to compliance. The first was a method whereby the heat loss was established based on proposed U-values for wall, window, floor, and roof elements and compared with a notional building with prescribed U-values and glazing ratios. The second method allowed the use of BPS+ to compare the energy use and CO_2 emissions resulting from the high internal gains and associated cooling requirements.

An initial study focused on a typical building section to evaluate the impact of design options such as the use of single or double glazing in the internal atrium wall and different insulation treatments applied to the existing facade. The 'big idea' was that the refurbished building might use less energy with lower insulation levels due to the enhanced heat loss counterbalancing the elevated cooling load. In addition, the performance of the atrium buffer between the old and new parts was unknown, requiring an assessment of the effective U-value of the facade and perimeter atrium combined.

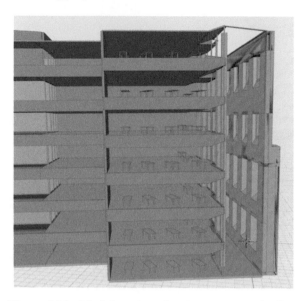

Figure 6.20   Model section showing atrium and offices

The ESP-r model is shown in Figure 6.20 – a 15m section of the atrium and 2,279 m² of office space. To this model were applied the variants of interest: three levels of insulation applied to the old facade and two fire-rated glazing systems placed between the atrium and offices. To reflect the transient nature of tenancy, one level was considered half-occupied and one unoccupied; the rest were assigned diversified occupancy patterns of interest to the client.

The proposed offsets shown between the windows in the old facade and the new accommodation raised questions about the quantity and quality of light within the offices. Initial simulations therefore focused on daylight availability and visual assessments. Figure 6.21 shows some typical outputs.

In relation to comfort conditions and energy use, simulations were performed to determine annual profiles of heating and cooling demand and the temperatures within the atrium. The offices were heated to 21 °C and the atrium to 10 °C during occupied periods. The office cooling set point was 24 °C. The base case model assumed that the historical facade was replaced with a code-compliant facade. Other variants added 25 or 50 mm of insulation. The inner atrium facade was defined with either double or single glazing.

Simulation results over a typical winter week and peak summer conditions were collated as shown in the example of Table 6.6.

The conclusion from the study was that altering the thickness of insulation at the inside face of the original facade had a minor effect on heating demands and on temperatures within the atrium during winter conditions. Furthermore, the use of single glazing between the offices and atrium reduced the overall demand for both heating and cooling. The reduction in cooling capacity during winter was

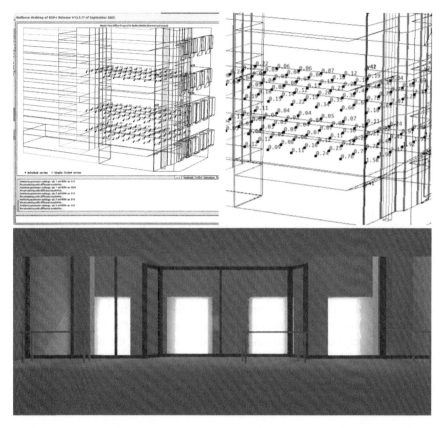

Figure 6.21   Daylight factors (upper) and visualisation of the atrium offset

Table 6.6   Simulation outcomes for office heating/cooling and atrium temperature

Facade U-value	Atrium glazing	Energy (kWh)		Capacity (kW)		Atrium peak temperature (°C)
		Heating	Cooling	Heating	Cooling	
0.3 (compliant)	Double	333	798	66.6	31.7	18.2
0.38 (50 mm insulation)	Double	343	797	64.9	31.5	17.2
0.55 (25 mm insulation)	Double	349	789	64.3	31.2	16.9
0.3 (compliant)	Single	257	529	68.1	25.8	19.2
0.38 (50 mm insulation)	Single	263	529	66.9	25.9	18.6
0.55 (25 mm insulation)	Single	267	518	66.3	25.4	18.3

significant. The simulations confirmed that the atrium would serve as an effective buffer in winter and that by maintaining the buffer at 10 °C, heat loss from the office spaces to the buffer resulted in a marked cooling load reduction throughout the year.

6.4   Housing assessments

This section describes a project to assist a developer to fine-tune a new product range. It demonstrates cooperation between monitoring and simulation and the ability to respond rapidly to technology options that were not initially considered – here replacing gas boilers with heat pumps. The project also contended with pro-blematic constructional detailing.

A high-resolution model was constructed for representative dwelling types recently constructed within a new estate and used to explore design modifications of interest to the construction company (Cowie *et al* 2020). Figure 6.22 shows a wireframe representation of one dwelling model.

The model was calibrated using data gathered from in situ pressurisation and pulse tests (see Section 6.24) and weather data from a local weather station. Furnishings were included in the model, and a family of two adults and two children was assumed to occupy the property. Occupancy-linked control of the heating system was imposed, and it was assumed that the occupants would open windows to maintain thermal comfort in summer.

A mechanical ventilation with heat recovery (MVHR) system was included according to design information received from the client, with a balanced extract and supply rate of 54 l/s and heat recovery efficiency of 90%. It was assumed that the MVHR always operated in basic mode (i.e., no use of the boost airflow rate). Without detailed information on the adaptive temperature control features of the ComfoAir Q350 unit, it was not possible to model this in detail. Instead, a constant supply temperature of 18 °C was imposed and subjected to a later sensitivity analysis.

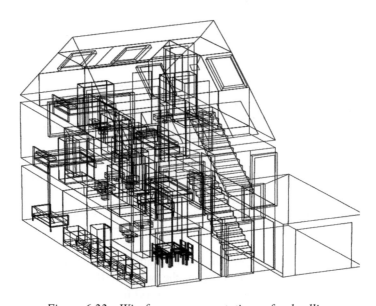

Figure 6.22   Wireframe representations of a dwelling

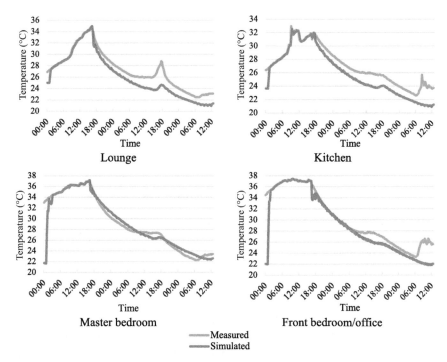

Figure 6.23   Comparison of predictions with measured results from a pulse test

Heating set points were obtained from the Chartered Institute of Building Services Engineer's (CIBSE) Guide A (Table 1.5), except in the case of the lounge where the set point was 21 °C (instead of 22 °C recommended by CIBSE) based on client instruction.

Figure 6.23 shows a comparison between simulated and measured results, the latter from a pulse test (see Section 4.4.1). Disparities in the cool-down profile were attributed to the lack of on-site weather data and brief periods when temperature sensors were in direct sunlight (and, of course, as yet undetected program deficiencies).

ESP-r simulations indicated an annual heating energy requirement of 57.3 kWh/m^2. This fell substantially outside the silver standard for space heating according to the Scottish Government's Building Standards technical handbook (SG 2017). Figure 6.24 gives a breakdown of the heating energy requirement by month and zone.

It was clear that the MVHR system accounts for the majority of the additional heating to bring the supply air to the set-point temperature of 18 °C after heat recovery. This high-energy demand was likely due to the disparity between the occupancy-related heating control and the always-on nature of MVHR, as well as the poor air tightness of the dwelling: when rooms are unoccupied, the MVHR extract is taking air that is likely to be below the set point, particularly in winter.

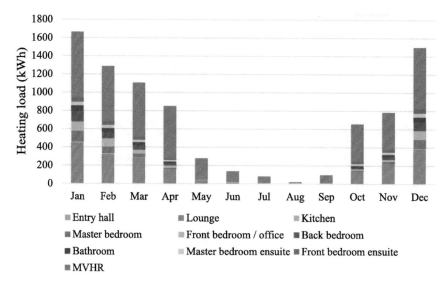

Figure 6.24   Breakdown of annual space heating energy requirement

Roof-mounted PV panels and Sunamp thermal batteries[14] were included in the model. Based on information received from the client, the system setup was as follows. Energy from the PV panels was stored in a 'UniQ dPV 3' heat battery with a 3.5 kWh capacity and heat loss rate of 18.7 W. Another 'UniQ HW+i 6' heat battery with a capacity of 7 kWh and a heat loss rate of 27 W was charged from the boiler, with a supplemental electric heating coil. The two heat batteries are connected in series to heat the domestic hot water (DHW). The PV panels were assigned an average efficiency of 15%. A simulated DHW profile was generated based on occupancy and the following assumptions:

- PV energy is transferred to the dPV 3 heat battery with 100% efficiency;
- DHW is heated from 10 °C to 50 °C;
- the heat battery can supply 50 °C temperature on its own; and
- the exchange efficiency of the heat battery is 95%.

Figure 6.25 shows annual simulation results for the charge state of the heat battery, the additional heat needed for DHW, and the residual energy from the PV that cannot be used to charge the heat battery.

Table 6.7 shows the aggregated energy utilisation over the year. It is stressed that these results are subject to the assumptions above, dependent on the simulated DHW profile, and neglect aspects such as pipework heat loss and variation in mains water temperature. Such issues were addressed via a sensitivity analysis that is not reported here.

The results confirmed that the PV array was sufficient to keep the heat battery charged from March to mid-October. Within this period, about half of the PV

---

[14]https://sunamp.com/en-gb/

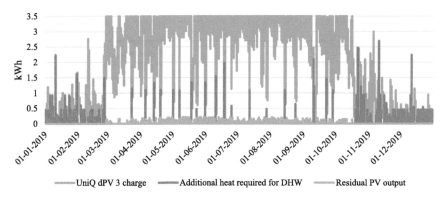

Figure 6.25   PV, heat battery, and DHW system performance profiles

Table 6.7   PV, heat battery and DHW energy (kWh)

Gathered by PV array	1110
PV used to charge dPV 3 heat battery	476 (43%)
Residual PV	633 (57%)
DHW	392
DHW from dPV 3 heat battery	125 (32%)
Additional needed for DHW	266 (68%)

energy was used to charge the heat battery, which is generally sufficient to heat the DHW during periods of light draw-off. Outwith this period, all PV output is used to charge the heat battery although with lower solar gain this is insufficient to keep the battery charged and supply the DHW. Overall, about a third of the heat for DHW is supplied from the PV array, and just less than half of the energy from PV is used for DHW. The discrepancy between the intermittent usage characteristics of DHW and the distributed generation profile of the PV limits the utilisation. A larger storage capacity would allow better utilisation provided the increased losses do not offset the benefit.

In the absence of information on the charge settings of the HW+i heat battery, it was not possible to assess its utilisation. However, assuming the remaining DHW requirement is supplied from the boiler without significant losses, the energy drawn from the grid at various boiler efficiencies was determined as shown in Table 6.8.

The predicted thermal comfort in all occupied zones was probed during occupied periods for the full year. Using the radiator specifications for upstairs rooms provided by the client, no periods of consistent underheating were observed, only brief heat-up periods at the beginning of occupancy because of the occupancy-linked control. No overheating according to the criteria in CIBSE Guide A was observed. Natural ventilation due to occupants opening windows and doors was found to be sufficient to cool the house in summer. Generally, the results showed

Table 6.8   Annual DHW energy use

Boiler	Grid energy	
efficiency (%)	kWh	kWh/m^2
85	306	2.1
90	293	2.0
95	279	1.9

that the property is able to maintain acceptable thermal comfort throughout the year. Air quality was not a concern: the maximum predicted CO_2 concentration in any zone was 1,045 ppm and concentrations above 1,000 ppm were rarely observed.

Three cumulatively applied upgrades were considered:

1. improved air tightness;
2. alternate geometry, constructions, and systems representing a proposed new product range; and
3. replacement of the gas boiler with an air-source heat pump.

For upgrade 1, an air leakage of 3.3 ACH at 50 Pa was reduced to 1.5 ACH. The predicted space heating energy requirement was 52.7 kWh/m^2, a reduction of about 8% on the base case. This figure is still substantially greater than the silver standard according to the Scottish Government's Building Standards technical handbook (SG 2017). Under the assumptions of the model, the dwelling does not operate efficiently. To investigate how much the MVHR design affects energy performance, additional simulations were commissioned with a reduced MVHR flow rate of 25 m^3/s and supply distribution weighted by room volume based on building regulation recommendations[15].

In this case, the predicted annual heating energy use was 30.5 kWh/m^2, a reduction of 47% relative to the base case, and slightly outside the gold standard. At this reduced flow rate, the maximum CO_2 concentration was 1,144 ppm, confirming that air quality requirements remain satisfied.

In upgrade 2, a model was constructed based on the base case geometry adjusted for a floor-to-ceiling height of 2.48m and a roof pitch of 35°. Constructions were updated in line with specifications provided by the client. Renewable energy systems in the model were PV and hot water heat recovery. The PV system comprised seven panels of 15% average efficiency. It was assumed that heating was delivered by a standard wet central heating system with intelligent, occupancy-linked control. The improved air tightness of 1.5 ACH at 50 Pa was assumed, and natural ventilation by trickle vents was activated. Figure 6.26 shows a wireframe representation of the new model.

When simulated for a year under the same conditions as the base case, this model resulted in an annual space heating energy use of 15.5 kWh/m^2; a 53%

[15]https://gov.uk/government/publications/ventilation-approved-document-f

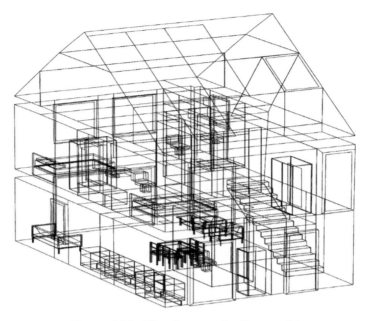

Figure 6.26   The alternate dwelling model

reduction relative to the base case and about half the gold standard referred to above. Figure 6.27 shows a monthly energy breakdown for the alternate model.

The slightly higher energy use in some spring and summer months is due to overcooling from open windows. Other than this, trends are similar to the base case. Some of the improvement in energy use may be due to differences in occupancy and casual gain patterns as the profiles are stochastically generated, but it is clear that the adapted model has significantly improved energy performance relative to the base case.

Some overheating was detected in the adapted model: predicted hours of exceedance was 3.2% in the kitchen and 3.1% in the back room ensuite, which is slightly above the limit of 3% in CIBSE TM52[16]. In addition, four time steps in the kitchen exceeded the absolute overheating criterion of 4 °C above the limit. However, this is likely to be an artefact of the assumption of constant conditions over each time step of the simulation; results clearly show that opening windows is sufficient to cool the house. Marginal underheating during heat-up periods was observed but nothing that is likely to compromise occupant comfort.

As may be expected when moving from mechanical to natural ventilation, air quality will become more of a concern. Maximum CO_2 concentration was 2,068 ppm in the front bedroom whilst the concentration in other zones did not exceed 2,000 ppm. As with the base model, the balance between air quality and thermal comfort is dependent on occupant behaviour. In general, if occupants are

[16]https://cibse.org/knowledge-research/knowledge-portal/tm52-the-limits-of-thermal-comfort-avoiding-overheating-in-european-buildings

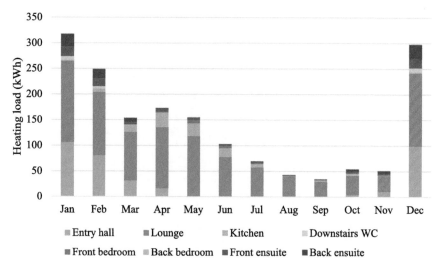

Figure 6.27   Monthly space heating energy breakdown for upgrade 2

particularly sensitive to stuffiness, energy use for space heating may be higher than the figure reported above.

The total energy generated from PV over the year was 1,795 kWh. Without specification of the hot water heat recovery systems in place, it was not possible to estimate exact energy saving. As for the base model, a simulated DHW usage profile was generated based on occupancy. Assuming water is heated from 10 °C to 50 °C, the total predicted energy required annually was 1,517 kWh. Assuming the hot water heat recovery system recovers 30% of this energy, this gives an annual energy saving of 455 kWh.

Upgrade 3 replaced the boiler in the adapted model with a heat pump. Two options were considered: replacement of the boiler only and replacement of the entire system including larger radiators. In place of an explicit heat pump model, a coefficient of performance (COP) profile was imposed as shown in Figure 6.28 and determined from an equation proposed by Baster (2017):

$$COP = 6.7 \, e^{-0.022\Delta T}$$

where ΔT is the difference between the heating system return and external temperatures (°C).

Where the existing distribution system is retained, the lower heat pump supply temperature approximately halves the radiator capacity, which results in compromised thermal performance. Underheated hours within occupied periods for the boiler and heat pump cases are given in Table 6.9.

The thermal performance is compromised to a lesser degree than in the non-upgraded base case and is unlikely to affect thermal comfort to a significant degree. In addition, and in contrast to the base case model, energy use for space

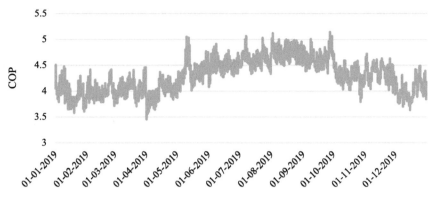

Figure 6.28   Annual heat pump COP profile

Table 6.9   Occupied period underheating (%)

	Boiler	Heat pump
Living room	2.0	3.3
Kitchen	0.0	0.0
Master bedroom	0.0	0.0
Second bedroom	0.0	0.0
Master ensuite	7.4	12.0
Second ensuite	3.2	18.7
Ground floor hall	0.0	2.0

Table 6.10   Net-at-the-meter heating energy

	Energy (kWh/m^2.y)
Boiler, 85% efficient	17.8
Boiler, 90% efficient	17.1
Boiler, 95% efficient	16.3
Heat pump retrofit	4.0
Heat pump + radiators retrofit	3.8

heating is increased, from 15.5 kWh/m^2 to 16.3 kWh/m^2 due to more time with the heating on. A comparison of the energy draw at the meter for the boiler and heat pump cases is shown in Table 6.10. It can be seen that despite the increase in space heating energy requirement, energy draw is still reduced by the improved COP of the heat pump.

Where the radiators are upsized, their heating capacity is similar to the base case so that the thermal performance would not differ. In this case, a reduction in energy draw at the meter results from the improved heat pump performance as shown in Table 6.10. In summary, the replacement of the boiler with a heat pump resulted in reduced energy draw, though this is dependent on the system being appropriately designed and operated.

The common tropes that housing issues are best dealt with via compliance tools, that BPS+ is overkill, and that environmental controls in housing are simple are fallacies. Compliance tools are not intended as design tools although they are often misused as such and, paradoxically, developing robust control for the domestic market can be more difficult than the commercial case due to inappropriate distribution system layout and the vagaries of occupants.

6.5   Building overheating

This section describes the application of BPS+ to an operational problem: severe building overheating. The building is complex (>1,000 surfaces) and its modelling involved network airflow and CFD. In addition, multiple simulations were necessary to accommodate highly variable occupant loadings.

The project (Hand and Samuel 2004) focused on problems with unacceptable summertime overheating in the high-seating gallery within the main auditorium of a theatre. The existing ventilation system was not providing adequate fresh air to cool the two galleries in the auditorium causing ceiling temperatures to approach 40 °C. A part of the problem was that the air extracts were at a low level preventing the less dense warm air from being extracted from the higher parts of the auditorium.

A high-resolution model of the theatre was created, including an airflow network to represent the HVAC system. Permutations of this airflow network were then applied (e.g., adding a roof-level air extract) to determine if the existing plant could be economically modified. Figure 6.29 (upper) shows the ESP-r model with some external surfaces removed, whilst the lower image shows the multi-zone nature of the model.

Figure 6.30 shows a symbolic representation of an airflow network established to represent the ventilation system connecting the foyer, auditorium, and stage for the existing situation in which 70% of the fresh air is supplied at ground level and 15% to each of the galleries. All air is extracted at the stage sides.

Several variant networks were created to represent possible remedial solutions. The following results correspond to the best performing of these variants: air supply apportioned according to seating distribution (40% to the high gallery and 30% each to the low gallery and ground level seating area), with equal extracts at ceiling level and at stage sides. Table 6.11 lists comparative peak ceiling surface temperatures under typically warm conditions with full occupancy. The problem is exacerbated by the high thermal mass.

Figure 6.31 shows the corresponding air temperature variations in principal spaces over the same period (beware of the different y-axis scales). The improvement is substantial, with the variant scheme reducing the peak temperature in the high gallery by more than 10 °C.

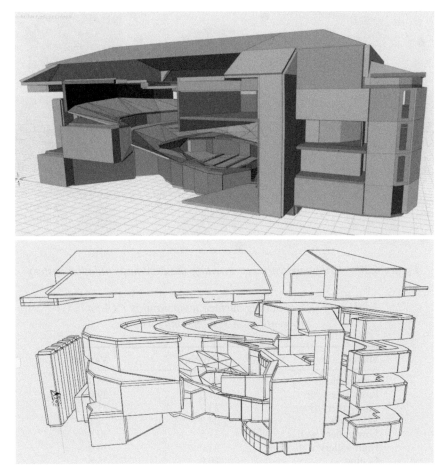

Figure 6.29 The ESP-r theatre model and its constituent zones

To examine the conditions in the high gallery more closely, a CFD domain model was defined for use with Ansys Fluent[17] based on boundary conditions exported from the ESP-r simulations during peak summer conditions. Figure 6.32 presents one snapshot result amongst many, depicting velocity vectors coloured by temperature for the above two cases.

It is clear from the left image that introducing the supply air at a low level and extracting it at the stage sides is an ineffective configuration. In the right image, where half of the extracted air takes place at the roof and the remainder at the stage sides, the distribution is uniform and the conditions improved throughout the theatre. These results were used to justify modifications to the existing ventilation

[17]https://ansys.com

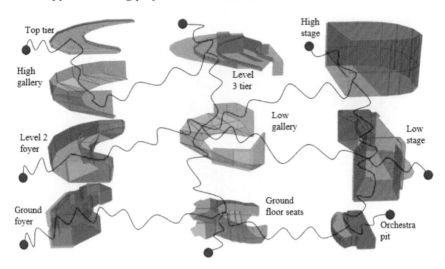

Figure 6.30   Airflow network representing the existing ventilation system

*Table 6.11   Peak ceiling surface temperatures (°C) for
two ventilation system configurations*

Surface	Existing	Variant
Front ceiling	40.0	36.0
Front side walls	31.0	29.0
Low gallery floor	23.7	23.7
Side walls	24.3	24.1
Low gallery walls	31.5	30.5
High gallery walls	33.0	28.0
Ceiling	40.0	35.8
Orchestra pit	21.4	20.0
Stage	26.0	25.0
Wings	30.6	28.8

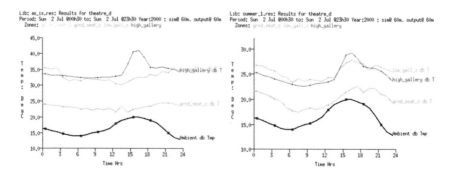

*Figure 6.31   Air temperatures in principal spaces, warm weather (left existing,
right variant)*

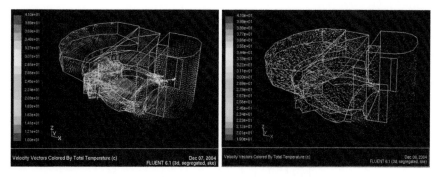

Figure 6.32   Conditions at peak summer, existing (left) and variant cases

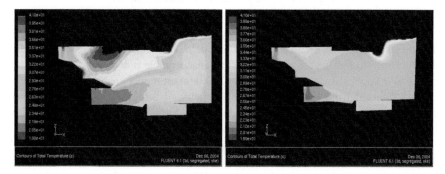

*Figure 6.33   Central vertical plane temperature distributions for existing (left)
and proposed ventilation schemes*

system, including ducting rerouting and control system adjustment. Figure 6.33 shows the temperature distribution associated with the existing and proposed ventilation schemes (the stage is on the right).

It is evident in the new scheme that the seating areas are cooler, with reduced radiant gains from the ceiling. The proposed alteration was deemed cost-effective because the existing plant and distribution system was retained.

This study was commissioned due to observations of untenable environmental conditions. Mitigation ideas emerged from using BPS+ to establish the context in which those conditions arose and to explore interventions in the days before a performance. It was demonstrated that the thermal inertia of the building could be used to time-shift discomfort and airflow redistribution was shown to mitigate the problem at full occupancy. The lesson is that thinking outside the box has the potential to generate low-cost options for facilities management.

In contrast to the above CFD approach, it is possible to conflate thermal and CFD models and thereby reconfigure the latter throughout a simulation in response to changing thermal conditions. This approach is strictly necessary where air movement corresponds to an unsteady, low Reynolds Number flow regime as typifies a building.

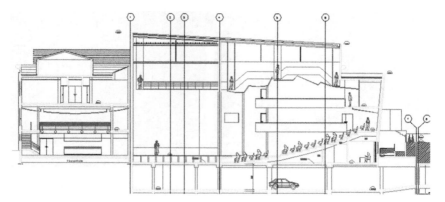

Figure 6.34   The AutoCAD theatre model

The project summarised here (Hand and Samuel 2004) undertook an assessment of a different theatre that was also experiencing unacceptable indoor conditions in summer. ESP-r was used with an embedded CFD domain model enabled.

As inferred from Figure 6.34, the geometry of a high-resolution model was created from an existing AutoCAD model and additional attributions were applied, including CFD domain gridding and building interaction information[18].

The HVAC system was configured to supply fresh air at low velocity through diffusers under the seats, on the stage, and in the technical void. At predetermined times of interest, the results from the building energy solution served as the boundary conditions for the CFD domain, which comprised 500+ cells and air inputs from 420 diffusers. The additional data required to define a CFD domain is extensive and technical as follows.

- Number of gridding regions, number of cell divisions in each, and rate of cell size reduction when nearing a boundary surface.
- Identification of source cells with inputs from people and equipment (heat, humidity, contaminant, etc.).
- Identification of cells to be considered non-participatory because they contain blockages associated with internal furnishings and fittings.
- Definition of airflow boundary conditions where the domain connects to ventilation inlets and outlets.
- The connection type indicates how information is shared between the thermal zone and CFD domain: one-way (to CFD only) or two-way.
- Selection of turbulence model from: none (laminar flow only), defined, or adaptive (determined automatically at simulation time). In the second case, the following options are available: $k - \varepsilon$ (recirculating flows), fixed eddy viscosity,

---

[18]The CFD domain can be coupled to the building and/or network flow domains and the CFD grid cells linked to the conditions prevailing at building surfaces and interactions with ventilation components.

Figure 6.35   Animated indoor conditions

zero-equation, or initial fixed eddy (after a stated number of iterations, $k - \varepsilon$ is resumed).

- Selection of buoyancy model: none, ideal gas, Boussinesq (fixed buoyancy at reference temperature).
- Selection of surface boundary-flow model: fixed or adaptive, log-law or Yuan wall function, local or global determination, and thermal or CFD side heat transfer coefficients determination.
- Required equation solution method, initialisation values for pressure, velocity, k, ε, and CO_2, relaxation factors, and iteration limit.
- Directives for cell monitoring during simulation and data capture resolution and frequency.

Whilst some of these data can be inferred from the input model or intelligently defaulted, others will require user input.

Simulations with the CFD domain disabled were undertaken for the summer months and the results were used to gauge the effect of design parameter changes on the cooling load. Focused simulations were then carried out to study indoor comfort and air movement under various occupancy loadings during summer conditions. In this case, the coupled CFD domain was enabled[19]. Figure 6.35 corresponds to snapshots of animated outputs relating to the distribution of temperature, airflow, and mean age of air during critical periods.

As with the first example, the results supported decisions on critical adjustments to the HVAC and air distribution system to eliminate the overheating problem.

6.6   Integrated thermal and lighting appraisal

This section describes two BPS+ applications that required combined thermal and visual simulation, with explicit sky modelling and enhanced resolution to enable

---

[19]Using an ESP-r compatible code that was under development at the time.

visual assessment of visual comfort. In one case, the model was enhanced by the inclusion of an explicit representation of a photocell for luminaire control.

The modelling of daylight-responsive buildings requires prediction of the time-varying internal illuminance distribution when influenced by blind movement and sky luminance changes. The Commission Internationale de l'Eclairage (CIE) daylight factor procedure (Tsang *et al.* 2022) is often used based on different methods to determine the daylight factors. These methods range from analytical approaches (Winkelmann and Selkowitz 1985), which are applicable to simple problem types (e.g. rectilinear geometry), to generalised lighting simulation (Ward 1993), which can handle problems of arbitrary complexity.

Daylight factors are subject to considerable variation even under overcast sky conditions, with variations of the order of 1:5 being reported (Teregenza 1980). Littlefair (1992) compared predicted and measured internal illuminance for different calculation methods. The conclusion was that the daylight factor method was not suited to situations that combine realistic sky types with complex building interactions.

Another area of concern relates to luminaire control. The hourly weather data normally employed is incompatible, in its frequency, with the requirements of realistic control algorithms. Some programs[20] offer statistical models to account for the sky luminance variation within the hour interval. However, because the results are provided as hourly integrations, it is not possible to use them to model specific control characteristics (such as photocell position, field of view, time constant, and switch-off delay). The use of hourly average data might misrepresent significantly the system response.

BPS+ can be applied to evaluate indoor visual comfort alongside other significant issues such as energy use, daylight utilisation, emissions reduction, and electricity grid interaction. In one study (Citherlet and Hand 2002) the potential of deploying low-cost, solar-tracking, internal blinds within an office environment was evaluated. Several approaches to control have been studied, and two of significance are summarised here:

- manually operated vertical blind and luminaire control;
- solar azimuth tracking vertical blind, with daylight-responsive luminaire control.

As shown in Table 6.12, the results from ESP-r/Radiance simulations indicated an energy saving of 38% for the tracking configuration when compared with manual operation.

Based on the conversion factors for gas and electricity prevailing at the time, these results corresponded to a substantial CO_2 reduction with the automatic control case.

The results also indicated a marked improvement in visual comfort as characterised by the Previsible Percentage Dissatisfied visual comfort index (JPPD; Guth 1966), which correlates the integrated luminance in the field of view to

---

[20]For example, the SUPERLINK program (Szerman 1994), which was extant at the time of the project.

Table 6.12   Results for the two blind/luminaire control cases

Control	Annual energy demand		Equivalent CO$_2$ emissions		Visual comfort
	kWh/m^2	% manual	kg/m^2	% manual	JPPD (%)
Manual	122	–	49	–	30
Automatic	75	−61	28	−57	15

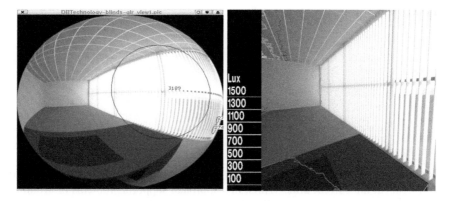

Figure 6.36   Office luminance (left) and illuminance distributions

subjective perceptions of visual comfort. The index was used to characterise the situation where the office module is viewed in a certain direction from a specific location – here the index was reduced by 50%. The study demonstrated that the introduction of an integrated blind and lighting control system within an office environment has the potential to reduce significantly the energy use (and thereby the associated CO$_2$ emissions) whilst improving occupant visual comfort.

The left image in Figure 6.36 shows a fish-eye representation of the luminance distribution as would be experienced by the desk user under clear sky conditions (units cd/m^2); the right image shows the corresponding illuminance distribution.

In another project (Clarke and Janak 1997), a higher fidelity approach was required to model variable sky luminance distribution, movable and light redirecting systems, multiple external/internal reflections, and artificial lighting control. Figure 6.37 summarises the approach taken based on a runtime coupling of ESP-r and Radiance.

At each simulation time step[21], ESP-r's luminaire control algorithm invokes Radiance to determine the internal illuminance distribution based on the transfer of data defining the current building state and weather condition. The returned data are used to determine, as a function of the photocell location and control algorithm,

---

[21]To minimise computing time, periods when handshaking is enabled can be defined by the user.

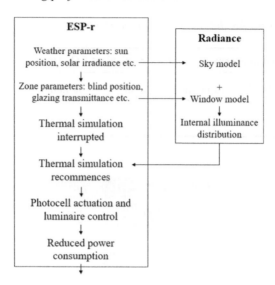

Figure 6.37   ESP-r/Radiance interactions at the time-step level

the luminaire status and hence the power consumption and casual gain associated with lights at the current time. Whilst this approach supports problems of arbitrary complexity, it is computationally demanding and therefore inappropriate for routine design application. An alternative approach was therefore researched and implemented.

Where problems may be characterised in terms of a finite number of discrete states (e.g., blind open, partially closed, and closed) an alternative approach is available whereby the internal illuminance calculation is based on daylight coefficients (Littlefair 1992a). This method discretises the sky vault into 145 elements as depicted in Figure 6.38.

The daylight coefficients for each photocell sensor point and building state are calculated by Radiance based on a unit luminance before commencing the ESP-r energy simulation. Then, during simulation, the internal illuminance distribution is determined by simple multiplications and additions to factor in the prevailing luminance of each sky patch. In this way, ESP-r supports luminaire and shading device control based on the changing value of model parameters such as internal air temperature, illuminance at specified locations, or external conditions such as facade solar irradiance. This allows the appraisal of different photocell options in terms of position, vision angle, controller set point, switch-off lux level, switch-off delay time, and minimum stop. Two dimming control algorithms are available within ESP-r as follows.

An *integral reset* controller adjusts the dimming level so that the measured photocell signal is kept at a constant reference value. This reference level is set during nighttime photocell calibration and represents the measured signal from the

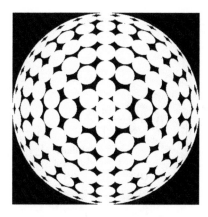

Figure 6.38   Sky discretisation

artificial lighting. The dimming level in the controller dynamic range is determined as follows:

$$f_{\text{dim}}^{\tau} = 1 - \frac{E_{\text{d,s}}^{\tau}}{E_{\text{e,s}}}.$$

A *closed-loop proportional* controller adjusts the dimming level so that it is a linear function of the difference between the photocell signal and the nighttime reference level. With this controller, a daytime calibration must be performed to determine the linear control function slope for use in the following expression.

$$f_{\text{dim}}^{\tau} = \frac{1 + m_{\text{slope}}\left(E_{\text{d,s}}^{\tau} - E_{\text{e,s}}\right)}{1 - m_{\text{slope}}E_{\text{e,s}}}$$

where f_{dim}^{τ} is the time-varying dimming level (–), m_{slope} the slope of the controller's linear function determined by daytime calibration, $E_{\text{d,s}}^{\tau}$ the time-varying daylight photocell signal (lux), and $E_{\text{e,s}}$ the artificial lighting photocell signal during nighttime calibration (lux).

Consider the problem shown in Figure 6.39: an office module of dimensions $4.5 \times 4.5 \times 3.2\text{m}$ with a combination window comprising an upper portion with an external light shelf and a lower portion with a movable blind.

The office is to be lit by wall-mounted, asymmetric luminaries designed to provide an average workplace illuminance of 320 lux. The lamp's luminous output can be regulated between 10% and 100% of the full light output. Daylight-responsive control is implemented via a ceiling-mounted photocell located at 2/3 of the room depth from the window. Within the study, the following control parameters were applied:

- daylight sensor set point, 320 lux;
- switch-off light reference level, 150% of set point;

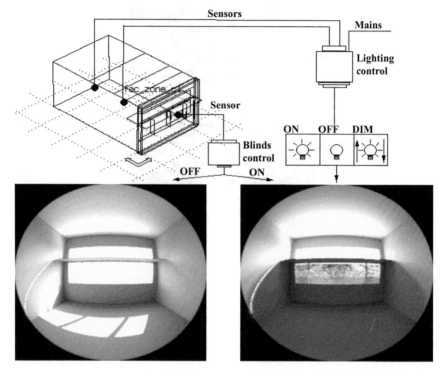

Figure 6.39   Thermal/lighting model conflation

- minimum light dimming, 10% of full light output;
- minimum power dimming, 10% of full circuit power;
- switch-off delay time, zero for 60-minute time-step simulations; 15 minutes for 5-minute time-step simulations;
- blind sensor location, vision window plane measuring vertical global irradiance; and
- set point for blind rotation to the shading position (45°), 300 W/m².

Figure 6.40 shows two photocell geometries: a partially shaded case on the left and a fully shaded case ($E_{e,s}$ = 44.5 and 14.1 lux, respectively). For the closed-loop, proportional control action applied, the linear control slope, m_{slope}, was set to −0.023 and −0.056 for the partial and fully shaded cases, respectively.

Table 6.13 lists the predicted lighting energy consumption for different controller/photocell combinations (including the difference relative to the ideal control case); Figure 6.41 shows the time-step variation in the contribution to average worktop illumination.

Of course, these results relate only to the problem studied here and are included only to indicate the potential of the BPS+ approach, which is relatively straightforward to apply.

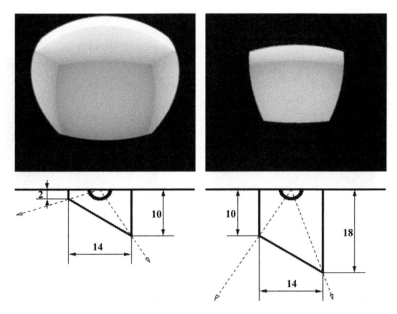

Figure 6.40 Partial and fully shading photocells

Table 6.13 Predicted lighting power consumption

Sensor characteristics	Time step (minutes)	Lighting energy (Wh/day)	Relative difference (%)
Integral reset, partially shaded	5	1168	−32.5
Integral reset, fully shaded	5	997	−42.4
Closed-loop, proportional, partially shaded	5	1670	−3.4
Closed-loop, proportional, fully shaded	5	1636	−5.4
Closed-loop, proportional, partially shaded	60	1792	3.6
No photocell	60	1730	−

The project demonstrated that to model realistic lighting control, short-term daylight availability must be considered. Different controllers resulted in large differences in power consumption (c.f. integral reset vs. closed-loop action). Insights into how best to model daylight-responsive luminaire control also emerged from the project. For example, a properly calibrated, closed-loop proportional controller with 5-minute time steps gave similar power consumption predictions as ideal control with hourly time steps, whilst an integral reset controller with ideal control failed to reproduce the dynamic behaviour and therefore the correct power consumption.

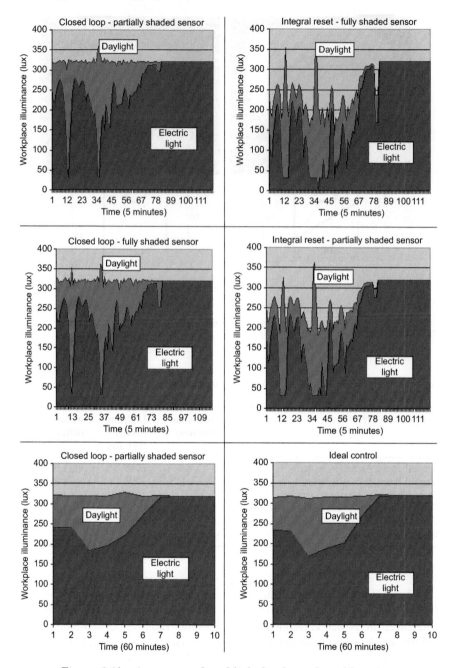

Figure 6.41   Average predicted light level at selected locations

It is noticeable that the models created in lighting studies are usually markedly different to those intended to support a thermal analysis. For example, the representation of window frame depth included in the former will be typically absent in the latter, excluding the possibility of analysing thermal bridge effects. In addition, a thermal analysis is often applied to an empty space that can then support only a cursory daylight factor analysis. It is surely better to address such issues in a high-resolution model that is able to support integrated thermal and lighting appraisals. Section 5.1.4 and Section 11.5 elaborate on this issue.

6.7   Wet central heating control

This section describes a focused study involving the modelling of heating system control with the aim of matching control approaches to dwelling archetypes. The results were judged in relation to boiler operational efficiency. The project was undertaken under DEFRA's Market Transformation Programme (Cockroft and Samuel 2012) with the aim of establishing a suite of models for use by an industry partner and others to appraise options for wet central heating system control. ESP-r models were developed for six heating systems and eight control regimes when applied to five typical dwelling types as summarised in Table 6.14. Each model comprised separately controllable living/non-living zones intended to emulate the standard SAP calculation protocol.

Table 6.14   Dwelling/control combinations evaluated

Dwelling type		Heating/control system
1. 104 m^2 detached, pre-1918, solid wall, single glazed, 100 mm loft insulation.		Non-condensing gas boiler (21.8 kW), non-modulating burner, radiators (living 7.5 kW, non-living 11.3 kW). Living zone mechanical thermostat, no TRVs.
2. 90 m^2 semi-detached, 1939–59, uninsulated cavity wall, single glazed, 100 mm loft insulation.		Condensing gas boiler (17.9 kW), modulating burner, radiators (living 5.6 kW, non-living 9.3 kW). Living zone mechanical thermostat, TRVs in other zone.
3. 90 m^2, semi-detached, EU stock average, double glazed, filled cavity, 100 mm loft insulation.		Condensing gas, combi boiler (14.2 kW), modulating burner, radiators (living 4.4 kW, non-living 6.8 kW). Living and non-living zones with mechanical thermostats.
4. 90 m^2 semi-detached, 1990–99, timber frame, double glazed, 100 mm loft insulation.		Condensing oil boiler (13.7 kW), non-modulating burner, radiators (4.3 kW, 6.3 kW). Outside temperature compensation, modulating supply water set point, living zone temperature compensation, TRVs in other zone.

(Continues)

Table 6.14   (Continued)

Dwelling type	Heating/control system
5. 79 m² mid-terrace, 2006 (Part L Regulations), filled cavity, 270 mm loft insulation.	Condensing gas boiler (8.8 kW), modulating burner, underfloor heating in living zone (2.2 kW), radiators in other zone (3.6 kW). Living zone sensor, modulating supply water set point, TRVs in other zone.
6. As 5.	Non-condensing gas boiler, modulating burner, radiators. Time proportional, modulating (TPM) thermostat in living zone, TRVs in other zone.
7. As 5.	As 6 but with TRVs in both zones.
8. As 5.	As 6 but with time proportional modulating thermostats in both zones.

Figure 6.42 shows a schematic of the central heating system as modelled. The boiler was modelled as a 'black box'[22], using the results of flue gas tests to derive the combustion efficiency as a function of return water and ambient temperature. Where the boiler is capable of modulation, the water supply temperature is controlled to a fixed or variable set point as determined by the control system. Below the modulation range, control is on/off at the lowest firing rate.

Figure 6.43 depicts the scheme for a modulating or on/off boiler. The logic element is represented by the left-hand box, and the thermal dynamics by the right-hand box. Here, L is load ($0<L<1$), ϕ water heat input, T_r water return temperature, T_{fs} water flow temperature set point, ϕ_{max} maximum boiler output, and η_b boiler efficiency. If $L<0.3$, $\phi =0$.

The 2-node thermal model, selected from ESP-r's plant components database, is linked to the distribution system and solved at each simulation time step. The logic element updates the thermal inputs at each plant time step, whilst secondary controls signal a water temperature set point to the boiler. In the case of a condensing boiler, heat recovery from the flue gas is represented by a combustion efficiency, η_f, derived from the empirical Siegert formula:

$$\eta_f = 100-(A/C + B)(T_{flue}-T_{amb})$$

where A and B are coefficients for a specific fuel, C the % concentration of CO_2 by volume in the flue gas after removal of water vapour, and T_{flue} and T_{amb} are the flue exit and ambient temperatures, respectively (°C).

---

[22]ESP-r can co-simulate with TRNSYS with plant networks comprising a mix of components from both programs (Beausoleil-Morrison *et al.* 2012). Alternatively, ESP-r component models can encapsulate TRNSYS type models to represent their internal state algorithmically rather than by numerical means (Aasem 1993).

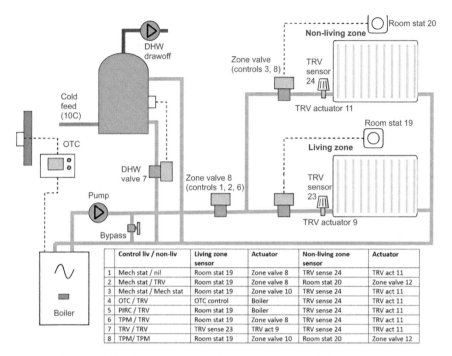

Figure 6.42 The wet central heating system with basic control

	Control liv / non-liv	Living zone sensor	Actuator	Non-living zone sensor	Actuator
1	Mech stat / nil	Room stat 19	Zone valve 8	TRV sense 24	TRV act 11
2	Mech stat / TRV	Room stat 19	Zone valve 8	Room stat 20	Zone valve 12
3	Mech stat / Mech stat	Room stat 19	Zone valve 10	TRV sense 24	TRV act 11
4	OTC / TRV	OTC control	Boiler	TRV sense 24	TRV act 11
5	PIRC / TRV	Room stat 19	Boiler	TRV sense 24	TRV act 11
6	TPM / TRV	Room stat 19	Zone valve 8	TRV sense 24	TRV act 11
7	TRV / TRV	TRV sense 23	TRV act 9	TRV sense 24	TRV act 11
8	TPM/ TPM	Room stat 19	Zone valve 10	Room stat 20	Zone valve 12

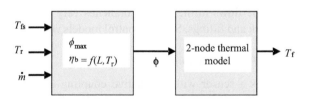

Figure 6.43 Boiler model

Regular and combi boilers incorporate additional logic to determine action when there is a call for domestic hot water. A regular boiler will simply react to the water flow rate when the DHW zone valve is opened on a call for heat from the cylinder thermostat. The domestic hot water storage cylinder is modelled as a separate two-node storage tank. When the domestic hot water call is satisfied (hot water temperature rises to the upper thermostat limit), the zone valve is closed.

When a boiler is activated from an off state, it carries out a sequence of pre-purge, ignition stabilisation, and anti-cycling operations. Table 6.15 lists typical values for condensing and non-condensing boilers.

A combi boiler will switch to DHW service mode when a draw-off is detected. In this case, the heat flux to the DHW, supplied via an internal fast-response heat

Table 6.15   Boiler start-up sequence

Operation	Condensing		Non-condensing	
	Gas rate (%)	Time (s)	Gas rate (%)	Time (s)
Pre-purge (no gas)	100	10	100	3
Ignition stabilisation (no gas)	70	3	–	0
Ignition and ignition stabilisation	70	10	30	10
Anti-cycle time	30	217	30	227
Rate of water temperature set point increase during anti-cycle time	3°C/minute			

exchanger is known. The heat exchanger is modelled in a similar fashion to the storage cylinder but with no thermal mass. The known heat flux to be supplied by the boiler enables the flow and return temperatures to be determined dynamically, and the boiler will modulate as necessary, mimicking the behaviour of a combi boiler maintaining a fixed DHW supply temperature.

The boiler model is a two-node representation, which ensures that the effect of boiler thermal mass is simulated, and jacket heat losses are calculated as a thermal input to the living zone. The supply water temperature is the output from the model.

The water distribution network, modelled by a flow network, incorporates a pump component assigned a pressure/flow characteristic. As individual valves open, close, or modulate, the flows in the various branches vary according to the flow resistances of pipework and fittings. Several control models were defined as follows.

On/off mechanical room thermostat

This model comprises a sensor with a radiative coupling to zone surfaces, convective coupling to the air, conductive coupling to the attached surface, intrinsic thermal mass, and anticipator heating effect. The dynamic output is the on/off status of the thermostat according to set point and mechanical differential settings. As shown in Figure 6.44, if the thermostat is 'on' then the water supply temperature is set to the maximum. If the thermostat is 'off' then the control temperature output is set to the minimum.

Weather compensation

Weather compensation is based on the outside temperature, which is used to derive the supply water temperature as shown in Figure 6.45.

Overall controls integration

The above control models are assimilated into the overall boiler control as shown in Figure 6.46.

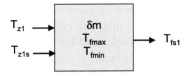

Parameters: T_{fmax}, T_{fmin}, δm

Inputs: T_{z1} zone 1 thermostat sensor temperature
T_{z1s} zone 1 temperature set point

If $T_{z1} > T_{z1s}$ then $T_{fs1} = T_{fmax}$
If $T_{z1} < T_{z1s} - \delta m$ then $T_{fs1} = T_{fmin}$

Figure 6.44   On/off thermostat logic

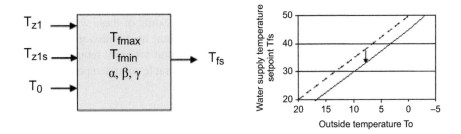

Parameters: T_{fmax}, T_{fmin}, α, β, γ
Inputs: T_{z1} zone 1 temperature
T_{z1s} zone 1 temperature set point
T_o outside temperature

$T_{fs} = 20 + \alpha + \beta(20 - T_o) + \gamma(T_{z1} - T_{z1s})$
$T_{fmin} < T_{fs} < T_{fmax}$

Figure 6.45   Weather compensated control

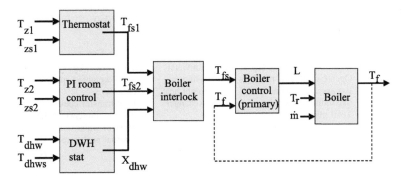

Figure 6.46   Assimilation of zone/DHW control into the boiler control model

The interlock ensures that boiler firing can only occur when at least one control module is calling for heat. Of course, it is possible with circuits controlled only by thermostatic radiator valves (TRVs) that the boiler will run when TRVs are closed as there is no feedback to the boiler control system.

If there is a simultaneous demand for zone heating and DHW then all zones calling for heat will be serviced, with water flow shared between the circuits according to the water network model. When the combi boiler is servicing DHW, heat is not supplied to any heating zone for the duration of the DHW call.

Unlike other thermostats, the electronic proportional-integral (PI) modulating thermostat does not operate a zone valve. When there is a call for DHW, the boiler has to increase the supply temperature to the maximum operating limit. To avoid living zone overheating during this period, a zone valve is included in the heating circuit that limits the room temperature, shutting off the water flow if the room temperature rises more than 1 °C above the set point.

Table 6.16 lists the user-adjustable parameters associated with the plant and control definitions included in the ESP-r model. The parameters listed are those corresponding to Dwelling 3, Plant 2 in Table 6.14. Some of these will be scaled, depending on the dwelling type, e.g., radiator nominal heat emission rate, boiler output rate, and maximum pump flow rate. Thermostat set-point temperatures are nominal values and may be adjusted to ensure comparable room temperature levels for each system.

Table 6.16   User-adjustable input model parameters

Component	Parameter	Component	Parameter
Boiler	Total mass (kg)	Living zone	Nominal heat emission (W)
	Mass-weighted specific heat	radiator	Nominal supply temperature (°C)
	(J/kg.K)		Nominal return temperature (°C)
	Full load heat delivered		Nominal environment temp. (°C)
	(nominal rating) (kW)		Heat transfer exponent (–)
	Gas heating value at STP		Mass (kg)
	(MJ/m³)	Non-living	Nominal heat emission (W)
	Upper boiler temperature limit (°C)	zone radiator	Nominal supply temperature (°C)
	Lockout time at upper limit (s)		Nominal return temperature (°C)
	Low limit of modulating		Nominal environment temp. (°C)
	range (%)		Heat transfer exponent (–)
	Low limit total differential (%)		Mass (kg)
	CO_2 and flue-return water		
	temperature vs. load		
	(8 polynomial coefficients)		
Circulating	Maximum flow rate (kg/s)	Hot water	Total mass (kg)
pump	Open circuit flow rate kg/s	cylinder/	Mass-weighted specific heat
	Bypass setting (fraction of	combi heat	(J/kg.K)
	maximum pump flow)	exchanger	Heat transfer coefficient
	Total mass (kg)		(to containment) (W/K)
	Mass-weighted specific heat		Mass of water in coil (kg)
	(J/kg.K)		Internal heat transfer area (m²)

(Continues)

Table 6.16   (*Continued*)

Component	Parameter	Component	Parameter
	Heat transfer coefficient (to containment) (W/K)		Internal heat transfer coefficient (W/m².K)
	Total absorbed power (W)		External heat transfer area (m²)
			External heat transfer coefficient (W/m².K)
PI room control (modulating)	Mass-weighted specific heat (J/kg.K)	Time control, heating	Enabled weekdays 7 h–9 h and 16 h–23 h; weekends 7 h–23 h
	Convective coupling to air (W/K)	Time control, hot water cylinder	Enabled weekdays 7 h–9 h and 16 h–23 h; weekends 7 h–23 h
	Radiative coupling to opposite wall (W/K)	Hot water cylinder thermostat	On set point (°C)
	Radiative coupling to mounting wall (W/K)		Off set point (°C)
	P proportional band (K)		
	I integral action time constant (s)		
	Cycle time (s)		
	Minimum on/off time (s)		
Mechanical thermostat	Total mass (kg)	TRV control (P control)	Valve fully closed (°C)
	Mass-weighted specific heat (J/kg.K)	Outside temperature-compensated control	Valve fully open (°C)
	Convective coupling to air (W/K)		Alpha (parallel shift) (K)
	Radiative coupling to opposite wall (W/K)	TPI room control (on/off)	Beta (ratio)
	Radiative coupling to mounting wall (W/K)		Gamma (room temperature compensation gain) (–)
	Accelerator heater (W)		Mass-weighted specific heat (J/kg.K)
	On set point (°C)		Convective coupling to air (W/K)
	Off set point (°C)		Radiative coupling to opposite wall (W/K)
			Radiative coupling to mounting wall (W/K)
			P proportional band (K)
			I integral action time constant (s)

The appraisal possibilities supported by the suite of models are many. Figure 6.47 shows a comparison between an electronic on/off thermostat (time proportional modulating) and an electronic modulating PI thermostat over a typical winter week.

The room temperatures are maintained at approximately similar levels in both cases whilst the boiler flow temperatures are also similar. The electronic PI controller directly modulates the boiler output until the demand drops below the boiler's lower limit of modulation. The boiler is then cycled on/off at the lower limit-firing rate. During a call for DHW, the electronic PI-controlled room temperature drops as this system cannot supply DHW and space heating simultaneously.

Another useful comparison is the impact of controller type on annual energy consumption as summarised in the example of Table 6.17.

The models also supported an assessment of boiler efficiency under various loads as indicated in Figure 6.48.

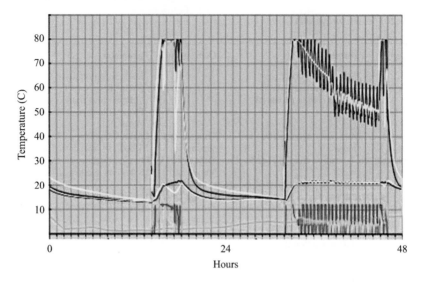

Figure 6.47   Comparison of controllers

Table 6.17   Annual heating energy consumption (MWh), condensing boiler

Controller	Space heating	DHW	Total	Heating saving (relative to TRV only)	Total saving (relative to TRV only)
TRVs only	13.1	2.5	15.6	–	–
Mechanical thermostat	10.9	2.5	13.4	17%	14%
TPM controller	11.2	2.6	13.8	15%	12%
Modulating PI controller	10.9	2.5	13.4	17%	14%

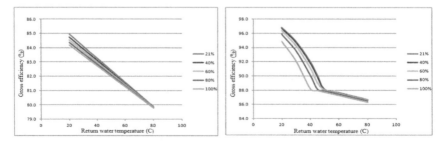

Figure 6.48   Efficiency of non-condensing (left) and condensing boiler

This level of heating systems modelling is most suited to the design and proving of heating system components before introduction to the mass market. It is also able to produce lower-level models for routine incorporation in downstream simulations corresponding to options appraisal directed at large housing stocks as summarised in Section 8.1.

6.8 Multi-zone control

Distributed control of buildings has a high potential to reduce energy use because different zones can be regulated separately and to different levels depending on usage. This section describes the application of BPS+ to assess multi-zone control of domestic heating[23], a market hitherto neglected in this regard. Only in recent years have even the simplest heating controls become a standard for new heating systems in the United Kingdom, the minimum installation requiring a single time and temperature control zone and TRVs on all radiators except in the rooms where thermostats are located. Even this basic level of control does not exist in 70% of UK homes. Recent developments in wireless TRVs offer a practical solution to this dilemma. What follows highlights the ability to automate simulation runs to evaluate the effectiveness of multi-zone control under a range of weather conditions and occupant behaviours.

The technology underpinning multi-zone control allows the thermostatic radiator valve head (the part that actuates the valve) to be replaced by a wirelessly controlled motorised actuator. By this means, every room can be controlled independently so that heating can be turned off in those rooms not in use during parts of the day, or temperatures increased in a room without affecting the heat supply to the rest of the building. A central control unit receives the demands from all radiators and switches the boiler on or off as required. This is a relatively low cost but effective retrofit option because it requires no pipework alteration and minimal additional wiring. By this means, energy savings should be achievable, compared to single location and predetermined time/temperature control of the heating system.

The premise under investigation in the study (Cockroft *et al.* 2017) was that multi-zone control offers a means to save energy in existing properties, as an alternative or in addition to conventional upgrade solutions. The purpose was to quantify the potential energy savings of multi-zone control for different house types, occupancy patterns, and locations. The study utilised ESP-r and, to ensure credibility, carried out an initial validation check using published data from a monitored site. Whilst detailed dwelling models were used, the plant components (thermostats, radiators, pipework, and boiler) were not modelled explicitly as it was expected that the type of heating control, radiator, and boiler would not materially affect the results in terms of temperatures achieved. Conversely, the airflow between zones was considered important, so an airflow network was included.

---

[23]The approach is equally applicable to commercial and public buildings.

Multi-zone control in a domestic property achieves energy savings because radiators in rooms that would otherwise be heated can be turned off or adjusted to reduce heat output. The achievable energy saving will depend on the extent to which the temperatures in such rooms fall before heating is again required. This, in turn, is dependent on room location, duration of the off period (or reduced temperature set point), thermal exchange with connected rooms, the overall insulation level of the property, internal gains, solar gains, and external temperature. For example, a room in a semi-detached property, with heated rooms on all sides and below, will not cool down as rapidly as a corner room in a bungalow, and therefore will deliver a lower energy saving if turned off for short periods. Turning the heating off in a room for a longer period will increase the obtainable energy saving per unit of time because a lower average temperature, and therefore lower heat loss to the external environment, will be experienced. A room with an open door to a neighbouring zone will gain heat from that zone as long as the temperature is lower, and this will reduce the energy saving in that room, and increase the energy required to maintain the temperature in the connected zone. The potential for energy saving will also be affected by the external fabric insulation levels. A well-insulated building will lose heat less rapidly than a room with no heating so the potential energy savings will be low compared with an older, unimproved property. A similar property in a cooler climate would also be expected to achieve greater energy savings (though not necessarily in proportion to its total energy consumption). The study therefore included variations in parameters that would allow observation of these various effects on the savings due to multi-zone control.

To highlight the effect of different zone interactions, two archetype dwellings were modelled: a typical UK semi-detached property on two floors, and a typical detached bungalow. Four occupancy patterns were imposed:

- 4-person family with two young children (YF);
- 4-person family with two teenagers (FT);
- elderly couple (EC); and
- young couple (YC).

Simulations were carried out for each dwelling with standard heating controls (temperature control in each room but only one temporal setting for the entire house), and with a zoned heating control system with independent time and temperature control in each room. These were termed the non-zoned and zoned control strategies, respectively. In all cases, heat was delivered via radiators with an equal radiant/convective split assumed. Radiators were sized at 1.75 kW per zone in the semi-detached dwelling, and 2 kW per zone in the bungalow, to ensure a rapid heat up to the set-point temperature (maximum 30 minutes), which were then maintained by an idealised control without on/off fluctuations or load-influenced deviations from set point. Thermostats in both the zoned and non-zoned cases were set to sense the spatially averaged resultant temperature as a proxy for comfort.

To assess the effect of insulation and airtightness on energy consumption, construction thermal properties were adjusted to create models that met minimum

Table 6.18   Dwelling age bands and construction properties

Band	Year range	Wall U-value (W/m².K)	Roof U-value (W/m².K)	Ventilation (ACH)
C	1930–49	2.0	2.0	0.5
F	1976–82	1.0	1.2	0.4
I	1996–2002	0.45	0.43	0.4
K	2007–on	0.3	0.3	0.3

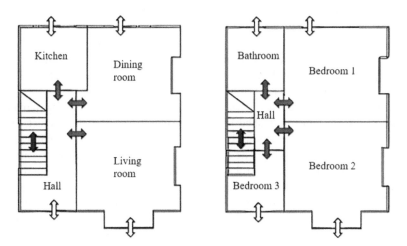

Figure 6.49   Airflow paths for the semi-detached house

building standards corresponding to four age categories as defined by the UK government's Standard Assessment Procedure (SAP 2023). The age bands and the corresponding fabric properties are listed in Table 6.18.

Two locations were considered corresponding to Heathrow (51.4680°N, 0.4551°W) and Glasgow (55.8700°N, 4.4345°W) airports and test reference year weather data for these locations applied (comprising hourly values of dry bulb temperature, direct normal radiation, diffuse horizontal radiation, wind speed, wind direction, and relative humidity).

The effects of wind and buoyancy on air movement and the air exchange between zones are important factors affecting energy consumption in a zoned house. The airflow paths included in the model of the semi-detached house are depicted in Figure 6.49.

For clarity, connections between internal zones representing cracks in solid constructions are not shown; these connections generally represent insignificant flows compared to the connections shown here. Bi-directional airflows may occur through doorways due to the combined effects of pressure and temperature differences (blue arrows). Windows were modelled as openable, and each was subject to proportional control such that the opening area was linearly proportional to the dry bulb temperature in the associated zone between 25 °C and 28 °C (white arrows). Windows were closed at and below 25 °C, and fully open (with an area of 2.5 m²) at and above 28 °C. This was intended to model occupant behaviour in preventing rooms from overheating. Two parallel, uncontrolled openings were included to allow air mixing on the staircase (red arrows). A similar flow network was constructed for the bungalow model. In the semi-detached house, air can also flow between the ground and upper levels, mainly driven by buoyancy effects. To assess the impact of airflow on the performance of the houses, four levels of door opening were modelled, corresponding to average opening areas of 0%, 10%, 50%, and 100% of full open door area applied to each room in each house.

To build confidence in the model, the work of Beizaee *et al.* (2015) was identified as significant because it reported simultaneous measurements of energy consumption in two semi-detached properties, one with conventional single-zone heating control, and one with a multi-zone system. Extrapolating the results to the UK climate, demonstrated that zonal control could reduce space heating demand by around 12% for the un-refurbished 1930s houses being tested in this project. Details of the two properties were acquired and used to define an ESP-r model to compare predictions with observations. The approach adopted was to calibrate a variant of the model with non-zoned control by adjusting the fabric thermal properties and leakage distribution to achieve reasonable agreement in terms of energy con-sumption and temperature statistics. The control parameters were then changed to represent the multi-zone controller. Simulations of the calibrated model were car-ried out for the periods for which the houses were monitored, using con-temporaneous weather data and internal gain profiles. The predicted savings resulting from the two control options were compared to the measured savings in the side-by-side experiment.

Figure 6.50 compares the differences in average temperatures during the heating periods in each zone, with and without zone control. In all cases, the differences are negative, due to the shorter heating periods in each zone with zoned control.

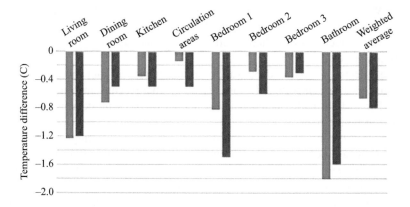

Figure 6.50   Temperature difference between predicted (blue) and measured cases

Table 6.19   Measured and predicted average energy consumption (kWh/d)

	Not zoned	**Zoned**	**Saving (%)**
Beizaee *et al.* measurement	62.4	53.6	14.1
ESP-r estimate	62.6	55.2	11.8
Estimate as % of measured	100.3	103.0	83.7

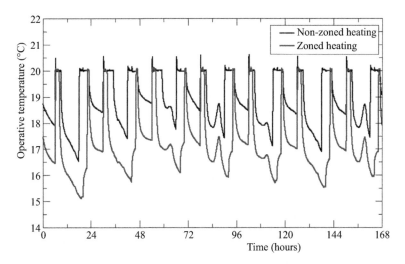

Figure 6.51   Operative temperatures, winter period, bungalow

Table 6.19 summarises the agreement obtained between predicted and measured energy consumption over the monitored period. As an initial indicator of annual saving potential, the ESP-r estimates equate to an average consumption of 62.6 kWh/day with zoning, a saving of 11.8% compared with the non-zoning case.

Annual simulations were then carried out for each house type, for each of four age bands, for each location (Glasgow and London), for each of four occupancy types, and for four door-opening percentages. Thus, 256 annual simulations were executed, each with a 24-day pre-simulation start-up period to eliminate initialisation assumptions and imposition of a 5-minute time steps to allow reasonable close control. The runs were automated using shell scripts in a Linux environment (see Section 4.5). Results comparing internal operative temperatures across all simulation runs were collated for a typical one-week winter period, whilst results comparing energy consumption were annual.

Figure 6.51 shows the operative temperatures in a bungalow zone for YC occupancy, age band K, Heathrow weather, doors closed, with and without zoning, and over a 7-day winter period. Providing heat only during the shorter occupied periods (zoned heating), resulted in lower temperatures during the unoccupied periods.

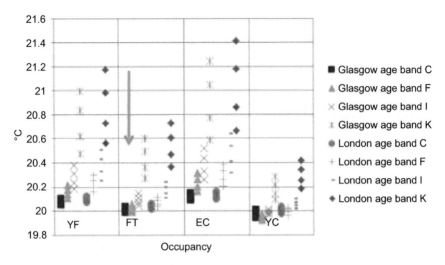

Figure 6.52   Effect of occupancy on living room temperature (semi-detached, winter, non-zoned)

Figure 6.52 shows the effect of occupancy (YF, FT, EC and YC as described above) on the average living room temperature during occupied hours in the semi-detached house for the non-zoned case. For each occupancy type, the average temperature is shown as a function of location and construction age band. For each case, a range of four values is plotted corresponding to the four door-opening percentages (0%, 10%, 50%, and 100%).

The arrow indicates the direction of the increasing door-opening area. When doors are fully open, air can pass freely between zones so heated zones will lose heat to unheated zones. This will tend to increase the heat demand of the heated zones and, because they will take longer to reach their set-point temperature at the start of each occupied period, will reduce their average temperature. Therefore, the lowest temperatures correspond to the 100% door-open case, and the highest temperatures to the door closed case. However, the range in each case is small with a maximum of 0.8 °C for the EC case. Two observations were reported as follows.

- The temperatures in the YC case are generally slightly lower due to reduced internal heat gains throughout the day and, to some extent, set-point temperatures are not being achieved due to shorter heating-on periods.
- There is a tendency towards higher temperatures as insulation standards improve and for the milder London climate. The highest temperatures are observed in the YF and EC cases, in the milder London climate, and for the best-insulated K age band houses. This is due to the longer heating periods and therefore shorter heating start-up times. The effect of door-opening percentage is greatest in these cases due to the larger impact of inter-zone heat transfer.

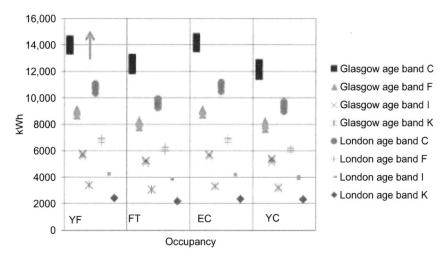

Figure 6.53   Effect of occupancy on change in living room temperature (semi-detached, winter, zoned)

Despite these variations, the overall spread of average temperature is 1.5 °C. Similar results were obtained for other zones, in both the semi-detached and bungalow cases, with a similar overall spread of average temperatures.

On comparing the results from the simulations with zoned controls, it is important that average temperatures during the occupied periods in each zone be maintained close to the set point. This is necessary to ensure a fair comparison of the energy performance of the non-zoned and zoned cases. Figure 6.53 shows the effect of occupancy on the average living room temperature in the semi-detached house during occupied hours for the zoned cases, minus that for the non-zoned cases.

For each occupancy type, the difference in average temperature is shown as a function of weather and construction age band. For each case, a range of four values is plotted corresponding to the four door-opening percentages. The arrow indicates the direction of the increasing door-opening area. The smallest temperature differences correspond to the 100% door-open case and the largest differences to the 0% door-open case. This reflects the lower inter-zonal heat transfers when the doors are closed. These differences, and the impact of door-opening percentage, are greater for the higher insulated and milder London climate cases due to the larger influence of inter-zonal heat transfer. However, the range in each case is small with a maximum reduction in average temperature of less than 0.4 °C.

Similar results were obtained for other zones, in both the semi-detached and bungalow cases, with a similar overall spread of temperature differences being obtained. This raised confidence that the energy consumption comparison is fair.

The annual energy consumptions for the semi-detached house, non-zoned, are shown in Figure 6.54.

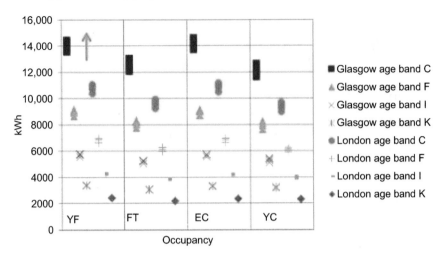

Figure 6.54   Annual energy consumption (semi-detached, non-zoned)

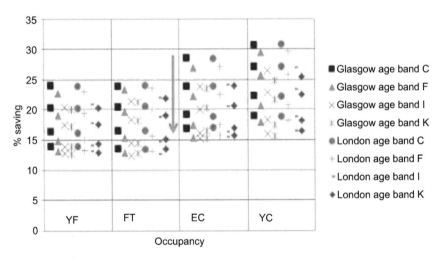

Figure 6.55   Energy savings (semi-detached, zoned)

The pattern is as expected, with consumption reducing as insulation levels improve, and less so for London than Glasgow. The consumptions for the FT and YC occupancy cases are slightly less than the YF and EC occupancy cases due to longer periods of absence. The effect of door opening is small as all zones are heated together, minimising the effect of inter-zone heat transfer. The arrow indicates the direction of the increasing door-opening area. Consumptions are slightly higher with 100% door opening due to greater inter-zonal heat transfer. The annual energy savings for the semi-detached house due to zoning are shown in Figure 6.55.

Savings as a percentage of the non-zoned energy consumption are hardly affected by climate or age band. EC and YC occupancy cases exhibit larger savings than the YF and FT cases. The differences are due to complex interacting factors, one of which is that both EC and YC 'couple' occupants have shorter evening bedroom heating periods than the YF and FT 'family' occupants.

By far the biggest influence on saving is the door-opening percentage (the arrow indicates the direction of increasing opening area), with 100% door opening almost halving the saving obtainable if all doors are closed. This indicates that an effective zoning strategy relies on isolating zones as far as possible. Although the house age band has only a minor effect on savings potential as a percentage, there is some tendency to reduce savings in better-insulated houses. Of course, the absolute savings will be lower, as the baseline non-zoned energy consumption is lower. Overall, the savings obtainable across all simulations range from 12% to 31%, with an average of around 19%.

A similar pattern of energy consumption for the bungalow was observed. Overall consumptions are slightly higher than for the semi-detached house, due to the lack of a party wall and a less compact layout. Again, the effect of door-opening percentage is minor.

Overall, the savings obtainable in the bungalow across all simulations range from 8% to 37%, with an average of around 20%. This is a wider spread of savings than was observed in the semi-detached dwelling. The main difference in savings between the dwelling types is due to the reduced inter-zone heat transfer from living, dining, and kitchen zones to the bedroom zones in the bungalow, compared to the semi-detached type where buoyancy effects in the stairway, and upwards heat transfer through ceilings are present. However, the overall increase in savings is small, being greatest for the EC and YC occupancy types with doors closed.

Table 6.20 summarises the energy savings for each door opening and each occupancy type, these being the most significant influences. The figures are averages for the two climates and four age bands.

In conclusion, the simulation results showed that significant energy savings are possible by adopting a multi-zone control strategy, whereby temperature and time-based control is applied independently in each room, compared with a non-zoned strategy, whereby all rooms follow a single time/temperature profile.

Table 6.20 Summary of energy savings relative to not zoned (%)

	Door opening (%)	Young family (YF)	Family with teenagers (FT)	Elderly couple (EC)	Young couple (YC)
Semi-detached	0	21.9	22.7	25.8	28.1
	10	18.6	19.4	21.8	24.5
	50	14.8	15.4	17.4	19.8
	100	13.2	13.0	15.8	17.3
Bungalow	0	23.0	25.0	27.3	33.3
	10	18.9	20.8	22.1	28.9
	50	13.8	15.4	15.9	22.5
	100	11.2	12.2	12.6	18.1

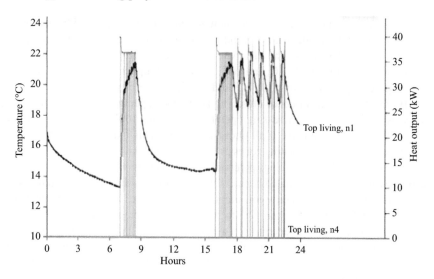

Figure 6.56   Operative temperature and boiler output

Given the range of parameters studied, a typical average saving of 20% was deemed possible across a range of house types, ages, and occupancy patterns. These savings are in line with the limited previously reported studies on the benefits of zonal control. This compares favourably with other demand reduction measures such as wall insulation or double glazing and in many cases can be applied to an existing central heating system at low cost and disruption. It is an attractive option where property architecture and conservation considerations make other options untenable.

The study demonstrates a typical approach to BPS+ usage: carry out a limited comparison with measured data (or some other credible data), determine the modelling detail required to replicate important performance indicators and expand the scope of the simulations as required by the topic being addressed. Whilst such studies can be undertaken on a project-by-project basis, a more productive option might be for professional bodies to commission generic studies and disseminate the outcomes thereafter.

The underlying message is that BPS+ is a useful tool to investigate innovative control concepts at the concept stage. This was further demonstrated with the chop-cloc[24] retrofit heating system controller that turns the heating off for a period each hour (as set by the occupant) to take advantage of released heat from the building fabric and contents. ESP-r was used to evaluate the impact on thermal comfort and energy saving in typical dwellings under various 'pause' intervals each hour (Cockroft and Samuel 2010). Figure 6.56 is an example output showing space operative temperature (°C) and boiler heat output (W) for a 30-minute pause per hour.

---

[24]https://nea.org.uk/publications/independent-evaluation-of-the-chop-cloc-product-cp941/

The study indicated a 13% energy saving against a modest drop in the average operative temperature over occupied periods. It also indicated a reduction in boiler switching because boiler control is prescribed rather than controlled by a thermostat. Positive results from the project encouraged the client to proceed to a more expensive field trial and demonstration stage.

The results of this section highlight the deficiency of compliance-based software that places a premium on abstracting a building into a few zones (e.g., living and non-living spaces, or a single zone in the case of PassivHaus) when the temperature variation between rooms is a dominant factor. This is a case where the remit of compliance limits the scope of enquiry and constrains any need to progress beyond the performance goals of the standard. The incentive to explore only the limited palette of options expressible in a compliance tool (because those are the only ones that deliver points) is sometimes perverse. Whether such incentives are more egregious than simplifying the complexity of the built environment to fit within a compliance framework is a discussion for the wider community. When it comes to control, the elephant in the room is time. All components have thermal inertia. The firing of a boiler does not instantaneously heat a radiator, or the stopping of a pump reduce heat emission to a room with no delay. Control instructions take time to change conditions in a space. Section 5.1.4 demonstrates the impact of neglecting such matters in the case of a thermostat.

6.9   Simulation-assisted control

This section describes the linking of BPS+ to a real control system to provide a look-ahead capability to give advice on the best course of action. The approach essentially implants a model of the process being controlled within the control system. It has been postulated that predictive control will result in the more efficient operation of building energy management systems. This prospect was explored in a project (Clarke *et al.* 2002) that used ESP-r as part of the control system to evaluate possible actions and return the one that matched the given criteria. Simulation-assisted control is likely to be of most use in the following circumstances.

- When significant look-ahead times are involved (hours rather than minutes).
- For high-level supervisory control (e.g., load shedding, where several alternatives and their implications for occupant comfort need to be evaluated).
- Where interaction is high (e.g., blinds/lighting/cooling).
- Where there are large variations in occupancy known in advance.

Table 6.21 lists the many possible targets and outcomes for a smart approach to control based on simulation look-ahead.

Figure 6.57 summarises the project's approach to evaluating the efficacy and impact of embedding BPS+ within a building energy management system (BEMS).

As with traditional BEMS, inputs are obtained from weather and building state sensors and an algorithm decides on the appropriate control action at a suitable frequency.

Table 6.21   Possible targets of simulation-assisted control

Application	Controlled component	Optimised output
Optimum start/stop	Heating/cooling system	Start/stop time
Nighttime cooling	Fans	Hours of operation
Set-back temperature	Heating system	Set-back temperature
Boiler sequencing	Boilers	Heating system efficiency
Load shedding	Heating system	Priority for heating
Combined heat and power	CHP Engine	Hours of operation
District heating	Heating system	Heat demand forecast
Underfloor heating	Heating system	Hours of operation
Mixed mode ventilation	Fans	Start/stop time
Charging of ice storage	Refrigeration	Hours of operation
Night-operated heat pumps	Water pump/compressor	Start time
Optimum control mode	Various	Control mode selection

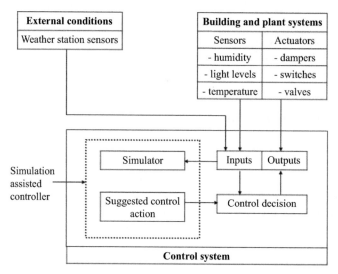

Figure 6.57   Embedding a BPS+ capability in BEMS

The new element added here is the BPS+ simulator that evaluates and rates the short-term impact of possible control actions and selects the most appropriate one. Within the project, the BEMS system was emulated using LabView's[25] Supervisory Control and Data Acquisition (SCADA) capability, with ESP-r used as the support simulator. The former is an environment that can produce virtual instruments on a computer display. The combination was then tested by application to a test room.

Within ESP-r, a control system comprises a set of open or closed control loops acting together or individually. Each loop comprises a sensor linked to an actuator

[25]https://ni.com/en.html

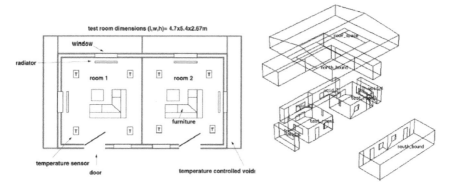

Figure 6.58   The Honeywell test rig and ESP-r exploded model

via an algorithm. ESP-r offers a library of sensors, actuators, and algorithms enabling control ranging from 'ideal', by which set points are imposed, to 'realistic' where sensor and actuator characteristics (offset, response, lag time, etc.) are modelled.

At each control decision point, the simulator receives information from LabView and looks ahead to determine the impact of the control action being considered. For example, in the case of optimum start, several simulations spanning the period around occupancy commencement are undertaken with the start time incremented. The results are then judged in terms of closeness to the required zone temperature at the start of the occupied period.

Figure 6.58 shows the test rig as established at Honeywell's Newhouse site in Scotland: two realistically dimensioned rooms are surrounded by temperature-controllable voids. LabVIEW was linked to the existing data acquisition system and an ESP-r model was established and calibrated.

The main experiment involved setting the optimum start time for the room 1 heating system. Data collection was at 1-minute intervals. At experiment commencement, the test rooms were allowed to free-float for 24 hours. The voids remained unconditioned throughout the experiment, whilst room 2 was maintained at 24 °C. The simulation controller was set to determine the switch-on time required to bring room 1 to a temperature of 25 °C with a nominal 1.2 kW heat input. Figure 6.59 (left) summarises the experiment outcome.

In the preceding 24 hours, the room temperature floated at around 21 °C. Given a 25 °C set point and target of 11:00, ESP-r predicted a heating system switch-on time of 10:20. Note that the room temperature was not at exactly 25 °C at this time as the simulated temperature was compared to the set point with a tolerance of ±0.5 °C. When the test room heating was switched on, the room reached 25 °C at 11:06. The actual room temperature coincided with the ESP-r room temperature prediction at 11:02. ESP-r slightly over-predicted the test room temperature, with prediction leading the monitored temperature. Measured and simulated temperatures coincided with a temporal error of 5%, a maximum deviation of 1 °C,

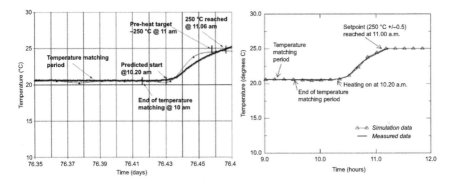

Figure 6.59   Test results before and after model refinement

and a 6-minute delay in reaching the set point. To address this discrepancy, the radiator model was refined, resulting in the close match shown in Figure 6.59 (right).

The ESP-r simulation time for six exploratory simulations per control assessment was of the order of 1–2 minutes on a (then) low-end Pentium PC. Given the advances in computer power since then it may be concluded that simulation-assisted control is feasible where the input model is relatively simple. Where not, the simulator could be brought into play regularly but infrequently, e.g., twice a day.

There is a general trend for manufacturers to embed control logic in their devices and the future may see machine learning whereby the relationship between sensors and control signals are highly mutable. This suggests a role for BPS+ in the design and operation of future control devices. It may even be possible to implement BPS+ within a black box controller (e.g., see Section 9.3.1).

6.10   Electrification of heating

The project described here demonstrates the use of BPS+ to evaluate the replacement of gas boilers within wet central heating systems with a connection to a heat pump – either directly or indirectly via a district-heating scheme. Because a heat pump will typically deliver a lower supply temperature, it is necessary to increase the size of radiators to maintain the same level of heat output or extend the heat input period. Alternatively, or in addition, it is possible to reduce the heat demand through the deployment of energy efficiency measures. Such actions are expensive and disruptive and should be avoided if possible. BPS+ was used in a feasibility study to confirm the best course of action in a specific case.

The assessment commenced with the identification of representative dwellings within a targeted estate. ESP-r input models were created and used to determine heat demand profiles for a defined level of comfort. The peak heat output required from each radiator was identified and simulations were rerun but with the maximum output restricted to correspond to the lower supply temperature associated with the district-heating scheme. This restriction was estimated from

$$P = P_{50} \left[\frac{(t_i - t_r)}{\ln\left(\frac{(t_i-t_a)}{(t_r-t_a)} 49.32\right)} \right]^{1.33}$$

where P is the new radiator heat output (W), P_{50} the heat output at a 50 °C radiator to room temperature difference (W), t_i the radiator inlet temperature (°C), t_r the return temperature (°C), t_a the room temperature (°C), and 1.33 an empirical exponent for a panel radiator.

Thus for a gas boiler P = 1139 W (when P_{50} = 1000 W, t_i = 70 °C, t_r = 50 °C, and t_a = 20 °C), whilst for a heat pump P is 50% less (P_{50} = 1000 W, t_i = 60 °C, t_r = 40 °C, t_a = 20 °C). The heat pump simulation will therefore result in lower comfort levels (i.e., unmet hours) at times when the heat demand is greater than the radiator can supply. The efficacy of heat pump use is then expressed in terms of the reduced heating energy requirement and loss of environmental control as measured by the total number of hours that the heating system is unable to meet the defined set-point temperatures during the heating season. Further simulations will establish to what extent adjustments to control regimes (e.g., an earlier start-up) and/or implementing retrofit measures (e.g., an insulation upgrade) can recover the situation. The attractiveness of the approach is that it does not require explicit modelling of the heat pump or the district-heating infrastructure.

In one study, this approach was applied to dwelling types comprising a community in Glasgow (Clarke and McGhee 2017). Simulations were undertaken for dwellings in their present state when serviced by a wet central heating (WCH) system and when reconfigured to connect to a heat pump, district-heating (DH) scheme delivering water at 60 °C. In the latter case, the WCH boiler is replaced by a heat exchanger connected to the low-temperature network resulting in a reduced radiator heat output due to the lower supply temperature. The performance of the latter system was then assessed relative to the former. The approach assumes that the heating system radiators are not upsized as part of the DH connection. Further simulations then focused on the ability of dwelling upgrades and control system timing adjustments to recover the lost comfort amenity associated with DH connection. Figure 6.60 shows two of the dwelling types considered in the study: a

Figure 6.60 Example of dwellings being considered for DH connection

2-bed apartment (68 m^2) located in a 4-in-a-block cottage flat (left) and a 2-bed apartment (60 m^2) located in a 4-storey tenement flat.

Table 6.22 summarises the heating system performance for one of the dwellings when serviced by the WCH system and the alternative DH connection with a 50% reduction in radiator heat output as indicated above.

As can be seen, a modest 2.7% reduction in heating energy requirement in the DH case is counterbalanced by increased hours when the heating system cannot maintain the indoor comfort levels of the WCH case. Table 6.23 gives the corresponding results for the same dwelling after the upgrade.

The effect of the upgrade is to (a) reduce the heating energy requirement by 59.8% (for the WCH case), (b) reduce to insignificance the difference between the heating energy required in the WCH and DH cases, and (c) marginally reduce the number of unmet hours for the DH case. Table 6.24 gives results corresponding to a DH connection and an extended start-up period.

Relative to the corresponding results in Table 6.22, this represents a 10.3% increase in annual heating energy requirements with a modest reduction in unmet hours. Undertaking simulation studies such as this is a cost-effective way to support rational decision-making that is likely to lead to a successful outcome when deployed.

Table 6.22   Heating system performance

Case	Heating energy (kWh/y)	Hours required	Unmet hours
WCH			
Living	1853	4074	0
Non-living	5088	5417	0
All	6942	–	–
DH			
Living	1799	4237	725
Non-living	4954	5433	927
All	6754	–	–

Table 6.23   Post-upgrade performance

Case	Heating energy (kWh/y)	Hours required	Unmet hours
WCH			
Living	775	2556	0
Non-living	2014	3367	0
All	2788		–
DH			
Living	740	2597	611
Non-living	2002	3387	396
All	2742		–

Table 6.24 DH + extended preheat time

Case	Heating energy (kWh)	Hours required	Unmet hours
DH			
Living	2216.3	5000	464
Non-living	5233.4	5722	842
All	7449.7		–

In another project, ESP-r results were used to develop a decision-support tool to enable an HVAC systems manufacturer to compare the economic and environmental benefits of an air-source heat pump with other technologies and fuels. The tool was implemented as a spreadsheet model incorporating building performance data derived from the simulation of different house types under conditions of continuous and intermittent occupancy. It enabled the effects of equipment efficiencies, fuel tariffs, and CO_2 emission assumptions to be immediately assessed.

The spreadsheet implemented hourly time-step calculations based on energy demand data derived from annual simulations for each combination of house type and heating schedule. To restrict the size of the spreadsheet, a two-dimensional frequency distribution method was implemented.

The performance of the heat pump in meeting the heat demand was compared to that of four alternative heat sources: a gas-fired condensing boiler, an oil-fired boiler, a coal-fired boiler, and direct electric heating. The fuel consumption, fuel cost, and CO_2 emissions for each of the four technologies for the six house types and occupancy scenarios were determined. Figure 6.61 shows the annual costs associated with meeting the heating requirement for each technology type and house type ([*] indicates continuous occupancy, else intermittent).

The results indicated that the deployment of air-source heat pumps would yield energy costs comparable with gas-fired condensing or oil-fired boilers and considerably lower than direct electric heating systems. CO_2 emissions for the heat pump were 40%–70% less than for the other fuel types.

6.11 Heat pump with thermal storage

This section describes projects that investigated the utilisation of building-integrated thermal energy storage (TES) when deployed using a variety of materials and at different scales. Kelly *et al.* (2017) have summarised the capabilities and potential applications of this technology, which provides a mechanism to bridge the temporal gap between supply and demand.

To match heat pump output to a varying heat load, a TES-based heat exchanger can be deployed to replace a boiler. The TES is then charged by a heat pump operating at off-peak times under favourable tariff and in day-ahead charge mode.

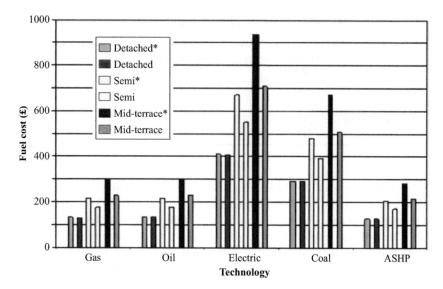

Figure 6.61   Annual fuel costs (£)

Mukhtar (2024), for example, used ESP-r and Fluent in a proof-of-concept study for such a configuration. Using information gleaned from the Scottish House Conditions Survey[26], representative dwellings were identified and simulated by ESP-r to emulate the case where an external agent delivers a next-day heat demand estimate for use to determine the day-ahead charge. The simulation results were also used to size a dwelling's phase change material (PCM) based TES cylinder comprising PCM elements as shown in Figure 6.62.

Within the project, the storage elements were converted to a Fluent[27] input model for use to study charge and discharge behaviour via flow rate regulation of a heat transfer fluid (water) in the latter case. Figure 6.63 shows time sequences of the temperature changes throughout a single PCM element during the charging and discharging process.

The conclusion was that a TES could be charged by a heat pump and then discharged in a manner that is an approximate match to a dwelling's next-day heat demand profile. Allison *et al.* (2017) investigated a variant of the above system whereby the heat pump pre-charges an underfloor heating system under the direction of a smart controller.

Kelly (2013) conducted a similar heat pump-based thermal storage project using ESP-r but with two PCM stores as shown in Figure 6.64 intended to allow ease of retrofit.

Based on an extensive parametric sensitivity analysis, it was concluded that for all building types and climates a PCM heat buffer of this type could facilitate heat

[26]https://gov.scot/collections/scottish-house-condition-survey/
[27]https://ansys.com/en-gb/products/fluids/ansys-fluent/

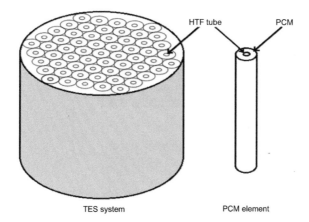

Figure 6.62   TES with PCM elements

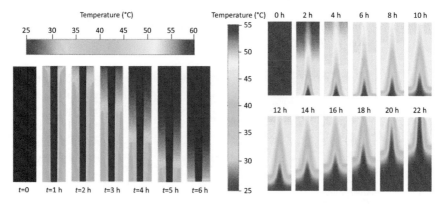

Figure 6.63   PCM element temperature change, charging (left) and discharging (credit: Mukhtar 2024)

pump load shifting to off-peak periods whilst still delivering comfort and hot water to the occupant.

Another approach to building-integrated TES is to use the fabric or foundation of a building as a heat store. This approach was researched within an Engineering and Physical Science Research Council funded research project (Allison *et al.* 2018) that used ESP-r to examine active and passive schemes (de Castro *et al.* 2018). Figure 6.65 shows one possibility: embedding piping in a well-insulated concrete foundation slab (in this case at the BRE Innovation Park in Scotland, which is now dismantled).

Options for daily, weekly, and seasonal storage were examined for a variety of dwelling types and occupancies. Diurnal storage provides support as described in the previous PCM-based example. Weekly storage, if grid connected, offers the

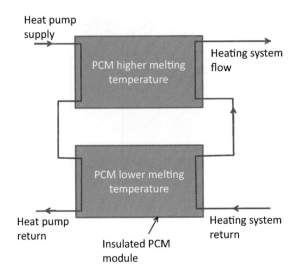

Figure 6.64   Integration of a PCM thermal store with a heating circuit (credit: Kelly 2013)

Figure 6.65   Charge/discharge piping within an insulated concrete slab

possibility to absorb surplus electricity during periods of excess wind power generation. Long-term storage would limit the large seasonal variation in demand that could undermine attempts to electrify heating at a large scale. The project method was as follows.

- ESP-r was used to determine the space heating demand profiles associated with key housing archetypes.

Table 6.25　Diurnal storage volumes for a detached house

Insulation	occupancy	Storage material (m^3)			
		Water	**Concrete**	**Magnetite**	**Salt**
Average	High	1.90	3.75	0.06	0.38
PH*	Low	1.93	3.83	0.06	0.39
Average	High	0.66	1.31	0.02	0.1
PH	Low	0.50	0.99	0.02	0.1

*Passive house standard

- Hot water demand profiles were generated using the OccDem stochastic occupancy predictor (Flett 2017).
- The combined heat demand profiles were integrated over the three seasonal periods (winter, summer, and transitional) to determine the required storage capacity.
- These capacities were converted to storage volumes for four materials (water, concrete, magnetite brick, and hydrated salt).
- The storage volumes where discretised and connected to an ESP-r plant model representing the storage charge and discharge processes.
- Finally, an assessment was made of the practicality of integrating such storage volumes into housing.

Table 6.25 shows an extract of the diurnal storage volumes for a detached dwelling when located in the northeast region of the United Kingdom.

The project conclusion, based on the combination of cases studied, was that only diurnal storage is practical because store volumes are then an acceptable fraction of the host building.

6.12　Daylight utilisation

This project considered the merit of complex fenestration linked to novel lighting control. It involved the use of thermal and lighting tools and the production of IPVs to assist with a comparison of overall performance.

Several technologies exist that can be used to improve the performance of a building's fenestration, including:

- evacuated or argon-filled cavities to reduce heat loss by blocking longwave radiation heat transfer;
- surface thin films to reflect incident radiation in the longwave or shortwave part of the solar spectrum;
- prismatic glass to redirect daylight to maximise its capture whilst eliminating glare;

Figure 6.66   The complex glazing arrangement

- adaptive coatings (thermochromic, photochromic, and electrochromic) that modify shortwave transmission as a function of temperature, solar irradiance, or an electrical control signal, respectively; and
- light shelves to enhance sky vault access.

Such options introduce technical complexity that places an added burden on the model definition and simulation procedures.

The project outlined here (Clarke *et al*. 1997) involved the application of ESP-r and Radiance to a school building located at Modane in France that embodied a complex glazing arrangement as shown in Figure 6.66. The building featured tilted windows with external light shelves, an internal window connecting peripheral classrooms with a central atrium, and a lighting system with timed control. The project was one of several case studies undertaken as part of the EC-funded Daylight Europe project (Kristensen 1996) intended to demonstrate innovative approaches to high-quality, low-energy building design.

A high-resolution ESP-r model was created based on architectural drawings and measured internal surface reflectance (Fontoynont 2017). To accommodate the daylighting focus of the project, the glazing layers were represented explicitly so that the intra-pane temperature and radiation processes could be studied. Table 6.26 summarises the thermal and optical characteristics of the window system in terms of total visible transmittance (TVT), direct solar transmittance (DST), and solar heat

Table 6.26   Optical and thermal properties of glazing

Glazing	TVT (–)	DST (–)	SHG (–)	U (W/m².K)
Double	0.76	0.61	0.71	2.75
Single	0.89	0.82	0.86	5.40
Translucent	0.65			
Surface reflectivity (%)				
Floor	40	Ceiling	89	
Wall	79	Roof	28	
Window frame	21	Ext. wall	65	

gain coefficient (SHG), all at normal incidence angle, and the overall thermal transmittance (U)[28]. The table also lists the opaque surface diffuse reflectivity values.

During the occupied period, the classroom lighting is automatically switched off every hour and, if required, must be turned on again by the occupants. Given the difficulties associated with interpreting results corresponding to stochastic processes, the explicit modelling of occupant behaviour was not attempted. Instead, at the start of each hourly teaching period, a controller determines whether the desktop illumination level is less than 300 lux. If it is, the lights are activated as a proxy of occupant behaviour according to best practice guidelines (CIBSE 1994).

The heating set-point temperature was 19 °C during weekdays when the rooms are occupied and 16 °C; otherwise, the set point was 8 °C at weekends. The switch from 19 °C to 16 °C occurs every hour and, as with lighting, has to be reversed intentionally by the pupils or teacher. During weekends and holidays, the heating set point was 8 °C. An MVHR system supplied 4,500 m³/h of fresh air to the classrooms during occupied periods with a 60% heat recovery efficiency. Model calibration was carried out by applying judicious adjustments to the input model until acceptable agreement was obtained with measured classroom daylight levels at various distances from the window wall. To obtain the agreement shown in Figure 6.67, it was necessary to include the obscuration effects of the sky vault by the surrounding mountains.

A reference model was established by removing the daylight-enhancing features (tilted glazing, light shelves, and internal windows between the classrooms and atrium) and changing the lighting control strategy to one with the lighting and heating on during occupied periods. In this way, the impact of the daylight capture features was quantified, with the results for the two models reported in IPV format as shown in Figure 6.68.

A comparison of the performance indicators represented in the IPVs gave rise to the following conclusions.

---

[28]ESP-r does not use such data but requires angle-dependent values of shortwave transmissivity, reflectivity, and absorptivity per layer, and overall visible transmissivity.

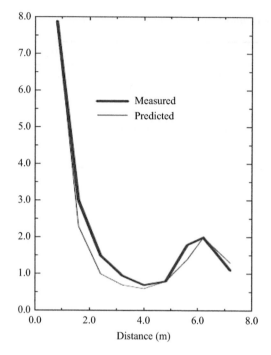

Figure 6.67   Daylight factor calibration outcome

Maximum capacity

The diversified total heating peak capacity (W/m^2) represents critical plant sizes and hence capital costs. The results indicated an insignificant increase in the heating capacity in the base case model.

Energy performance indicators

The normalised annual energy requirement for the base case was 142 kWh/m^2.y, whilst the reference case value was 158 kWh/m^2.y. When compared to normalised performance indicators (NPI) for schools, this placed the base and reference cases in the good and average categories, respectively. The base case delivered an overall energy consumption reduction of 11% relative to the reference case due mainly to the borrowed daylight from the atrium reducing the need for artificial lighting.

Typical seasonal energy demand profiles

The delivered energy data are expressed as cumulative daily profiles for each season. In both cases, the heating energy consumption is dominated by ventilation losses (note the consumption peaks after mechanical ventilation start-up). On the other hand, there is a relatively high lighting energy consumption in the reference case.

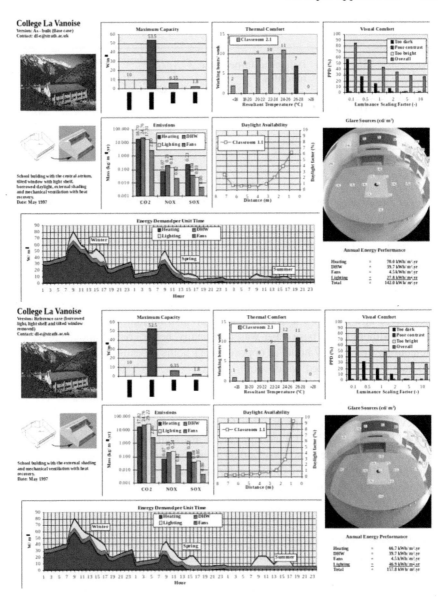

Figure 6.68   IPVs for the base and reference cases

Environmental emissions

The annual energy performance data were converted to primary energy units and then to equivalent gaseous emissions by applying the conversion factors of Table 6.27.

Table 6.27   EU-average generation and emission conversions

Fuel	Use	Delivered-to-primary modulus	Emission modulus (g/kWh)		
			CO_2	NO_2	SO_2
Gas	Heating	1.25	100	0.2	Nil
Electricity	All	1.73	360	3.0	0.57
Oil	Heating	1.67	160	0.6	2.0

The base case emissions were reduced by 14.7%, 18.5%, and 3.2% for CO_2, NO_2, and SO_2, respectively.

Thermal comfort

Optimising the daylighting performance of a building will often adversely affect thermal comfort. To examine this possibility, the resultant temperatures in a south-facing classroom on the first floor, during occupied periods over a typical summer week, were arranged into a binned frequency distribution. Both models perform similarly although the base case was marginally better, rising above 26 °C less often during the week (7 hours as opposed to 11 hours) because of the reduced solar heat gain due to the tilted windows.

Daylight availability

The level and distribution of daylight factors within a space is a reasonable indicator of daylight availability and therefore artificial lighting requirements. The daylight factors in the base case model were higher by a factor of 4 or more, especially at points remote from the external facade. This pattern was apparent in each classroom regardless of location.

Glare sources

The IPV output highlights potential glare sources within a 3D view, with luminance values given in cd/m². As indicated, a relatively bright classroom perimeter constitutes a potential glare source, with the glare sources in the reference case more pronounced because the background luminance is lower due to the limited daylight availability at the back of the classroom.

Visual comfort

The Previsible Percentage Dissatisfied index (JPPD; Meyer 1993, Compagnon 1996) relates visual discomfort to any excess/lack of light or inadequate contrasts in the field of view of a person performing a reading task. This output gives results in terms of the range of luminance values encountered. The base case performed better in relation to the 'lack of light' and 'overall visual comfort' categories.

Table 6.28   Parametric analysis results

Category	Base case	No borrowed light	No light shelf	Luminaire dimming control
NPI (kWh/m^2.y)	142.0	158.2	142.8	129.4
Heating (kWh/m^2.y)	70.0	69.0	69.7	72.8
Lighting (kWh/m^2.y)	27.8	45.0	28.9	12.4
Htg. capacity (W/m^2)	53.5	53.5	53.4	54.3
CO$_2$ (kg/m^2.y)	63.5	74.0	64.2	54.8
NO$_x$ (kg/m^2.y)	0.4	0.5	0.5	0.4
SO$_x$ (kg/m^2.y)	0.3	0.3	0.3	0.3

Having confirmed its superiority, the base case model was used to quantify the contributions of the light shelves and borrowed light, and the potential to improve performance through a different approach to luminaire control. Three design variants were explored: removing the external light shelf, removing the borrowed light feature, and actuating luminaires based on proportional dimming control. The results are summarised in Table 6.28.

Based on a comparison of the IPV performance entities, it was concluded that the as-built design offered

- no significant reduction in maximum heating capacity and therefore plant capital cost;
- a significant primary energy reduction, particularly in relation to electrical power consumption;
- slightly improved performance with respect to thermal comfort;
- significantly higher daylight levels, particularly at the back of the classrooms; and
- improved visual comfort in relation to perimeter glare.

Overall, the results indicated that the building provided significant performance improvements when compared to a reference design without the special daylighting features. In terms of the contribution of the different systems:

- the borrowed light feature was the principal contributor to the enhanced daylight levels;
- the light shelf's contribution was marginal; and
- proportional dimming control provided significant additional energy savings.

A takeaway message is that a focus on daylight utilisation, which is now routine using BPS+ tools, offers significant scope for electrical energy reduction in building.

6.13   Phase change materials

This section considers the use of BPS+ to explore options for deploying phase change materials (PCM) encapsulated in a building's construction. The use of PCM in the built environment has evolved slowly as new products have emerged. Whilst some BPS+ tools provide PCM analysis capability, it remains a niche topic for many in the design and simulation community. Acquiring data from manufacturers remains problematic and there is a paucity of expertise within design teams on how and where such a quirky product might contribute. BPS+ can play a role here by helping practitioners to identify cost-effective deployments.

Thermal energy can be stored as sensible or latent heat. In the former case, the temperature of the medium changes during charging or discharging of the storage medium, in the latter case the temperature of the medium remains more or less constant since it undergoes a phase transformation. Such an isothermal phase change can be employed to provide passive space temperature control. This project demonstrated the use of BPS+ to evaluate the potential of innovative constructions in which traditional building materials are combined with an inner PCM layer. Essentially, the latent heat of fusion is used to increase the thermal capacity of the construction.

Heim and Clarke (2004) incorporated a phase change model within the ESP-r system and demonstrated its use to appraise the performance of a PCM–gypsum composite board when applied to a portion of a multi-storey office building as a way to avoid overheating. Table 6.29 lists the properties of a 1.2 cm composite lining as shown in Figure 6.69.

Table 6.29   PCM–gypsum composite properties

Property	Value
Latent heat of phase change (kJ/kg)	45.0
Phase change temperature range (°C)	1.0
Conductivity (W/m².K)	0.35
Density (kg/m³)	1000.0

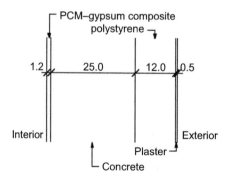

Figure 6.69   Cross-section through an external wall with a 1.2 cm inner PCM–gypsum panel

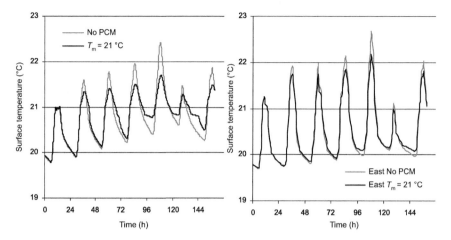

Figure 6.70 Wallboard surface temperature (left) and zone resultant temperature with and without PCM

Annual simulations at 15-minute time steps were carried out for the office with and without the PCM panel under a Warsaw climate. In each case, zone resultant temperatures, internal surface temperatures, and PCM temperatures were extracted and compared.

For the case of no composite, the maximum resultant temperature recorded in the west- and east-facing zones was 33.6 °C and 31.8 °C, respectively, with overheating observed from June through September when internal temperatures exceeded 25 °C. Based on the resultant temperature history obtained from an initial simulation, the melting temperature (T_m) was varied throughout the year: 21 °C for 6–12 March, 24 °C for 27 March to 2 April, 27 °C for 23–29 May and 30 °C for 19–25 July. The solidification temperature (T_s) was assumed 1 °C higher than the melting point and each composite was assigned the same value of latent heat of fusion (45 kJ/kg). In this way, the PCM composite's performance was compared with the traditional gypsum board. Figure 6.70, for example, shows the internal surface temperature comparison for a PCM melting temperature of 21 °C.

The latent heat stored in the PCM panel is expected to cover partially the energy demand at the beginning and end of the heating season. The most important feature of the material that determines its usefulness is the solidification temperature, which should be above but relatively close to the required internal air temperature. Therefore, the first composite, with a solidification temperature of 22 °C was deemed the most suitable in this case. Figure 6.71 shows the significant impact on the zone heating load during a one-month period in winter. These profiles integrate to 48.7 kWh and 4.2 kWh for the cases without and with PCM, respectively.

Further information on PCM materials and their numerical modelling is available elsewhere (Hawes *et al.* 1993).

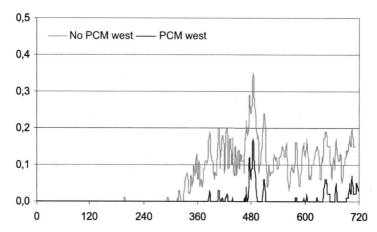

Figure 6.71   Heating loads in a west-facing zone in November with and without PCM wallboard

6.14   Advanced facades

This section considers the use of innovative approaches in the design of building facades that progress beyond the traditional. In some cases, this will require measurements to determine a material's thermal and optical properties.

Energy action at the urban scale requires attention to be directed at the energy demand-supply interface in an attempt to enhance the temporal match. It is likely that demand profile reshaping will be as important as overall demand reduction because the former can facilitate the integration of energy sources that are both intermittent and stochastic. Advanced facades[29] can play a significant role in this regard given the potential to blend design options that affect both energy demand and local supply (Selkowitz *et al.* 2003).

On the demand side, several construction options exist to reshape/reduce energy demand. Insulation and thermal capacity can be enhanced to a level beyond which the law of diminishing returns will apply. Novel materials can be deployed, such as breathable insulation (Imbabi and Peacock 2004) whereby heat loss reduction is enhanced by utilising the solar energy captured in the surface boundary layer, or transparent insulation material (Wittwer and Platzer 2000) to allow the deep transmission of the incident solar flux. Advanced glazing (Hutchins 1998) can be used to change demand through switchable properties (via thermo-, photo-, and electro-chromic coatings), reduce heat loss (via low emissivity coatings and low conductivity inter-pane gas), redirect light (via prismatic coatings and attached light shelves), or control solar penetration (via shading devices, shutters, or holographics). Double envelopes (Loncour *et al.* 2004) can be introduced to provide solar ventilation, air pre-heating, and/or passive cooling in addition to lowered thermal transmission and daylight control.

---

[29]Advanced Building Skins Conference series, https://abs.green/home/.

On the supply side, technologies can be integrated within the facade to provide thermal and electrical energy from renewable sources. Photovoltaic cells can be deployed to convert the incident solar flux to electrical power and, where used in hybrid mode, to produce heat (this mode of operation has the advantage in the case of mono-crystalline cells of reducing cell temperature and so raising the energy conversion efficiency). Advanced glazing can also be used to enhance/control the penetration of daylight to the interior; a photocell-based control system may then be deployed to regulate artificial lighting use in a complementary manner to displace the consumption of high-grade electricity. Ducted wind turbines (Grant *et al.* 2008) can be deployed at a roof parapet to produce electrical power by reducing the pressure at the leeward side of the duct through the action of an integrated spoiler, thereby enhancing inward airflow at the windward side (see Section 6.16). Where a double facade is employed, the solar chimney effect can be used to enhance the airflow at the turbine entry.

Just as the options are many, so too are the conflicts that arise because of the interactions between systems technologies when deployed in combination. The question then is how can the multi-variate impacts of alternative deployment combinations be assessed at the design stage to support decision-making that respects the implicit performance trade-offs. For example, will a photovoltaic facade harvesting solar energy to provide electricity give rise to a more cost-effective solution than mutually exclusive daylight capture devices deployed as a means to reduce electricity usage?

BPS+ is an apt technology for appraising such options in whatever combination suits the context. Clarke and Johnston (2010), for example, reported the use of ESP-r to evaluate the performance of a solar facade retrofitted to student residences at the University of Strathclyde. Figure 6.72 shows the retrofit target: a 1,040 m^2 solar facade encapsulating polycarbonate honeycomb transparent insulation material (TIM) and automated solar shading.

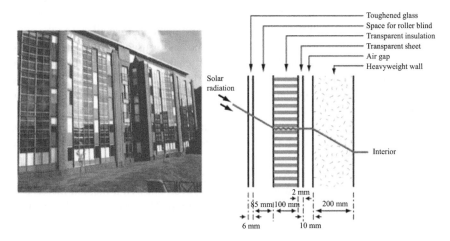

Figure 6.72   The target of an ESP-r model of a facade with transparent insulation

The inner heavyweight wall absorbs the transmitted solar radiation whilst the adjacent TIM reduces heat loss and allows 'deep' solar penetration. The effective thermal transmittance of the structure is substantially improved, and overheating is avoided by incorporating a movable, reflective roller blind. Results from simulations with the original facade and with the replacement solar facade indicated that the latter resulted in a 40% reduction in heating energy use from 280 kWh/m².y to 168 kWh/m².y. This result was subsequently confirmed by post-installation monitoring.

Encapsulated in the facade of Figure 6.72 is an automated, reflective roller blind to control overheating. This blind is guided vertically by U-shaped channels on either side. Unfortunately, the weight of the blind gave rise to mechanical problems. To address this issue and improve further the performance of advanced facades, holographic optical elements (HOEs) were examined as a way to diffract useable frequencies of sunlight and concentrate that energy on the inner wall. Alternatively, reflection holograms can be used as a printed shading device to avoid overheating problems.

The requirements for a lightweight HOE shading device were explored in another EC-funded project (Willbold 1995), and in a companion project, an ESP-r model was established for part of a low-energy office building with a TIM facade (Kuhn *et al.* 2011). This model was simulated to quantify the transmission parameters of an HOE that would be required to maintain acceptable internal comfort conditions, namely high transmission coefficients during the low sun angles of winter months and low transmission (high reflection) during the higher sun angles of summer months. With most transparent materials, transmission is dependent on the solar angle of incidence. With HOE, both solar altitude and solar azimuth must be considered in relation to the material orientation.

Bi-angular transmission data was obtained as measured by a project partner. This included transmission versus axial angle of incidence for bandwidths 500–1000 nm and 500–1500 nm, and versus solar altitude in steps of 2° and wall azimuth in steps of 5°. Before using ESP-r, a code-level intervention was necessary[30] (Frontini *et al.* 2009) to read these data and make them available to the solar processing algorithms in place of the values normally used. ESP-r calculates the solar altitude and azimuth angles prevailing at each time step. The relative angle between the solar azimuth and the azimuth of the plane containing the holographic element could then be calculated and linear interpolation used to determine the associated transmission data.

Figure 6.73 shows the predicted HOE direct radiation transmission factor for offices facing west and east of south. Similar data were produced for diffuse radiation transmission.

As can be seen, the transmission is high in winter and low in summer as required. The offices facing west of south, however, showed a marked increase in transmission during the day in the spring and autumn. The reverse was observed for

---

[30]A capability available to all users due to the open source nature of this program.

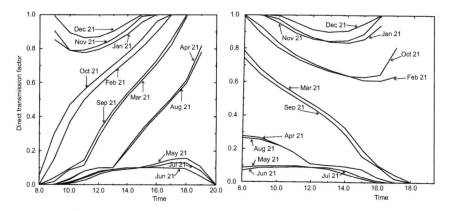

Figure 6.73   Direct transmission factor for west (left) and east of south

offices facing east of south. This unwanted behaviour can be modified by applying a structure rotation to the HOE.

At each time step, the calculated direct and diffuse transmission factors were used to modify the solar radiation transmitted and absorbed within the TIM. The amount of direct radiation passing through the transparent layers and the absorption in these layers (with the exception of the glass on the outside of the HOE) were multiplied by the calculated direct transmission factor. Similarly, the transmitted and absorbed diffuse radiation were multiplied by the diffuse transmission factor. Based on the predicted temperature profiles at various positions within the facade during summer, it was concluded that the HOE would work at least as well as a movable, reflective blind.

The search for higher-performance facades has seen any number of fashions over time. What was missing from that formative period were techniques to envisage how designs with multiple parts would work and how they might be maintained or repaired. As evidenced in Section 11.5 and Section 12.4, BPS+ has evolved to include design detail and support product design.

The permutations for smart facade design are many, including the use of transparent insulating materials as described above, phase change materials to enhance thermal capacity, transparent infrared absorbers to absorb infrared radiation, thermochromic materials to control light transmission and reduce heat loss, and the incorporation of PV within double envelope designs to capture heat and power. This last option is considered in the next section.

6.15   Hybrid PV facades

This section describes a project to assess PV facade configurations when operated in hybrid mode to produce both heat and electricity. This allowed the characterisation of PV cell efficiency under realistic weather conditions and enabled the production of performance maps for use elsewhere.

Figure 6.74   A building with PV facade configurations applied

The aim of encapsulating PV components within a facade is to enable cavity heat recovery whilst reducing the operational temperature of the PV cells and thereby increasing their efficiency. The expectation is that this will enhance the thermal performance of the building envelope by providing an economic delivery of electric power and heat and BPS+ tools have been applied to this end (Clarke *et al.* 1996).

The study described here was undertaken as part of the EC IMPACT project (Hand and Strachan 2000). It used ESP-r to investigate how the electrical performance of PV modules integrated into the facade of commercial buildings might be affected by the enhanced cooling effect of cavity airflow. Several facade variants were compared when located in warm and temperate climates (Milan and London, respectively). These variants addressed natural and forced ventilation PV facade arrangements as depicted in Figure 6.74.

The building is served by a displacement ventilation system supplied by a central air-handling plant. To account for structural thermal storage, the models included separate zones for floor voids and ceiling plenums. Office lighting was controlled based on daylight availability, with on-off control maintaining a desk-level illumination of 300 lux. Each floor comprised an east and west office zone and a core services zone. The facade-mounted PV modules and their enclosures were represented as distinct thermal zones to capture the radiant and convective exchanges affecting module performance. The following variants were considered.

- Base case (upper left) – the building with no PV applied.
- V1 (upper right) – as the base case but with 99 PV modules mounted in a horizontal spandrel enclosure and naturally ventilated via grills in the upper and lower surfaces. A convective heat transfer coefficient (h_c) of 3.0 $W/m^2.K$ is applied at the back face of the PV module.
- V2 (bottom left) – as V1 but with the PV modules mounted on a tilting enclosure that is altered depending on the season.
- V3 (bottom right) – as the base case but with the PV modules mounted in vertically aligned enclosures (again with an h_c of 3.0 $W/m^2.K$).

Assessments were carried out for each variant under the two climates. Annual and seasonal simulations were conducted and performance metrics for plant capacity, energy demand, comfort, temperatures, flow rates, and power production were extracted. Table 6.30 summarises the average capacities and energy demand

Table 6.30   Predicted equipment capacity and energy use

Variant/item	London		Milan	
	Capacity (W/m^2)	Energy (kWh/m^2.y)	Capacity (W/m^2)	Energy (kWh/m^2.y)
Base case				
Heating	71.3	49.1	69.1	45.0
Cooling	40.5	13.6	62.5	50.4
Lighting	12.1	5.9	12.1	4.0
Fans	5.8	15.6	5.8	15.4
Small power	4.5	24.2	4.5	24.2
PV	0.0	0.0	0.0	0.0
Total	–	108.3	–	138.9
V1				
Heating	71.2	49.3	69.1	45.0
Cooling	40.3	13.6	62.5	50.4
Lighting	12.1	5.9	12.1	4.0
Fans	5.8	15.6	5.8	15.4
Small power	4.5	24.2	4.5	24.2
PV	9.7	10.5	9.8	10.5
Total	–	108.6	–	138.9
PV efficiency (%)		10.7		10.3
V2				
Heating	71.2	49.4	70.0	45.4
Cooling	40.3	13.6	62.5	50.8
Lighting	12.1	5.9	12.1	4.0
Fans	5.8	15.6	5.8	15.6
Small power	4.5	23.2	4.5	23.2
PV	13.7	14.6	14.8	24.1

(Continues)

Table 6.30   (Continued)

Variant/item	London		Milan	
	Capacity (W/m^2)	Energy (kWh/m^2.y)	Capacity (W/m^2)	Energy (kWh/m^2.y)
Total	–	108.7	–	139.8
PV efficiency (%)		10.7		10.8
V3				
Heating	71.7	50.5	70.1	46.3
Cooling	39.8	13.1	61.9	49.5
Lighting	12.1	5.0	11.1	3.0
Fans	5.8	15.6	5.8	15.4
Small power	4.5	24.2	4.5	24.2
PV	9.7	10.4	10.0	14.6
Total	–	108.3	–	138.3
PV efficiency (%)		10.8		11.0

normalised to floor area. To express performance in terms of PV module face area, these data can be multiplied by 7.24.

These and other results gave rise to several conclusions as follows. There is little difference between the performance of the spandrel mounted (V1) and vertical (V3) PV designs. The differences in natural ventilation airflow caused by the form of the enclosure resulted in the flow rate in V1 being roughly half that of V3, with the air temperatures up to 5 °C higher in V1. PV output in V3 was 0.4% higher with module temperatures 0.7 °C lower. Enhanced heat transfer resulted in increased flow rates but the temperatures and efficiencies of the modules were not markedly different. It appears that enhanced heat transfer would require enclosure redesign to enhance airflow.

The use of tilting PV modules enhances PV capacity and power delivery over time, especially in the Milan climate. Although module efficiency is marginally higher with tilting PV modules, the primary benefit is increased radiation capture. For example, for London in summer, mid-day absorbed radiation on one group of cells was 419 W in V1 and 626 W in V2. For the Milan climate, increases from 442 W to 863 W were observed. Typical values for transition seasons are 534 W to 630 W for London and 644 W to 918 W for Milan. For a London climate, the enclosure air temperatures were 5–8 °C above ambient and the module temperatures were 2–3 °C above that in the V1 case. Enhancements to natural ventilation design details will therefore have a limited effect on module efficiency. Conversely, with the Milan climate, the enclosure air temperatures are 8–12 °C above ambient and enhancements to natural ventilation design details will be worth considering.

Although the above data indicate an increase in cooling demand for Milan, this is due to a few occurrences when solar ventilation preheat delivers air above 24 °C. Additional control logic to dump the tempered air when the building does not

require heating would correct this anomaly. A summary of the project outcomes
follows.

- The PV electrical outputs are in line with expectations. Enhanced heat transfer
  results in only a small improvement in module temperature and therefore PV
  electrical output.
- Tilting PV modules offers the best prospect for increased power output. It will
  then be easier to enhance heat transfer by increasing the exposure of the rear of
  the modules.
- The PV thermal output in the case of forced ventilation resulted in modest
  reductions in heating requirements. The problem is one of utilisation – the heat
  output is maximum during summer afternoons when demands are low or zero.
  Heat storage with release early the next morning would be beneficial but its
  cost-effectiveness is unlikely.
- Strategies for heat and power utilisation in niche cases have been reported
  elsewhere (e.g., Strachan *et al.* 1997).

In another project, EC SOLINFO (Lewis 1995), ESP-r was used to generate a
performance map for the rapid assessment of the output from PV components
when operating under non-standard conditions. The first step was to calibrate an
ESP-r PV cell model using data from a PASSYS outdoor test cell as shown in
Section 2.6 (Vandaele and Wouters 1994). The data required by the model are the
short-circuit current, the open circuit voltage, the voltage and current at the
maximum power point, the reference values at which these are measured (normally 1,000 W/m² irradiance and 25 °C containment), and the number of series
and parallel connected cells in each panel. In addition, an empirical constant is
required to account for the change in output with a change in temperature
(Buresch 1983). A base case model of the hybrid facade was created, and a corresponding reference model was formed by adding the PV component to facilitate
comparison.

After model calibration using test cell data, the performance of a hybrid
facade was simulated under typical UK conditions. Figure 6.75 shows the predicted PV cell temperatures plotted against incident irradiation: note that these
conditions are significantly removed from those comprising a standard test,
demonstrating the inappropriateness of using manufacturers' published data for
calculation purposes.

Figure 6.76 shows the predicted facade performance under summer conditions
corresponding to an average PV cell temperature of 32.5 °C.

Finally, Table 6.31 summarises the operating performance, by season, for the
electrical power only, and combined heat and power cases.

The study demonstrated that the PV efficiency could be improved significantly
if the cells were cooled and heat recovered. Even during winter operation, when
thermal demands are at their highest, the combined efficiency increases from
11.7% to 33.2%. The utilisation of the thermal energy during the mid-heating
season could result in the building's thermal demands being largely met from the
PV system.

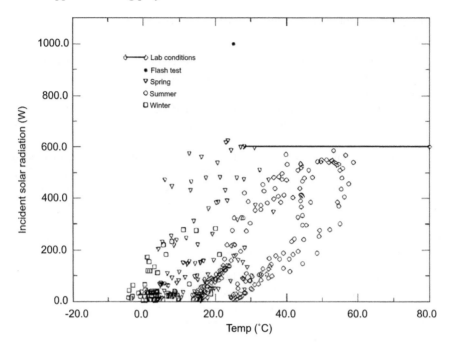

Figure 6.75   PV cell temperature vs. incident solar radiation

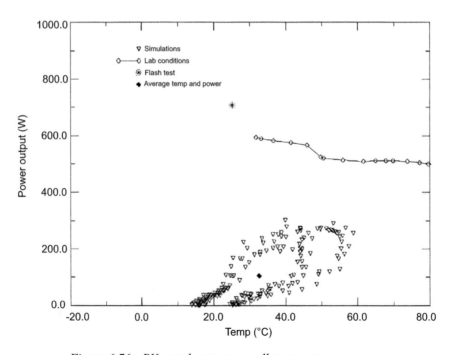

Figure 6.76   PV panel output vs. cell temperature, summer

Table 6.31 Seasonal hybrid PV facade performance

	Winter week	Spring week	Summer week
Insolation (kWh)	26.5	132.2	211.6
Electrical power (kWh)	3.1	16.3	25.8
Electrical efficiency (%)	11.7	12.3	12.2
Heat recovered (kWh)	7.3	42.0	94.6
CHP efficiency (%)	33.2	44.1	56.9
CO_2 reduction (kg)	4.3	23.9	34.9

6.16 Building-integrated renewable energy systems

This section addresses the use of BPS+ to blend building-integrated micro-generators and demand-side measures in a manner that allows a building to operate autonomously. This requires the matching of demand and supply profiles where possible and battery storage to meet any residual.

Network connection is the preferred configuration for building-integrated renewable energy systems. The generated electricity is exported to the local distribution network, whilst the building's electricity demands are met from separate imports. The additional capital and operating costs that such schemes place on network operators to maintain acceptable network performance are limited at present due to the low number of connected schemes and the availability of government subsidies. In the longer term, an increase in the number of intermittent, single-phase supplies using the electricity network as a buffer device will introduce network imbalance and power quality problems. The increasing costs associated with correcting these problems will not be absorbed by the network operators but passed on to the generators (the consumer in the case considered here). This additional cost will reduce the financial viability of small-scale schemes.

Non-exporting systems – hereinafter referred to as 'building-embedded' – have the potential to contribute significantly to the attainment of targets for clean energy systems. For this to happen, new approaches to component sizing, system configuration and power utilisation are needed to ensure the best overall match between demand and supply with local energy storage to accommodate temporal mismatches.

PV is widely deployed whilst, although available, small wind turbines attached to buildings have generally been disappointing due to the relatively poor wind resources in urban areas. That said there is a market for small turbines in niche applications, chiefly in remote areas without access to mains electricity, and special urban installations as described below and elsewhere (Grant and Dannecker 2000).

To illustrate the role of BPS+ in devising building-embedded solutions, consider the installation of new and renewable energy (RE) systems within part of

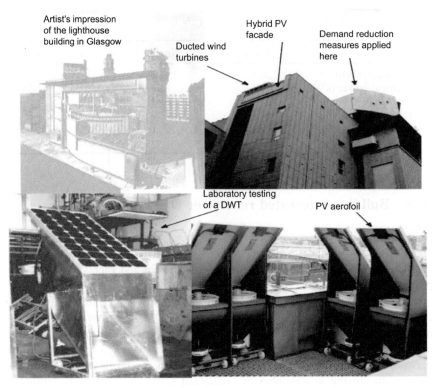

Figure 6.77   The refurbished Charles Rennie Mackintosh building in Glasgow

the refurbished Charles Rennie Mackintosh Lighthouse Building in Glasgow[31] as shown in Figure 6.77 (Clarke *et al.* 2000).

The project, which was funded by the EC RE-Start project (Burton *et al.* 1996) and supported by Glasgow City Council, used BPS+ to design a hybrid solution, comprising a blend of demand reduction measures and new supply options, and then retrofit these to the building as part of a monitored demonstration facility. Two applications were used in the project, ESP-r and Merit[32].

The first task was to create an ESP-r model of the portion of the building being targeted for refurbishment. Several variants of this model were defined corresponding to possible demand reduction measures and embedded supply options. The ESP-r predicted, disaggregated energy demand/supply profiles for each variant were passed to Merit, which applied statistical techniques to rank-order the combinatorial matches. Based on the outcome, the following deployments were deemed suitable.

- Advanced glazing to minimise heat loss without significantly reducing daylight penetration.

[31]https://thelighthouse.co.uk
[32]https://www.esru.strath.ac.uk/Applications

- A facade with transparent insulation and automated blind control to time-shift incident solar energy to offset later heating requirements.
- Dynamic indoor temperature and illuminance set-point adjustment as a function of occupancy.
- An array of parapet ducted wind turbines (DWT) to meet a portion of the power demands during the winter and transitional seasons[33].
- A PV array to meet a portion of the power demand during the transitional and summer seasons (this array was subsequently incorporated within the spoiler of the DWT).
- A hybrid PV facade to meet a portion of the power (PV_e) and heat (PV_h) demands during the transitional and summer seasons[34].
- A battery storage system to meet the temporal mismatch between demand and supply.

Figure 6.78 shows (in the form of IPV extracts) the predicted cumulative impacts of the demand reduction measures and the final match with supply: a 68% reduction in annual energy demand – comprising a 58% reduction in heating and an 80% reduction in lighting – and a close quantitative match between demands and supplies.

Without local energy storage, the temporal mismatch would result in the renewable supplies meeting only around 65% of the total demand. This figure would fall to around 20% were the energy efficiency measures not adopted.

These predictions were transformed into a project plan by the design team[35], the components procured and installed within the building as shown in Figure 6.77. The DWTs were manufactured locally, whilst the PV systems and advanced glazing were supplied by international manufacturers. Since the direct current (DC) operating voltage of the battery/inverter system was 48 V, it should be noted that:

- the primary charging of the batteries comes from the combined DWT and PV_e systems, with the backup charge controller only operational after a sustained period of low or no supply availability;
- the DWT supply voltage is regulated by the 12 V DC sector of the battery bank across which it is wired;
- a balanced distribution of DWTs across each 12 V sector of the battery bank is required to ensure equal charging of the battery cells to maximise their operational lifespan;
- the PV modules, wired three in series using blocking diodes, supply the PV controller at a V_{max} operating voltage of 54 V, the PV controller then regulates the charging voltage to 52 V across the 48 V battery terminals;

---

[33]The power output from a DWT is roughly proportional to the cube of wind speed. Below 8 m/s, the output is very small, rising rapidly for higher speeds. In this case the power output was enhanced by using a PV panel as the spoiler.

[34]The use of a hybrid facade increases the PV cell efficiency and gives a useful thermal output in the form of pre-heated air to the ventilation system.

[35]https://pagepark.co.uk

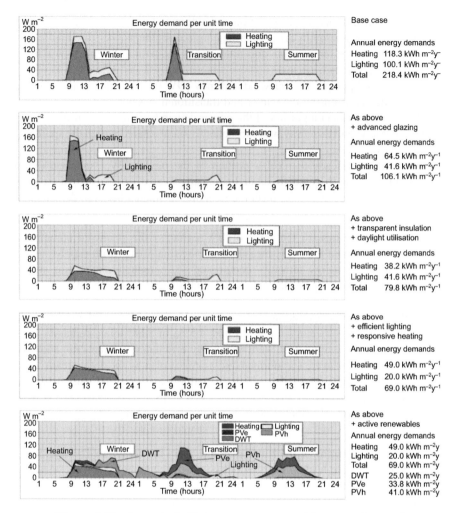

Figure 6.78   Impact of efficiency measures and best supply match

- the backup battery charge controller is only operational when the battery terminal voltage falls below 43 V; and
- battery overcharging is prevented by dump load activation when the terminal voltage rises above 44 V.

As shown in Figure 6.77, the hybrid PV component was subsequently incorporated within the south-facing facade, whilst the DWTs were mounted on the south- and west-facing roof parapet giving significant wind exposure.

To determine the accuracy of the approach, the building was monitored over a heating season. Figure 6.79 shows the monitored DWT and PV outputs. The performance of the south-facing PV components is close to the initial predictions

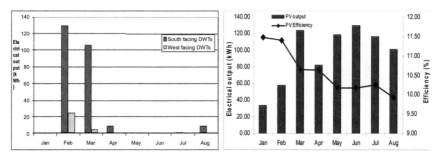

Figure 6.79 Monitored performance of the DWT (left) and PV components

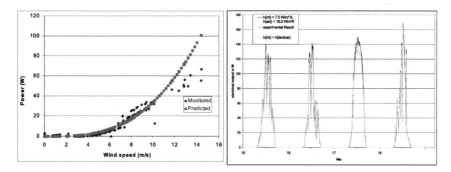

Figure 6.80 Monitored and predicted performance of the DWT (left) and PV

with low-energy yields during the winter period and higher energy yields during the transitional and summer periods. The mean monthly efficiency of the modules varies inversely as a function of the operating temperature.

For a given solar irradiation, the modules experience a specific temperature rise. Therefore, because the ambient temperature varies by season, for a given irradiation level the PV efficiency will be different throughout the year. This results in the PV achieving its highest efficiency during the winter period and its lowest level during the summer.

Figure 6.80 shows the agreement between monitored and predicted data. From these results, it was concluded that the final case predictions of Figure 6.78 were an acceptable representation of the systems as deployed.

The DWT and PV systems are complementary when integrated within a building under Glasgow weather conditions as the majority of the power generation occurs during the winter/spring period for the DWT system and the spring/summer period for the PV system. It was estimated that the south-facing DWTs should be able to meet the energy demands associated with the winter period. This is because the predominantly southerly winds are greater than 7 m/s for prolonged periods. Whereas the south-facing turbines have open exposure across a wide arc to the

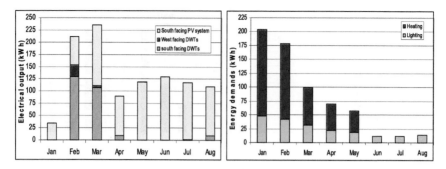

Figure 6.81   The monthly demand/supply match

south, the local upwind obstructions to the west compromise the performance of the west-facing DWTs. Figure 6.81 shows the energy supply and demand over 8 months.

The magnitudes associated with the combined electrical power generation from the active renewable systems highlight the importance of demand-side reduction measures (Arteconi *et al.* 2014) if renewable energies are to make a significant contribution to meeting the building's energy demands. After such reductions are applied the active renewable technologies can be deployed to satisfy the majority of the residual demand as shown.

6.17   Estate upgrade planning

This section demonstrates the use of BPS+ to explore upgrade strategies for large estates. It involves the use of a tool that has been preloaded with information derived from thermal simulations applied to models diversified in relation to the many possible design parameter settings.

Models of representative dwellings can be used to compose an estate model for upgrade options assessment. In one project (Clarke and McGhee 2017) a HUE[36] estate model was created comprising 40 × 2-bed apartments located within 10 four-in-a-block cottage flats, and 40 × 2-bed apartments located within 5 four-storey conjoined tenement flats. The model was then adapted to represent possible upgrades applied separately and together: air tightness at five levels (0.1, 0.25, 0.5, 0.75, and 1.0 ACH), boiler efficiency at three levels (65%, 80%, and 90%), and insulation at five levels as summarised in Table 6.32.

Table 6.33 gives the results for a single apartment[37] when configured as indicated (the base case against which other cases are compared).

---

[36]https://www.esru.strath.ac.uk/Applications. HUE is an application based on pre-simulated models representing building stocks and possible upgrades (see Sections 7.7 and 8.1).
[37]Such results were integrated to obtain whole estate values.

Table 6.32   Level U-values (W/m^2.K)

Level	1	2	3	4	5
Wall	2.0	0.60	0.31	0.31	0.13
Roof	2.0	0.35	0.16	0.14	0.11
Floor	0.88	0.46	0.25	0.22	0.13
Glazing	0.56	3.4	2.1	1.7	1.1
Door	3.0	3.0	2.0	1.7	1.0

Table 6.33   Apartment results

Floor area	68 m^2
Insulation	@ Level 3
Air leakage	@ Level 3
Boiler efficiency	@ Level 2
Space heating	200 kWh/m^2.y
Cost of energy	6.2 £/m^2.y
Emissions	40 kgCO$_2$/m^2.y

Table 6.34   Impact of insulation changes

Insulation level	1	2	3	4	5
Space heating (kWh/m^2.y)	508	302	200	176	118
Improvement on base case (%)	154	51	–	−12	−41
Cost of energy (£/m^2.y)	15.7	9.4	6.2	5.5	3.7
Improvement on base case (%)	153	52	–	−11	40
Emissions (kgCO$_2$/m^2.y)	100	60	40	35	23
Improvement on base case (%)	150	50	–	−13	−43

These results are in alignment with benchmark values for a building with an energy efficiency rating of C.

Table 6.34 summarises the impact of degrading/improving the insulation standard across the five discrete levels (whilst retaining air leakage and boiler efficiency at the base case levels). Here, yellow indicates the base case, red a worsening of the situation, and green an improvement.

Table 6.35 summarises the corresponding results for air leakage reduction (whilst retaining insulation and boiler efficiency at base case levels).

Table 6.36 summarises the corresponding results for boiler efficiency improvement (whilst retaining insulation and air leakage at base case levels).

Finally, Table 6.37 summarises the impact of applying combined retrofits.

Such results were used by the design team to identify appropriate upgrades in terms of measure combinations and their rate of rollout as a function of likely operational budgets.

Table 6.35   Impact of air leakage changes

Air leakage level	1	2	3	4	5
Space heating (kWh/m^2.y)	317	249	200	144	123
Improvement on base case (%)	59	25	–	−28	−39
Cost of energy (£/m^2.y)	9.8	7.7	6.2	4.5	3.8
Improvement on base case (%)	58	24	–	−27	−39
Emissions (kgCO$_2$/m^2.y)	63	49	40	29	24
Improvement on base case (%)	58	23	–	−28	−40

Table 6.36   Impact of boiler efficiency changes

Boiler heating efficiency level	1	2	3
Mains gas (kWh/m^2.y)	247	200	178
Increase on base case (%)	24	–	−11
Cost of energy (£/m^2.y)	7.7	6.2	5.5
Improvement on base case (%)	24	–	−11
Emissions (kgCO$_2$/m^2.y)	49	40	35
Improvement on base case (%)	23	–	−13

Table 6.37   Impact of combined retrofits

Retrofit[*]	Space heating (kWh/m^2.y)	Cost of energy (£/m^2.y)	Emissions (kgCO$_2$/m^2.y)
Ins (2) + Air (2) + Boiler (2)	350	10.9	70
Ins (2) + Air (2) + Boiler (3)	314	9.7	62
Ins (3) + Air (2) + Boiler (2)	249	7.7	49
Ins (3) + Air (2) + Boiler (3)	222	6.9	44
Ins (2) + Air (3) + Boiler (2)	302	9.4	60
Ins (3) + Air (3) + Boiler (2)	200	6.2	40
Ins (3) + Air (3) + Boiler (3)	179	5.5	35
Ins (4) + Air (4) + Boiler (2)	122	3.8	24
Ins (4) + Air (4) + Boiler (3)	109	3.4	22
Ins (5) + Air (5) + Boiler (2)	49	1.5	10
Ins (5) + Air (5) + Boiler (3)	45	1.4	9

*Ins ( ) = Insulation Level; Air ( ) = Air leakage Level; Boiler () = Boiler efficiency Level.

Although the number of parametric excursions (75) is limited in this case, the planning of what variants to impose and performance data to capture is not essentially different from a project supporting a much larger matrix. In contrast to the evolution of a BPS+ model in response to feedback from a client, parametric studies must imagine 'what if' questions. Whereas a high-resolution model can be fine-tuned to answer specific questions posed by the client, the models used in parametric studies tend to be more abstract (in the past to fit with limited computing resources). The challenge now is to reframe how models used in parametric studies are designed: tools developers need to be bolder by supporting the imposition of uncertainty and optimisation techniques on high-resolution models.

6.18 Swimming pool refurbishment

This section presents a niche application of BPS+ entailing the cooperative use of tools and the use of outcomes to undertake a cost analysis. The project (Cockroft *et al.* 2008) utilised ESP-r and TRNSYS[38] operating in tandem to assess energy-saving options being considered as part of the refurbishment of a community swimming pool. As summarised in Figure 6.82, a base case model was established comprising an ESP-r model of the building and a TRNSYS model of the gas-fired boiler providing heat to air-handling units, pool water, and domestic hot water. Pool covers were deployed at night to reduce evaporation losses and fresh air requirements.

Table 6.38 summarises the simulation results for the base case model. These results correspond to the release of 480 tonnes of CO_2 and an annual running cost of £77,900 (at 2008 prices).

Variant system models were then defined and simulated as summarised in Figure 6.83.

Table 6.39 summarises the results for each variant. In all cases, evacuated tube solar collectors are used to provide heat for pool water and domestic hot water.

Several recommendations were made based on these results and some were taken forward for inclusion in the refurbishment.

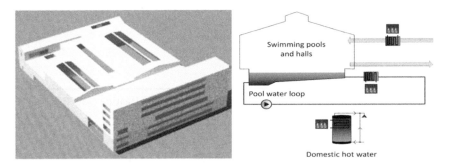

Figure 6.82   Base case model of the community swimming pool

[38]https://trnsys.com/

*Table 6.38   Annual energy (MWh) breakdown for
the base case model*

Space heating	204
Pool ventilation heating	1155
Pool water auxiliary heating	584
Domestic hot water auxiliary heating	133
Total heating (met by gas boiler)	2076

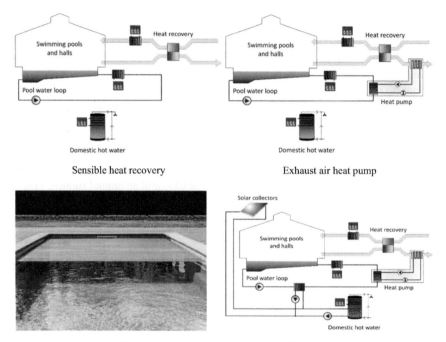

Sensible heat recovery

Exhaust air heat pump

Pool covers & ventilation control

Solar thermal collectors

Figure 6.83   System variants

Table 6.39   Summary of results

Variant	CO_2 saving (%)	Heating cost saving (%)	Renewable fraction (%)
Sensible heat recovery sized for a design effectiveness of 65%	33	33	–
Sensible heat recovery sized for a design effectiveness of 85%	42	42	–
As 1 with 60 kW heat pump on exhaust air	42	42	20
As 2 with 60 kW heat pump on exhaust air	51	50	21
As 4 with improved pool cover and ventilation control strategy	56	56	24
As 5 with 200 m^2 solar thermal collectors providing heat to the pool water	60	60	31

6.19   Telecommunication shelters and electrical substations

This section describes an atypical application of BPS+ to a ubiquitous building type with high energy-saving potential. It involves the use of zonal airflow modelling applied to explore approaches to electronic equipment cooling and heat recovery. The structures containing such equipment are numerous and widely dispersed.

There are around 45,000 telecommunication shelters in the United Kingdom (ACIS 2020) and approximately 400,000 final distribution substations transforming electricity from 11 kV to 230 V (EMFS 2023). Most of these structures are 'lights out' facilities that are rarely staffed and satisfying the needs of the equipment they host is the main aim. The equipment is continually evolving to match demands. Much of the generated heat is lost so improving the insulation and ventilation standards would both improve energy efficiency and increase the potential for heat recovery (Lagoeiro *et al.* 2023). Although any captured heat would be low grade (<50 °C), it could be used to supply nearby facilities by contributing to a local heat network. The energy-saving potential is substantial with a short payback period associated with the types of energy-saving measures that are likely to be deployed.

BPS+ support for improving the performance of such buildings is demonstrated by a project focused on a mechanically ventilated telecommunications shelter (Hand and Cockroft 2002). The project involved simulation support for an electrical engineer, who was evaluating options for reducing running costs and carbon emissions for a telecommunication company. Figure 6.84 shows a typical factory-assembled shelter.

The facades of such shelters are typically concrete or metal and are designed primarily for security and low maintenance. As shown in Figure 6.85, the electronic equipment is tightly packed and must be cooled.

Figure 6.84   A telecommunication shelter

Figure 6.85   Shelter interior with equipment racks

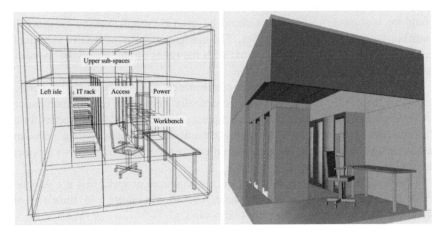

Figure 6.86   ESP-r model of the telecommunication shelter

A survey indicated that whole-space mechanical ventilation was often used to this end. Could an approach be implemented that facilitates direct equipment rack cooling as a way to reduce energy use, perhaps based on a hybrid natural/ mechanical ventilation system? Because the project aim was to assess the risk over time of high-temperature conditions within the equipment racks, a zonal airflow modelling approach was adopted representing position-dependent heat inputs.

A typical case of a 3m × 7.2m shelter was assessed, large enough to accommodate eight racks, each housing equipment between 4 kW and 8 kW and with power controls and backup batteries arranged along the right wall. The facade and ceiling were steel-faced panels with rigid insulation. As shown in Figure 6.86, an

ESP-r model was constructed to represent the shelter as several thermal zones to demarcate the spaces between and above the equipment racks.

The model comprised a left aisle, IT racks, central access, power/battery rack, and staff working space; additional thermal zones were included corresponding to upper air spaces. In one option studied, a rear-mounted mechanical ventilation unit was ducted to grills near the equipment racks, with the unit and ducting represented as distinct thermal zones.

As illustrated in Figure 6.87, a zonal airflow network was created to represent the ventilation system including a frost protection unit, supply ducting, and air intake grills at the rear. This was then merged with the multi-zone building model as shown.

Orifice and bi-directional flow components were used to represent the airflow between thermal zones with the latter component type assigned to the horizontal connections between the aisles and equipment racks so that airflow was temperature and pressure dependent.

The left-rear grill supplied the access aisle behind the main rack, which disperses air through the mesh of the equipment rack enclosure and around its periphery. The right-rear grill supplies the power conditioning and battery rack and from there through the rack mesh to adjacent spaces. The air then exhausts via the staff area, high-level exhaust grill, and gaps around the entry door (lower right). Air rising from the equipment racks mixes at the upper level and exits via an upper-level exhaust grill.

This model allowed a first-order assessment of temperature distribution and heat recovery potential without incurring the computational burden of resolving a CFD domain model repeatedly over thousands of time steps. An annual simulation

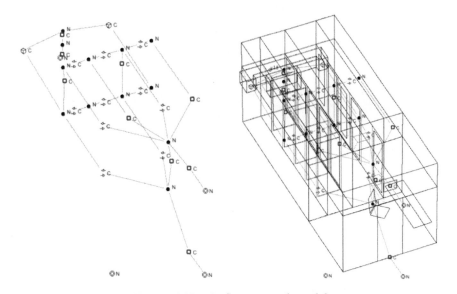

Figure 6.87   Airflow network model

at 6-minute time steps executed in approximately 2 minutes thus facilitating rapid studies for various design configurations and for different UK locations. Initial assessments examined the impact on rack temperatures of fixed and variable ventilation flow rates. Figure 6.88 shows two typical results.

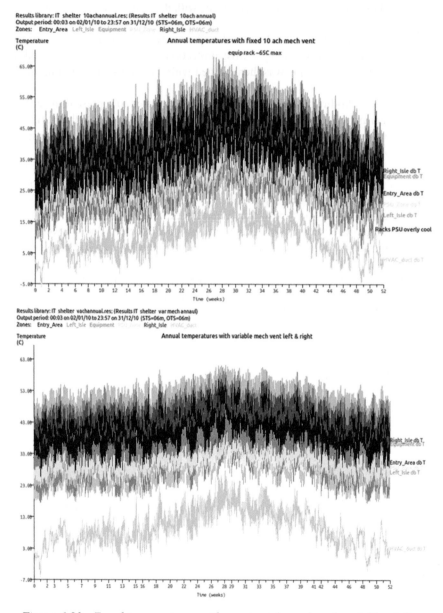

Figure 6.88   Zonal temperatures with constant (upper) and variable airflow

The upper graph corresponds to the flow rate required to limit the equipment rack (dark green) to 70 °C. If applied all year, this resulted in the left and right cold aisles (light brown and yellow) dropping to 5 °C and the equipment racks dropping below 15 °C. It also incurred high fan running costs. In the lower graph, the fan flow rates are linked to the temperatures in the equipment racks, reducing up to 50% and boosting to 150%. This moderates the range of temperatures in the equipment racks with only brief periods on boost. It also tended to track better any changes in equipment demands.

The changes in environmental conditions during a shorter timescale in February are shown in Figure 6.89. The fixed ventilation needed to limit summer temperatures on the left results in excess cooling of the equipment rack. Allowing the flow to be scaled still manages to control peak temperatures and reduce running costs further.

Having established variable airflow as an effective cooling option, construction retrofits were assessed as an alternative to complete shelter replacement in future. This involved exploring compositions that could be readily implemented by adding material to what existed. The results indicated that for communication shelters with a constrained equipment provision (4–8 kW) there were minor benefits to upgrading the fabric. For fully provisioned shelters, higher shell heat retention required higher ventilation.

For the case of an electricity substation, the after-diversity maximum demand (ADMD) index, as used by the industry to size wires in a network, has been shown to be substantially greater in design guidelines than observed in practice (Tait 2014). This means that the ability to connect new power supplied to existing infrastructure is being underestimated. To address this issue, it is possible to add an electrical network representation to the model to enable the determination of the impact on ADMD of community energy actions involving PV and heat pump retrofits.

6.20 Outdoor air quality

This section describes the use of CFD to assess emission dispersal and give advice on pollution control. It reports a project that used PHOENICS[39] to model the extraction of gases from fume cupboards in a new clinical research facility (CRF) on the grounds of the Western General Hospital in Edinburgh (Grant and Dannecker 1998). Concerns related to the venting of fume cupboards and the purpose of the investigation was to estimate the minimum height of the chimney that would ensure that gases were carried clear of surrounding buildings.

The site included a number of taller buildings of irregular shape in close proximity and to simplify the simulation process the buildings were modelled as combinations of rectangular boxes as illustrated in Figure 6.90.

The modelling of all buildings as combinations of rectangular boxes was necessary for computational reasons. One result of this is that all buildings in the model have flat roofs, which is correct for some, but not all buildings on site. If a

[39]https://www.cham.co.uk/phoenics.php

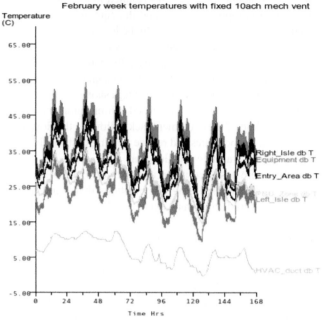

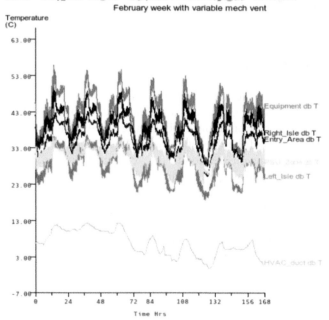

Figure 6.89   Zoom-in on the Figure 6.88 graphs

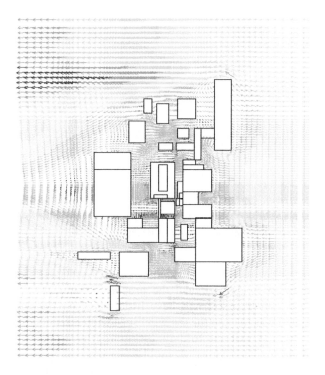

Figure 6.90   Flow pattern on a horizontal plane 2.6m above ground level, wind from the south (right)

building with a pitched roof is modelled with a flat-roofed building of the same height, the turbulence and eddying in the wind will be over-predicted. For the purposes of this investigation, the effect was thought to be conservative.

The approaching wind was represented as a vertical profile typical of an urban situation (i.e., speed increasing with height above ground). A nominal wind speed of 10 m/s at a height of 10m was imposed. The shape of the flow pattern around the buildings is independent of wind speed, and local velocities for lighter or stronger winds will vary in proportion. After consideration of the local topography and prevailing winds, five wind directions were investigated. To simplify computation, these were referenced to the grid of the site plan shown in Figure 6.90. This grid is not exactly aligned with the points of the compass and the actual wind directions modelled were as follows.

Nominal direction	NE	S	SW	W	NW
True bearing	33°	168°	213°	258°	303°

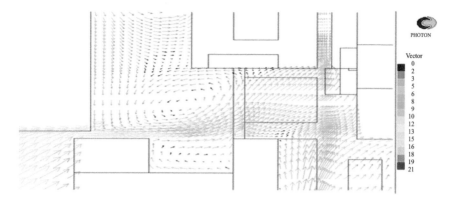

Figure 6.91   Flows in the region of the CRF building viewed on a horizontal plane, 15m above ground, NW wind

No attempt was made to model the proposed chimney. A chimney is a slender structure and by itself will have little effect on the wind flow. The procedure adopted was to examine the wind speed and direction in the region of the proposed chimney exit and to draw the appropriate conclusions about the likely path of the exhaust gases. A problem is deemed to exist if the wind passing through the point of exit subsequently becomes trapped in a zone of recirculating air. This is perhaps conservative in that the buoyancy and upward exit velocity of the exhaust gases would be expected to move them upwards relative to the surrounding air, but it was considered prudent to have this margin of safety.

The 3D wind flow patterns can be viewed in two dimensions at any plane of interest. For example, Figure 6.90 shows the effects of a southerly wind on a horizontal plane a few metres above the ground. The funnelling of the wind through spaces between buildings is clearly illustrated. At a higher level, tall buildings influence the flow over the roofs of lower structures, producing complex flow patterns as seen in Figure 6.91. Wind velocity vectors are colour coded corresponding to wind speed in m/s. Due to model complexity, computation was a lengthy process, each run requiring around 3 days to reach convergence with the computer technology available at the time.

To study the performance of the proposed dispersal system, wind velocity vectors in the vertical plane are more critical. It is important that the gases rise clear of surrounding buildings and are not trapped between them. For the proposed chimney location, wind vectors in both N–S and E–W planes were examined. An alternative proposal to attach a chimney to the adjacent part of the CRF building was investigated using the same E–W plane, with a new N–S plane located a few metres to the east of the original.

A recurring pattern in the predicted wind flow was that the presence of a tall building downwind of a chimney is likely to cause problems. With a wind from the

NE, there is an escape route downwind from the CRF building over relatively low buildings to the SW. But for all other directions modelled matters are more problematic. Winds from the S or SW give rise to a large recirculating region between the CRF building and planned buildings as shown in Figure 6.92 for a worst-case SW wind. If it is to exhaust into the air that remains clear of this region, a chimney must be at least as tall as the latter buildings. This constituted one of the two worst cases detected in the analysis.

For a chimney attached to an adjacent building, there is the possibility of an S or SW wind of gases concentrating in the region immediately to the N where a space is enclosed on three sides by high walls. Figure 6.93 illustrates this situation by looking at an N–S plane through the region. A chimney flush with the roof of the building might cause problems, although an exit 3m above the roof should carry the gases clear.

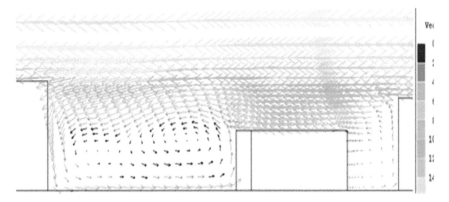

Figure 6.92   Flows in the region of the CRF building, N-S plane, SW wind

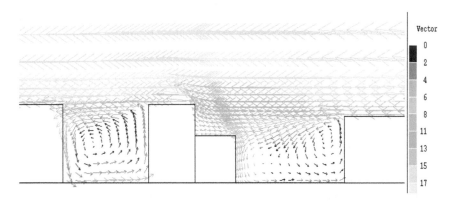

Figure 6.93   Flows in the region of the CRF building, SW wind

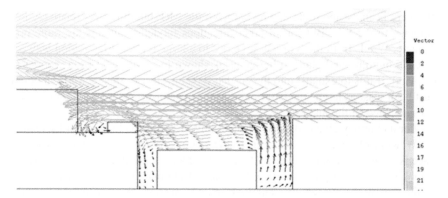

Figure 6.94   Flows in the region of the CRF building, W wind

As shown in Figure 6.94, a second difficult case arose in a W wind, when air flowing over the roof of the CRF building feeds into a recirculating region in front of another adjacent building. To avoid this completely, a chimney must be at least as tall as the upwind building, and preferably higher.

The severity of the case illustrated here is similar to that in Figure 6.92: for both, a chimney exiting at a level 3m above the roof of the surrounding buildings was shown to avoid problems. It was therefore recommended that a chimney of this height be installed.

In another project (Tang 2002), an ESP-r compatible CFD research code that was under development at the time was applied to investigate microclimate changes due to the discharge of pollutants from rooftop stacks on a university chemistry building. Simulations were undertaken for the four cardinal wind directions, each at 2 wind speeds (normal and a worst case), plus a zero wind speed condition. Figure 6.95 shows different views of the model, which was created by importing an existing AutoCAD file. The modelled CFD domain was approximately 25m × 300m, with different wind conditions imposed on this virtual wind tunnel.

Figure 6.96 shows typical outputs corresponding to airflow vectors over the site under a west wind condition with a contaminant plume. Such outputs are typically animated to emphasise the variable direction and speed.

The results identified locations where the risk of air pollution was high, the reasons why such contaminations occur, and the magnitude of the pollution in relation to wind speed and direction. This information helped the design team to develop measures to mitigate the risk.

In yet another project with similar aims (Tang *et al.* 2003), the site geometry was mapped to a rectilinear grid and the domed roof was discretised into prisms that preserved the profile. CFD outputs were selected to show the spread of contaminant at various concentration levels as depicted in the examples of Figure 6.97.

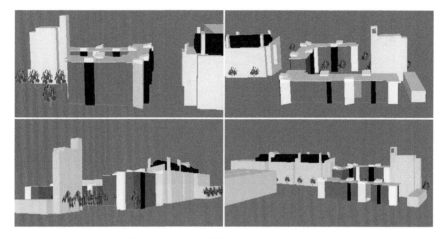

Figure 6.95   CFD model viewpoints

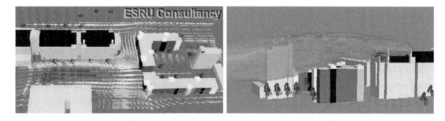

Figure 6.96   Typical CFD outputs

Figure 6.97   Coverage of contaminant at 1% and 0.5% of exit concentration

The results showed that under the worst-case wind condition, an adjacent building was within the spread of the contaminant at 0.25% of the exit concentration level as shown in Figure 6.98.

The outputs shown in this section may be classed as experiential in that they readily communicate complex information to all project stakeholders. This should be an aspiration in all projects irrespective of the issue being addressed. For

Figure 6.98   Contaminant spread

example, in the case of energy data, information displayed via a pie chart might be more effective than a table of numbers.

6.21   Retrofit analysis

BPS+ is often used to assess building retrofit options and this section describes the use of archival drawings and the generation of stochastic occupancy patterns when preparing models. Ultimately, the approach taken will depend on the nature of the uncertainties associated with the construction, heating/cooling system, air leakage paths, and occupancy mix.

The project described here (Hand and Kelly 2020) relates to a 1961 tower block retrofit undertaken as part of the EC Ruggedised[40] project. Past upgrades included the application of external insulation, window replacement, and the introduction of electric storage heaters.

To minimise the required modelling effort, a typical floor of the tower block was selected. As is often the case with such projects, information was only available in legacy formats. Figure 6.99 shows one of many drawings available in hardcopy requiring transformation to a digital format.

The approach taken was to treat each room as a separate thermal zone, with the digitised points aligned to internal surfaces to preserve space volume and surface area. Constructions were then added, extending 'outward' from the resulting wireline representation. The hardcopy plan was scanned and processed via a 'click-on-bitmap' facility in ESP-r. To capture the architectural elements, the bounds of the room were initially defined followed by the left and right edges of the door and windows. Figure 6.100 (upper) shows the identification of the doors in the hall area and the supporting menu of options.

[40]https://ruggedised.eu/legacy/

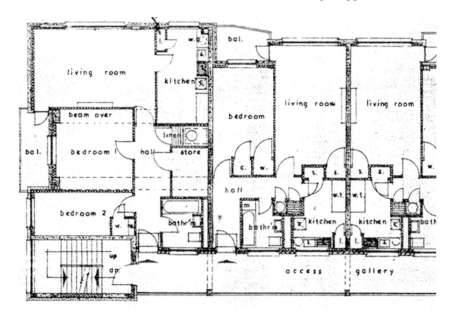

Figure 6.99   A hardcopy plan for a portion of the tower block

The middle portion of the figure shows the situation after three zones have been captured. Once the digitising session is complete, the surfaces are attributed in terms of their construction, window/door details, thermal bridges, and boundary conditions. The outcome is the 3D model of the lower portion of the figure.

The next step was to augment the model with explicit representations of the existing storage heaters as shown in the collage of Figure 6.101. The approach taken was to represent each storage heater as multiple thermal zones – core zones representing the brick thermal store into which heat is injected, and an outer zone representing the casing void where air circulates via controlled flow openings (see Section 9.2 for further details on this modelling approach).

The previously installed storage heaters had limited control capability and a core aim of the assessment was to compare the cost-effectiveness of installing modern replacements against retaining the existing units but with upgraded control.

Occupancy patterns and air leakage characteristics presented significant modelling uncertainty. To address the former, the OccDem stochastic profile generator[41] (Flett 2017) was employed, resulting in realistic internal heat gain profiles such as those shown per dwelling zone in Figure 6.102 (see also Section 11.2).

Here the gains in the kitchen from cooking (purple line) are seen to dominate. In comparison with prescribed occupancy, the OccDem method introduces

[41]https://www.esru.strath.ac.uk/Applications/

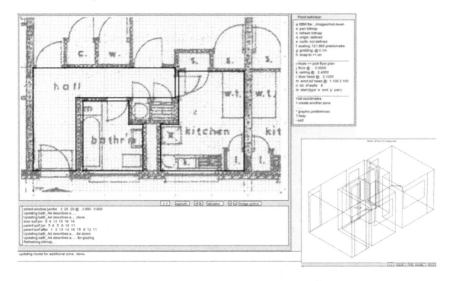

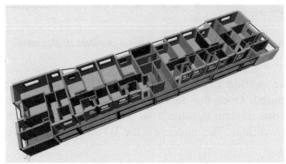

Figure 6.100   An ESP-r digitising session and final model

variability with occasional 'everyone over for pasta' events. To see the gains of lesser magnitude the vertical axis has been zoomed. Table 6.40 contrasts the significant difference between prescribed and stochastic casual heat gains over a January period.

In the absence of field measurements, defining the air leakage distribution of a 60-year-old building is highly uncertain. As the focus of the study was to explore different control approaches, simulations were undertaken corresponding to prescribed infiltration rates and an uncertainty analysis was commissioned to bracket low and high cases.

Model calibration focused on the heat emission from the storage heaters. Adjustments were made to the model to alter the decay of temperatures in the heater core and room-facing surface. As the existing heaters were manually controlled, a guesstimated regime was implemented in the model.

Figure 6.101   Addition of storage heater models

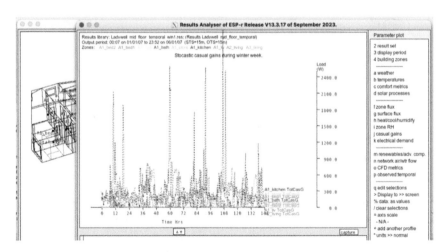

Figure 6.102   OccDem generated internal gains for each zone

Table 6.40   January internal gains (kWh) by different methods

Source	Prescribed	Stochastic
Occupants	958	784
Lights	1227	918
Small power	888	1099

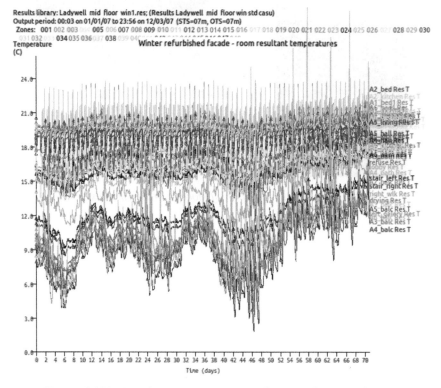

Figure 6.103   Resultant temperatures with original storage heaters

Wintertime simulations were conducted for the pre- and post-upgrade designs. In relation to thermal comfort, Figure 6.103 shows the variation of zone resultant temperature within multiple zones. This indicated that the heating system was failing to maintain acceptable indoor conditions when required.

The next step was to explore better storage heater design and smarter control. Changes to the model included improved case insulation (to limit uncontrolled heat emission), fan control, and smart charge control as available in the latest generation of storage heaters (e.g., Quantum Heaters[42]). Incorporating these aspects led to an improvement in charge/discharge control and less discomfort events thus increasing their viability within an all-electric future.

6.22 Inter-program comparison

Progress in the field is often accomplished when one tool is used to help refine another. An inter-model comparison is frequently used to scrutinise a new program or one that is subject to change. This is especially helpful where the benchmark program has the ability to emulate the assumptions of the target program. In ASHRAE Research Project 472 (Sowell 1988) ESP-r was one of three programs used to cross-check the DOE-2 program's zone weighting factor calculation. The aim was to determine the amplitude and time delay of the zone cooling load resulting from a 24-hour period, unit amplitude solar heat gain. Configurations of zones with different geometry and construction arrangements were processed by ESP-r based on its finite difference method as summarised in Table 6.41.

Before simulation could commence, it was necessary to configure ESP-r to emulate the response function approach of DOE-2 as follows.

- All constructions were defined as adiabatic by setting adjacent air temperatures equal to the associated surface temperature for all times. External long-wave radiation exchanges were set identically zero.
- Internal and external surface heat transfer coefficients were assigned time invariant values, first as combined convective/radiative values and then as separated values.
- ESP-r's solar algorithm was bypassed and replaced by a unit pulse generator stimulating internal surfaces in specified proportions.

The first analysis corresponded to the simplest case: combined surface radiative and convective heat transfer coefficients. That is, ESP-r's internal long-wave algorithm is disabled and combined resistances link the surfaces to the air point. Table 6.42 lists the results.

Considerable disagreement was evident in both amplitude and time delay, with the ESP-r amplitudes being less by 10–20% and the time delays greater by up to 60%.

ESP-r's theoretical model was then modified and Table 6.43 gives the results when the internal convective part of the surface coefficients was prescribed and ESP-r's internal long-wave radiation exchange algorithm was enabled.

The effect is to worsen consensus by moving the ESP-r amplitudes away from those predicted by DOE-2 and increasing the time delays: the best case (H2) being 10% greater; the worst case (N) being 586% greater!

---

[42]https://dimplex.co.uk/quantum/

Table 6.41   Zone parameter diversification

Case	Zone size	Glazing[#] %	Floor finish	Floor weight	Ceiling type	External wall	Drapes %
A[*]	LG	0	CP	2C	NC	EE	HD
B[*]	LG	0	CP	2C	NC	AA	FD
C	LG	50	CP	1W	CL	CC	FD
D	SM	99	NC	8C	NC	EE	ND
E	LG	99	NC	8C	CL	GG	ND
F	LG	99	NC	8C	NC	EE	ND
G	SM	99	NC	8C	NC	EE	HD
H[*]	SM	0	NC	8C	CL	AA	HD
I	LG	99	NC	2C	NC	GG	ND
J	SM	99	NC	2C	NC	AA	ND
K	LG	99	NC	2C	CL	CC	ND
L	SM	50	NC	2C	CL	EE	FD
M[*]	LG	0	NC	2C	NC	CC	FD
N	SM	99	NC	1W	NC	GG	ND
A2	SM	50	CP	2C	CL	EE	ND
B2	LG	50	CP	2C	CL	CC	ND
C2	SM	50	CP	1W	CL	AA	ND
G2	LG	0	CP	8C	NC	GG	ND
H2	LG	0	CP	8C	NC	CC	ND
L2	SM	0	NC	1W	CL	CC	ND
M2	SM	0	CP	8C	NC	CC	ND

[*]Not processed due to the ambiguity of zero glazing with drapes.
LG – large zone, SM – small zone.
[#] % of the south-facing external wall.
CP – carpet, NC – no carpet.
1W – 1 in. wood floor, 2C – 2 in. concrete floor, 8C 8 in. concrete floor.
CL – suspended ceiling, NC – no suspended ceiling.
AA, CC, EE, and GG – ASHRAE wall types A, C, E, and G.
ND – no window cover, HD – half window cover, FD – full window cover.

The next step was to modify ESP-r's adiabatic wall treatment to emulate better the DOE-2 assumption of no wall heat flux[43]: for the external construction, the outside air temperature was set identical to the internal air temperature; for internal constructions, the 'other side' air temperature was set identical to the corresponding inside surface temperature. Whilst it was not certain that this treatment is appropriate in the context of ESP-r's numerical model, it is interesting to observe the effect on predictions as shown in the following table for the F and N cases as shown in Table 6.44.

[43]The response function technique employs the principle of superimposition in which each heat transfer process is considered independently.

Table 6.42 Inter-program comparisons

Case	DOE-2 amplitude (–)	ESP-r amplitude (–)	DOE-2 delay (hrs)	ESP-r delay (hrs)
A[*]	0.672	n/a	0.784	n/a
B[*]	0.647	n/a	0.829	n/a
C	0.825	0.858	0.655	2.42
D	0.438	0.408	1.992	3.98
E	0.293	0.277	2.13	4.13
F	0.331	0.277	2.94	4.12
G	0.526	0.508	1.26	3.15
H[*]	0.524	n/a	1.08	n/a
I	0.641	0.475	1.99	5.27
J	0.696	0.535	1.65	4.68
K	0.513	0.465	2.83	5.35
L	0.739	0.858	0.91	2.42
M[*]	0.629	n/a	1.33	n/a
N	0.860	0.712	3.55	0.53
A2	0.686	0.573	0.77	2.62
B2	0.617	0.540	0.79	1.68
C2	0.838	0.678	0.66	2.68
G2	0.577	0.471	0.91	0.98
H2	0.534	0.486	0.95	1.05
L2	0.833	0.690	0.86	3.58
M2	0.606	0.531	0.95	2.22

Table 6.43 Split heat transfer coefficients

Case	DOE2.1C amplitude (–)	ESP-r amplitude (–)	DOE2.1C delay (hrs)	ESP-r delay (hrs)
A[*]	0.672	n/a	0.78	n/a
B[*]	0.647	n/a	0.83	n/a
C	0.825	0.594	0.66	2.85
D	0.438	0.291	1.99	4.22
E	0.292	0.260	2.13	4.42
F	0.331	0.211	2.94	4.52
G	0.526	0.250	1.26	3.22
H[*]	0.524	n/a	1.08	n/a
I	0.641	0.378	1.99	6.02
J	0.696	1.651	0.43	5.45
K	0.513	0.430	2.83	5.55
L	0.739	0.511	0.91	3.25
M[*]	0.629	n/a	1.33	n/a
N	0.860	0.594	0.52	4.38
A2	0.686	0.473	0.77	3.15
B2	0.617	0.445	0.79	2.35
C2	0.838	0.542	0.66	3.38

(Continues)

Table 6.43   (Continued)

Case	DOE2.1C amplitude (–)	ESP-r amplitude (–)	DOE2.1C delay (hrs)	ESP-r delay (hrs)
G2	0.577	0.294	0.91	1.72
H2	0.534	0.299	0.95	1.92
L2	0.833	0.513	0.86	3.95
M2	0.606	0.351	0.95	2.68

Table 6.44   Alternative adiabatic wall treatment

Case	DOE-2 amplitude (–)	ESP-r amplitude (–)	DOE-2 delay (hrs)	ESP-r delay (hrs)
F	0.331	0.238	2.94	4.95
N	0.860	0.574	0.52	2.85

Now, the ESP-r amplitude moved closer to the DOE-2 value in case F but further away in case N, whilst the time delay worsened in the former case and improved considerably in the latter case (although still some 82% greater).

The complete set of ESP-r results was submitted for processing along with the results from the other programs participating in the study. Figure 6.104 shows two of many results, here for cases C and F as reported by Sowell (1988).

These two cases show poor and good agreement between the participating programs; in the former case, ESP-r is the greatest outlier depending on the modelling assumption.

Sowell (1988) summarised the project thus:

> *A cross-check was performed in which DOE-2.1C custom weighting factor results were compared to dynamic response predictions of three other internationally recognized programs. The results of the cross-check show that whilst the programs do not predict identical response, cooling load amplitude agreement was within ±10% of the heat gain excitation, and delay agreement was within ±0.6 hours. Although this agreement was judged to be adequate, the discrepancies for certain zones suggest that there are unresolved issues regarding load calculations that should be addressed.*

DOE-2.1C was subsequently refined in relation to its treatment of boundary conditions and the modelling of radiative transfer. Whilst the specific modifications are unknown, it is interesting to observe that the program subsequently offered users the choice of zone models based on response function or finite difference methods.

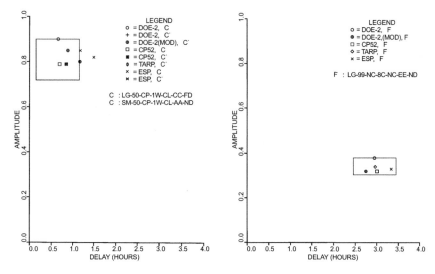

Figure 6.104 Result for cases C and F (from Sowell 1988)

6.23 Assessing the impact of algorithm simplification

This section describes the systematic degradation of a BPS+ tool to emulate the methods and assumptions of a simplified tool as a means to determine routes for improvement. This was the aim of a project commissioned by the (then) UK Energy Technology Support Unit (ETSU)[44] (Lomas 1992) to:

- use ESP-r to determine the effect of different glazing types and areas when applied to a direct gain, passive solar house; and
- compare the findings with those obtained by the SERI-RES program (Littler and Haves 1986) to confirm the suitability of this tool for use in ETSU's passive solar buildings programme (Clarke 1994).

The advantage of benchmarking against a first principles simulation program is that it is possible to degrade systematically the underlying methods to emulate aspects of the simplified program. Furthermore, although the simulation program does not use aggregated input parameters such as construction U-values, it is possible to output such parameters to ensure input model equivalency during the inter-program comparison.

An ESP-r model of a passive solar house was established as shown in Figure 6.105 as a copy of a previously established SERI-RES model. The ESP-r model included an airflow network as shown.

---

[44]https://discovery.nationalarchives.gov.uk/details/r/C16470/

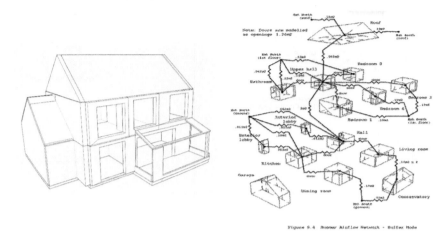

Figure 6.105   Summary of the study model

Annual simulations focused on the south facade glazing area and type. The areas were varied between 0 and 1.4 times the initial design value whilst the glass included single clear float, double clear float, and double Kappafloat™ (now called Pilkington Optitherm™). Window curtains were closed overnight, with operation times depending on the time of year. The results demonstrated a marked disagreement between the two programs.

The project brief was to apply ESP-r in SERI-RES emulation mode in an attempt to understand the reasons for divergence. These required adjustments to ESP-r related to the following issues.

- SERI-RES uses environmental temperature as the control point, whereas ESP-r can use any user-defined reference temperature. For initial ESP-r simulations, the reference point was defined as air temperature. The use of environmental temperature will lead to higher predictions of auxiliary heating demand. ESP-r simulations were re-run assuming environmental temperature as the reference point.
- The initial modelling specification required convective heat input to zones, which was not applied in SERI-RES due to the use of environmental temperature.
- Casual gains were assumed 20% radiant and 80% convective in ESP-r whilst the use of environmental temperature implies a different split. ESP-r was changed accordingly.
- Inter-zone airflow in ESP-r is a vector quantity meaning that heat is carried to or removed from only the receiving zone, the zone from which the air originates being affected in turn by its replacement air. In SERI-RES, an equivalent conductance was being used to represent inter-zone air coupling, a mechanism that permits heat exchange in both directions even where there would be advection in only one direction. This treatment was imposed on ESP-r.

- In the initial ESP-r simulations, ground temperatures were set to typical monthly values for the United Kingdom. SERI-RES assumed that the ground temperature for any month was equal to the mean external air temperature for the previous month. This assumption was imposed on ESP-r.
- ESP-r calculates surface convective and radiative heat transfer coefficients at each computational time step. The constant values assumed by SERI-RES were imposed on ESP-r.

The annual simulations for double and single glazing were repeated but with ESP-r operating in SERI-RES emulation mode. The results demonstrated that the main reason for the initial disagreement was due to the use of environmental temperature in SERI-RES. This in turn forced the use of fixed heat transfer coefficients and dictated an unrealistic treatment of casual gains and zone heat input. Other contributing factors related to the building physics representation.

- SERI-RES used a constant 'solar to air fraction', whilst ESP-r determines this heat flux from the window's spectral absorptance at the prevailing solar incidence angle.
- SERI-RES used a constant 'solar lost' fraction, whilst ESP-r determined this heat flux as a function of solar incidence angles and system thermophysical properties.
- ESP-r determines the spectral transmittance, absorptance, and reflectance of each window and at each angle of incidence. All solar processes correspond to an anisotropic sky model operating in tandem with a solar ray-tracking algorithm. SERI-RES used shading coefficients that relate normal incidence solar gain to a reference glazing type.

The outcome of the analysis was that ETSU switched to the use of a BPS+ tool in its passive solar design programme.

6.24   Characterising background infiltration

One problematic aspect of BPS+ model is the representation of infiltration. This section describes the use of a pressurisation test to identify this parameter.

A BPS+ tool can be used to emulate an intended real-world monitoring intervention to inform decisions on issues such as sensor placement and the frequency of data capture. Not only does this improve the veracity of the monitoring scheme, but it also ensures that the collected data can be used to recalibrate the model before simulations aimed at performance improvement or to extrapolate outcomes to other situations.

The example described here relates to the monitoring of low-energy dwellings with layout and system variations applied. An initial task was to determine airtightness via a 50 Pa pressurisation test involving the use of a blower door fan (Martín-Garín *et al.* 2020). Whilst this is a conventional approach, other methods exist or are under development:

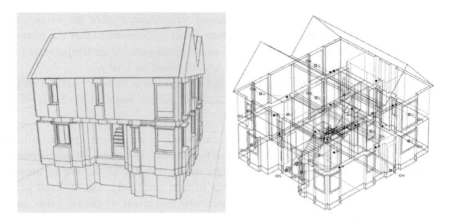

Figure 6.106   Dwelling with blower door attached and a final airflow network representation

- a low-pressure (∼4 Pa) pulse technique involving the release of a pulse of air and the measurement of the decay in building pressure over a few seconds;
- an air tightness test that switches on and off the ventilation system enabling airtightness to be calculated from measured changes in pressure and volume flow rate; and
- a method that uses infrared and acoustic measurements to locate air leakage paths directly.

The last method is interesting in that it can be used to identify leakage distribution and the parameters of a corresponding airflow network. This has the potential to make airflow modelling routine in future. Figure 6.106 shows a dwelling with a blower door highlighted.

An initial ESP-r model was configured to match the test by including a simple airflow network comprising leakage paths connecting each room at an air pressure of 50 Pa to the outside via a single orifice component. The leakage paths associated with windows, doors, and ventilation units were not included at this stage as these are disabled during a pressurisation test.

The pressurisation test result – the whole-dwelling infiltration rate corresponding to a 50 Pa internal pressure – was then used to identify the parameter of the orifice component that gives the same volume airflow in the simulation as measured in the test.

The previously disabled leakage paths are now enabled, and the imposed 50 Pa internal pressure constraint is removed. The ESP-r model (Figure 6.106, right) now has a reasonable representation of infiltration and is ready to be used to explore the impact of design changes under realistic operational conditions.

6.25 Chapter summary

BPS+ models are highly adaptable and may be applied to a wide range of problem types from new build and refurbishment to component design and air quality appraisal. This chapter has summarised problem types and scales that characterise the potential. In most cases, the projects were serviced by ESRU software applications that were under development at the time and so could be adapted rapidly in response to exposed deficiencies. Because these applications serve as a platform to research the issues underpinning the computational approach to design, they remain objects of ongoing refinement. They are made available at no cost under an open-source licence in the hope that others will contribute to the refinement process and application in practice.

References and further reading

Aasem E O (1993) 'Practical simulations of buildings and air-conditioning systems in the transient domain', *PhD Thesis*, ESRU, University of Strathclyde, Glasgow.

ACIS (2020) 'Telecoms cabin monitoring', *Case Study Report*, airedale.com/wp-content/uploads/2020/01/Telecoms-Monitoring-Case-Study.pdf.

Allison J, Bell K, Clarke J A, *et al.* (2018) 'Assessing domestic heat storage requirements for energy flexibility over varying timescales', *Applied Thermal Engineering*, 136, 602–616.

Allison J, Cowie A, Galloway S, Hand J, Kelly N and Bruce S (2017) 'Simulation, implementation and monitoring of heat pump load shifting using a predictive controller', *Energy Conversion and Management*, 150, 890–903.

Anderson J G A, Clarke J A, Kelly N, McLean R C, Rowan N and Smith J E (1996) 'Development of a technique for the prediction/alleviation of conditions leading to mould growth in houses', *Final Report for Contract Number 68017*, Scottish Homes, Edinburgh.

Arteconi A, Costola D, Hoes P and Hensen J L M (2014) 'Analysis of control strategies for thermally activated building systems under demand side management mechanisms', *Energy and Buildings*, 80, 384–393.

Baster E (2017) 'Modelling the performance of air source heat pump systems', *PhD Thesis*, ESRU, University of Strathclyde.

Beausoleil-Morrison I (2000) 'The adaptive coupling of heat and air flow modelling within dynamic whole-building simulation', *PhD Thesis*, ESRU, University of Strathclyde.

Beausoleil-Morrison I, Kummert M, Macdonald F, Jost R, McDowell T and Ferguson A (2012) 'Demonstration of the new ESP-r and TRNSYS co-simulator for modelling solar buildings', *Energy Procedia*, 30, 505–514.

Beizaee A, Allinson D, Lomas K J, Foda E and Loveday D L (2015) 'Measuring the potential of zonal space heating controls to reduce energy use in UK homes: the case of un-furbished 1930 dwellings', *Energy and Buildings*, 92, 29–44.

Buresch M (1983) 'Photovoltaic energy systems—design and installation', *Solar Energy*, 33(3–4), p. 388.

Burton S, Doggart J and Grace M (1996) 'Community planning for Glasgow', *Renewable Energy Strategy in European Towns Report*, EC RE-Start Project Report, Brussels.

Chiesa G and Grosso M (2019) 'Python-based calculation tool of wind-pressure coefficients on building envelopes', *Journal of Physics: Conference Series*, 1343, 012132.

Chow T (1995) 'Air-conditioning plant components taxonomy by primitive parts', *PhD Thesis*, ESRU, University of Strathclyde, Glasgow.

Citherlet S and Hand J (2002) 'Assessing energy, lighting, room acoustics, occupant comfort and environmental impacts performance of building with a single simulation program', *Building and Environment*, 37(8–9), 845–856.

Clarke J A (1994) 'Assessing the credibility of SERI-RES for buffer mode conservatories', *Final report for project S/D3/00190/00/00*, Energy Technology Support Unit.

Clarke J A and Hensen J L M (1991) 'An approach to the simulation of coupled heat and mass flow in buildings', *Indoor Air Quality and Climate*, 1, 283–296.

Clarke J A and Janak M (1997) 'Integrated thermal and lighting appraisal of daylight utilisation technologies in buildings', *Proc. BEPAC '97*.

Clarke J A and Johnston C (2010) 'Assessing and benchmarking the performance of advanced building facades', Chapter 24 in *Materials for Energy Efficiency and Comfort in Buildings and Occupied Spaces*, Woodhead Publishing Energy Series, 484–502.

Clarke J A and McGhee R (2017) 'Assessment of estate upgrade options', Consultancy report, ESRU, University of Strathclyde.

Clarke J A, Cockroft J, Conner S, *et al.* (2002) 'Simulation-assisted control in building energy management systems', *Energy and Buildings*, 34, 933–940.

Clarke J A, Dempster W M and Negrao C (1995) 'The implementation of a computational fluid dynamics algorithm within the ESP-r system', *Proc. Building Simulation '95*, University of Wisconsin, Madison, USA.

Clarke J A, Hand J W, Janak M and MacDonald I (1997) 'Simulation case study: college La Vanoise, Modane, France', *Report of the Daylight-Europe Simulation Team*, ESRU.

Clarke J A, Hand J W, Johnstone C M, Kelly N and Strachan P A (1996) 'Photovoltaic-integrated building facades', *Renewable Energy*, 8(1–4), 475–479.

Clarke J A, Hand J W, Johnstone C M, Kelly N, and Strachan P A (1996) 'The characterisation of photovoltaic-integrated building facades under realistic operating conditions', *Proc. 2nd European Conference of Architecture*, Berlin, 26–29 March.

Clarke J A, Hand J W, Mac Randal D F, and Strachan P A (1995) 'The development of an intelligent, integrated building design system within the European COMBINE project', *Proc. Building Simulation '95*, University of Wisconsin, Madison, USA.

Clarke J A, Johnstone C M, Macdonald I A, *et al.* (2000) 'The deployment of photovoltaic components within the lighthouse building in Glasgow. *Proc. 16th European Photovoltaic Solar Energy Conf.*, Glasgow.

Cockroft J and Samuel A (2010) 'Energy saving potential of Chop-cloc device in UK housing', *Consultancy Report E401*, ESRU, University of Strathclyde.

Cockroft J and Samuel A (2012) 'Benefit of adding room temperature control to a basic central heating system', *Consultancy Report* E412, ESRU, University of Strathclyde.

Cockroft J, Cowie A, Samuel A and Strachan P (2017) 'Potential energy savings achievable by zoned control of individual rooms in UK housing compared to standard central heating controls', *Energy and Buildings*, 126, 1–11.

Cockroft J, Kokogiannakis G and Kummert M (2008) 'Govanhill Baths Development Project: Energy Saving Opportunities', *Consultancy Report E359*, ESRU, University of Strathclyde.

Compagnon R (1996) 'The JINDEX visual discomfort analysis program', *Daylight Europe Project Report*, LESO-PB, EPFL.

Cowie A, Clarke J A and Hand J W (2020) 'Project 2020, Type C House Initial Simulation Study', *Consultancy Report*, ESRU, University of Strathclyde.

de Castro M, Sharpe T, Kelly N and Allison J (2018) 'A taxonomy of fabric integrated thermal energy storage: a review of storage types and building locations', *Future Cities and Environment*, 4(1).

EMFS (2023) *Electric and magnetic fields and health*, emfs.info/sources/substations/final/.

Flett G (2017) 'Modelling and analysis of energy demand variation and uncertainty in small-scale domestic energy systems', *PhD Thesis*, University of Strathclyde.

Fontoynont M (Ed.) (2017) *Daylight Performance of Buildings*. London: Routledge.

Frontini F, Kuhn T E, Herkel S, Strachan P and Kokogiannakis G (2009) 'Implementation and application of a new bi-directional solar modelling method for complex facades within the ESP-r building simulation program', *Proc. Building Simulation '09*, Glasgow.

Grant A D and Dannecker R (1998) 'Dispersal of gases from fume cupboards in the new Wellcome Millennial Clinical Research Facility, Edinburgh', *Technical Report*, ESRU, University of Strathclyde.

Grant A D and Dannecker R (2000) 'A hybrid PV/wind energy module for integration in buildings', *Proc. 16th European Photovoltaic Solar Energy Conf.*, Glasgow, May.

Grant A D, Johnstone C M and Kelly N (2008) 'Urban wind energy conversion: the potential of ducted wind turbines', *Renewable Energy*, 33(6).

Guth S K (1966) 'Computing visual comfort ratings', *Illuminating Engineering, October*, 634–642.

Hand J W (1998) 'Removing barriers to the use of simulation in the building design professions', *PhD Thesis*, University of Strathclyde.

Hand J W (2020) 'Analysis of design options for Home Group Inovation Village', Report E11/20, ESRU, University of Strathclyde.

Hand J W and Cockroft J (2002) 'Analysis of communication shelter ventilation strategies', *Consultancy Report E211*, ESRU, University of Strathclyde.

Hand J W and Hensen J L M (1995) 'Recent experiences and developments in the training of simulationists', *Proc. Building Simulation '95*, University of Wisconsin, Madison, USA.

Hand J W and Kelly N (2020) 'Analysis of supervisory control options for storage heaters in high rise housing', *Ruggedised Project Task Report*, ESRU, University of Strathclyde.

Hand J W and Samuel A A (2004) 'Performance assessment of Municipal Theatre of Besancon', *Consultancy Report E227*, ESRU, University of Strathclyde.

Hand J W and Strachan P A (2000) 'Modelling study of the integration of PV modules into an office building', *Report for IMPACT project Work Package 8*, ESRU, University of Strathclyde.

Hand, J W (2020) *Strategies for Deploying Virtual Representations of the Built Environment*, https://contrasting.onebuilding.org/Strategies/Index.html.

Hawes D W, Feldman D and Banu D (1993) 'Latent heat storage materials', *Energy and Buildings*, 20, 77–86.

Heim D and Clarke J A (2004) 'Numerical modelling and thermal simulation of PCM-gypsum composites with ESP-r', *Energy and Buildings*, 36, 795–805.

Hensen J L M (1991) 'On the thermal interaction of building structure and heating and ventilating systems', *PhD Thesis*, Eindhoven University of Technology.

Hutchins M (1998) 'Advanced glazings: a status report', *Sunworld*, 22(1), 5–11.

Imbabi M S and Peacock A (2004) 'Allowing buildings to breathe', *Renewable Energy*, 85–95.

Kelly N J (1998) 'Towards a design environment for building-integrated energy systems: the integration of electrical power flow modelling with building simulation', *PhD Thesis*, ESRU, University of Strathclyde.

Kelly N J (2013) 'Performance of PCM-based thermal storage for load shifting of heat pumps', *Consultancy Report E448*, ESRU, University of Strathclyde.

Kelly N J, de Castro M, Sharpe T, Shea A and Strachan P (2017) 'Domestic thermal storage requirements for heat demand flexibility', *Proc. 4th Sustainable Thermal Energy Management Conference*, Alkmaar, Netherlands.

Kristensen P E (1996) 'Daylight Europe', *Proc. 4th European Conference on Solar Energy in Architecture and Urban Planning*, Berlin, March.

Kuhn T E, Herkel S, Frontini F, Strachan P and Kokogiannakis G (2011) 'Solar control: a general method for modelling of solar gains through complex facades in building simulation programs', *Energy and Buildings*, 43(1), 19–27.

Lagoeiro H, Davies G, Marques C and Maidment G (2023) 'Heat recovery opportunities from electrical substation transformers', *Energy Reports*, 10, 2931–2943.

Lewis J O (1995) 'SOLINFO, innovative support for European building designers', *Report to EC DG XII JOU2-CT92-0219'*, Contractors meeting, Barcelona, 26-27 September.

Littlefair P J (1992) 'Modelling daylight illuminance in building environmental performance analysis', *Journal of the Illuminating Engineering Society*, 25–34.

Littlefair P J (1992a) 'Daylight coefficients for practical computation of internal illuminance', *Lighting Research and Technology*, 24(3), 127–135.

Littler J and Haves P (1986) 'Development of SERI-RES within the UK passive solar programme', *Proc. 10th National Passive Solar Conference*.

Lomas K J (1992) 'Applicability study I: executive summary', *ETSU Report* S1213.

Loncour X, Deneyer A, Blasco M, Flamant G and Wouters P (2004) *Ventilated Double Facades – Classification & Illustration of Facade Concepts*, Belgian Building Research Institute.

Martín-Garín A, Millán-García J A, Hidalgo-Betanzos J M, Hernández-Minguillón R J and Baïri A (2020) 'Airtightness analysis of the built heritage – field measurements of nineteenth century buildings through blower door tests', *Energies*, 13, 6727.

McDougall S and Hand J W (2003) 'Using simulation to move from daylighting issues to first use of the new UK carbon emissions calculation method', *Proc. Building Simulation '03*, Eindhoven.

Meyer J J (1993) 'Visual discomfort: evaluation after introducing modulated light equipment', *Proc. Energy-Efficient Lighting*, Arnhem, The Netherlands, pp. 348–357.

Mukhtar U A (2024) 'A computational approach to the sizing of heat pump integrated thermal energy storage systems for wet central heating', *PhD Thesis*, ESRU, University of Strathclyde.

Nakhi A (1995) 'Adaptive construction modelling within whole building dynamic simulation', *PhD Thesis*, ESRU, University of Strathclyde, Glasgow.

Negrao C O R (1995) 'Conflation of computational fluid dynamics and building thermal simulation', *PhD Thesis*, ESRU, University of Strathclyde, Glasgow.

Samuel A and Cockroft J (2008) 'Thermal performance of steel window to determine condensation risk', *Consultancy Report E377*, ESRU, University of Strathclyde.

SAP (2023) 'UK Standard Assessment Procedure for Energy Rating of Dwellings', *https://gov.uk/guidance/standard-assessment-procedure/*.

Selkowitz S, Aschehoug O and Lee E S (2003) 'Advanced interactive facades – critical elements for future green buildings', *Proc. GreenBuild*, Lawrence Berkeley National Laboratory.

SG (2017) *https://gov.scot/publications/building-standards-2017-domestic*.

Sowell E F (1988) 'Cross-check and modification of the DOE-2 program for calculation of zone weighting factors', *ASHRAE Transactions*, 94(2), 737–753.

Strachan P A, Johnstone C M, Kelly N, Bloem J J and Ossenbrink H (1997) 'Results of thermal and power modelling of the PV facade on the ELSA Buildings, Ispra', *Proc. 14th European Photovoltaic Solar Energy Conference*, Barcelona, pp. 1910–1913, July.

Szerman M (1994) 'Auswirkung der Tageslichtnutzung auf das energetische Verhalten von Burogebauden', *Doctoral Thesis*, Stuttgart.

Tait L (2014) 'Impact on network load profiles of 2016 building standards for new buildings and energy efficient retrofits', *Final Report for KTP Project KTP008752*, Scottish Power Energy Networks.

Tang D (2002) 'Environmental impact assessment, Belmont Hall, University of Dundee', *Consultancy Report*, ESRU, University of Strathclyde.

Tang D, Kelly N and McElroy L B (2003) 'Environmental impact assessment of flue stacks located at the centre for inter-disciplinary research, University of Dundee', *Technical Report E205*, ESRU, University of Strathclyde.

Teregenza P R (1980) 'The daylight factors and actual illuminance ratios', *Lighting Research and Technology*, 12(2), pp. 64–68.

Tsang E K W, Li D H W and Li S (2022) 'Predicting daylight illuminance for 15 CIE standard skies using a simple software tool', *Sustainable Cities* 4, doi:10.3389/frsc.2022.792997.

Vandaele L and Wouters P (1994) 'PASSYS services: summary report', *Publication No. EUR 15113 EN*, European Commission, 1994.

Ward G J (1993) *Radiance 2.3 Imaging System*, Lawrence Berkeley Laboratory, Berkeley.

Willbold-Lohr G (Coordinator) (1995) 'Passive sun control with holographic optical elements', *Final Report for Contract JOU2-CT92-0227*.

Winkelmann F C and Selkowitz S (1985) 'Daylighting simulation in the DOE-2 building energy analysis program', *Energy and Buildings*, 8, 271–286.

Wittwer V and Platzer W (2000) 'Transparent thermal insulation materials and systems: state of the art and potential for the future', *High Temperatures, High Pressures*, 32(2) 143–58.

## Chapter 7

Urban energy schemes

This chapter considers the application of BPS+ in support of community energy schemes whereby technologies for energy generation are co-located with demands as a means to attain improved resource utilisation. Whether and where this approach is generally viable has yet to be proven.

Whilst hydro stations, biogas plants, and wind/solar farms are the typical technologies used to capture renewable energy, these technologies introduce complexities in network management, which (under traditional operation at least) will limit renewable energy deployment to an estimated 25%–30% (EA 99) of installed capacity. These limitations are due to the intermittent nature of renewable resources, which require reserve capacity and energy storage to maintain network stability and power quality during periods of fluctuating outputs.

As an alternative to the strategic connection of renewable energy sources to the electricity grid, it is possible to deploy small-scale systems in a manner that offsets local energy demand, as demonstrated in Section 6.16. The analysis of proposals will require a model comprising multiple buildings and the various technologies that can be used to service them – such as roof-mounted solar panels, local solar/ wind farms, individual or communal heat pumps, and smart overall control. Because renewable energy sources are stochastic, geographically dispersed, and of low/variable energy density, their ability to match demand at the urban scale generally requires prior action to reduce the demand to a level commensurate with the locally available resource. Consider the following scenario.

For a temperate climate characterised by a mean annual solar irradiance of 175 W/m^2, the mean power production from a photovoltaic panel with a 15% conversion efficiency is approximately 25 W/m^2. For a mean wind speed of 5 m/s, the power produced by a micro wind turbine will be of a similar order of magnitude but with a different profile shape. The point is that such power densities are significantly lower than a building's energy demand. In the UK, for example, a typical office building will have an annual energy demand of between 200 and 400 kWh/m^2 corresponding to naturally ventilated and air-conditioned cases, respectively. These data translate to a typical power demand of around 70 and 140 W/m^2 respectively – some three to five times the available renewable resource! This situation changes dramatically where aggressive energy efficiency measures are deployed: marginally surpassing best practice standards in the natural ventilation case, for example, would reduce the annual energy demand to 110 kWh/m^2 (or 35 W/m^2 average power demand). This suggests the possibility of a useful quantitative match where systems are first made energy efficient. The challenge is to overcome the temporal mismatch without recourse to importing/exporting from/to the public supply networks: this is the aim of dynamic demand management.

Performance simulation provides the means to appraise the performance of low-energy communities featuring cooperating technologies for demand management and sustainable heat/power delivery. Only by addressing such communities in a holistic and dynamic manner can the operational characteristics of the individual technologies be discerned and overall, well-found solutions established.

Based on a review of urban energy system models, Keirstead *et al.* (2012) concluded that there was significant potential to move beyond single disciplinary approaches towards an integrated perspective that captures the theoretical complexity of the domain. This can be achieved by applying several models in cooperative mode. The example application that follows is based on cooperation between the ESP-r and Merit (Born 2001) programs. The aim is to inform the user of the mix of technologies (by type and capacity) that best matches the demand without imposing arbitrary external factors – such as the available roof area in the case of PV – which serve only to inappropriately constrain the solution space. This requirement has been met by implementing a 'chunking' procedure in the Merit program, as explained below.

7.1   Formulating an exploratory model

The supply technologies of interest are initially identified. This list can be highly speculative and even include technologies that are only of marginal interest. The starting point for the purposes of the project described here is that the demand profile is known. Where the community exists, this profile may well be available in the form of utility data. Where it corresponds to a new build situation, estimates of demand may be available from the design process. If no data were available, it would be necessary to estimate the demand via an extended energy audit or modelling study (perhaps based on the semi-automated approach to estate model formulation as described in Section 8.1). The important point is that the method commences with full knowledge of the community energy demands that are to be serviced via a hybrid supply system.

The capacity of each technology option is set in Merit to an initial value equal to or greater than that required to match the peak demand if that technology were deployed alone (and, of course, the renewable source was simultaneously available, which is unlikely to occur in practice). Note that the eventual outcome is independent of these initial capacity value assignments.

The capacity of each technology is now 'chunked' as a function of the minimum feasible contribution. For example, 1 MW of wind capacity might be composed by appropriate selections from an arbitrary list of turbine capacities (e.g., 50, 150, 250, 350, 450, 550, 650, 750, 850, and 950 kW, and 1 MW) because the rated power will affect the turbine's capacity coefficient (see Section 13.2). On the other hand, 1 MW of PV might be represented as 100 separate modules, each of 10 kW, because outputs are additive. The user has control of the chunk size, which will depend on the scale of the scheme being processed: as with the discretisation process applied within a numerical representation of a building, chunk size can vary as a function of the required accuracy. This initial specification is designed to be as broad as possible, with the subsequent BPS+ analysis used to establish the operational feasibility and efficacy of the best-fit supply technology mixes to result from the Merit analysis as described below.

Merit undertakes a demand/supply profile match assessment involving a large number of combinatorial possibilities due to the chunking process (wind chunk 1 with wind chunk 2 etc., PV chunk 1 with PV chunk 2 etc., wind chunk 1 with PV chunk 1 etc., and so on). The result is a rank-ordered set of matches based on three criteria, as follows.

A Rank Correlation Coefficient (RCC; Scheaffer and McClave 1982) describes the correlation between demand and supply by calculating the degree to which the profile variables fall on the same least square line:

$$\text{RCC} = \frac{\sum_{t=0}^{n} (D_t - d)(S_t - s)}{\sqrt{\sum_{t=0}^{n} (D_t - d)^2 \sum_{t=0}^{n} (S_t - s)^2}} \tag{7.1}$$

where D_t is the demand at time t, S_t the supply at time t, d the mean demand over time period n, and s the mean supply over time period n. RCC describes the trend between the time series of two datasets and does not consider the relative magnitudes of the individual variables. Thus, if a supply system were doubled in size, RCC would remain the same even though the excess supply would be greater. Additionally, two profiles perfectly in phase with one another, but of very different magnitudes, would result in a perfect correlation, but not a perfect match.

An Inequality Coefficient (IC; Williamson 1994) describes the magnitude of the inequality due to three sources – unequal tendency (mean), unequal variation (variance), and imperfect co-variation (co-variance). IC ranges between 0 and 1, with 0 indicating a perfect match and 1 denoting no match.

$$\text{IC} = \frac{\sqrt{\frac{1}{n}\sum_{t=0}^{n} (D_t - S_t)^2}}{\sqrt{\frac{1}{n}\sum_{t=0}^{n} (D_t)^2} + \sqrt{\frac{1}{n}\sum_{t=0}^{n} (S_t)^2}} \tag{7.2}$$

A Percentage Match (PM) is then given by

$$PM = (1 - IC)100 \tag{7.3}$$

where PM indicates the overall match between demand and supply, ranging from excellent (90%–100%) to very poor (0%–10%).

The rank-ordered matches that emerge from the process, as depicted in Figure 7.1, represent the hybrid system supplies that deliver the best quantitative match to the demands. However, the above criteria do not reflect the feasibility of the matches so identified. This requires that BPS+ be used to assess the operational performance of schemes of interest based on heat and power usability considerations.

A scheme selected from the matching stage is transformed into a model suitable for simulation. In the present case, the target program was ESP-r, requiring the explicit definition of multiple buildings – in terms of geometry, construction, operation, and control parameters – and multiple supply technologies – in terms of system layout, building interaction, and control parameters. Whilst this ESP-r input model could be created in a conventional manner, with each building/system model explicitly defined, the scale of an embedded generation community scheme, perhaps comprising several hundred dwellings, will render this approach infeasible in most cases. To reduce the model creation workload, an alternative approach can be used based on the parametric diversification of characteristic archetype models (Clarke and Samuel 2011) as further elaborated in

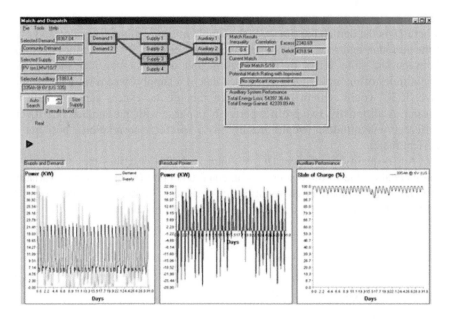

Figure 7.1   A Merit demand/supply matching outcome

Sections 7.7 and 8.1 for the non-domestic and domestic sectors, respectively. These prototype models represent conceivable building design configurations now and after all possible upgrades. The approach, as encapsulated in the Building Stock Modeller (BSM[1]) procedure can be used to generate automatically a multi-building ESP-r model based on user inputs specifying the design parameters to be diversified and the required level of granularity.

Within the BSM approach, it is assumed that the energy performance of a building is a function of a finite number of principal parameters, such as exposure, insulation level, air tightness, thermal capacity location, solar ingress level, occupancy level, and living area fraction in the case of a dwelling. The outcome in this case would represent the housing stock as a distinct model differentiated by the unique combination of its parameters. Application of the approach to the Scottish Housing Stock, as explained in Section 8.1, resulted in 18,750 models corresponding to five levels for each of the parameters indicated above, with the exception of solar ingress, which had three parameters, and thermal capacity position, which had two. Application to the Seoul commercial building estate, as explained in Section 7.7, resulted in 122,472 models (representing the existing buildings and all likely upgrades).

An ESP-r estate model can then be formed by selecting and combining models based on descriptive parameters such as type, age, construction, use, etc. A model calibration exercise is then carried out to ensure that predictions are aligned with the initial demand profile of the community. Alternatively, this estate model can be formed at the beginning to determine the demand profile of the community prior to the matching process. In this case, the calibration exercise might utilise benchmark data.

Finally, to confirm operational resilience, the community model can be simulated, in part at least, alongside the supply technologies identified as the best candidates in the Merit demand/supply matching procedure. The next section considers an approach that can be applied when the details of the supply system are known and the buildings to be connected already exist – here a proposed district heating system.

7.2   District heating

The scheme considered here is the connection of a community to a district heating system based on conventional or renewable energy sources, as summarised in the schematic of Figure 7.2 (Wiltshire *et al.* 2014).

At the design stage, it is appropriate to consider demand-side management to reduce and reshape energy demands to present a favourable load profile to the centralised supply. On the supply side, it is possible to consider a single energy source or a hybrid design comprising a mix of available technologies (e.g., conventional/biomass boiler, ground source heat pump (GSHP), geothermal heat extraction, and waste incineration).

[1]https://www.esru.strath.ac.uk/Applications

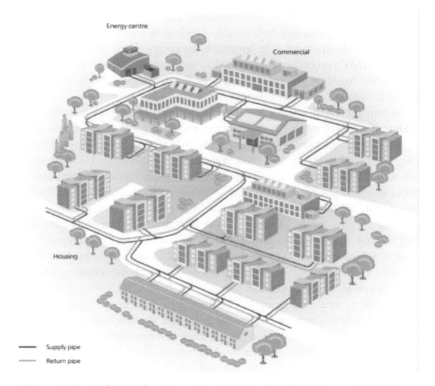

Figure 7.2   A district heating schematic (credit: Wiltshire et al. 2014)

To support such an analysis, a BPS+ model would need to include the heat distribution network, the heat generation plant, the connected buildings, and, perhaps, a heat store to accumulate heat when demand is low for use later when the supply is insufficient. Of course, the building models could be separately used to explore energy efficiency measures, and then the aggregate load profile could be imposed on a district heating model with the building entities excluded.

As a real-world example, consider the Combined Heat and Power (CHP)-based district energy scheme commissioned by the University of Strathclyde as summarised in Figure 7.3 and modelled using ESP-r as part of the EC-funded Ruggedised project[2] (Ruggedised-Glasgow 2019).

The intention was to provide energy to supply 75% of the needs of a well-established, city-centre campus. This level of supply was estimated as equivalent to an annual saving of 875 tonnes of CO_2 compared to the existing heating system (Morales and Roberts 2021). As an aside, initial research found that 62% of the embedded carbon associated with a district heating scheme is associated with the distribution network. This suggests the use of pipes constructed from recycled

[2]https://ruggedised.eu/legacy

Figure 7.3   Heat network within the University of Strathclyde, Glasgow (credit: Morales and Roberts 2021)

materials or low-carbon steel in the future. Section 3.1.5 describes how such a consideration can be explicitly included in a BPS+ model.

Figure 7.4 summarises an ESP-r model corresponding to part of the scheme: a central gas boiler and flow network to deliver hot water to the campus buildings. The purpose of the model was to determine the building energy efficiency measures required to enable low-temperature boiler operation.

Such a model can be used to study a range of issues affecting scheme performance, such as optimum supply temperature, dwelling comfort, locus of control, operating cost, and equitable billing. The model can also be adapted to appraise alternative energy sources and distribution strategies (e.g., two-pipe and parallel networks).

Figure 7.5 shows two typical outputs from ESP-r showing supply and return temperatures from/to the boiler and the total heat loss from the piping network. At the time of writing, the project had not concluded.

Another project example is shown in Figure 7.6, which depicts a district-heating scheme for Queens Quay in Glasgow (Clean Heat at the Water's Edge 2019).

The intention was to provide space heating and hot water to residential and commercial buildings. The heating network consisted of a series of pipes running below the roads of the development. A modelling study indicated that utilising a supply temperature range of 55 °C–75 °C could reduce emissions to 348 tCO_2 from 1,719 tCO_2 at a higher initial temperature, a saving of 79.7%.

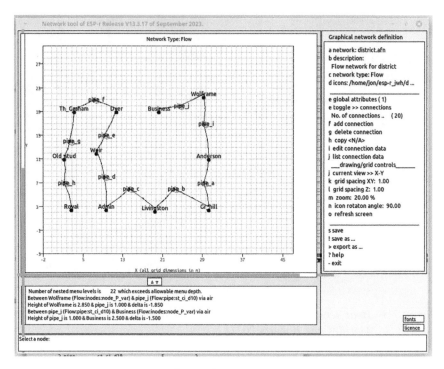

Figure 7.4   ESP-r district heating network model creation

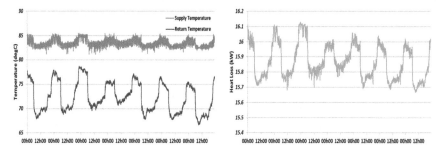

Figure 7.5   Typical simulation outputs

A final project example addressed an existing district heating network serving the West Whitlawburn community in South Lanarkshire[3] (West Whitlawburn 2017). This comprised 644 dwellings divided between six high-rise tower blocks, five low-rise terraces, and 100 new-build properties. The towers and terraces were built in the 1960s. An Energy Centre heats water using a 685 kW wood chip biomass boiler. This is pumped around the district heating network, providing heating and hot water to

[3]https://wwhc.org.uk/district-heating/

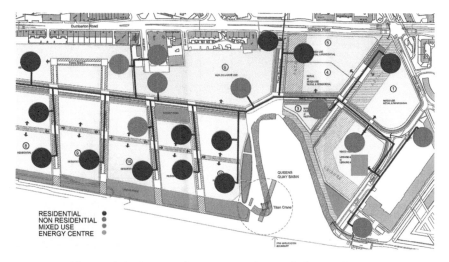

Figure 7.6   Queens Quay masterplan with district heating

individual dwellings. There are three gas boilers providing top-up, and the system incorporates a 50,000-litre thermal store. ESP-r was used to undertake an assessment focused on the tower blocks. The results indicated that a 21% reduction in heat loss could be achieved if the central supply temperature was lowered from 85 °C to 70 °C.

In summary, after a heating network model has been established, it can be used to explore issues such as

- the connectability of existing houses before and after refurbishment;
- the optimum control of central and/or distributed heat pump solutions to maximise COP;
- the phased deployment of the network;
- the connection of disparate consumer types as a means to level out load; and
- the adoption of a hybrid approach to heat supply to utilise local renewable energy resources.

It is then possible to compare outcomes to counterfactual cases such as individual home heating, conventional, electrified, or otherwise.

7.3   Hydrogen filling stations

Hydrogen has widespread support as a clean replacement for fossil fuels. For example, the Scottish Hydrogen and Fuel Cell Association[4] has over 180 commercial and academic members and organises many events, including an annual conference. There is much academic activity in the field and some of this has focused on the development of assessment methods (e.g., Ete 2009).

[4]https://www.shfca.org.uk

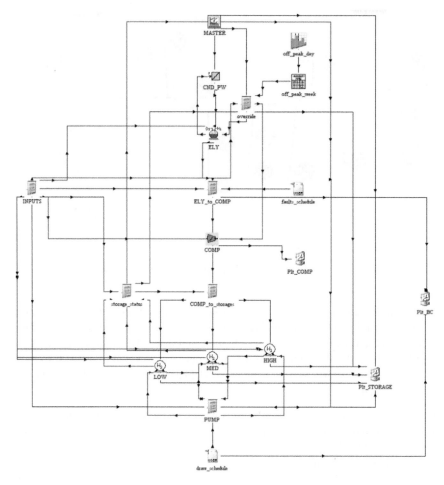

Figure 7.7   TRNSYS model of a hydrogen filling station

The application example summarised here (Kelly and Monari 2019) used TRNSYS[5] and GENOPT[6] to determine the volume of storage and number of electrolysers required to meet typical hydrogen demands placed on a city filling station. Figure 7.7 shows the filling station model, the main components of which are:

- a master controller to dictate the operation of the electrolysers and compressor as a function of storage tank pressure;
- three electrolysers defined by parameters such as the area of the electrode, the number of cells, the number of stacks per electrolyser, and the operating temperature and pressure;

[5]https://trnsys.com/
[6]https://simulationresearch.lbl.gov/GO/

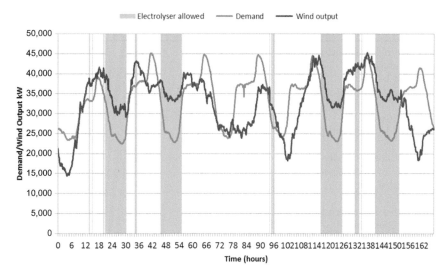

Figure 7.8 Electrolyser operating periods for the constrained capacity case

- a compressor to raise the pressure of the hydrogen leaving the electrolyser and transfer this to the storage tank; and
- three storage tanks each of 4 m^3 volume holding hydrogen at 500 bar.

The model was used to optimise component sizes to supply different daily volumes of hydrogen whilst minimising the station's 10-year capital and running costs. Three filling station operating scenarios were considered:

- a base case where the electrolysers can operate at any time in order to refill the hydrogen storage;
- off-peak, where the electrolysers can operate only during low electricity tariff periods; and
- capacity constrained where the electrolysers operate during periods when the electricity demand is lower than the wind supply in order to absorb surplus wind power (as indicated by the green bands in Figure 7.8).

Simulations were undertaken over a 1-week period, giving rise to the results shown in Table 7.1 for the three cases.

The clear trend is that as more constraints are placed on the operation of the filling station, the larger the volume of storage and the number of electrolysers required.

7.4 Car park PV canopies

Altering the carbon balance of utility grids can be done using PV arrays. In urban environments, this involves challenges requiring more than electrical engineering skills for their resolution: inputs are also required from architects, structural

Table 7.1   Results for the three filling station cases

Daily hydrogen demand (kg)	288	384	480	576	672
Base case					
Storage volume (m³)	7.3	7.5	12.3	11.2	13.1
Number of electrolysers	4	5	5	7	9
Number of pumps	2	2	3	3	4
Capital Cost (£M)	5.0	8.2	8.3	11.5	14.8
Indicative 10-year cost (£M)	83.1	113.8	136.1	163.2	191.5
Off-peak					
Storage volume (m³)	8	9.4	14.6	19.1	21.3
Number of electrolysers	4	6	6	7	8
Number of pumps	2	2	3	3	4
Capital Cost (£M)	6.6	9.8	9.9	11.6	13.3
Indicative 10-year cost (£M)	84.1	119.4	139.0	171.2	198.1
Capacity constrained					
Storage volume (m³)	67.5	67.8	102.5	122.1	133.8
Number of electrolysers	8	13	14	17	21
Number of pumps	2	2	3	3	4
Capital Cost (£M)	13.8	21.8	23.9	28.9	35.5
Indicative 10-year cost (£M)	88	143	154	187	231

engineers, and urban planners, in addition to utility personnel, equipment manufacturers, and end users. BPS+ studies will often involve models that can address
multiple concerns. This was demonstrated in a project (Ruggedised-Glasgow 2019)
to assess the feasibility of installing PV canopies on the roofs of multi-storey car
parks to support car-charging hubs and the possibility of delivering power to nearby
buildings. Figure 7.9 shows the project case study: a multi-storey car park with
nearby highrise flats to the north and northeast and student residences to the south
and southeast.

The design issues to be considered included:

• viable area and layout of the PV arrays consistent with vehicle access and
  potential loss of parking space revenue;
• the impact of the arrays and supports on the building structure;
• the location and type of charging stations consistent with the needs of users
  (especially local taxi drivers with regular charging demands);
• provision of local battery storage for load shifting and flexibility;
• the environmental conditions near charge points and battery banks; and
• the disruption of sight lines from nearby buildings.

The assessment used an ESP-r model as a core repository of all project information.
This model included the following four elements.

(1) A low-resolution model of the car park, ground topography, and surrounding
    buildings is used to support visual impact assessments (Figure 7.10).
(2) A high-resolution model of the car park and surrounding obstructions to
    support an energy analysis (Figure 7.11) and deliver temperatures and air

Figure 7.9   Car park PV canopy case study

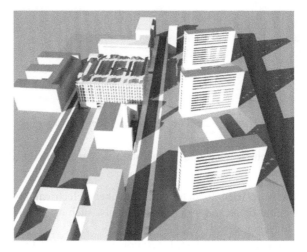

Figure 7.10   ESP-r model of car park and surroundings

movement patterns at multiple points, along with detailed data on roof solar irradiation.

(3) Representations of the PV canopy arrangements and associated power inter-connections. Two options are depicted in Figure 7.12.

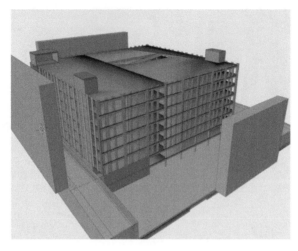

Figure 7.11   High-resolution model of the car park

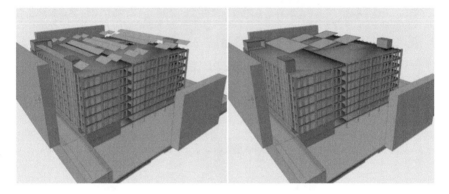

Figure 7.12   Two PV array layout options

(4) Schedules of vehicle charging to facilitate appraisals of the demand/supply match over time.

The models were used to determine the likely electrical power output for different PV module types and array arrangements, such as shown in Figure 7.13 for one layout in May.

The simulation process required iteration to accommodate volatile factors. For example, feedback from taxi drivers suggested a need for more fast charging capabilities as well as adjustments to the charge times and durations.

Adaptations were also required relating to the installed battery capacity and charge station locations. An interesting result concerned the inclination of the PV array, with a near horizontal angle (with just enough inclination to facilitate rain run-off) producing more power output due to the vastly increased deployment area (because inter-array spacing to avoid shading was not then an issue).

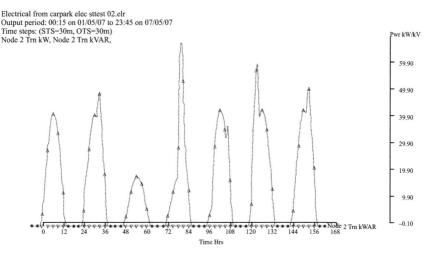

Electrical from carpark elec sttest 02.elr
Output period: 00:15 on 01/05/07 to 23:45 on 07/05/07
Time steps: (STS=30m, OTS=30m)
Node 2 Trn kW, Node 2 Trn kVAR,

Figure 7.13   Electrical output for an array/module variant for May

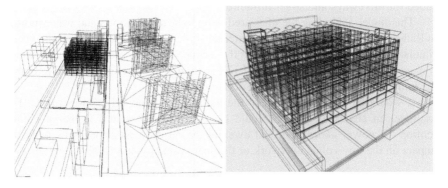

Figure 7.14   The mixed-resolution ESP-r model

At first glance, the model is atypically complex for the needs of the PV array and battery sizing. However, the decision to begin with an urban scale model designed to support a range of assessments and then iterate on that base to address concerns as they arose in the project proved a good choice. As highlighted in Figure 7.14, the resolution of the sightlines model is markedly different from the car park model.

The up-front cost of providing a richer site model paid off as it allowed structural and urban planning staff to be kept fully informed as the EV aspects of the project evolved.

7.5   Smart street lighting

Building performance simulation is typically applied to individual buildings or parts thereof. Less frequently, it might be used for stock modelling, involving the processing of thousands of buildings. Between these extremes lie assessments at the neighbourhood scale, where the goal might be to check for equitable sky access or design a heat-sharing network. The project described here set out to confirm that new LED street lighting would generate acceptable and compliant nighttime light distribution. The scale of the project was restricted to 12 houses along a residential street. A mix of dwelling types was included, representing different plan forms and facade arrangements. The performance metrics required by the client included illuminance levels throughout the street and visualisations from various locations along the street and from rooms within dwellings.

As with most numerical assessments, the project accessed different data sources. The light distribution of the lamps was obtained by laboratory measurement, and high-resolution housing models were drawn from recent research and consulting projects. ESP-r was the primary repository for project data because its data model could accommodate experimental data, building models, and visual entities such as gardens, pavements, lampposts, and roads.

Figure 7.15 shows a neighbourhood model where each dwelling has detail that would support studies beyond lighting distribution, and visual entities have been added for driveways and street furniture.

The model, including the measured lamp light distribution data, was exported to Radiance, with textures added to external surfaces. The outcome from the visual simulation included renderings of the scene for the proposed LED lamps (Figure 7.16) and lux maps (Figure 7.17); such images were interrogated to obtain the luminance/illuminance at points of interest.

It would be possible to extend the model to support other assessments. For example, an electrical network could be added to represent EV charge points, rooftop PV power generation, and dwelling power draw as a means to study the impact on the low voltage network or the spare capacity at a local substation.

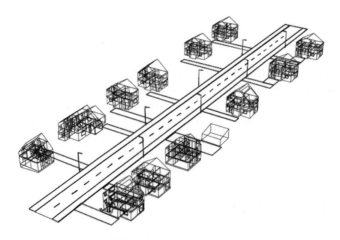

Figures 7.15   An ESP-r street with 12 dwellings

Figure 7.16   Aerial and driver view

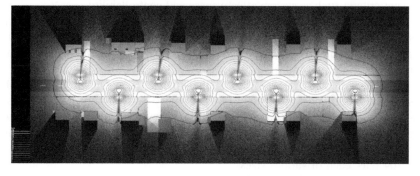

Figure 7.17   Illuminance (lux) contours

7.6   Matching demand and supply within hybrid schemes

To maximise the potential of a hybrid energy scheme, two actions need to be supported at the design stage, as follows.

- Determining the mix of supply technologies (by type and capacity) that best match the community demand profiles. This task equates to a rational sizing

Figure 7.18   The Upton low-carbon community development

procedure for hybrid supplies as opposed to the arbitrary selection of a technology mix based on exogenous factors such as available roof area or peak
demand.
• Generating a model comprising a combination of the supply technologies and
the buildings to be serviced. This step needs to be automated because it is
unlikely that practitioners would invest the time to create one-off, detailed
community models.

This model is then used to study and refine the proposed hybrid scheme in terms of
relevant criteria such as system effectiveness and controllability.

This two-part procedure is now demonstrated through application to a real
case: a community at Upton in Northampton, UK as depicted in Figure 7.18. The
objective was to design a hybrid renewable energy supply system to meet existing
demands.

The development was regarded at the time as an exemplar of sustainable
design due to the novel application of the Enquiry-by-Design method for master
planning (EST 2006) and the typicality of its size and structure. The case study
considered here targeted a residential neighbourhood occupying 3.7 hectares
accommodating 214 dwellings subdivided into 110 mixed semi-detached houses
and 104 flats, each of various sizes.

The total measured annual electricity and heating energy demand of the site
was 817,187 kWh and 3,106,098 kWh, respectively, with corresponding peak
demands of 311 kW and 1,278 kW. The hourly electricity and heating profiles for a
winter week are shown in Figure 7.19.

The low-carbon supply technologies considered were PV and wind turbines for
electricity, and solar thermal and heat pumps for heating. The chunking procedure
described in Section 7.1 was applied. Taking electricity as an example, the PV
was chunked into 700 × 500W capacity units, whilst wind power was chunked
into multiple turbines of increasing capacity: 10, 30, 50, ... 290, 310, and 330 kW.

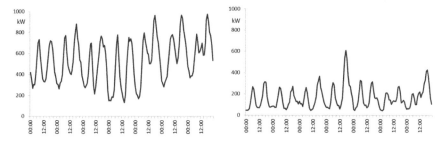

Figure 7.19   Site heating (left) and electricity profiles

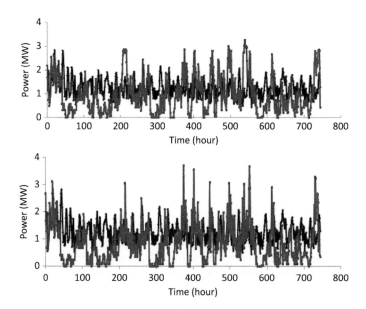

Figure 7.20   Top 2 demand (black) and supply matches over a month

This resulted in 131,771 supply/demand permutations to be searched by Merit, the matching tool used in the study.

Figure 7.20 depicts the top two rated matches for electricity resulting from the Merit search, and Table 7.2 summarises the match statistics and required hybrid capacities for both cases.

At this stage, and as described in Sections 7.7 and 8.1, an estate model was constructed by selecting prototype models based on available dwellings and connecting these to hybrid supply models as required (energy centre, building-integrated, etc.).

This estate model can be selectively refined to study specific issues. For example, within the ESP-r system, whilst heat-flow network models are typically used to represent HVAC systems within individual buildings, the approach can be

Table 7.2   Match statistics and hybrid supply capacities

Statistic (see Section 7.1)	Match 1	Match 2
RCC	0.07	0.07
IC	0.34	0.37
PM	65.7	63.4
Solar PV capacity (kW)	100	50
Wind turbine capacity (kW)	350	300

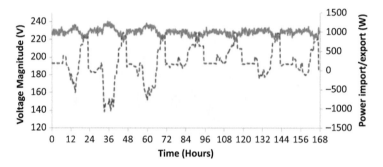

Figure 7.21   Voltage (upper) and net power import/export

extended to the modelling of community heating networks. Similarly, an ESP-r electrical model can be used to analyse the performance of electrical networks of varying scales and levels of complexity: from distribution inside a building to local microgrids. Two options exist for the creation of a supply model: as a community scheme by which the supply is connected to aggregate load profiles, or as a building-integrated scheme by which the hybrid supply is apportioned to individual dwellings. In the case study reported here, an ESP-r network model was established to represent the low-voltage distribution system associated with a part of the estate, and this was used to analyse the impact on the low-voltage electricity network of building-integrated electrical supply technologies. The network model comprised a series of supply cables, each of which accommodated approximately 50 dwellings connected radially to a local supply substation, which was the boundary of the analysed system.

The following discussion considers an individual dwelling selected from the multi-dwelling community model: a 2002 standard dwelling with 140 m^2 of floor area. Attributing this dwelling to a portion of the generation suggested by the Merit match results in a 6.6 m^2 PV installation (0.5 kW$_e$ capacity) and a 1m diameter wind turbine (1.3 kW$_e$ capacity). The focus of interest was the potential impact of this level of microgeneration on the local supply network.

Figure 7.21 illustrates the net electrical power flows and voltage levels associated with the dwelling with hybrid supply over a springtime week.

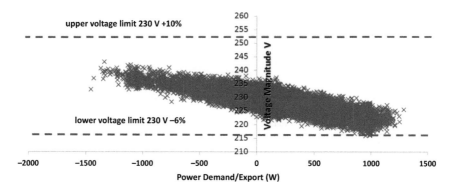

Figure 7.22   Voltage variation against power import and export (system sized by Merit)

Negative power values indicate that power is being exported to the local network from PV; positive values indicate that the building is drawing power from the network. The voltage level is highly influenced by the performance of the dwelling's generation and demand. As the demand increases, the local voltage drops below the nominal network supply voltage of 230V. Conversely, as the supply from the local generation increases the voltage rises.

Figure 7.22 shows the predicted variation in the local network voltage plotted against the power exchange between the dwelling and the local network over the simulated week. Superimposed on the graph are the voltage constraints for electricity supply in the UK (230V +10%/−6%).

The voltage at a substation usually has a mean value of 230V but is subject to random perturbation, with a normal distribution about the mean and a standard deviation of 1.9V as determined from monitored low voltage network data (Thomson and Infield 2007). The random variation represents the influence of the wider electricity system on the local network.

These results indicate that for the suggested Merit generation scheme, the voltage excursions would be relatively minor and the supply voltage would remain within the prescribed limits. The suggested hybrid supply mix is feasible and does not result in specific problems. Whilst this is a positive outcome, affirming the feasibility of the Merit match, this may not be the case for other local generation schemes.

Figure 7.23, for example, shows the voltage variation with power exchange for an alternative electricity match possibility comprising a PV-only scheme of 30 roof-mounted panels (3 kW$_e$ capacity).

Now the upper voltage constraints for the dwelling are frequently breached, and this would be the case for other dwellings connected to the same branch of the low-voltage network.

Whilst this example has focused on the impact of hybrid power generation on a single building in an estate model, it is possible to expand the scope to issues at a larger scale, such as phase balancing and cable overloading. Further, BPS+ can be

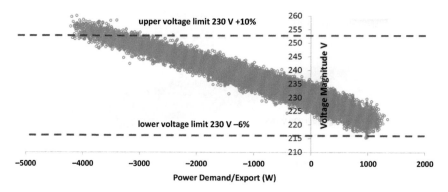

Figure 7.23   Voltage variation against power import/export (PV-only system)

applied to assess the integration of other technologies and actions such as heat pumps (Hong *et al.* 2013) and thermal load shifting. It can also be used to assess the impact of changes in load profiles on electricity network design as expressed, for example, via the After Diversity Maximum Demand (ADMD) index used by the industry to size wires in a network (Tait 2014).

7.7   Urban stock profiling

Assuming that 'internet-of-things' technology becomes ubiquitous in the future, it is possible that high-frequency, disaggregated energy use data will be available for most buildings. Pending this situation, is it possible to use BPS+ to generate such data for substantial parts of a city? Reinhart and Davila (2019) summarised the task of creating an urban model for energy simulation, stressing the need to combine several datasets. To be practical, the process needs to be automated because it involves a large number of entities. Other researchers (e.g., Walter and Kämpf 2015) have sought to reduce the input data burden by employing a simplified thermal model.

This section describes a simulation-based approach developed within a project funded by the Korean Institute of Energy Research[7] and led by the Woorizen software company based in Seoul[8] (Clarke *et al.* 2015). A building stock modelling tool for Parametric Energy Profile Analysis (PEPA[9]) was used to generate a stock model for commercial office buildings located in Seoul city centre, with ESP-r applied to produce hourly energy use profiles for heating, cooling, hot water, lighting, appliances, and elevators. The intention was to make the generated database available to assess direct energy management interventions throughout the estate. Table 7.3 summarises the project procedure.

[7]https://www.kier.re.kr/eng/
[8]https://woorizen.com/ko/
[9]https://www.esru.strath.ac.uk/Applications

Table 7.3   A procedure to generate disaggregated load profiles for a large building estate

1.	Survey building estate.
2.	Identify the major determinants of building energy use.
3.	Specify representative increments in the design parameters.
4.	Define base model(s) for building type(s).
5.	Calibrate model(s) using selected monitored data.
6.	Specify heating, ventilation, and air conditioning (HVAC) system configurations and control regimes.
7.	Use PEPA to generate a building stock model.
8.	Use ESP-r to simulate the stock model to produce a database of disaggregated load profiles.

The project commenced with a review of the Seoul commercial building stock based on available design data, site visits, practitioner interviews, and monthly gas and electricity consumption data over a 2-year period (2012–2013) for ~2000 buildings. A 3D map service called Vworld (Lee and Jang 2019) was used to determine basic information on building geometry, window type, and surrounding environment. The major design and operational parameters determining energy use were identified and assigned representative range values corresponding to various standards. PEPA was then used to generate a stock model automatically by applying these design parameter states, in various combinations, to a base model of a typical office layout. A selection of these models was calibrated against monitored data before all models were simulated to produce disaggregated load profiles.

As illustrated in Table 7.4, building facades were grouped into four categories according to window-to-wall ratio and window shape: curtain wall (all glazed except the spandrel area), belt type (windows continuous in an external wall), grid type (single window unit installed in a criss-cross pattern), and punch type (windows located randomly).

Small-to-medium-sized office buildings are located mainly in dense urban areas and are therefore affected by adjacent buildings in terms of solar access and wind pressure. These issues were included in the stock model.

The building stock was classified into six groups based on build year as a proxy for the envelope insulation standard corresponding to prevailing regulations. The structure of the office stock is heavyweight (i.e., steel frame construction or reinforced concrete construction). Only 0.1% of buildings are of wood construction. Table 7.5 shows a breakdown by construction type.

The *U*-value standard has been established at 0.58 $W/m^2.K$ since 1981, which represents a radical change from the 1.05 $W/m^2.K$ required since 1979. Table 7.6 summarises the change in the requirement for envelope insulation over time.

Fan coil units are typically installed within perimeter zones, and constant air volume (CAV) or variable air volume (VAV) systems are applied in interior zones. In general, heating systems are supplied by a central boiler plant. The cooling plant comprises compression (68%) or absorption (21%) chillers. Occasionally, a

Table 7.4   Classification of office building facades

Type	Example
Curtain wall	
Belt	
Grid	
Punch	

Table 7.5   Breakdown of building stock by construction structure

Structure	Count
Reinforced concrete	1724
Not identified	91
Steel frame	57
Steel frame and reinforced concrete	168
Wood	1
Total	2041

building is equipped with an ice-storage system, taking advantage of low tariffs during the night. Table 7.7 gives the breakdown of chiller types.

Based on these and other analyses, the representative parameter types and diversification levels were agreed as listed in Table 7.8.

Table 7.6   Mandated U-values over time

Renewal period	External wall (W/m^2.K)	Window (W/m^2.K)
09/79 to 12/80	1.05	3.49
01/81 to 07/87	0.58	3.49
07/87 to 01.01	0.58	3.37
01.01 to 01/11	0.47	3.84
02/11	0.36	3.84
02/11 to 09/13	0.27	3.84

Table 7.7   Breakdown by cooling plant type

Type	Count
Absorption chiller	371
Compression/turbo (electricity)	1181
Unidentified	184
Total	1736

Table 7.8   Design parameter diversification

Parameter	Level	Criteria
S (Operating schedule)	1	Offices and facilities that operate 9 hours a day, 5 days a week.
	2	Neighbourhood facilities that operate 13 hours a day, 6 days a week.
	3	Neighbourhood facilities that operate 24 hours a day, days a week.
D (Occupant density)	1	High: 4 m^2/person.
	2	Normal: 9 m^2/person.
	3	Low: 14 m^2/person.
E (Electrical use)	1.	High: 36 W/m^2 peak (10 W/m^2 lighting, 26 W/m^2 small power).
	2	Normal: 27 W/m^2 peak (10 W/m^2 lighting + 17 W/m^2 small power).
	3	Low: 16 W/m^2 peak (4 W/m^2 lighting + 12 W/m^2 small power).
W (Window/wall ratio)	1	70%
	2	50%
	3	30%
	4	10%
I (Solar ingress)	1	Solar transmission factor: 0.7
	2	Solar transmission factor: 0.5
	3	Solar transmission factor: 0.2
U (Insulation)	1	Poor: External wall = 1.25 W/m^2.K, window = 4.5 W/m^2.K
	2	Normal: External wall = 0.53 W/m^2.K), window = 4.5 W/m^2.K
	3	High: External wall = 0.36 W/m^2.K, window = 1.8 W/m^2.K, low-e glazing.

(Continues)

Table 7.8   (Continued)

Parameter	Level	Criteria
V (Ventilation)	1	High: 1.6 ACH infiltration, 8.4 l/s.person mechanical fresh air intake.
	2	Medium: 0.1 ACH infiltration, 8.4 l/s.person mechanical fresh air intake.
	3	Low: 0.1 ACH infiltration, no mechanical fresh air intake (8.4 l/s. person recirculated), 0.5 ACH controlled natural ventilation (set point 24 °C).
F (Fan power)	1	CAV
	2	VAV
H (Heating plant)	1	Boiler
	2	Heat pump
	3	Boiler and heat pump
C (Cooling plant)	1	Electrically driven chiller
	2	Absorption chiller
	3	Heat pump
	4	Electrically driven chillers
	5	Absorption chiller + heat pump
	6	Electrically driven chiller + ice storage
	7	Electrically driven chiller + ice storage + heat pump

The procedure of Figure 7.24 is now followed, with PEPA used to diversify the parameters of a base model to generate a set of models corresponding to the buildings comprising the Seoul office building estate now and in the future after an upgrade.

Figure 7.25 gives an example of a base model comprising four perimeter zones, one interior zone, and one circulation zone. This model is not a functioning ESP-r model; in the model files, the data values are replaced by keywords that are assigned values during the diversification procedure as a function of the related parameter combination.

Each model is therefore unique, being created from a combination of the first seven parameters with the final three parameters post-processed based on the simulation outcomes. Seven of the 10 parameters have three levels, one has four levels, one has seven, and one has two, so there are a total of $3^7 \times 4 \times 7 \times 2 = 122,472$ models created. It is important to note that only a small portion of these models will correspond to existing buildings. Most will define a future state after refurbishment: the ability to use the profiles database to identify suitable interventions for any building was the aim of the project.

The PEPA procedure assigns names to models that reflect the underlying parameter permutation. For example, S1D2E2W2I2U2V3F1 signifies a model with an operating schedule at level 1, occupant density at level 2, electrical use at level 2, window/wall ratio at level 2, solar ingress at level 2, insulation at level 2,

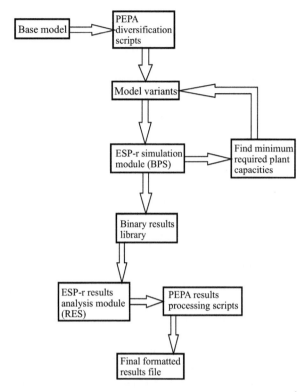

Figure 7.24   Generating energy profiles with auto diversification of a building stock model

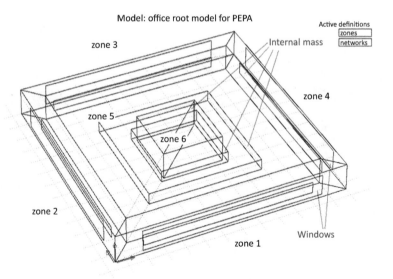

Figure 7.25   A base office model

ventilation at level 3, and fan power at level 1. The remaining parameter levels, corresponding to heating and cooling plant types, are applied at simulation time.

Simulation of the set of models resulted in a large database of disaggregated hourly demand profiles. This database has several potential uses, most notably stock profiling, the production of estate benchmarks, and the identification of refurbishment actions for buildings individually and collectively. For example, in the second case, a scan of the database suggested that buildings with energy use:

- <80 kWh/m^2.y are the very best of the stock but correspond to only 2.1% of the estate;
- 80–260 kWh/m^2.y can be considered good, but further significant energy savings are likely to be difficult to achieve;
- 260–520 kWh/m^2.y can be considered mediocre, with significant energy savings relatively easy to achieve; and
- >520 kWh/m^2.y can be considered the worst of the stock and has substantial room for improvement.

By considering the parameter levels of a model matched to a particular building, and comparing these with the levels of another model considered a reasonable target for improvement, both the actions to be taken and the likely energy-saving outcome are immediately apparent.

Before simulation, however, it is necessary to undertake model calibration using monitored data for sampled buildings. Figure 7.26 shows one of these cases corresponding to a 'S1D2E2W2I2U2V3F1' model as defined in Table 7.8.

Figure 7.26   A monitored building used for model calibration

7.8 Chapter summary

Locating the means of energy supply alongside energy demand within urban settings has the potential to improve operational efficiency through the avoidance of transmission losses and by enabling smart control whereby the supply/demand match can be nudged over time. BPS+ can be applied to identify opportunities at the individual and community scales whilst balancing energy and environmental issues. It can also be applied to large building stocks to provide insight into present performance and help identify building-specific upgrade actions.

References and further reading

Born F J (2001) 'Aiding renewable energy integration through complimentary demand-supply matching', *PhD Thesis*, ESRU, University of Strathclyde.

Clarke J A, Cowie A and Kim J (2015) 'Development of an energy profile generator for small-to-medium scale office buildings in Seoul', *Project Report*, ESRU, University of Strathclyde.

Clarke J and Samuel A (2011) 'Housing upgrades policy development', *Proc. eSim 2011*, Halifax, Canada.

Clean Heat at the Water's Edge (2019) https://esru.strath.ac.uk/EandE/Web_sites/18-19/cleanheat/index.html.

EST (2006) 'Creating a sustainable urban extension – a case study of Upton, Northampton', *Energy Saving Trust Report*.

Ete A (2009) 'Hydrogen systems modelling, analysis and optimisation', *MPhil Thesis*, ESRU, University of Strathclyde.

Hong J, Kelly N J, Thomson M and Richardson I (2013) 'Assessing heat pumps as flexible load', *Power and Energy*, 227(1), pp. 1–13.

Keirstead J, Jennings M and Sivakumar A (2012) 'A review of urban energy system models: approaches, challenges and opportunities', *Renewable and Sustainable Energy Reviews*, 16, pp. 3847–3866.

Kelly N and Monari P (2019) 'Assessing sizing and operation of Aberdeen filling station components', *Report WP2-022-0.1*, ESRU, University of Strathclyde.

Lee A and Jang I (2019) 'Spatial information platform with VWorld for improving user experience in limited web environment', *Electronics*, 8(12), 1411.

Morales A M and Roberts J J (2021) 'Is the University of Strathclyde's Combined Heat and Power (CHP) District Energy Scheme compatible with its carbon reduction targets? A Life Cycle Emissions Assessment of the Strathclyde's Energy Centre and Implications for its Expansion', University of Strathclyde, Glasgow.

Reinhart C and Davila C C (2019) 'Building performance simulation – challenges and opportunities', In Hensen J L M and Lamberts R (Eds.), *Building Performance Simulation for Design and Operation* (2nd Edn.), Chapter 21, pp. 696–722, Routledge, Milton Park, ISBN 9781138392199.

Ruggedised-Glasgow (2019) Ruggedised project, Glasgow component, https://ruggedised.eu/cities/glasgow/.

Scheaffer R L and McClave J T (1982) *Statistics for Engineers*, PWS Publishers.

Tait L (2014) 'Impact on network load profiles of 2016 building standards for new buildings and energy efficient retrofits', *Final Report for KTP Project KTP008752*, Scottish Power Energy Networks.

Thomson M and Infield D (2007) 'Impact of widespread photovoltaics generation on distribution systems', *Renewable Power Generation*, (1), pp. 33–40.

Walter E and Kämpf J H (2015) 'A verification of CitySim results using the BESTEST and monitored consumption values', *Proc. IBPSA-Italy Conf.*, Bozen-Bolzano, 4–6 February.

West Whitlawburn (2017) https://esru.strath.ac.uk/EandE/Web_sites/17-18/west-whitlawburn/index.html.

Williamson T J (1994) 'A confirmation technique for thermal performance simulation models', *Technical Report*, University of Adelaide.

Wiltshire R, Williams J and Woods P (2014) 'A technical guide to district heating', Building Research Establishment, Garston, UK.

Chapter 8

Regional/national scale energy action

This chapter takes the Parametric Energy Profile Analysis (PEPA) procedure of Section 7.7 and generalises it further to make it applicable to very large building stocks (e.g., national) where specific property details are not available. Reducing energy demand on a large scale and replacing conventional energy sources requires that policymakers comprehend the large variation in systems and behaviours across the stock as well as the uncertainties underpinning upgrades and energy supply options. This chapter explores the application of BPS+ to devise strategies for building stock upgrades at the regional or national scale. This requires the generation of BPS+ input models in a manner that bypasses the need to consider buildings individually, given the unacceptable data processing effort that would otherwise be required. It also requires the application of BPS+ to explore innovative solutions for sectors such as agriculture and water management.

8.1   National building stock upgrading

Taking Scottish housing as an example, there are around 2,699,000 dwellings in the country, of which 3.4% are vacant and 1.7% are due for demolition (NRS 2023). The majority of dwellings are either houses (46%) or flats (54%) with around 15% of all dwellings built within the last 22 years (SG 2022). The 2021 Scottish House Condition Survey (SHCS 2023) identified seven predominant house types: Detached, Semidetached, Terraced, Tenement Flat, Four-in-a-Block, Conversion, and Tower/Slab Block. It reported a mean National Home Energy Rating (NHER) rating of 7.9 for the stock (on a scale of 0/poor to 10/good), with an associated mean Standard Assessment Procedure (SAP) rating of 66.7. Annual CO_2 emissions for an average Scottish dwelling were estimated at 6.5 tonnes, resulting in a production of around 17.5 million tonnes of CO_2 per year for the stock. In comparison, the 1996 House Condition Survey established a mean NHER rating of 4.1 and a mean SAP rating of 43, indicating a 10% improvement between surveys, with 12% of all dwellings achieving an NHER rating of 7–9 and no dwellings attaining a rating of 10.

From the 2017 survey (NRS 2021), around 77% of dwellings had main gas central heating (8% had partial central heating). This represents an 11% improvement since 1996, with the number of dwellings with no central heating down from 13% to 2%. This still significant figure gives rise to concerns about fuel poverty and the health-related problems associated with hypothermia and mould growth in a cold climate. Although around 76% of houses have loft insulation, in only 48% of cases does this meets the 2000 Building Standards insulation thickness.

Energy and carbon calculation methods in current use, such as SAP[1] and SBEM[2] procedures in the UK, are based on energy balance methods that do not account for the dynamic characteristics of buildings. Whilst the application of BPS+ to inform changes to national building stocks has great potential, the challenge is to a priori establish input models that represent all buildings now and after future upgrades.

8.1.1   Formulating a stock model

Although it is a straightforward task to identify building types from an architecture and construction (AC) viewpoint, such a categorisation is unhelpful when viewed thermodynamically. Two separate buildings, each belonging to the same AC group (i.e., archetype), may have substantially different energy consumption patterns because of dissimilar energy efficiency measures that have been previously applied, different location characteristics, and different occupancy. Likewise, two buildings corresponding to different AC groups may have the same energy consumption (after normalisation relative to floor area) because the governing thermodynamic-related design parameters are essentially the same.

The strategic application of BPS+ is enabled by an approach that operates in terms of thermodynamic classes (TC) whereby different AC types may belong to the same TC (Clarke *et al.* 2004). As with urban stock modelling (see Section 7.7), a representative model is formed for each TC and its energy performance is determined by simulation using long-term weather data. Any actual building may then be related to a TC via the present level of its governing design parameters. Should any of these parameters be changed as part of an upgrade, that building would be deemed to have moved to another TC.

The simulation results for a set of representative models, scaled by their proportion of the overall population, then define the possible performance of the entire housing stock, present and future, for the exposure, occupancy, and system control assumptions made within the simulations. By varying these assumptions and re-simulating, scenarios such as future climate change and improved living standards may be assessed. It is important to note that the aim is not to predict future performance but to enable a comparison of outcomes as an essential input to policy-level decision-making and energy action planning.

Consider the following rudimentary example of TC model formulation in relation to the housing stock. If it is assumed that the determinants of dwelling energy demand are insulation level, thermal capacity level, capacity position, air permeability, window size, exposure, wall-to-floor ratio, and system efficiency, this gives 3,240 potential TCs representing the universe of possibilities depending on the number of levels allowed for each parameter (here 6, 2, 3, 3, 3, 5, and 2, respectively)[3]. That is,

---

[1]https://gov.uk/guidance/standard-assessment-procedure/
[2]https://www.uk-ncm.org.uk/
[3]Set to values that give the required modelling resolution.

any possible dwelling, existing or planned, will correspond to a unique combination of these parameters and therefore belong to one, and only one TC. Note that most of these TCs will not yet exist where the housing stock is poor but will represent designs that will result from the application of upgrades in various combinations when implemented over time. Where the level of granularity is increased by adding more parameters or parameter levels, the number of TCs representing the stock will increase. Each TC model is realised by applying parameter diversification to a base BPS+ input model (a starter seed) comprising living, eating, and sleeping areas for dwellings (or another typical configuration in non-domestic cases). The assumptions underlying a base model might correspond to an average house as suggested in the literature (e.g., Bartholomew and Robinson 1998, CIBSE 1999, Shorrock and Utley 2003). One way to enact this procedure is to use the PEPA program as elaborated in Section 7.7 but here with no requirement to survey the building stock.

Two options for using such a stock model exist. Selections can be made for a given estate in support of explicit simulation, or the entire TC collection can be subjected to long-term simulation, the predicted energy demands normalised by floor area to facilitate intercomparison, and the results encapsulated in a simplified tool for rapid upgrade evaluation by designers, estate managers, or policymakers. The latter approach underpins the HUE tool[4] for housing upgrade evaluation, the interface of which is shown in Figure 8.1.

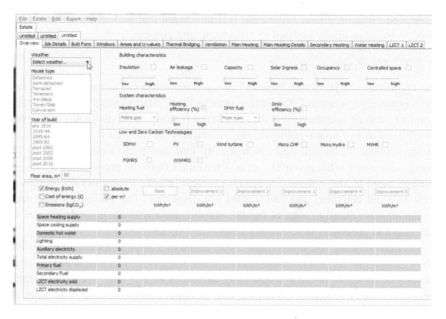

Figure 8.1   The HUE tool interface

[4]https://www.esru.strath.ac.uk/Applications

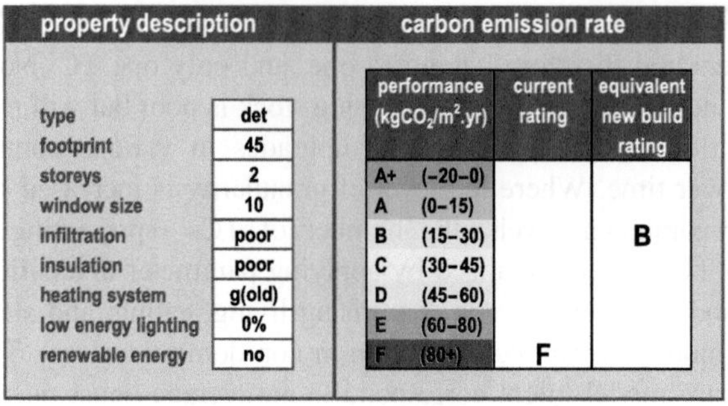

property description		carbon emission rate		
		performance (kgCO$_2$/m^2.yr)	current rating	equivalent new build rating
type	det			
footprint	45			
storeys	2	A+ (−20−0)		
window size	10	A (0−15)		
infiltration	poor	B (15−30)		B
insulation	poor	C (30−45)		
heating system	g(old)	D (45−60)		
low energy lighting	0%	E (60−80)		
renewable energy	no	F (80+)	F	

Figure 8.2   A carbon rating

HUE is designed to be flexible in its application: a context is defined, pragmatic property data is gathered and input, a representative TC is identified, upgrade technologies are selected, and outputs are generated based on pre-simulation results. The outputs comprise energy/carbon/cost estimates, and a carbon rating as shown in Figure 8.2. This rating is determined by the UK standard calculation method as encapsulated in SAP.

The following HUE application example relates to the Scottish housing stock. To apply the tool to another estate, its underlying TC models would need to be generated using the PEPA procedure. It is envisaged that such stock models would need to be sporadically updated to reflect ongoing changes to the estate.

8.1.2   Upgrade appraisal

An application of HUE will typically proceed as follows. The context of the analysis is set, e.g., 'current UK building standard' or '2050 climate with high indoor comfort', and simple describing data is entered relating to the dwellings in an estate of arbitrary size. The governing design parameters are inferred from this input, and an appropriate TC is selected automatically and assigned to each dwelling. The energy demand for the given TC set is determined from the pre-simulated TCs and used as input to the system-side appraisal. Next, planned upgrades are suggested, new TCs are activated and new results are displayed. The process is essentially instantaneous. Finally, outputs for the starting case and all upgrade phases are collated and presented in terms of energy, carbon, and cost.

At the time of the project, the 2002 Scottish House Condition Survey data showed that the 2,278,000 dwellings in Scotland translated to a total annual space heating demand of 14.5 TWh and CO$_2$ emissions of 5.5 MT. The space heating energy demand equated to 17% of the total Scottish energy demand at the time. An analysis revealed that the entire national stock could be classified into three TC groups, as listed in Table 8.1. The largest portion of the stock is contained within Group 1, which includes several TCs associated with unimproved dwellings constructed prior to 1981. This grouping accounted for 11.1 TWh of annual heating energy.

Table 8.1   Digest of Scottish dwellings

1: high thermal mass, poor insulation, large air change rate
   Number of dwellings: 1,594,600
   Heating demand (kWh/m^2.y): 87
2: standard insulation, large air change rate
   Number of dwellings: 660,620
   Heating demand (kWh/m^2.y): 47
3: high insulation, standard air change rate
   Number of dwellings: 22,780
   Heating demand (kWh/m^2.y): 26

Practical considerations dictate that any upgrade strategy should focus on low-cost technologies initially to maximise the return on investment, and be phased over time thereafter to accommodate technical advances and budgets. Reducing fabric and ventilation heat loss were assessed in this study.

An initial HUE analysis indicated that the upgrade strategy should be to target Group 1 dwellings by improving their air tightness (e.g., applying draught proofing to windows and doors) and insulation (via cavity, internal or external insulation plus double-glazing and loft insulation), both to standard levels. This would shift these properties to a Group 2 category with an associated saving of 40 kWh/m^2.y. A second upgrade phase could then be carried out to move Group 2 dwellings to Group 3, again by improving insulation and infiltration (in this case, improvements would attain compliance with all elements of the 2002 regulations).

The implementation of the first phase of improvement measures was predicted to result in savings in the annual space heating energy demand of 4.7 TWh (or 33.2% of the national space heating energy demand). This would be achieved by focusing solely on basic upgrades to dwellings in Group 1. In the second phase of the programme, the annual space heating energy savings would rise to 7.36 TWh (51.6%) by targeting Group 2 dwellings. Overall, by improving Group 1 to Group 2 and the original Group 2 to Group 3, a phased programme would reduce the annual space heating energy demand of the Scottish housing stock from 14.5 TWh to 7.14 TWh (i.e., a 51.6% reduction).

In another project (Tuohy *et al.* 2006), a Local Authority housing stock comprising 7,876 dwellings was evaluated using HUE to determine the impact on the carbon footprint of a range of upgrades. The housing stock was decomposed into TCs, and feasible upgrades were identified from Energy Savings Trust publications (EST 2007) and a site survey. The considered upgrades are listed in Table 8.2.

Figure 8.3 illustrates the impact of each upgrade option on the carbon footprint. As can be seen, the current footprint per dwelling is 4.9 tonnes of CO_2/y whilst future scenarios can have less than 1 tonne.

HUE can also be used to provide energy performance ratings. In this case, the Environmental Index (EI) and Energy Band (EB) are calculated in accordance with the standard UK method.

Table 8.2   Proposed upgrade options

1. Current stock, no upgrades.
2. Low-cost fabric improvement (loft/underfloor insulation, double-glazing and draught proofing).
3. Upgrade 1 + roof and wall *U*-value improvements.
4. Gas, electric, and solid fuel heating systems upgraded to condensing combi-boiler, air source heat pump with radiators, or wood boiler with 2007 regulation efficiencies.
5. Upgrades 1+2+3.
6. Upgrades 1+2+3 plus solar water heating (920 kWh/y useful energy).
7. Upgrades 1+2+3 plus local generation by PV (650 kWh/y useful energy).
8. Upgrades 1+2 plus CHP (Stirling engine except where >2 storeys then community CHP).
9. Upgrades 1+2 plus biomass (comprising individual or community wood boiler systems).

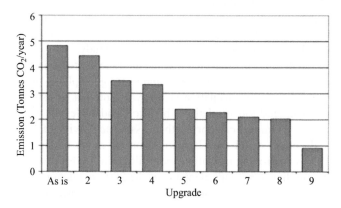

Figure 8.3   Upgrade impact on CO_2 emissions

Table 8.3   Upgrade options appraisal outcomes

Upgrade	kg CO_2/y	EI	EB
1. As is.	3391	57	D
2. Fabric.	2778	66	D
3. 2 + gas condensing boiler.	1679	81	B
4. 2 + ground source heat pump.	1515	83	B
5. 2 + community biomass.	817	93	A
6. 2 + community gas CHP.	1000	98	A
7. 2 + 3 + PV + solar thermal.	1454	84	B

Table 8.3 gives an example result for an electrically heated 1980s top-floor flat, which had previously been upgraded with cavity wall insulation, double-glazing, and 200 mm of loft insulation. The rating of the base property is 'D'. A number of improvement scenarios were assessed aimed at raising the fabric to

2002 standards and then applying system upgrades: a condensing boiler, ground source heat pump, community biomass heating, community CHP, a condensing boiler combined with solar water heating, and PV panel (providing 920 kWh heat and 650 kWh electricity annually, respectively).

In this specific case, two options achieved an 'A' rating: upgraded fabric with either community biomass heating or a community CHP system.

8.2 Solar access

Landforms and the layout of buildings influence the timing and magnitude of useful solar energy and daylight. Planning procedures attempt to limit the degree to which proposed schemes diminish sky vault access. This project (Hand 1999) examined a proposal to redevelop a portion of a community located on a north-facing hill above the town of Greenock in Scotland. The approximately 1-km^2 site was constrained in terms of winter sunlight and had high wind exposure. Evidence was required on the patterns of building facade shading at different times of the year. The approach taken was to establish an ESP-r model of the community, including the site topography as indicated in Figure 8.4.

The site model was exported to Radiance to support visual assessments at various times of day and across seasons to characterise the solar access. This identified the most promising areas of the site for dwelling placement. Dwelling models (semidetached, short terraces, and longer terraces) were then added and shifted in position to represent different site layouts.

To determine 'windows of opportunity' for winter solar access, orthogonal projections from the sun were made to determine site shadow patterns over time. Figure 8.5 shows four images spanning the period from 11:00 (upper left) to 14:00 (lower right) in January. The views are from the top of the hill, looking southwest down the slope. It was apparent that the low sun angles resulted in long shadows until 14:00 when the hill to the South blocked the sun altogether.

After reviewing the animations, the design team adopted a configuration of short terraces to provide alternating periods of light and shade. In the second phase

Figure 8.4 The ESP-r community model

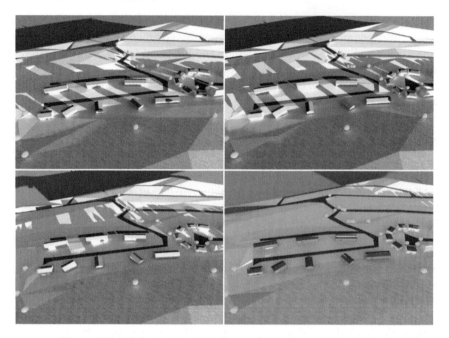

Figure 8.5   Solar access 'windows-of-opportunity' in January

of the project, dwelling descriptions following the designer's revised site layout were incorporated into the model and used to assess the impact on heating.

The findings of the study were that the site is disadvantaged when compared with the conditions normally associated with housing. However, with selective site planning and choice of dwelling types, the extent of the reduced sunlight could be mitigated, resulting in heating cost savings and improved indoor daylight levels.

8.3   Waste heat for horticulture

Greenhouses are widely used for crop growing and correspond to various typologies, as reported by Brunetti (2022), who also summarised a number of applicable design tools, including BPS+. The study summarised here (Hand and Cockroft 2013) examined possible improvements to a double-skinned, polyethylene-lined horticultural tunnel with underground heating. The objective was to determine whether proposals for improved skin and the use of sub-soil heating would reduce energy use and still support the growing of a mix of local produce. The intention was to take waste heat from a nearby facility, although this would require a flexible arrangement to accommodate times of supply shortage or lower temperature availability. There were also decisions to be made on the composition of the growing medium and the depth of the sub-soil heating pipes. The client, a horticultural expert, required air and earth temperature profiles throughout the year for a matrix of design options.

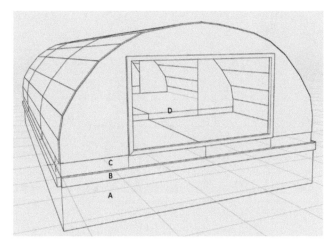

Figure 8.6   ESP-r model of a polytunnel

The approach taken was to separate the polytunnel into three thermal zones with an access door at one end and a ventilation opening at the other end, as shown in Figure 8.6.

An initial oval polygon was extruded to form the tunnel, and the end details were added. The model comprised a 7m wide section, with a total floor area of 100 m² split into three zones. Layer A in the figure is 1m of earth, layer B the heat injection layer with embedded pipes, layer C the growing medium, and layer D a thicker growing medium in the centre of the tunnel. Some BPS+ tools have the capability to generate automatically a curved surface as a multi-faceted representation. The conditions were assessed at 10-minute time steps over a year for each design option.

The sides of the polytunnel were insulated to a depth of 1.5m below ground level to minimise peripheral heat loss. The temperature in the central zone was controlled to 12 °C by an on/off controller. Normally, the access doors would remain closed. Overheating was controlled by opening the access door and rear vent according to the following rules:

- <10 °C, doors open 50 mm;
- 10 °C–20 °C, doors open 100 mm;
- 20 °C–24 °C, doors open 400 mm; and
- >24 °C, doors open 600 mm.

Enactment of these rules was facilitated via an airflow network, as shown in Figure 8.7, which responded to wind effects that were significant at the site.

Several heating scenarios were modelled and Figure 8.8 presents the result for a 10 kW (100 W/m²) supply with a 40 °C layer temperature limit. This resulted in polytunnel air temperatures in the range of 5 °C–12 °C. There are brief drops in the heat supply temperature when the air temperature set point is reached or exceeded.

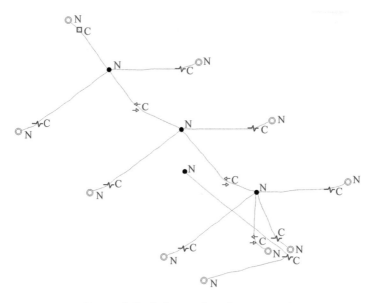

Figure 8.7   Polytunnel airflow network

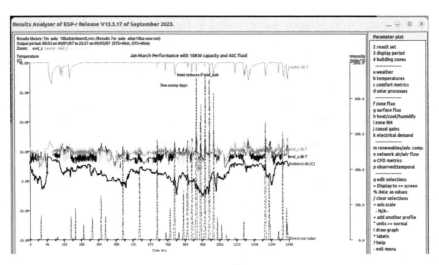

Figure 8.8   Zone temperatures, 100 W/m² heat supply and controlled airflow

Figure 8.9 (upper) is focused on the centre of the polytunnel where the growing medium temperatures (upper line) are in the region of 19 °C–25 °C. The distribution of temperatures within the growing medium is shown in the lower image over the course of a day.

Such results were critical for assessing stress in the root systems and the growth prospects of the intended crops. The assessments highlighted the inertia of

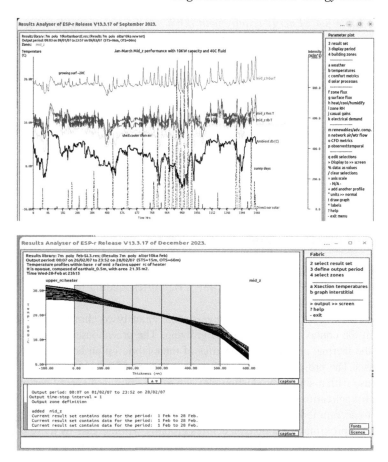

*Figure 8.9   Centre zone temperatures (upper) and growing medium temperature
profile over time*

the system and supported the conclusion that it would not be possible to control
polytunnel zone temperatures more precisely by controlling the heat input rate.
However, it also indicated that some disruption in heat supply could be tolerated.
Table 8.4 shows the energy requirement for the three heating levels considered in
the project.

Overall, it was concluded that the temperatures within a double-skinned polytunnel could be usefully raised by several degrees during winter periods by providing
heat input to the soil below. Maximum heat injection rates of 100–150 W/m^2 are
appropriate: lower rates would limit the useful temperature rise during the coldest
weather, risking frost damage, whilst higher rates do not offer significant additional
benefits where the heated ground layer temperature is limited to 50 °C.

Assuming a heating intensity of 150 W/m^2, a 1-acre area (4047 m^2) with
polytunnels arranged to cover 80% of the ground surface would require a peak heat

Table 8.4   Energy requirement (kWh/m^2) for polytunnel heating

Month	100 W/m^2	150 W/m^2	200 W/m^2
Jan	49.6	63.6	64.7
Feb	56.5	69.4	72.3
Mar	47.1	60.0	67.6
Apr	29.9	37.3	42.6
May	11.1	14.5	17.5
Jun	5.8	7.7	10.3
Jul	8.7	11.5	14.2
Aug	2.9	3.9	5.2
Sep	18.2	21.5	24.8
Oct	27.4	32.5	36.8
Nov	55.9	73.0	77.7
Dec	66.5	82.7	84.9
Annual	379.6	477. 6	518.6

input of 4047 × 0.8 × 150/1000 = 486 kW, corresponding to an annual energy consumption of 1548 MWh.

In another ESP-r application, Lee *et al.* (2019) analysed climate-adaptive greenhouses in the context of Dutch tomato growing. The greenhouses had the ability to change the thermal and optical properties of their fabric on an hourly, daily, or seasonal basis. The project reported a net profit increase of 20% compared with a static greenhouse with a similar production rate due to the optimised energy performance.

Such applications are likely to grow in the future with the expansion of data centres to accommodate the growing demand for digital services. According to the IEA[5], the approximately 8000 data centres globally will increase by more than 70% by 2026 and consume 800 TWh annually (cf. the ~300 TWh of total electricity consumption in the UK in recent years).

8.4   Water desalination

A large proportion of the available water on the earth's surface is saline, and there is limited access to potable water. Solar distillation is one method of producing fresh water from saline sources. In the project described here, BPS+ was used to optimise the performance of a solar still (Madhlopa 2009) and demonstrate how BPS+ can be used to undertake detailed appraisals. Figure 8.10 shows a solar still design comprising a thin layer of water in a horizontal basin with a single or double-sloped transparent cover.

The single-slope (SS) still has a back wall that acts as an internal reflector, whilst the double-slope (DS) still (El-Swify and Metias 2002) has no back wall. The transparent covers in a DS, which may be symmetrical or asymmetrical, have a gable along each end. In both cases, saline water in the basin is heated by solar radiation passing through the cover and being absorbed by the water and bottom

[5]https://www.iea.org/energy-system/buildings/data-centres-and-data-transmission-networks

part of the still basin. Vapour flows upwards from the heated water and condenses when it encounters the cooler inner surface of the cover. The condensate (clean water) is collected in a channel fitted along the lower edge. The distillate output is influenced by weather and operational factors, with an SS intercepting a higher proportion of solar radiation than a DS at high and low latitudes (Garg and Mann 1976). The solar energy capture is principally dictated by the aspect ratio of the still base, $R = L/B$, where L is the length and B is the breadth. Conversely, the DS is more economically viable than the SS (Mukherjee and Tiwari 1986).

ESP-r was used to identify the value of R for optimum performance. An input model was created in which the basin liner was defined as 0.001m thick steel and the covers as 0.004m clear float glass inclined at 55.9° and 19.9° to the horizontal when the model was located in a temperate and warm climate, respectively. The sidewalls were triple-layered with expanded polystyrene (0.05m thick) sandwiched between 0.005m thick plywood. The base of each still was triple-layered with external 0.005m plywood, 0.05m expanded polystyrene, and 0.001m steel inner layer. The surface area of the base remained constant for different values of the aspect ratio. Further, the height of the lower vertical sides was fixed at 0.05m above the still base, whilst the height of the higher vertical sides varied with R.

The DS was simulated with the covers oriented in east-west and north-south directions, whilst the SS faced south at both locations. Figure 8.11 shows the

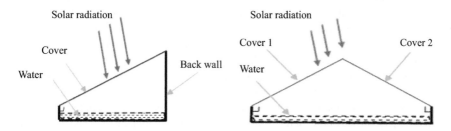

Figure 8.10   A basic solar still with a single and double slope

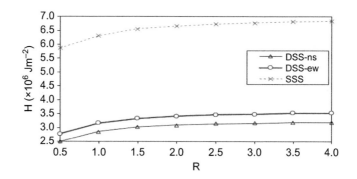

Figure 8.11   Variation of H with R, temperate climate

variation of the mean annual daily effective insolation (H) on the base of the still when the DS is oriented north-south (DSS-ns) and east-west (DSS-ew).

The DS collects less solar radiation than the SS for a given value of R due to the presence of the back wall in the latter, which reflects part of the incoming solar radiation onto the base. The DS captures more solar radiation in the east-west orientation due to the effect of shading from the gables. The effect of shading is significant at low solar altitudes in the morning and afternoon, which accounts for the observed effect of orientation on the optical efficiency of the DS as reported by others (e.g., Dwivedi and Tiwari 2010). It is also apparent that H increases with R to an optimum level for both designs. This is due to a reduction in the self-shading arising from the wall along the breadth of the still. It should be noted that R increases as B decreases, leading to a decrease in the height and area of the slanted wall of the still (for a constant slope) and its shading effect on the internal part of the still base in both designs. Under the imposed meteorological conditions, the optimum value of R was around 3.0 for both designs.

Figure 8.12 shows the corresponding variation of H against R for a warm climate.

Again, the DS collects less solar radiation than the SS and captures more solar radiation in the east-west orientation than in the north-south. For the east-west orientation, one of the gables would cast a shadow on the still base when the sun is due north or south, except when the sun traverses the sky over the local latitude. It should nevertheless be noted that the sun traverses the sky overhead, to the south and north in the warm climate location. In this case, a gable would cast a shadow on the still base when the sun is not crossing the sky overhead, but the beam radiation would be able to reach the still base directly even at low solar altitudes in the morning or afternoon. For the north-west orientation, one of the gables would cast a shadow on the still base in the morning or afternoon, except at local solar noon. Solar radiation would be unable to reach the still base directly at low solar altitudes during certain times in the morning or afternoon, which accounts for the observed effect of orientation on the optical efficiency of the DS. Figure 8.12 shows that H increases with R to an optimum level for both

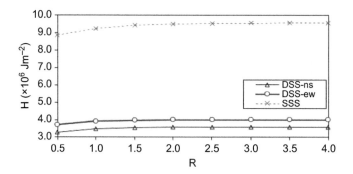

Figure 8.12   Variation H with R, warm climate

designs again due to a reduction in self-shading from the walls along the still width. For the imposed meteorological conditions, the optimum value of R was around 2.0 for both designs. It should be noted that the zenith angle (θ) is low around solar noon on most days at low latitude sites. Low values of θ tend to reduce the effect of shading, and optimal values of R. In contrast, the corresponding value of θ at high latitude sites is relatively high around solar noon, and this will increase the effect of shading and the optimal values of R. This indicates that $R = 2.0$ is not a universal optimum value.

In summary, ESP-r simulations showed that the optical performance of a DS is lower than that of an SS for two contrasting site locations. The DS collected more solar energy in the east-west than north-south orientation for a given value of R. In addition, effective insolation increased with R, but the increase was insignificant for values of $R > 3.0$ for both designs at high latitudes. Similarly, effective insolation increased with R, but the increase was insignificant for values of $R > 2.0$ for both designs at low latitudes. It was concluded that R significantly affects the collection of solar energy by a DS, the optimum value of R is sensitive to site parameters, the orientation of the DS does not affect this optimum, and the optimum value of R is approximately the same for both designs at a given site.

8.5   Chapter summary

BPS+ can be applied in a conceptual manner to large building stocks via the parametric diversification of 'seed' models corresponding to domestic and non-domestic archetypes, followed by automated long-term simulations to generate performance curves for the stock including possible upgrades. At the other end of the scale, the approach can be applied to assess issues such as solar access, waste heat utilisation in horticulture, and solar still design.

References and further reading

Bartholomew D and Robinson D (Eds) (1998) *Building Energy and Environmental Modelling: Applications Manual 11*, CIBSE, London, ISBN 0-900953-85-3.

Brunetti G L (2022) *Design and Construction of Bioclimatic Wooden Greenhouses Volume 4: Architectural Integration and Quantitative Analyses*, Wiley, New York, ISBN 978-1-78630-854-2.

CIBSE (1999) *Environmental Design, CIBSE Guide A*, Chartered Institution of Building Services Engineers, London, ISBN 0-900953-96-9.

Clarke J A, Johnstone C M, Kondratenko I, *et al.* (2004) 'Using simulation to formulate domestic sector upgrading strategies for Scotland', *Energy and Buildings*, 36, pp. 759–770.

Dwivedi V K and Tiwari G N (2010) 'Experimental validation of thermal model of a double slope active solar still under natural circulation mode', *Desalination*, 250, pp. 49–55.

El-Swify M E and Metias M Z (2002) 'Performance of double exposure still', *Renewable Energy*, 26, pp. 531–547.

EST (2007) *UK Government Energy Efficiency Best Practice Program*, https://est.org.uk/bestpractice.

Garg H P and Mann H S (1976) 'Effect of climatic, operational and design parameters on the year round performance of single-sloped and double-sloped solar still, under Indian arid zone conditions', *Solar Energy*, 18, pp. 159–164.

Hand J W (1999) *Private Communication with Michael and Sue Thornley Architects*, Glasgow.

Hand J W and Cockroft J (2013) 'Assessment of potential to use heat from a biomass CHP plant for horticultural purposes', *Consultancy Report* E458, ESRU, University of Strathclyde.

Lee C, Hoes P, Cóstola D and Hensen J L M (2019) 'Assessing the performance potential of climate adaptive greenhouse shells', *Energy*, 175, pp. 534–545.

Madhlopa A (2009) 'Development of an advanced passive solar still with separate condenser', *PhD Thesis*, ESRU, University of Strathclyde.

Mukherjee K and Tiwari G N (1986) 'Economic analysis of various designs of conventional solar stills', *Energy Conversion and Management*, 26, pp. 155–157.

NRS (2021) *National Records of Scotland*, https://scotlandscensus.gov.uk/media/hdrduafj/housing_and_accommodation_central_heating_topic_report.pdf.

NRS (2023) *National Records of Scotland*, https://nrscotland.gov.uk/statistics-and-data/statistics/statistics-by-theme/households/household-estimates/2022.

SG (2022) *Local Authority Housing Stock by Type*, Scottish Government, https://statistics.gov.scot/data/local-authority-housing-stock-by-type.

SHCS (2023) *Scottish Housing Condition Survey 2021*, https://gov.scot/publications/scottish-house-condition-survey-2021-key-findings/documents/.

Shorrock L D and Utley J I (2003) *Domestic Fact File*, BRE Publications, ISBN 1860816231.

Tuohy P G, Strachan P A and Marnie A (2006) 'Carbon and energy performance of housing: a model and toolset for policy development applied to a local authority housing stock', *Proc. Eurosun*, Glasgow.

Chapter 9

Smart grids with active demand management

This chapter addresses the concept of a smart grid (Kamran 2022) defined as an electricity supply network that uses digital communications technology to detect and react to local changes in demand and supply. The issues to be considered within a BPS+ simulation include demand management/response, supply asset dispatching, active network control, accommodating behaviour change, appliance selection, and design for operational resilience. BPS+ has evolved to a level that can address these issues and thereby enable a balance to be struck between the requirements of consumers and network operators.

9.1   The Lerwick case study

In principle, a smart grid can use high-capacity backup batteries to balance intermittent renewable supplies with stochastic demand in order to maintain network stability over short periods. An alternative, potentially cheaper option is to distribute storage throughout the building estate (Farid and Chen 1999, Foote *et al.* 2005, Qureshi *et al.* 2011), thereby displacing the need for backup and improving the load factor for renewables (Strbac *et al.* 2012). Such distributed capacity exists at present: electric storage heaters are used in areas without mains gas and, in conjunction with tele-switched control, could be utilised by a utility to level demand throughout the day. At the time of the project described here, approximately 1.7 million households in the UK had electric storage heating, and 1.3 million had electric hot water tanks (DECC 2012). Together, these accounted for 16% of domestic electricity consumption (DUKES 2011).

Storage heaters produce heat from electric resistance elements and store it in a high-heat capacity material covered by a layer of insulation. In modern appliances, heat transfer to the room occurs via two operation modes: uncontrolled natural convection from the heater surface, and fan-assisted air movement through the core (Hasnain 1990). Such devices are inexpensive to install, require no maintenance, and deliver heat without distribution losses (Oughton *et al.* 2008, Which Review 2013). On the negative side, older units have excessive uncontrolled output, leading to overheating, and charge forecasting is an uncertain science. Based on the monitoring of space heating consumption in houses with storage heaters, Hayton (1994) reported that the heaters typically provided up to 86% of the total requirement, with the balance coming from direct heating. Room temperatures in a house with storage heaters are typically highest in the morning, rather than the evening maximum observed in houses with direct heating systems (Yohanis and Mondol 2010).

The project considered here[1] was funded by Ofgem[2] and coordinated by Scottish Hydro Electric Power Distribution (SHEPD 2012). It involved a trial of dispatchable space and water heaters within a smart grid established in Lerwick in the Shetland Islands, including the use of BPS+ to assess algorithms for appliance charge/discharge control (Clarke *et al.* 2014). Because the island is not connected to the UK grid, the distribution network operator has to balance demand and supply across a population of 22,000. Network stability constraints dictated that only a small 3.7 MW wind farm could be accommodated. The smart grid trial was intended to reduce fossil fuel use, support an increase in wind generation capacity, and improve the reliability and quality of the electricity supply. An Active Network Manager (ANM) dispatches wind generators in response to demand fluctuations and controls the charging of domestic space and water heaters with a total electrical capacity of 2.1 MW distributed across 235 dwellings. The heaters, manufactured by Glen Dimplex (2013)[3], receive a centrally generated charge schedule from the ANM and relay back data on thermal store status with updates every minute in both directions (Gill *et al.* 2013). The charge schedule specifies the desired heater input power for each 15 minute interval in the upcoming 24 hour period based on anticipated demand, supply, and network status. The charge can be varied between discrete 1/3rd levels. An additional feature is that the charge can respond automatically to short-term changes in grid frequency, shutting down when the frequency drops below an acceptable level and increasing when the frequency rises. In this way, the space and water heaters support direct load shifting rather than relying on differential pricing to influence the customer's actions as in traditional demand side response (Ofgem 2013).

Figure 9.1 shows the space heater and hot water storage devices (Lacroix 1999) and indicates the extent of the monitoring scheme deployed in the project.

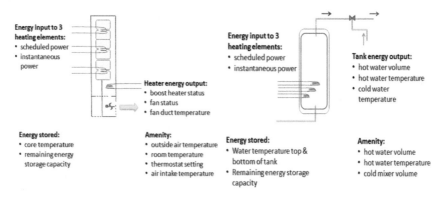

Figure 9.1   Space and water heater monitoring scheme

[1]Northern Isles New Energy Solutions (NINES), https://ssen-innovation.co.uk/nines/
[2]https://ofgem.gov.uk
[3]https://glendimplex.com/en-ie

Data was collected for 18 devices, each with 12–14 data channels, at 1–5 minute intervals. The full monitoring scheme was in operation from March to October 2012 covering weather that was typical of conditions in Shetland.

Within the project, BPS+ was employed to model the performance of the distributed storage solution, with monitored data used to calibrate input models and assure the veracity of predictions. The project explored how a controllable storage system would respond to different charge schedules. Performance was assessed from the point of view of both the grid operator (controllability, storage utilisation, overall demand, and wind curtailment avoidance) and the occupants (comfort, demand, and cost).

9.2 Formulating a smart charge model

ESP-r (Clarke 2001) was used to study the effectiveness of approaches to charge control. In an early phase of the project, a local controller in each of six monitored houses communicated with a central controller established to emulate the ANM. Five were built in the late 1990s with insulated timber construction; the sixth was within a 2004 conversion of a stone-walled public building into flats. Each dwelling had two or three storage heaters with a total capacity of 4.3 to 6.2 kW. These were located in the hall and living areas, with direct panel heaters in other rooms. Occupancy ranged from single working persons, through families with two children, to an elderly, all-day resident. Table 9.1 summarises the dwellings and occupancy regimes.

The dwellings were modelled as described in Chapter 5 based on data gathered from various sources, including site surveys. The models included detailed internal layouts to allow storage and panel heater outputs to be controlled separately. Figure 9.2 summarises the deployment of the storage heaters in one dwelling.

Table 9.1 Monitored houses in the initial trial

Dwelling	Form	Construction	Occupants	Floor area (m²)	Controllable input power (kW)	Effective storage (kWh/ day)	Direct heating (kW)
1	Flat	Stone, retrofit insulation	1	40	4.35	25.6	3.25
2	Semi-detached	Timber, well-insulated	4	95	6.30	37.1	4.00
3	Bungalow	Timber, well-insulated	4	54	4.35	25.6	3.25
4	Flat	Timber, well-insulated	1	48	2.40	13.3	2.00
5	Bungalow	Timber, well-insulated	3	54	4.35	25.6	3.25
6	Flat	Timber, well-insulated	2	48	4.35	25.6	2.00

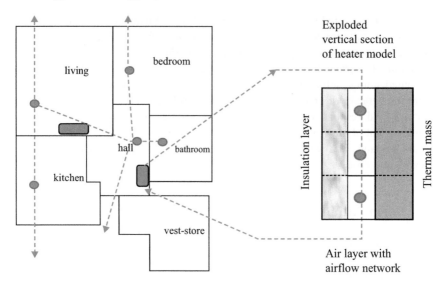

Figure 9.2   Simulation model for a dwelling with heaters

The space heaters were highly insulated and offered flexible control of output through a timer and room thermostat linked to fan-assisted mode. They were equipped with a built-in controller that maintained occupant comfort by providing a heat reserve at all times through automatic charging when the store reached that reserve, regardless of the imposed charge schedule. In the event of a conflict with ANM instructions, the internal device controller had priority.

Large heaters had four stacks of bricks sandwiching 3 × 800W heating elements; smaller versions had three stacks and 650W elements. The heater model comprised nine thermal zones, representing the following processes:

- variable energy input to each of the three sections of the core during charging;
- heat transfer from the core to the intra-heater air stream;
- heat transfer through the insulation to the room air (uncontrolled output); and
- heat transfer from the intra-heater air stream to the room air (fan-assisted output).

The models utilised performance data provided by the manufacturer and as measured in laboratory tests:

- specific heat capacity was derived from the temperature rise recorded whilst charging;
- airflow rates around the brick core were estimated from heat outputs under uncontrolled discharge and fan-assisted operation; and
- convection coefficients were estimated from empirical correlations corresponding to natural convection in the uncontrolled discharge case (Alamdari *et al.* 1984) and forced convection when operating in fan-assisted mode (Fisher and Pederson 1997).

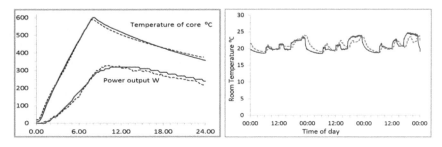

Figure 9.3   Storage heater predictions vs. measurements

To ensure that the model was well-founded, it was calibrated using data collected from laboratory tests and field monitoring. Figure 9.3 compares simulated and measured performance.

In the laboratory test (left image), full charge was applied for 8 hours and then discontinued. Temperature and power output rose rapidly, then fell slowly: after 24 hours, the thermal store was two-thirds full. Output peaked around 2 hours after the power input stopped as the store temperature equalised. The predicted temperature (red) is close to that measured in both shape and magnitude, with a Spearman's rank correlation coefficient of 0.99 and an inequality coefficient of 0.02, respectively (Born 2001). The corresponding agreement for power output was slightly lower but still acceptable at 0.97 and 0.03. The most significant model adjustment to achieve this alignment was the convective heat transfer coefficient between the heater and room air. The right image shows the comparison between predicted room temperature and monitored data over a 3-day period. The input to the model in this test was the measured charge profile and outside air temperature. The predictions are generally within 15% of the measured. The mean room temperatures agreed to within 0.1 °C. The predicted fluctuation range was 6.3 °C compared to 5.1 °C as measured, indicating that casual gains and infiltration rates were more complex than modelled. The rank correlation coefficient was 0.7 and the inequality coefficient was 0.03, indicating moderate agreement between profile shapes but good agreement in magnitude.

Three significant lessons emerged from the calibration exercise, as follows.

- Since the storage heaters were intended to heat the whole house, an airflow network was needed, including the kitchen extractor fan.
- Given the high thermal capacity of the heater cores, an extended pre-simulation period was required to ensure an acceptable heater starting state.
- Because the heater fan can switch at one-minute intervals, simulations should ideally be constrained to that frequency to avoid misrepresenting the heat input to the room. However, because the project required simulations over extended periods this would incur a time penalty. A sensitivity study indicated that a time step of five minutes would yield results that were within 10% of those using one minute, so this was applied in the charge scheduling study.

Within the field trial, charging schedules were kept within existing tariff windows, and input power was held at a minimum level. During cold spells, the heater core was not hot enough to deliver the required output and frequently hit its programmed minimum reserve level, at which point the heater would charge at full power irrespective of the received charge instruction from the ANM. Under warmer conditions, the automatic energy cap prevented full delivery of the scheduled charge. This approach to charging therefore worked badly for both the occupants and the utility.

Simulations of alternative charging schedules were undertaken to search for better solutions:

- *Teleswitching* – the traditional approach where heaters charge at maximum rate during reduced tariff hours if physically possible;
- *Exact scheduling* – storage of the next day's forecast energy requirement each day;
- *Approximate scheduling* – storing the average monthly demand each day to mimic inaccurate forecasting or deliberate storage of small amounts of wind; and
- *Flexing with wind speed* – storing up to twice the next day's demand when the wind speed is high; minimum power is then needed to top up when the wind is low, taking into account the forecast fill level. Whilst not identical to the proprietary (unknown) schedule produced by the ANM, it generated a similar charging pattern.

Each schedule was then applied at different times of day, and at varying power levels, giving 15 permutations as listed in Table 9.2.

The heater's internal controller could not be fully modelled as the algorithm for calculating the adaptive control cap was intellectual property. The cap was therefore emulated by running each schedule with four maximum heater fill settings

Table 9.2   Charging regimes investigated

Schedule	Energy delivered	Power setting	Time of day
Te	Teleswitch, as demanded by the heater	Max	Tariff hours
De	Average heater demand for the month	Min	Tariff hours
Hs	Exact heater demand each day	Max	Early, starts 04:00
En1	Exact house demand each day	Min	Early, starts 04:00
Ex1	Exact house demand each day	Max	Early, starts 04:00
Ex2	Exact house demand each day	Max	Late, starts 18:30
An1	Average heater demand for month	Min	Early, finishes 08:00
An2	Average heater demand for month	Min	Late, finishes 00:00
Ax1	Average heater demand for month	Max	Early, finishes 08:00
Ax2	Average heater demand for month	Max	Night, starts 00:30
Ax3	Average heater demand for month	Max	Daytime, starts 14:00
Ax4	Average heater demand for month	Max	Late, starts 18:30
Sn1	Demand flexed with wind speed	Min	Early, starts 04:00
Sx1	Demand flexed with wind speed	Max	Early, starts 04:00
Sx2	Demand flexed with wind speed	Max	Late, starts 18:30

as observed in the trial: unconstrained, 55%, 14%, and 4%. The scheduling of water heating (Moreau 2011) was considered in the project but is not reported here.

Simulations modelled a regular daily routine, with prescribed thermostat settings (Shipworth *et al.* 2012) and ventilation rates rather than the real vagaries of occupant behaviour. Even so, the predicted temperature variation throughout the day was similar to that observed, although peak temperatures in the evenings were higher. In May and June, room temperatures exceeded 25 °C in the evenings, even with the living room and kitchen heaters switched off. Individual occupants might react to this in a variety of unpredictable ways – accepting higher temperatures, opening windows, or switching heaters off. Figure 9.4 shows monitored daily temperatures over 6 months in one of the dwellings.

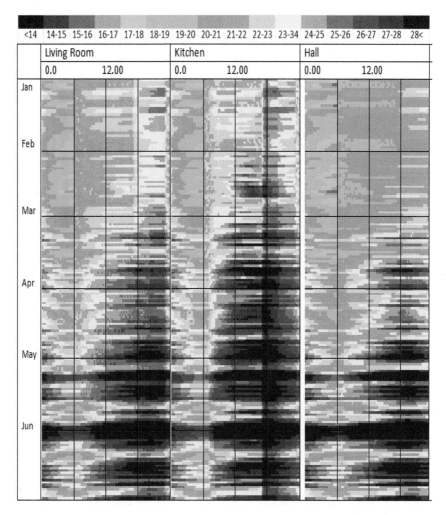

Figure 9.4   Room temperature (°C) variation in a test dwelling

The hall is cooler than the living area because the set point is lower and the heater is heating the first floor via the stairwell. The highest temperatures in the evening, with gains from cooking, can exceed 30 °C. The average living room temperature when occupied is above the set point. A similar pattern can be seen in the high thermal mass flat, although being less insulated, this does not overheat to the same extent. This profile remained almost constant for all the active scheduling regimes in both dwellings. However, with teleswitching, temperatures were significantly higher in both.

9.3   Active network management

The following ESP-r results relate to two dwellings over a January to June period with observed occupancy patterns imposed. Other casual gains and ventilation were estimated based on observations and judiciously adjusted to bring temperatures roughly in line with measurements. Figure 9.5 shows the predicted energy consumption and schedule for two dwellings with different charging regimes.

Neither case shows a significant difference between any of the active schedules, even though dwelling 1's heat demand of 115 kWh/m^2.y is 2.3 times that of dwelling 2. The highest consumption in dwelling 2 is 7% more than the lowest; in dwelling 1, where thick walls provide additional storage, the difference is 2%. However, with teleswitching, both dwellings consume considerably more energy, 66% and 32%, respectively.

For the Utility, the degree to which the schedule instructions are followed is of key importance. A schedule following index, S, was defined as

$$S = 1 - \frac{unscheduled\ charge + scheduled\ charge\ not\ drawn}{total\ scheduled\ charge}.$$

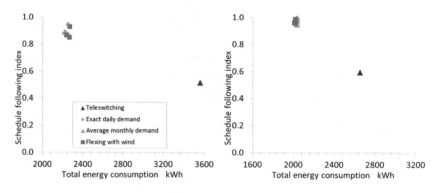

Figure 9.5   Heat consumption and schedule following for dwelling 1 (right) and 2 (left)

With active scheduling, the index ranged from 0.95 to 1.0 for the stone flat and 0.81 to 0.95 for the timber house. Here again, teleswitching gives dramatically worse performance.

Timing and power input level have a small secondary effect. Delivering energy at high power over a shorter time results in 2%–3% higher unscheduled consumption than the same amount scheduled at low power over an extended period. Scheduling late in the day – afternoon or early overnight – consistently gave rise to unscheduled demand in the mornings. The effect of capping the heater fill level can be seen in Figure 9.6 (note the scale of the axes: the small dotted boxes represent the total range covered in Figure 9.5).

In all cases, total consumption decreased as the maximum fill level decreased: the uncontrollable proportion increased, and dramatically so when the cap was low. This replicated the observed situation with high unscheduled charging if the cap is applied. Room temperatures are on average 1 °C lower when the fill level is constrained to 14% or lower.

With unconstrained or 55% fill, the stone flat required almost no direct heating due to its compact layout and the effective storage capacity increase from the walls. The better-insulated house required more heating because the hall heater was delivering heat to the entire upper floor. At the same time, the living room and kitchen were overheated by having two heaters in a relatively small area. Teleswitching is essentially demand-driven, so there is less unscheduled charging and use of panel heaters, even at lower cap levels.

Figure 9.7 shows storage capacity utilisation for the two houses under various unconstrained charging regimes. Scheduling over more than a single day increases capacity utilisation at the cost of around 1% more consumption than a charging regime delivering exactly the day's forecast. In dwelling 2, the storage capacity is 2.2 times the average daily demand, whilst in dwelling 1, with a 1.5 times greater thermal capacity, the demand has less headroom. The fact that the heaters never

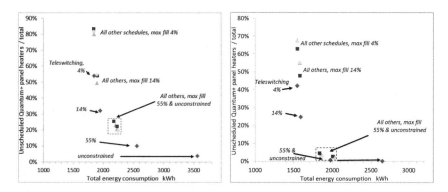

Figure 9.6   Uncontrollable and total heat consumption with different levels of storage capacity for dwelling 1 (right) and 2 (left)

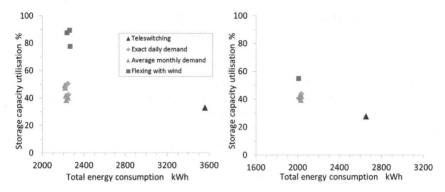

Figure 9.7   Storage capacity utilisation in dwelling 1 (right) and 2 (left)

reach 100% capacity utilisation illustrates the limitations of this kind of static scheduling.

Improving heater insulation to a level that halves the uncontrolled heat loss, reduces energy consumption by about 6%, and increases capacity utilisation, both without reducing charge schedule following or room temperatures significantly. However, reducing the uncontrolled loss by 90% or more resulted in a 0.5 °C–1.0 °C lower room temperature and increased reliance on panel heaters.

A similar set of schedules was applied to dwelling 2 – a poorly insulated bungalow with timber frame and block walls typical of older dwellings in Scotland. The heaters performed much as before: a maximum 3% difference in total energy consumption, and a schedule following index of 0.85–0.9 across all active schedules. However, with a specific heat demand around double that of dwelling 1, the four storage heaters struggled to maintain set-point temperatures in January and February, even though they drew 5% more energy than scheduled. There was also less headroom for storing excess wind as the storage capacity utilisation was over 75%.

For the network operator, heaters offering smart external control are demonstrably better than teleswitched storage heating. The schedule following is higher and the total demand lower provided the scheduled energy delivery is roughly in line with demand. It was observed that it is slightly better to charge early in the day at a low power level. Scheduling in the afternoon or early overnight caused an unscheduled load shortly after the heating came on in the morning: this could be compensated by a slightly higher total energy delivery.

The new technology also benefits occupants: the heaters are more controllable, so they use less electricity and cause less overheating. Occupants should see only minimal differences in room temperature, energy used, or costs with different charging schedules, again provided the charge is roughly in balance with demand. This applies even when double the demand is delivered on some days, none on others, and is not influenced by house construction. However, for a given thermostat setting, room temperatures will be above the set point in well-insulated dwellings and lower in poorly insulated ones, whatever the schedule. The simulation also highlighted a counter-intuitive feature of heater control: since both fan-assisted and uncontrolled output depends on the level of fill, turning up the thermostat can lead to a downward spiral where the heater fan switches on more often, lowering the core temperature and reducing the heat output, which then requires even more use of the fan. Eventually, this has to be compensated for with costly direct heating.

Benefits to both utility and occupant can be wiped out if the heater's adaptive control cap is operational. In this case, heaters and central controllers compete, with the net effect of pushing demand from scheduled to unscheduled periods and from storage to direct heating, increasing costs for customers and the peak load for the utility. Overall, however, the simulations demonstrated the efficacy of the device storage cap in reducing energy consumption and regulating temperature where the heater would otherwise fill to a maximum at every opportunity.

Storage heaters use more energy than direct heating because the uncontrolled output is continuous. If heat output between midnight and 6 a.m. is considered mostly unnecessary, the best-insulated house consumed 25% more electricity than needed, and the worst-insulated 14%. This is the same or less than the energy penalty reported when using a hot water buffer to shift heat pump operation (Hong *et al.* 2013): 16% where it is shifted outside a 3 hour early evening peak (Arteconi *et al.* 2013) and over 60% if the pump operates under an Economy 10 tariff[4]. However, storage heaters allow much greater flexibility in timing.

The storage capacity considered in the project allowed scheduling over 48 rather than 24 hours. In all houses assessed, it was possible to do this without increasing indoor temperatures and incurring only a small penalty in demand. Future improvements in heater insulation that reduce uncontrolled output by 50% would reduce even that penalty.

9.3.1   *Building-integrated BPS+*

In order to work reliably, the ANM scheduling system needs a method to forecast the heating and domestic hot water (DHW) demand of the buildings within the smart grid. BPS+ can assist on a recurring basis if embedded within representative dwelling types. Hand (2015) enabled this possibility by installing ESP-r on an ultra-low-cost Raspberry Pi[5] as shown in Figure 9.8. This single-board computer mounted behind an LCD monitor offers wireless, Bluetooth connectivity, 1.2 GHz CPU, and 1 GB RAM for around £28 (+£35 for the monitor) at 2024 prices.

An ESP-r 13-zone annual simulation at four time steps per hour required around 40 minutes on a 2017 Raspberry Pi 3 whilst a 3-zone simulation at 20 time steps per hour required 10 minutes.

A practical prerequisite of the smart grid approach is robust communications. Within the Lerwick field trial, the system was set to exchange data between houses and centre at 1-minute frequency using a cellular network. This proved impractical,

---

[4]https://en.wikipedia.org/wiki/Economy_10
[5]https://raspberrypi.org

Figure 9.8   A Raspberry Pi 3 with 5" display

with many communications outages and a loss of central control. Since the charge schedule is enacted in 15 minutes blocks, data exchange at 5 or 10 minutes intervals would place less onerous demand on communications without compromising functionality. The small computer solution would support signal accumulation/ processing and robust transmission via WiFi.

Several findings emerged from the project based on simulations of a network with about 20% of all Lerwick households equipped with smart heaters (Gill and Kockar 2013):

- the increased load factor enabled an additional 15 MW of wind to be operated economically;
- the 2.9–4.2 GWh/y reduction in fossil fuels has high replication potential; and
- the peak load reduction of 2 kW/dwelling could underpin new electricity network guidelines.

The overall conclusion was that a distributed storage solution that permits controllable load shifting over 6 hours would be economically advantageous (Strbac *et al.* 2012).

9.4   Microgrid trading

A core strength of BPS+ is the ability to explore solutions that do not yet exist. The case considered here is microgrid power trading. The rationale of the microgrid approach has been widely recognised (Ofgem 2007). Electricity can be generated locally from new and renewable sources, thus increasing the overall efficiency of generation, eliminating the losses incurred in transmitting power to the point of use, and reducing the overall environmental impact. Because the approach is community-based, it has the potential to represent a better value proposition for citizens by ensuring their greater engagement in energy issues and providing economic benefit. The approach also has the potential to contribute significantly to the security of supply at all scales.

As always, there are associated issues. In the absence of storage, electricity supply must be equal to demand. Normally, generators slow down with rising demand, and the supply frequency drops. Load frequency controllers sense this change and increase the fuel supply. However, in the case of renewable energy-based generators within a microgrid, the 'fuel' supply is weather-related, and it cannot be controlled in this manner. Instead, frequency control needs to be recovered by scheduling non-renewable energy generators to satisfy any deficit demand. Similar issues exist with voltage regulation. Fault detection and isolation are other crucial issues. Where the fault is within the national grid, the protection system could isolate the microgrid so that it may function normally. The islanded microgrid would then drift out of phase with the national grid in the absence of a phase-locking mechanism.

Conner (2003) studied the issues of control and coordination and proposed a dispatch management system based on centralised control. The project demonstrated successful trading within a simulation environment and an experimental test bed. Jiayi *et al.* (2008) proposed a multi-agent, distributed control system on the basis that the approach is better able to deal with the randomness of energy demand and supply within individual microgrids. Trichakis *et al.* (2008) also proposed the agent-based approach as best able to overcome the constraints associated with a low voltage distribution network: voltage stability control, cable thermal protection, and reverse power flow avoidance. In the project reported here, it was assumed that each microgrid operates autonomously and does not allow any external agent to direct its available resources. All micro-generators within a microgrid are primarily connected to the local loads. Surplus energy is sold to a central pool, and deficit energy is bought from the pool if available; otherwise, the public supply is used, as depicted in Figure 9.9. In this approach, control is autonomous and embedded within each microgrid.

If they develop in the future, it is likely that such microgrids will vary in size, and as the number of microgrids within heterogeneous trading networks grows, so too will the issue of complexity management: is the transient response and stability of hundreds or thousands of embedded generators manageable? A key issue is economic viability: in the medium term, a microgrid can only compete with the retail price of energy where other advantages are maximised, such as the utilisation of waste heat, the exporting of energy surpluses, and the capitalisation of

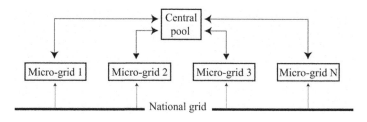

Figure 9.9   Microgrid trading via a central pool

socio-economic benefits. Where microgrids are interconnected, the possibility of trading exists. The question then is how can trading be controlled to incentivise all players, regardless of size or location.

The Micro-Grid Trading (MGT) program[6] (Ali 2009) was used to explore this issue. This software application does not possess models of demand and supply side entities; instead, it is given temporal demand and supply profiles as input, and these are used as a seed to generate a user-specified number of microgrids against directives relating to size, type, and operational variability. In this way, future network models can be established at any scale, comprising constituent microgrids of similar or dissimilar characteristics. The seed profiles may be informed inventions, field observations, or the results of BPS+ simulations (here ESP-r and Merit). In the study summarised here, it was assumed that the trading price of surplus microgrid electricity is always less than the retail electricity price. In addition, this price is fixed as opposed to fluctuating, as would happen in a competitive bidding regime.

MGT caters to three levels of microgrid size variation to reflect the likely diversity of scale: low, medium, and high. When set to low, the generated microgrids randomly fall within the range 0.5–1.5 times the seed capacity value, when set to medium, 0.3–3 times this value, and when set to high, 0.1–10 times. The program caters to the possibility of a microgrid being disconnected from the trading network due to equipment failure or planned maintenance by resetting the output from randomly selected microgrids to zero over randomly selected periods as a simulation proceeds. The intermittency of renewable energy sources is modelled by implementing similar random restrictions on randomly selected micro-generators. Based on user selections, MGT generates a trading network comprising the required number of microgrids with the required size variability and level of reliability.

The subsequent trading simulation has two principal constraints. First, a microgrid can only import energy from the public supply if the surplus energy from all other microgrids is less than its demand. Second, the simulation proceeds based on user-supplied prices for centrally supplied and microgrid surplus energy, with the latter constrained to be lower. Different rates may be imposed for each hour of a day to reflect the different possible tariffs. At each simulation time step, the energy available from all microgrids in surplus is added to a central pool. A random function is invoked at this point to decide the order in which microgrids in deficit may import from this pool. In this way, the random nature of trading in the real world has been emulated. Where there is insufficient energy in the pool to satisfy a selected microgrid's deficit in its entirety, it may import from the public supply at the prevailing price. Note that the simulations do not differentiate between real and reactive power as required where microgrid power factors vary. Choi *et al.* (1998) have demonstrated that the real-time pricing of real and reactive power can act to clip the peak power demand and raise the minimum demand, thus reducing the investment in power system equipment and potentially improving the supply/demand match.

---

[6]https://www.esru.strath.ac.uk/Applications

The results available from simulations include statistics on the temporal variation in the energy available in the central pool, the ability of the pool to satisfy microgrid demand over time, and the overall cost of trading to each microgrid. Three aggregate performance indicators are derived from these results to support an overall assessment of trading efficacy.

Energy wasted (E_w) defines the portion of the surplus microgrid energy transferred to the pool that is not utilised:

$$E_w = 1\left(\frac{E_u}{E_p}\right)$$

where E_u is the pool energy that is utilised and E_p the total surplus energy transferred to the pool by all microgrids.

Monetary loss (M_l) relates only to the seed microgrid and defines the difference between the amount paid for energy imports from the grid (if the microgrid operated alone) and the amount paid within the trading scheme for pool imports:

$$M_l = 1 - \left(\frac{i_g - i_t}{i_g}\right)$$

where i_g is the income when the microgrid imports only from the national grid, and i_t where it trades with other microgrids. A negative M_l indicates that the microgrid trading imparts a benefit (i.e., the microgrid receives income after paying for all imports).

Energy imported (E_i) defines the portion of the energy imported from the national grid by all microgrids (energy is imported from the national grid whenever the trading pool is unable to meet a microgrid's demand at any time):

$$E_i = \frac{E_g}{E_t}$$

where E_g is the energy imported from the grid and E_t the total energy consumed by all microgrids.

ESP-r and Merit were used to generate the demand and supply profiles for the seed mircogrid, and then MGT was applied to several scenarios, each relating to the impact of a principal network attribute or operational characteristic on microgrid tradability as measured by the above performance indicators. Within each scenario, a principal parameter was varied whilst all other parameters were held constant, as summarised in Table 9.3.

The trading simulation outcomes are listed in Table 9.4. These results gave rise to the following observations.

- All trading performance indicators improve as the number of participating microgrids increases: E_w indicates an increasing utilisation of pool energy, although a large portion (78%) remains unallocated within the largest network considered here – M_l indicates a reducing economic burden on the seed

Table 9.3   Operational parameter variation within each scenario

	Parameter value assumed within all scenarios except where varied as stated below:			
Number of microgrids	11			
Size variability	Medium			
Equipment failure	No			
Renewable energy intermittency	No			
Variable energy tariff	No			
Grid price (p/kWh)	10			
Microgrid trading price (p/kWh)	7			
Scenario	Operational parameter variation:			
	Case 1	Case 2	Case 3	Case 4
Number of microgrids	4	11	51	101
Size variability	Low	Medium	High	
Equipment failure	No	Yes		
Renewable energy intermittency	No	Yes		
Variable energy tariff	No	Yes		

Table 9.4   Microgrid trading simulation results

Scenario	Case	E_w	M_l	E_i
Number of microgrids	1	0.96	0.60	0.86
	2	0.86	0.48	0.51
	3	0.82	0.46	0.36
	4	0.78	0.42	0.28
Size variability	1	0.91	0.59	0.49
	2	0.86	0.48	0.51
	3	0.81	0.35	0.44
Equipment failure	1	0.86	0.48	0.51
	2	0.82	0.24	0.53
Renewable intermittency	1	0.86	0.48	0.51
	2	0.79	0.10	0.63
Variable energy tariff	1		0.48	
	2		0.10	

microgrid and E_i indicates a marked reduction in the energy imported from the national grid.

- Where a performance indicator improves as size variability is increased, this improvement is marginal, whilst the impact on the energy imported from the national grid (E_i) changes in a non-linear manner, with an initial rise followed by a substantial fall as the variability moves from low to high.
- Equipment failure at the levels considered here results in a marginal improvement in pool energy utilisation and has a beneficial impact on unaffected microgrids in monetary terms.

- As expected, high renewable energy intermittency results in an overall increase in the importing of energy from the national grid. Paradoxically, this can result in a financial benefit to some microgrids where, as here for the seed microgrid, they are a net exporter of energy to the trading pool.
- Also, as expected, a variable tariff has a major impact on the operational cost of an individual microgrid.

It was concluded that large microgrid trading networks would result in a more effective trading regime. Where a large network is further comprised of highly variable microgrid sizes, there will be more frequent trading opportunities, resulting in less interaction with the public supply network. Equipment failure and the intermittency of renewable energy sources can act as a stimulus to trading, although the net overall impact, beneficial or otherwise, could not be deduced from the present study. Finally, the differential cost of electricity between the trading pool and the public supply will be the principal determinant of successful trading. It is unlikely that an unregulated market will maintain a sufficient differential.

9.5   Chapter summary

BPS+ is able to evaluate the conflicting aspects underpinning smart grid operations relating to the use of external control agents to orchestrate the charging of distributed storage. Whilst the approach considered in this chapter involved domestic storage heaters, it could equally well be the excess battery capacity of EVs. By implementing embedding BPS+ on ultra-small and low-cost computers, the ability to embed a simulation model within a smart control regime has been demonstrated. Simulation is also an apt technology to generate data for the analysis of future projects such as trading microgrids.

References and further reading

Alamdari F, Hammond GP and Melo C (1984) 'Appropriate calculation methods for convective heat transfer from building surfaces', *Proc. UK National Conf. on Heat Transfer*; 2, pp. 1201–1211.

Ali (2009) 'Electricity trading among micro-grids', *MSc Thesis*, ESRU, University of Strathclyde.

Arteconi A, Hewitt NJ and Polonara F (2013) 'Domestic demand-side management: Role of heat pumps and thermal energy storage systems', *Applied Thermal Engineering*, 51, pp. 155–165.

Born F J (2001) 'Aiding renewable energy integration through complementary demand-supply matching', *PhD Thesis*, University of Strathclyde.

Choi J Y, Rim S-H and Park J-K (1998) 'Optimal real time pricing of real and reactive powers', *IEEE Transactions on Power Systems*, 13(4), pp. 1226–1231.

Clarke J A (2001) *Energy Simulation in Building Design*, Butterworth-Heinemann, Oxford.

Clarke J A, Hand J, Kim J M, Samuel A A, and Svehla K (2014) 'Performance of actively controlled domestic heat storage devices in a smart grid', *Power and Energy*, 229(1), pp. 99–110.

Conner S J (2003) 'Distributed dispatching for embedded generation', *PhD Thesis*, University of Strathclyde.

DECC (2012) *Energy consumption in the UK, Domestic Data Tables*, Dept. Energy and Climate Change, https://gov.uk/government/publications/energy-consumption-in-the-uk.

DUKES (2011) 'Energy consumption by final user (energy supplied basis) 1970 to 2012', Digest of UK Energy Statistics, https://gov.uk/government/publications/energy-chapter-1-digest-of-united-kingdom-energy-statistics-dukes/.

Farid M M and Chen X D (1999) 'Domestic electrical space heating with heat storage', *Proc. IMechE Part A: Power and Energy*, 213, pp. 83–92.

Fisher D E and Pederson C O (1997) 'Convective heat transfer in building energy and thermal load calculation', *ASHRAE Transactions*, 103(2), pp. 137–148.

Foote C E T, Roscoe A J, Currie R A F, Ault G F and McDonald J R (2005) 'Ubiquitous energy storage', *Proc. Int. Conf. Future Power Systems*, Amsterdam, November.

Gill S and Kockar I (2013) 'Unit scheduling simulations report in support of Shetland repowering', Dept. of Electrical and Electronic Engineering, University of Strathclyde.

Gill S, Ault G, Kockar I, Foote C and Reid S (2013) 'Operating a wind farm in the future smart grid - lessons from designing and deploying a smart grid on Shetland', *Proc. European Wind Energy Conference*, Copenhagen, February.

Glen Dimplex (2013) *The Quantum Energy System*, https://quantumheating.co.uk/the-system-explained.php/.

Hand J W (2015) 'Prospects for building simulation on small scale computing platforms', *Building Simulation'15*, Hyderabad.

Hasnain S M (1990) 'Review on sustainable thermal energy storage technologies. Part I: heat storage materials and techniques', *Energy Conversion and Management*, 30, pp. 1127–1138.

Hayton J (1994) 'Heating consumption measurement in electrically heated dwellings', *Building Serv. Eng. Res. Technol.*, 15, pp. 19–24.

Hong J, Kelly N J, Richardson I and Thomson M (2013) 'Assessing heat pumps as flexible load', *Proc. IMechE Part A: J Power and Energy*, 227, pp. 30–42.

Jiayi H, Chuanwen J and Rong X (2008) 'A review on distributed energy resources and MicroGrid', *Renewable and Sustainable Energy Reviews*, 12(9), pp. 2472–2483.

Kamran M (2022) *Fundamentals of Smart Grid Systems*, Elsevier, Amsterdam, ISBN 978-0-323-99560-3.

Lacroix M (1999) 'Electric water heater designs for load shifting and control for bacterial contamination', *Energy Conversion and Management*, 40, pp. 1313–1340.

Moreau A (2011) 'Control strategy for domestic water heaters during peak periods and its impact on the demand for electricity', *Energy Procedia*, 12, pp. 1074–1082.

Ofgem (2007) *Review of Distributed Generation*, https://ofgem.gov.uk/sites/default/files/docs/2007/05/review-of-distributed-generation.pdf.

Ofgem (2013) 'Demand side response', Office of Gas and Electricity Markets, https://ofgem.gov.uk/Sustainability/Documents1/DSR%20150710.pdf.

Oughton D R, Hodkinson S and Faber O (2008) 'Electrical storage heating', *Heating and Air-conditioning of Buildings* (10th edn.), Taylor & Francis, Milton Park.

Qureshi W A, Nair N K C and Farid M M (2011) 'Impact of energy storage in buildings on electricity demand side management', *Energy Conversion and Management*, 52, pp. 2110–2120.

SHEPD (2012) 'Proposals for the development of the integrated plan for Shetland', Ofgem https://ofgem.gov.uk/Networks/ElecDist/Policy/Documents1/Phase%201%20Consultation%20Aug%202011.pdf.

Shipworth M, Firth S K, Gentry M I, Wright A J, Shipworth D T and Lomas K J (2012) 'Central heating thermostat settings and timing: building demographics', *Building Research & Information*, 38(1), pp. 50–69.

Strbac G, Aunedi M, Pudjianto D, *et al.* (2012) *Strategic Assessment of the Role and Value of Energy Storage Systems in the UK Low Carbon Energy Future*, Report for Carbon Trust, June.

Trichakis P, Taylor P C, Coates G and Cipcigan L M (2008) 'Distributed control approach for small-scale energy zones', *Power and Energy*, 222(2), pp.137–147.

Which Review (2013) 'Home heating systems: electric central heating pros and cons', https://which.co.uk/energy/creating-an-energy-saving-home/guides/home-heating-systems/electric-central-heating/.

Yohanis Y G and Mondol J D (2010) 'Annual variations of temperature in a sample of UK dwellings', *Applied Energy*, 87, pp. 681–690.

Chapter 10

Urban energy systems deployment

BPS+ tools are normally applied by professionals to provide information in support of design decision-making. In this mode of operation, outputs are scrutinised and considered before being passed on as recommendations to clients. The impact can be amplified where outputs are used to inform policy and urban action planning. This chapter considers the extension of BPS+ to include the deployment of sustainable energy systems within urban environments requiring site selection that respects local policy and technical constraints. The factors underlying policy and technical decision-making, and the scoring and weighting factors to be applied, can be elicited from city planners and utility personnel and used to identify urban sites that are policy unconstrained and technically feasible. This chapter reports the form of a Geospatial Opportunity Mapping tool, named GOMap[1], which operates alongside BPS+ to quantify the energy potentials of urban energy systems deployment.

10.1 City development plans

City councils in the United Kingdom and elsewhere are increasingly interested in urban renewable energy technology (RET) for its potential to bring economic benefits (Mirzania *et al.* 2019). Local RET deployment is considered an apt contribution to net zero carbon aspirations, allowing energy to be generated on vacant and derelict land and then consumed locally or fed to a local substation (Pillot *et al.* 2020). The main challenge lies in identifying land that is both policy unconstrained and technically feasible. RET projects typically focus on the latter issue in terms of resource potential and access to grid transmission lines (Amjad and Shah 2020).

Glasgow City Council (GCC), for example, has considered novel uses for vacant and derelict land (VDL), which at the time of writing comprised approximately 880 ha across 644 sites, and accounted for 9% of Scotland's total VDL area (GCC 2022). The council is interested in local power generation and with support from Innovate UK[2] through the Future City Glasgow Demonstrator Project[3] (Innovate UK 2015), commissioned a sub-project to assess the potential for VDL usage in Glasgow (Clarke *et al.* 2020). The GOMap geospatial mapping tool was

---

[1]https://www.esru.strath.ac.uk/Applications
[2]https://www.ukri.org/councils/innovate-uk/
[3]https://www.gov.uk/government/news/glasgow-becomes-a-world-leading-smart-city

developed and applied to this end (McGhee 2022). GOMap identifies city areas where the deployment of community-scale RET schemes is most feasible whilst highlighting the policy and technical constraints affecting other possible locations.

To identify potential sites, the policy aspects that underpin city development, such as urban expansion, natural resource management, and regional integration, must be considered. This information is typically held by local authorities (Omitaomu *et al.* 2012). The planning process in Scotland has recently been redesigned (SG 2023), with numerous spatial principles and national policies highlighted as follows.

- Strategy for Economic Transformation Principle – to provide economic benefit.
- Compact Urban Growth Principle – to make efficient use of land and provide services and resources such as carbon storage and flood risk management.
- Housing to 2040 Principle – to provide access to affordable, quality homes to tackle the housing shortage.
- National Climate Change Plan – to achieve net zero emissions by 2045 by encouraging reduced travel and sustainable transport.
- Natural Places Policy – to protect the environmental landscape, including cultural and natural heritage and green space; and
- National Energy Strategy – to promote resource recovery and waste reduction.

The planning process focuses on policy regulations to promote city development and aims to protect areas with significant environmental and biodiversity presence. The process typically excludes the technical factors that govern the suitability of land to accommodate RET deployment.

Within the project summarised here, an evaluation procedure based on geographic information system (GIS) and BPS+ procedures was established to identify the available land in the light of technical and policy issues – such as ground condition and barriers to network connection in the former case, and environmental and biodiversity impacts in the latter (Georgiou and Skarlatos 2016). The project proceeded in two stages.

The first stage focused on combining policy and technical aspects to enable site selection for RET. Existing tools mainly addressed the regional scale or included only a subset of issues (Cunden *et al.* 2020, Finn and McKenzie 2020, Günen 2021). A more comprehensive evaluation method was needed to balance all policy and technical factors in the urban context. This resulted in the GOMap opportunity mapping tool. The second stage demonstrated the use of GOMap and BPS+ (here ESP-r) to undertake scenario appraisals at the city scale, here Glasgow.

The procedure allows policymakers and developers to examine building performance and RET deployment within a GIS framework, and explore energy planning strategies that match local energy demand with new energy supply through the coordinated deployment of building retrofit packages and RET (Yamagata *et al.* 2020, Anand and Deb 2023).

10.2 Balancing policy and technical issues

The functionality required to address the first stage was encapsulated in GOMap based on information acquired from the planning authority and utility provider. The RET target was set as Photovoltaic Power Station (PVPS) – see Section 13.3 – which was of interest to the local authority at the time.

Consultations were held with personnel from GCC's Land and Environmental Services department. Five policy aspects were identified, each comprising a number of factors affecting site suitability based on definitions from the following sources.

- GCC's City Development Plans (GCC 2017, 2017a);
- Scottish Planning Policy (SG 2023); and
- UK Legislation (2024).

The first policy aspect is *Biodiversity* and encompasses factors related to creature habitats within natural sites such as parks and forestry. The Land and Environmental Services department holds information on species in hotspot areas although this is confidential to prevent hunting (GCC 2001). GOMap, as released publicly, highlights only general areas where creature habitats are likely to be present, without providing precise locations.

The second policy aspect concerns *Development* and focuses on sustainable and economic growth. It prioritises support for local areas through new housing, facilities, businesses, and travel links. Community growth masterplans address projected housing land shortfalls over the next decade to meet future needs. The city's green belts are assessed to determine if boundary alterations are needed or if land can be removed for development. Green networks link natural, semi-natural, and constructed open spaces for physical activity and increased accessibility. Housing land supply is a major factor for city development, with areas selected for transformational regeneration for large-scale housing-led projects. Master plan factors provide physical environments that support strong communities and quality of life and create places of distinction.

The third policy aspect addresses *Environment* to protect natural green areas and preserve landscapes. It includes conservation areas designated under the Listed Buildings and Conservation Areas (Scotland) Act 1997, historic garden landscapes, and scheduled ancient monuments (SG 2010, SG 2011, HES 2014). These policies aim to protect, enhance, and manage heritage sites for current and future generations. Sites of importance for nature conservation provide guidance on the natural environment, including protected sites and species, and the enhancement of biodiversity. Sites of special landscape importance are identified by criteria including scenic quality, landscape quality, and natural/cultural heritage features. Two factors are deemed to have high priority. One is sites of special scientific interest, which designates areas with special flora, fauna, geology, or geomorphological features; and is designed to safeguard the longevity of rare and/or special natural features (SG 2008). The other high-priority factor is the Antonine Wall World Heritage Site, a famous landmark of the Frontiers of the Roman Empire. The Antonine Wall possesses a buffer zone to protect the site and its immediate surroundings.

The fourth policy aspect concerns *Visual Impact* and comprises factors based on the proximity of residential properties to a RET deployment. Because a PVPS structure is not high, it is unlikely to restrict the view from nearby communities, especially as there is often a natural barrier of trees. Where such a barrier is not present, a separation distance of 500m is recommended (Castillo *et al.* 2016).

The fifth policy aspect is *Visual Intrusion* and comprises a factor addressing the susceptibility to glare caused by reflections from a RET deployment. Research into temporary vision loss has been carried out by aviation agencies where flash blindness for a period of 4–12 seconds can be caused by 7–11 W/m^2 reaching an observer's eye (US Federal Aviation Authority 2010). In the United Kingdom, it is considered unlikely that this intensity level would be experienced from a large distance although 20 W/m^2 is possible when in close proximity (UK Civil Aviation Authority 2010).

Consultations were held with Scottish Power Energy Networks (SPEN) to obtain details on the distribution network and substation location/capacity throughout the city. The Ordnance Survey[4] and Centre for Environmental Data Analysis (CEDA 2021) were contacted to obtain building heights and terrain data, respectively. Four technical aspects were identified, each comprising factors that affect RET in general and one that is specific to PVPS.

The first technical aspect is *Overshading* caused by structures such as buildings and trees. It required pre-processing using GIS tools to generate a polygon shapefile using building height data (OS 2021). The efficiency of a PVPS depends on the incident solar irradiation, and overshading can significantly reduce power output. The GOMap/ESP-r procedure is able to determine shading based on sun position, obstructions geometry, PV panel inclination, and PV array arrangement on site (Vulkan *et al.* 2018).

The second technical aspect concerns substation *Proximity,* comprising factors that enable the determination of substation distance from a grid connection point as a function of the 11 kV network configuration.

The third technical aspect concerns substation *Congestion,* comprising factors that identify connection suitability based on issues such as existing fault levels, risk of reverse power flow, and voltage rise (SPEN 2015).

The fourth technical aspect concerns *Terrain,* comprising land topography and related vulnerabilities. Although most locations will not prevent accessibility, issues such as foundations, flood risk, and steep slopes may prove problematic. Terrain information stored in digital terrain models is provided by the Centre of Environmental Data Analysis. A digital terrain model is made available in raster or image format, where each pixel has values representing an attribute such as building height, soil concentration, and vegetation type.

---

[4]https://www.gov.uk/government/organisations/ordnance-survey

10.3 Identifying city-wide opportunities

To accommodate the information collated during the consultations, GOMap includes mechanisms for gridding, scoring, and weighting the factors underpinning the policy and technical aspects as described above.

The grid mechanism distinguishes active factors from others by assigning unique ID numbers to each grid cell. Variable grid dimensions can be set to suit geographical requirements. Because grid resolution issues can occur when cells overlap irrelevant parts of land, an embedded rule allows cells to gain information from active factors if 50% or more of the cell's total area overlaps a factor. If a cell fails the rule, no factor information is passed to the cell. A study by McGhee (2022) deduced an ideal grid resolution for Glasgow as 10 m × 10 m, giving 1.76 million cells: going beyond this would not yield significant new information. Each grid cell possessed a unique ID allowing for an overall cell score to be calculated.

The next step was to determine the scoring method to determine the overall policy and technical aspect scores for each grid cell. From the psychometric literature (Nunnally and Bernstein 1994), a 3-point scoring method was found to be effective in the present case as it reduces subjectivity by assigning a simple 'Good', 'Moderate,' and 'Bad' score (Preston and Colman 2000). This scoring system leads to scores according to the criteria shown in Table 10.1. In rare cases, a 'showstopper' score is included where a factor underlying any policy aspect cannot be mitigated, such as those related to the environment or World Heritage Sites. As each city can have its own local requirements, the GOMap tool allows these scores to be modified.

The scores for each policy and technical factor were agreed through consultation with local authority planners and utility specialists. The spatially varying factor scores were installed in GIS map layers representing the geographical data as a visual representation of the hierarchy shown in Tables 10.2 and 10.3 relating to policy and technical aspects, respectively, and their associated factors and scores. Each map layer can be independently enabled/disabled depending on the given application scenario.

A scoring mechanism was developed to calculate the overall score for a grid cell based on individual scores of overlapping policy and technical factors. The mechanism consists of two use cases: 'lenient' and 'stringent'. The lenient method calculates the median value of individual factor scores for each aspect. The overall

Table 10.1 Policy and technical aspect scores

Score	Policy aspect	Technical aspect
1	Possible	Favourable
2	Intermediate	Likely
3	Sensitive	Unlikely
4	Showstopper	–

Table 10.2   Policy aspect factors and scores

Aspect	Rating	Factor	Score
Biodiversity	Possible	No species on the protected list is believed to be present.	1
	Intermediate	UK protected species possibly resident, which requires an environmental survey and mitigation measures.	2
	Sensitive	European protected species possibly resident, which requires an environmental survey and serious mitigation measures.	3
Developmental	Possible	Master plan area; strategic economic investment locations; transformational regeneration areas.	1
	Intermediate	Community growth masterplan area; Economic policy areas; Green belt; Green network opportunity areas; Housing land supply; Industrial-business marketable land supply; Local development framework; Network of Centres; Strategic development framework; Strategic development framework – river.	2
	Sensitive	Housing land supply with consented developments.	3
Environmental	Possible	Green corridors; Local nature reserves.	1
	Intermediate	Conservation areas; Listed buildings; Ancient woodlands; Tree preservation orders; World Heritage site buffer zone.	2
	Sensitive	Sites of special landscape importance; Gardens and designed landscapes; Scheduled ancient monuments; Sites of importance for nature conservation.	3
	Showstopper	Sites of Special Scientific Interest; World Heritage site (Antonine Wall).	4
Visual impact	Possible	No residential areas overlook the site.	1
	Intermediate	Residential areas overlook the site.	2
Visual intrusion	Possible	All other areas.	1
	Intermediate	Between 1 and 5 km radius to the south of an airport or heliport or within 100 m of a motorway.	2
	Sensitive	Within a 1 km radius to the south of an airport or heliport or within 100 m of a motorway.	3

score is then calculated using the median value of all aspect scores. The stringent method is stricter and assigns the highest overlapping factor score. Because GCC wished to encourage development, the lenient scoring method was set as the default.

The introduction of an aspect weighting system in GOMap was necessary to allow planners to explore the impact of future policy change and infrastructure development (Kumar *et al.* 2017). This involved applying the Analytical Hierarchy Process method (Saaty 1980) to determine the ideal weightings of all aspects for a given scenario whilst alleviating subjective bias. Based on consultations and planning documents, biodiversity and environmental factors are emphasised greatly due to the possible dangers imposed on surrounding wildlife, flora, and fauna, and

Table 10.3   Technical aspect factors and scores

Aspect	Rating	Factor	Score
Overshading	Favourable	Falls outside the estimated annual shaded footprint.	1
	Unlikely	Falls within the estimated annual shaded footprint.	3
Substation	Favourable	Combined heat map score under 10.	1
congestion	Likely	Combined heat map score equal to 10.	2
	Unlikely	Combined heat map score greater than 10.	3
Substation	Favourable	Within 100 m of a substation connection line.	1
connection	Likely	Between 100 m and 200 m of a substation connection line.	2
distance			
	Unlikely	Further than 200 m from a substation connection line.	3
Terrain	Favourable	Flat ground, no access issues or risk of flooding.	1
	Likely	Heavily sloping or broken ground; restricted access; unsafe buildings; medium risk of river or coastal flooding; high risk of surface water over a large area.	2
	Unlikely	No direct access; site underwater or with a high risk of river or coastal flooding.	3

the local ecological system from project deployment. As a result, these aspects are given the highest level of significance to ensure protection and preservation emphasis is made for natural areas of landscape importance. This is followed by the visual impact on residents in close proximity to a PVPS and then city development due to economic impact. Visual intrusion has the least significance due to the low risk of glare.

On the technical side, the overshading aspect is given the highest level of significance. This is followed by substation connection distance due to the logistics and cost involved in the installation of new connection cables. Next is substation congestion, which indicates whether a new generation can be attached to a substation without overloading circuits. The terrain has the least significance as broken ground or flooding risk can usually be readily mitigated.

With this information, the weightings of each policy and technical aspect were set for Glasgow as shown in Table 10.4.

These weightings allow the final score of a grid cell to be determined by multiplying the scores of each aspect by its weight and summing the result (Parihari *et al.* 2021). Taken together, the policy and technical factor scoring and weighting identify the land available for RET (here PSPV) deployment. This land area is passed to ESP-r to determine the optimum parameters for the PV installation and the likely power output over relevant periods (Kalogirou 2014).

10.4   Photovoltaic power station potential

When deploying a PVPS, a key issue is the determination of the minimum distance between arrays of PV panels to avoid shading from neighbouring rows. This minimum distance is a function of panel orientation, inclination, and solar incidence angle

Table 10.4   Policy and technical aspect weightings

Policy		Technical	
Aspect	**Weighting**	**Aspect**	**Weighting**
Biodiversity	0.326	*Overshading*	0.484
Developmental	0.114	*Sub. congestion*	0.168
Environmental	0.326	*Connection dist.*	0.231
Visual impact	0.148	*Terrain*	0.117
Visual intrusion	0.086		
Σ	1.000	Σ	1.000

Table 10.5   PVPS output, Glasgow

Parameter	**Value**
Panel azimuth	205°
Panel tilt angle	37°
Panel area	2 m^2
Array spacing	7.4 m
Energy yield	172.8 kWh/m^2.y

and can be determined from simulations corresponding to increments in the value of these parameters. A site utilisation factor is introduced by GOMap to define the portion of a site that can be usefully utilised for PV array deployment given the need for maintenance access and other, perhaps unrelated site activities. Araki *et al.* (2017) have reported a land utilisation factor for PVPS between 47% and 51%. In GOMap, the factor is user-set with a default value of 50%.

After determining the available land, GOMap invokes ESP-r and receives back the predicted PVPS annual energy output corresponding to PVPS parameters determined from a parametric sensitivity analysis as typified by the data of Table 10.5 for a single panel.

GOMap delivers its findings as an opportunity map highlighting areas suitable for PVPS deployment as shown in the example of Figure 10.1.

The green areas depict land that is policy unconstrained and technically feasible. The focus selection panel (upper-left) contains all spatial information including policy, technical, and supplementary data. The weightings (upper right) for each policy and technical aspect are listed with default weightings set as equal. Individual factors can be disabled and aspect weightings changed to explore alternative planning strategies. The BPS+ returns are displayed bottom left.

Outwith GOMap's main function, the tool allows custom scripts to be introduced. One such script facilitates the identification of buildings possessing south-facing rooftops for possible building-integrated PV installation (Saretta *et al.* 2020). With reference to Figure 10.2, this process requires two sets of information:

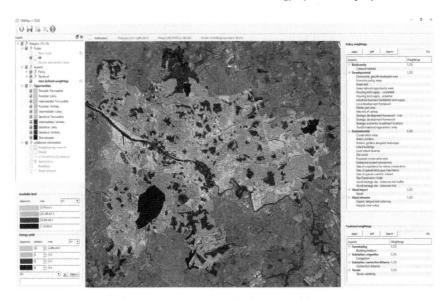

Figure 10.1 GOMap applied to Glasgow City

Figure 10.2 Procedure for identifying south-facing rooftops

building polygon information with associated location and occupancy status details (OS 2020) and a 2 m high-resolution digital surface model (upper-left) containing heights for tall structures such as trees and buildings (CEDA 2021).

The digital surface model is processed into an aspect raster file ('aspect' here is not to be confused with previous policy/technical designations). An aspect raster calculates and depicts the slope direction for each pixel to its neighbouring pixels, with values for each pixel representing the compass direction the surface faces (0° classified as due north and 90° as east, upper right). The pixels are reclassified to Boolean values where 1 is equivalent to the southern-facing range of 135° and 225° from the north, and 0 is equivalent to all other compass values. This reclassification allows the raster file to be filtered by extracting all south-facing pixels. As some south-facing pixels may originate from fabricated or natural structures, these pixels are filtered by first converting them into polygons. Then, using the building polygon file, which contains the perimeter of the rooftop as the top-down view of a 2D polygon, the south-facing polygons are spatially intersected with the building polygons (lower-left). The result is a new shapefile containing vector polygons representing south-facing rooftops of inhabited buildings (lower right).

10.5   Case study: Glasgow

GOMap was applied to Glasgow city in support of GCC's interest in repurposing some of the city's VDL sites to tackle fuel poverty – households who spend more than 10% of annual income on energy (SG 2020). Glasgow contains approximately 314,500 occupied dwellings (NRS 2018) of which around 107,000 (34%) are socially owned and of these, 26,000 (24%) have no wall insulation and are in the 'hard-to-heat' category.

Glasgow encapsulates 115 electoral districts containing 644 VDL sites with a combined land area of 880 ha. The energy demand of a dwelling is on average 14,000 kWh/y (BEIS 2023a, 2023b). As 75% of this energy is used for space heating (SG 2023), the typical heating demand is 10,500 kWh/y or around 3,300 GWh/y for the city. As part of the Scottish Government's Energy Strategy (SG 2017), there is a commitment to electrify home heating. GOMap was applied to explore future PVPS deployment scenarios as follows.

S1: base case with all policy and technical aspects active and with the scope set to VDL sites only.
S2: relaxes the potential glare and social-subjective perspective aspects against the assumption that future communities will be more accepting of RET.

Table 10.6   Scenario outcomes for Glasgow

Scenario	Area (ha)	Energy yield (MWh/y)	Dwellings equivalent (% of S1)	Hard-to-heat dwellings equivalent
S1	94.5	57,598	4430 (−)	21%
S2	109.9	67,038	6385 (+16)	25%
S3	183.9	111,938	10,661 (+94)	41%
S4	58.2	31,560	3006	12%

S3: relaxes substation-related aspects to emulate future infrastructure upgrades.
S4: investigates the potential of building-integrated PV upgrades.

The results are presented in Table 10.6. With all policy and technical aspect information active and equally weighted, S1 releases 94.5 ha of available land corresponding to an energy supply potential of 57,598 MWh/y, enough to supply 4,430 dwellings or 21% of the city's dwellings in the hard-to-heat category.

Figure 10.3(left) shows a site within the S1 category. Here, the brighter green area in the middle unconstrained land is designated as policy 'Possible' and technically 'Favourable'. The constrained land is coloured red. As this analysis focuses on sites designated VDL, the green areas to the east (a local Rugby pitch) and west (land which may be designated VDL in future) are not offered up for development.

The unconstrained land is passed to ESP-r to determine the optimum arrangement of PV arrays within the PVPS taking into account the land utilisation factor, with the outcome superimposed on the site (right).

S2 explored the potential for policy-assisted development by relaxing the two visual-related aspects. Compared to the base case, this released 15.4 ha of additional unconstrained area bringing the total to 109.9 ha. This has the potential to generate 67,038 MWh/y, which could meet the needs of 6,385 dwellings, a 16% increase on S1 and equivalent to 25% of hard-to-heat dwellings.

S3 explored the potential for technical-assisted development by relaxing the two substation-related aspects. Compared to the base case, this doubled the available unconstrained land area resulting in 183.9 hectares. This corresponds to an annual energy yield of 111,938 MWh/y, which could supply 10,661 dwellings, an increase of 94% on S1 and equivalent to 41% of hard-to-heat dwellings.

Finally, S4 explored the identification of occupied dwellings with south-facing rooftops for possible PV deployment as shown in Figure 10.4 (see also Section 7.4): this identifies a dwelling to the left with an existing installation and three possible new installations.

Of the 314,500 occupied dwellings in Glasgow, 11,530 dwellings were identified with south-facing rooftops with a minimum of 20 m^2 and therefore able to provide reasonable PV coverage (GreenMatch 2023). A total roof area of 58.2 ha was identified, the equivalent of 31,560 MWh/y or 3,006 dwellings, which accounts for 12% of hard-to-heat dwellings.

Figure 10.3   Identified unconstrained land (left) and PVPS deployment

Figure 10.4   Identifying rooftops for PV deployment

In summary, tools like GOMap extend support for the decision-making process by consolidating different information types within an opportunity mapping process that can identify unconstrained locations for RET deployment and provide this additional information to a BPS+ tool.

10.6   Chapter summary

The potency of BPS+ is considerably amplified when integrated with the development control process. This can be achieved by conjoining the technology with GIS tools that incorporate the means to score land availability against technical and environmental constraints. Additionally, the versatility of coupling BPS+ with GIS allows for individual buildings or communities to be identified for retrofitting or other energy-related purposes. The outcome is a powerful procedure to support rational planning whilst enabling a constructive dialogue between those who propose a project and those who appraise its acceptability.

References and further reading

Amjad F and Shah L A (2020) 'Identification and assessment of sites for solar farms development using GIS and density based clustering technique – a case study in Pakistan', *Renewable Energy*, 155, pp. 761–769.

Anand A and Deb C (2023) 'The potential of remote sensing and GIS in urban building energy modelling', *Energy and Built Environment*, in press, 5(6), pp. 957–969, Elsevier.

Araki K, Nagai H, Lee K H and Yamaguchi M (2017) 'Analysis of impact to optical environment of the land by flat-plate and array of tracking PV panels', *Solar Energy*, 144, pp. 278–285.

BEIS (2023a) 'Sub-national electricity consumption statistics 2005 to 2021', Department of Business, Energy and Industrial Strategy, www.gov.uk/government/statistics/regional-and-local-authority-electricity-consumption-statistics.

BEIS (2023b) 'Sub-national gas consumption statistics 2005 to 2021', Department of Business, Energy and Industrial Strategy, https://www.gov.uk/government/statistics/regional-and-local-authority-gas-consumption-statistics.

Castillo C P, Silva F B E and Lavalle C (2016) 'An assessment of the regional potential for solar power generation in EU-28', *Energy Policy*, 88, pp. 86–99.

CEDA (2021) *The CEDA Archive*, Centre for Environmental Data Analysis, archive.ceda.ac.uk/.

Clarke J A, McGhee R and Svehla K (2020) 'Opportunity mapping for urban scale renewable energy generation', *Renewable Energy*, 162, pp. 779–787.

Cunden T S M, Doorga J, Lollchund M R and Rughooputh S D D V (2020) 'Multi-level constraints wind farms siting for a complex terrain in a tropical region using MCDM approach coupled with GIS', *Energy*, 211, p. 118533.

Finn T and McKenzie P (2020) 'A high-resolution suitability index for solar farm location in complex landscapes', *Renewable Energy*, 158, pp. 520–533.

GCC (2001) *Local Biodiversity Action Plan*, Glasgow City Council, https://glasgow.gov.uk/CHttpHandler.ashx?id=31719&p=0.

GCC (2017) *IPG6: Green Belt & Green Network*, Glasgow City Council, https://glasgow.gov.uk/CHttpHandler.ashx?id=36884&p=0/.

GCC (2017a) *SG 4 Network of Centres, Glasgow City Council, https://glasgow.gov.uk/CHttpHandler.ashx?id=36886&p=0.*

GCC (2022) 'Glasgow continues trend in the reduction of Vacant and Derelict Land in the city', Glasgow City Council, https://glasgow.gov.uk/index.aspx?articleid=29534.

Georgiou A and Skarlatos D (2016) 'Optimal site selection for sitting a solar park using multi-criteria decision analysis and geographical information systems', *Geoscientific Instrumentation, Methods and Data Systems*, 5, pp. 321–332.

GreenMatch (2023) https://greenmatch.co.uk/solar-energy/solar-system/3kw-solar-panel-system.

Günen M A (2021) 'A comprehensive framework based on GIS-AHP for the installation of solar PV farms in Kahramanmaraş, Turkey', *Renewable Energy*, 178, pp. 212–225.

HES (2014) *Historic Environment Scotland Act 2014*, https://legislation.gov.uk/asp/2014/19/contents/.

Innovate U K (2015) *Future Cities*, https://innovateuk.blog.gov.uk/tag/future-cities/.

Kalogirou S A (2014) *Solar Energy Engineering* (2nd edn.), Academic Press, San Diego, CA.

Kumar A, Sah B, Singh A R, *et al.* (2017) 'A review of multi criteria decision making (MCDM) towards sustainable renewable energy development', *Renewable and Sustainable Energy Reviews*, 69, pp. 596–609.

McGhee R (2022) 'Utilising GIS mapping to identify areas of opportunity for Photovoltaic Power Station deployment in an urban environment', *PhD Thesis*, ESRU, University of Strathclyde.

Mirzania P Ford A, Andrews D, Ofori G and Maidment G (2019) 'The impact of policy changes: The opportunities of community renewable energy projects in the UK and the barriers they face', *Energy Policy*, 129, pp. 1282–1296.

NRS (2018) *National Records of Scotland*, https://nrscotland.gov.uk/files//statis-tics/household-estimates/2018/house-est-18-publication.pdf.

Nunnally J C and Bernstein I H (1994) *Psychometric Theory*, McGraw-Hill, New York.

Omitaomu O A, Blevins B R, Jochem W C, *et al.* (2012) 'Adapting a GIS-based multicriteria decision analysis approach for evaluating new power generating sites', *Applied Energy*, 96, pp. 292–301.

OS (2020) *Ordnance Survey – AddressBase*, https://ordnancesurvey.co.uk/business-government/products/addressbase.

OS (2021) *Ordnance Survey – OS Terrains*, https://ordnancesurvey.co.uk/business-and-government/products/os-terrain-5.html.

Pillot B, Al-Kurdi N, Gervet C and Linguet L (2020) 'An integrated GIS and robust optimization framework for solar PV plant planning scenarios at utility scale', *Applied. Energy*, 260, p. 114257.

Parihari S, Das K and Chatterjee N D (2021) 'Land suitability assessment for effective agricultural practices in Paschim Medinipur and Jhargram districts, West Bengal, India', Chapter in *Modern Cartography Series*,10, pp. 285–311, Elsevier.

Preston C C and Colman A M (2000) 'Optimal number of response categories in rating scales: reliability, validity, discriminating power, and respondent pre-ferences', *Acta Psychologica*, 104(1), pp. 1–15.

Saaty T L (1980) *The Analytic Hierarchy Process*, McGraw-Hill, New York.

Saretta E, Caputo P and Frontini F (2020) 'An integrated 3D GIS-based method for estimating the urban potential of BIPV retrofit of facades', *Sustainable Cities and Society*, 62, p. 102410.

SG (2008) *The Sites of Special Scientific Interest Regulations 2008*, https://legis-lation.gov.uk/ssi/2008/221/contents/made.

SG (2010) *The Town and Country Planning (Tree Preservation Order and Trees in Conservation Areas) (Scotland) Regulations 2010*, https://legislation.gov.uk/ssi/2010/434/contents/made.

SG (2011) *The Ancient Monuments and Archaeological Areas (Applications for Scheduled Monument Consent) (Scotland) Regulations 2011*, https://legisla-tion.gov.uk/ssi/2011/375/contents/made.

SG (2017) *Scottish Energy Strategy: The Future of Energy in Scotland*, https://www2.gov.scot/energystrategy/.

SG (2020) *Scottish Fuel Poverty Advisory Panel: Annual Report – 2019*, https://gov.scot/publications/scottish-fuel-poverty-advisory-panel-second-annual-report-2019/documents/.

SG (2023) *National Planning Framework 4*, Scottish Government, https://gov.scot/publications/national-planning-framework-4/.

SPEN (2015) *Distributed Generation Heat Maps Overview*, SP Energy Networks, https://spenergynetworks.co.uk/userfiles/file/DG_Heat_Maps_Overview.pdf.

UK Civil Aviation Authority (2010) *Interim CAA Guidance – Solar Photovoltaic Systems*, AARDVaRC Ltd, Sudbury.

UK Legislation (2024) *The National Archive*, https://legislation.gov.uk/browse.uk.

US Federal Aviation Authority (2010) *Technical Guidance for Evaluating Solar Technologies on Airports*, Federal Aviation Administration, Washington.

Vulkan A, Kloog I, Dorman M and Erell E (2018) 'Modeling the potential for PV installation in residential buildings in dense urban areas', *Energy and Buildings*, 69, pp. 97–109.

Yamagata Y, Yang P P J, Chang S, *et al.* (2020) 'Urban systems and the role of big data', Chapter in *Urban Systems Design: Creating Sustainable Smart Cities in the Internet of Things Era*, pp. 23–58, Elsevier.

Chapter 11

Tackling the performance gap

There can be many reasons why BPS+ predictions do not match reality and consequently fail to deliver design stage expectations – the so-called performance gap (de Wilde 2014). This chapter considers some principal causal factors that can be accommodated within a high-resolution model as required in specific cases. These include design parameter uncertainty, occupant behaviour, operational events, and construction defects.

11.1　Accommodating parameter uncertainty

The creation of virtual representations of buildings and their service systems requires the translation of architectural and engineering information (plans, sections, constructions, plant specifications, control schedules, etc.) to the input model syntax of a particular tool. Opportunities for errors and omissions are plentiful. The intent of construction documents may not be realised on-site with exactitude, materials may be substituted, the weather may have degraded the integrity of a protective membrane, or a retrofit may have introduced thermal bridging and unintended air leakage paths. Models at best are imprecise digital twins just as the buildings they represent are imprecise implementations of design intent. Each of these divergences introduces an element of uncertainty.

Methods have been incorporated within BPS+ tools to address uncertainties. The most popular methods are Differential Sensitivity Analysis (DSA) and Monte Carlo Analysis (MCA) as described elsewhere (Judkoff and Neymark 1995, Macdonald 2002). Both methods have the advantage of being relatively easy to apply to existing programs because the simulation is treated as a black box and the parameters to be considered are contained in the input model.

DSA requires a base case simulation in which input parameters are set to a best estimate. The simulation is then repeated with the value of one input parameter changed from P to $P+\delta P$ and the effect on output(s) noted. This is done for each parameter in turn, giving $N+1$ simulations to analyse the effects of N uncertain parameters. The underlying assumption is that the effect of uncertainty is linear over the perturbance – an assumption that can be tested to a limited degree by carrying out further simulations with the parameter values set to $P-\delta P$; if the effect on the output parameters is the same (but in the opposite directions) then linearity is confirmed. The DSA method is not optimised for the number of simulations required and does not identify any parameter interactions. However, it does inform the user which of the analysed parameters have the most influence on the output(s). When applied to simulations with many input parameters the technique becomes cumbersome and time consuming.

MCA generates an estimate of the overall uncertainty in the predictions due to all uncertainties in the input parameters, regardless of interactions and quantity. A probability distribution is assigned to each input parameter under consideration. For all parameters, values from within their probability distribution are randomly selected and a simulation is undertaken. Simulations are repeated with new values randomly selected. Given a large enough number of simulations, the uncertainty in the output parameters will have a Gaussian distribution irrespective of the input parameter probability distributions. It has been shown (Judkoff and Neymark 1995) that the number of simulations required by this technique is 60–80, after which there are only marginal gains in accuracy. The main application difficulty is in identifying the input parameter probability distributions. In practice, it is usual to assume that parameters have a Gaussian distribution although any distribution is possible. A disadvantage is that the method, unlike DSA, does not distinguish individual parameter sensitivities.

To demonstrate the imposition of uncertainty considerations on a simulation, consider the case of the viewing gallery portion of the Lighthouse building in Glasgow (see Section 6.16) when subjected to ESP-r simulation (Clarke *et al.* 2000). The following three tables list uncertainties defined for material thermophysical properties, weather parameters, and casual gains, respectively.

The entries of Table 11.1 represent the use of the widest possible sources of thermophysical properties in use for building simulation (Hong *et al.* 2000).

The entries of Table 11.2 define only small uncertainties in the weather parameters, as typical building performance is required.

The entries of Table 11.3 can be interpreted in two ways: an uncertainty in the magnitude of individual items or an uncertainty in the total number of items. For example, the occupancy gain relates to the uncertainty of the magnitude of a person's metabolic rate but could also relate to the number of occupants in the viewing gallery at the same time.

MCA was applied for three periods (winter, summer, and transition season) with a focus on heating energy demand. The maximum demand and associated uncertainty were determined as shown in Figure 11.1.

The upper and lower profiles correspond to one standard deviation above and below the average. The standard deviation varies with time (it tends to be larger in the afternoon), and therefore the maximum heating demand at one standard deviation may not occur at the same time as the maximum average heating demand.

A further analysis was undertaken with the maximum heater load restricted to 45 W/m^2. This value represents insufficient heater power to achieve the set-point temperature: from the initial analysis, only 2/3 of the maximum average load, equating to approximately 0.5% of probable maximum heating demands, would be satisfied with this capacity. This example is used to demonstrate how underheating risk can be assessed. Table 11.4 shows the annual heating energy demand (kWh/m^2) for the unlimited and limited heater power cases. There is a 10% energy saving on

Table 11.1 Thermophysical property distributions

Parameter	Property	Average value	Standard deviation
Heavy mix concrete	k	1.68	0.376
	ρ	2304	146
	C	869	65
Copper	k	333	89
	ρ	8858	177
	C	398	15
Steel	k	46	5.6
	ρ	7800	–
	C	497	29
Aluminium	k	211	17
	ρ	2733	58
	C	880	–
Plywood	k	0.16	0.04
	ρ	622	72
	C	1718	430
Slate	k	1.72	0.4
	ρ	2150	778
	C	1110	509
Gypsum plasterboard	k	0.28	0.2
	ρ	950	134
	C	882	102
Cement screed	k	0.9	0.36
	ρ	1452	382
	C	910	152
EPS	k	0.035	0.003
	ρ	28	9
	C	1328	103
Plate glass	k	0.95	0.13
	ρ	2515	148
	C	828	55
Glass fibre quilt	k	0.035	0.004
	ρ	32	23
	C	851	43

k – conductivity (W/m.K); ρ – density (kg/m^3); C – specific heat capacity (J/kg.K).

Table 11.2 Weather parameter distributions

Parameter	Variation
Dry bulb temperature (°C)	±0.2
Global solar radiation (%)	±3.0
Diffuse horizontal solar radiation (%)	±3.0
Wind speed (m/s)	±0.5
Wind direction (° from North)	±10

Table 11.3   Operations profile distributions

Parameter	Variation (%)
Occupancy total gain	±50
Convective portion of occupancy gain	±10
Lighting sensible gain	±20
Convective portion of lighting gain	±10
Infiltration rate	±100

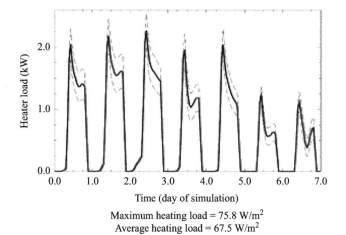

Maximum heating load = 75.8 W/m²
Average heating load = 67.5 W/m²

Figure 11.1   Heating demand with error bands

Table 11.4   Annual heating energy demand

Heater power	Maximum	Average	Minimum
Unlimited	80.6	66.0	51.4
Limited	65.9	59.4	53.0

average by adopting the limited heater power (not including any gain in running the boiler at higher part load efficiencies). The uncertainty in the heating energy consumption has also decreased.

The temperature profile of the occupied space during a typical winter week was analysed to identify periods of underheating. The results are shown in Figure 11.2: the average temperature does not reach the control set point of 20 °C during the first three days and only in the afternoons of the fourth and fifth days. The error bands have been drawn at one standard deviation. With this degree of confidence, the temperature is always above 17 °C and for the majority of occupied hours is within 1 °C of the set-point temperature.

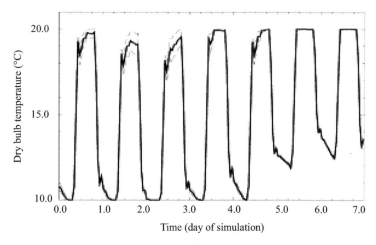

Figure 11.2 Temperature (°C) with error bands

The designer is now able to make an informed decision on plant size taking the possible energy savings (one-third smaller maximum load and 10% annual energy savings) and the risk of potential underheating (slight) into consideration.

Introducing uncertainty consideration into BPS+ enables risk assessment, and this, in turn, will improve designer confidence in simulation outputs. Where uncertainty analysis is automatically applied, risk assessment can be enabled as a routine procedure. Facility managers might even use BPS+ proactively to explore adaptation strategies rather than firefighting staff complaints. The challenge is to reduce the friction in the process so that uncertainty becomes a normal part of the workflow and thus is able to provide indications of where effort should be concentrated.

11.2 Incorporating occupant behaviour

Occupant presence and behaviour play a crucial role in determining building performance. The representation of this behaviour within BPS+ tools is essential to bridge the gap between design stage aspiration and performance in use (Yan *et al.* 2015, Rida *et al.* 2018, O'Brien 2020). Mathematical models have been developed for occupant-related aspects such as window usage (e.g., Fritsch *et al.* 1990, Yun and Steemers 2008, Tuohy *et al.* 2009, Haldi and Robinson 2009), luminaire switching (e.g., Hunt 1979), and appliance use (Flett 2017). Other researchers (e.g., Reinhart (2004) have combined occupant presence and behaviour within methods that also track inter-room movement (e.g., Page *et al.* 2008, Wang *et al.* 2011). Algorithms are usually stochastic and seek to capture the diverse perceptions and behaviours within a population in response to environmental stimuli.

BPS+ tools have been slow to integrate explicit occupant modelling, preferring pre-constructed schedules defining occupant presence, activity-related casual heat gain, luminaire switching, and thermostat setting adjustment. Explicit occupant modelling removes the need for such prescriptions and places the representation of occupants on the same dynamic footing as the other technical domains.

Efforts to incorporate explicit occupant modelling have been inconsistent, often blending deterministic and stochastic rules, or offering pre-constructed models based on unrepresentative empirical data. Pollutant emission, door and window opening, window blind adjustment, lights switching, small power usage, and thermostat setting adjustment are all stochastic occupant impacts. Before enabling the co-simulation link with the Occupant Behavior Functional Mockup Unit (obFMU[1]) as described below, ESP-r represented these impacts using pre-defined schedules, event-based input model adaptations, or by imposing field measurements. The program also offered a limited number of occupant presence/behaviour models based on stochastic considerations as follows.

The Sub-Hourly Occupancy Control (SHOCC) model (Bourgeois *et al.* 2006) targets occupancy-related phenomena within a building through the prediction of population mobility, occupancy-activated control, and individual/group behaviour modelling. It employs a stochastic occupancy predictor based on the Lightswitch 2002 algorithm (Reinhart 2004). Figure 11.3, for example, shows the daily update of occupant mobility: 'LSTrueAD' and 'LSTrueEvents' are stochastic functions defined in Bourgeois (2005, Appendix A).

Some ESP-r projects employed freestanding presence models to generate diversified occupancy profiles. For example, Richardson *et al.* (2008) utilised UK Time Use Survey data (ONS 2003), comprising 24 hour diaries completed at a 10 minute resolution, to devise a Markov chain, Monte Carlo method to generate occupancy data with the same transition probabilities. Figure 11.4 shows outputs from four runs of the method for a two-person house over a typical weekday. Note that the outputs are different in each case due to the random numbers used in the stochastic generation although each is based on the same transition probability matrices and thus exhibits similar characteristics.

Many algorithms exist for the prediction of luminaire switching, some restricted to switch-on probability on arrival (e.g., Hunt 1979) and others attempting comprehensive coverage of all scenarios such as the Lightswitch 2002 algorithm (Reinhart 2004) as summarised in Figure 11.5.

Tuohy *et al.* (2009) developed a comfort-driven, adaptive algorithm for window opening in offices based on survey data expressed by a logit function (Figure 11.6) describing the probability of window opening and closing.

Rijal *et al.* (2008) extended the approach to include door opening, ceiling fan use, heating/cooling system operation, and night cooling according to the algorithm depicted in Figure 11.7.

In a domestic context, ESP-r uses the OccDem[2] probabilistic occupant behaviour-modelling tool (Flett 2017), which uses enhanced occupancy modelling methods to generate appliance and hot water usage schedules.

---

[1]https://behavior.lbl.gov/?q=obFMUdownload
[2]https://www.esru.strath.ac.uk/applications

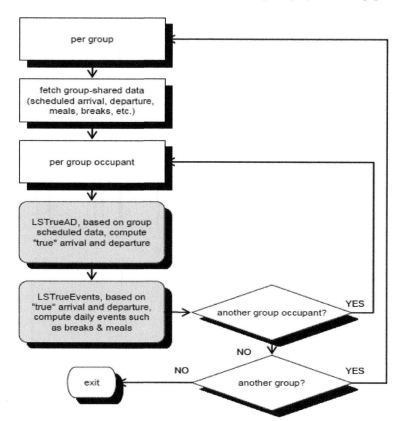

Figure 11.3   SHOCC daily update of occupant mobility data in ESP-r (credit: Bourgeois 2005)

Whilst the above algorithms are a marked improvement in the use of pre-scriptions, integrating them within BPS+ programs is undesirable for several reasons. First, they are generally empirical and relate only to the specific cases underlying their development. A model will give misleading results if applied in another context and such a caveat is typically lost after BPS+ implementation. Second, Gunay *et al.* (2015) reported that different models give significantly different results and these could be used to game the system. Third, if BPS+ tools implement algorithms independently, with no mechanism for harmonisation, inconsistencies in the capabilities of BPS+ will result (Cowie *et al.* 2017).

The IEA-EBC Annex 66 project[3] (Yan *et al.* 2017) addressed this issue by creating an occupant behaviour toolkit called obFMU (Hong *et al.* 2016), which is based on the functional mock-up interface (FMI) co-simulation standard[4]. Whilst

[3]https://annex66.org/
[4]https://fmi-standard.org

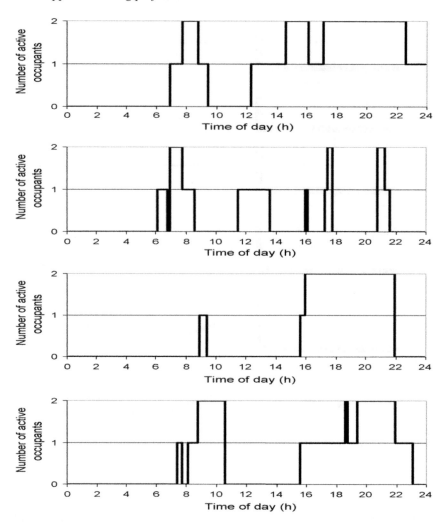

Figure 11.4   Stochastic occupancy presence prediction (credit: Richardson et al. 2008)

this represents a significant step forward, it places an additional burden on the BPS+ tool user, who must define the links with obFMU and interpret simulation outcomes under the stochastic influence. As a contribution to the Annex 66 project, the ESP-r program was adapted to enable co-simulation with obFMU (Cowie *et al.* 2017).

The obFMU framework

The obFMU approach provides a framework for implementing occupant behaviour models defined by the obXML schema, which implements the DNAS ontology (Drivers, Needs, Actions, and Systems) for occupant behaviour modelling

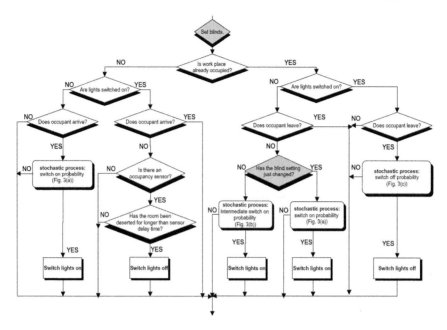

Figure 11.5   Lightswitch algorithm for luminaires and blinds in ESP-r (credit: Reinhart 2004)

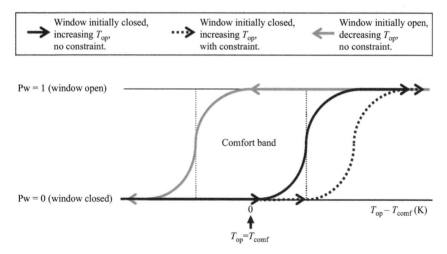

Figure 11.6   Logit functions representing window opening in ESP-r (credit: Tuohy et al. 2009)

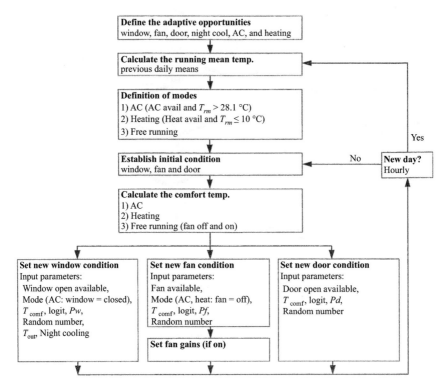

Figure 11.7   Occupant controls interaction algorithm in ESP-r (credit: Rijal et al. 2008)

(Hong *et al.* 2015, 2015a). Occupant presence prediction is based on a Markov chain movement model with a transition probability matrix used to determine the probability of an occupant moving from one location to another (Hong *et al.* 2016). Belafi *et al.* (2016) implemented ten sets of occupant behaviour models from the literature in obFMU[5]:

- blind and window opening and closing (Haldi and Robinson 2008, 2009);
- light activation (Hunt 1979);
- light activation (Love 1998);
- blind opening and light activation and deactivation (Newsham 1994);
- heater activation (Nicol 2001);
- light activation models from Reinhart and Voss (2003);
- air conditioning activation and deactivation models (Ren *et al.* 2014);
- window-opening models (Yun and Steemers 2008);
- window-opening models (Zhang and Barrett 2012); and

---

[5]Available from https://behavior.lbl.gov/obxmldownload.

- light activation and window opening and closing (Gunay *et al.* 2015), based on the models of Reinhart (2004), Hunt (1979), Haldi and Robinson (2009), and Yun and Steemers (2008).

Although this library contributes to harmonisation, it has imitations because it contains only a subset of the available models and no guidance is given on model applicability, placing an additional burden on the user when making selections. This situation may be expected to improve in future.

Running a co-simulation

Linking obFMU to a BPS+ program is based on the FMI tool-independent standard[6] that supports model exchange and co-simulation using a combination of xml files and compiled C-code. The standard defines the inputs, outputs, and various standard functions that are pre-compiled into a Dynamic Link Library (DLL), packaged with metadata in XML format, and delivered as a Functional Mock-up Unit (FMU). The master program, the BPS+ tool, calls the functions contained in the FMU. The master waits for the slave to finish an assigned task before continuing, which removes issues relating to CPU time synchronisation.

The open-source FMI Library[7] (FMIL) is a resource for implementing FMI in the master program. FMIL provides a programming interface for interacting with all parts of FMUs, including function definitions and data structures. It also provides utilities such as unzipping the FMU, loading the FMU DLL at run time, and parsing XML metadata files.

To explore and demonstrate the approach, obFMU was connected to ESP-r (Clarke and Cowie 2019). An ESP-r simulation involves a series of time steps with the building state captured at the present time row of any current time step. Data characterising each thermal zone's environment are passed from ESP-r to obFMU, selected behaviour models of obFMU are evaluated, and data characterising occupant actions are passed back to ESP-r. This information is then processed alongside control system actions before moving to the next simulation time step.

ObFMU has its own input data to define instances, occupants, and behaviour models. ESP-r requires linkage directives to associate obFMU instances with thermal zones and define how input and output variables of the behaviour models be handled. This necessitated an extension of the user interface as depicted in Figure 11.8, which also serves to summarise the input requirements to enable co-simulation.

The menu on the left shows linkage directives for obFMU inputs, whilst the menu on the right shows linkage directives for obFMU outputs. Four items of information are required: an association with an ESP-r thermal zone, an ESP-r variable reference, an association with an obFMU instance, and a variable reference for obFMU. These directives define associations between entities and data in ESP-r and obFMU.

---

[6]https://fmi-standard.org
[7]https://fmi-library.org

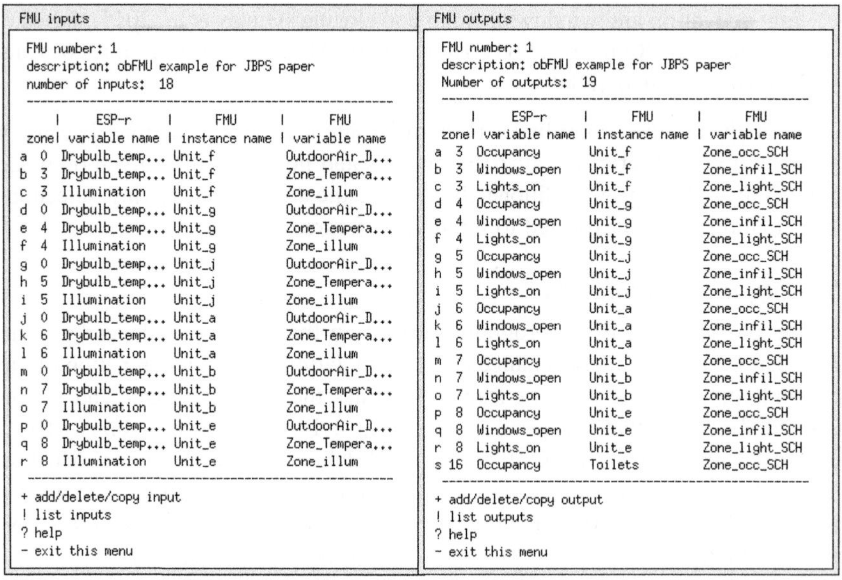

```
FMU inputs                                      FMU outputs

FMU number: 1                                   FMU number: 1
description: obFMU example for JBPS paper       description: obFMU example for JBPS paper
number of inputs: 18                            Number of outputs: 19
-------------------------------------------     -------------------------------------------
     I    ESP-r   I     FMU    I     FMU             I    ESP-r   I     FMU    I     FMU
   zoneI variable name I instance name I variable name   zoneI variable name I instance name I variable name
a  0  Drybulb_temp... Unit_f    OutdoorAir_D...   a  3  Occupancy      Unit_f     Zone_occ_SCH
b  3  Drybulb_temp... Unit_f    Zone_Tempera...   b  3  Windows_open   Unit_f     Zone_infil_SCH
c  3  Illumination    Unit_f    Zone_illum        c  3  Lights_on      Unit_f     Zone_light_SCH
d  0  Drybulb_temp... Unit_g    OutdoorAir_D...   d  4  Occupancy      Unit_g     Zone_occ_SCH
e  4  Drybulb_temp... Unit_g    Zone_Tempera...   e  4  Windows_open   Unit_g     Zone_infil_SCH
f  4  Illumination    Unit_g    Zone_illum        f  4  Lights_on      Unit_g     Zone_light_SCH
g  0  Drybulb_temp... Unit_j    OutdoorAir_D...   g  5  Occupancy      Unit_j     Zone_occ_SCH
h  5  Drybulb_temp... Unit_j    Zone_Tempera...   h  5  Windows_open   Unit_j     Zone_infil_SCH
i  5  Illumination    Unit_j    Zone_illum        i  5  Lights_on      Unit_j     Zone_light_SCH
j  0  Drybulb_temp... Unit_a    OutdoorAir_D...   j  6  Occupancy      Unit_a     Zone_occ_SCH
k  6  Drybulb_temp... Unit_a    Zone_Tempera...   k  6  Windows_open   Unit_a     Zone_infil_SCH
l  6  Illumination    Unit_a    Zone_illum        l  6  Lights_on      Unit_a     Zone_light_SCH
m  0  Drybulb_temp... Unit_b    OutdoorAir_D...   m  7  Occupancy      Unit_b     Zone_occ_SCH
n  7  Drybulb_temp... Unit_b    Zone_Tempera...   n  7  Windows_open   Unit_b     Zone_infil_SCH
o  7  Illumination    Unit_b    Zone_illum        o  7  Lights_on      Unit_b     Zone_light_SCH
p  0  Drybulb_temp... Unit_e    OutdoorAir_D...   p  8  Occupancy      Unit_e     Zone_occ_SCH
q  8  Drybulb_temp... Unit_e    Zone_Tempera...   q  8  Windows_open   Unit_e     Zone_infil_SCH
r  8  Illumination    Unit_e    Zone_illum        r  8  Lights_on      Unit_e     Zone_light_SCH
                                                  s 16  Occupancy      Toilets    Zone_occ_SCH
-------------------------------------------     -------------------------------------------
+ add/delete/copy input                         + add/delete/copy output
! list inputs                                   ! list outputs
? help                                          ? help
- exit this menu                                - exit this menu
```

Figure 11.8   ESP-r's FMI interface

For example, the first three entries are ambient dry bulb temperature, zone dry bulb temperature, and zone illumination, all associated with a single obFMU instance called 'Unit_f'. The FMU variable names associate these variables with the appropriate data for this instance in obFMU. The three outputs are occupancy, window open fraction, and lights switching state, all associated with the same obFMU instance. These linkage directives represent data describing the environment required by behaviour models in obFMU and data returned from the behaviour models describing occupant actions. In this case, outdoor temperature, indoor temperature, and zone illuminance are transferred to obFMU from ESP-r. ObFMU then evaluates its behaviour models using these values as inputs and returns decisions on whether there are occupants in the room, whether they have opened or closed windows, and whether they have switched lights on or off. These decisions are then imposed on the ongoing ESP-r simulation.

The process of setting up ESP-r and obFMU for co-simulation is crucial, with the focus here on the added burden placed on the user. ObFMU requires two XML files: obXML.xml and obCoSim.xml, which define occupants and behaviour models as well as instance mappings and simulation settings. Two key resources are the Web-based GUI, 'Occupancy Simulator,' and a library of occupant behaviour models implemented in obXML format. Both are freely available online. However, the effort required to create the obFMU input is significant, around 2 person-hours. For large input models, the effort will be greater and a planned development of ESP-r is to generate this input file automatically based on information already contained in the ESP-r model.

A key consideration for the user is to ensure that the input data for ESP-r and obFMU is consistent. The building space definitions in the obXML file and the instance mappings defined in the obCoSim file must be consistent with the corresponding ESP-r model, whilst the input and output definitions for ESP-r must be consistent with the behaviour models as defined in the obXML file. The recommended workflow is as follows.

- Create an ESP-r model, ensuring that the prerequisites for all required obFMU inputs and outputs are satisfied.
- Generate the obXML and obCoSim files for obFMU, using the resources mentioned previously (automated in future).
- Define directives for the required inputs and outputs of obFMU in the ESP-r model.

The procedure for commencing a co-simulation with obFMU is no different from running any other ESP-r simulation. ESP-r offers a command-line interface whereby user interaction is not required to control the path of a simulation and this is helpful when many simulations are needed.

The computational burden of initialising obFMU and running behaviour models at each time step can be substantial. For example, a model with three thermal zones and two occupants simulated over a week showed an increase of around 80% in run time. For a model with 50 thermal zones and 63 occupants, simulated over 1 year, the increase is around 45%. The use of stochastic occupant behaviour models renders simulation predictions stochastic, requiring multiple simulations to yield a robust picture. However, extracting results from an ESP-r simulation remains unchanged by co-simulation due to the automatic encapsulation of obFMU outputs in the ESP-r model equations.

Impact of occupant behaviour

To demonstrate how a behaviour-enabled simulation can give a more realistic outcome, consider the 2,880 m^2 floor area office depicted in Figure 11.9 as established to explore natural ventilation approaches to summer overheating alleviation.

The model comprises eight open-plan offices, six of which are sparsely occupied at 20 m^2/person (nine other ancillary zones in the model represent stairwells, toilets, entry corridors, and ceiling plenums). As the focus of the study was occupant behaviour, the model is simplified in terms of facade details, fittings, and environmental controls. Simulations were undertaken at 5 minute time steps over a typical warm week in July.

An initial simulation used prescribed occupant and lighting schedules and a prescription that opened windows when the office air temperature exceeded 21 °C during occupied hours. Subsequent simulations replaced the occupant schedule with an obFMU stochastic occupant movement model. Usual arrival and departure times were taken from the existing profile but allowed a 30 minute variation either way. Directives added a mid-day lunch event and introduced the requirement for occupants to spend 5% of their day randomly in another office. These directives

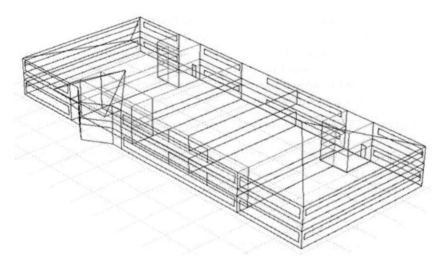

Figure 11.9   The case study office building

were generated via the Occupancy Simulator whilst ensuring that the peak casual gain was the same in both simulations. The lighting schedule was replaced by the behaviour model of Newsham (1994), and the window-opening model was replaced by the stochastic model of Haldi and Robinson (2009).

The sensitivity of the results to the added stochasticity was explored: over five simulations, the average variation in room temperature was less than 1%, and the average variation of the heat load associated with infiltration was approximately 4%. Given these small variations, only a single set of results is reproduced here: Figure 11.10 (left) for a modelled zone as produced by the non-stochastic simulation and (right) for a corresponding case incorporating stochasticity.

A comparison of the temperature profiles indicates that the building is more resistant to overheating with stochastic behaviour, principally because the available daylight levels do not necessitate the use of artificial lighting. Furthermore, although peak cooling from window opening is lower in the stochastic case (due to differences in inter-zonal airflow), the windows are left open overnight leading to peak temperatures that are more than 2 °C lower than the non-stochastic case.

Such results demonstrate how the use of occupant behaviour modelling can give rise to different outcomes. Whilst a skilled user would likely explore further and may well identify daylighting control and nighttime cooling as effective strategies, this would require time and effort. By allowing the occupants to interact with the building according to stochastic principles, such passive measures would likely emerge naturally from the simulation.

Although office spaces gain the most attention, it is in transition spaces such as corridors and stairwells that stochastic assessments might up-end the use of prescriptive schedules. The extreme variability in the numbers and persistence of occupants, coupled with neglect in environmental controls, results in marginal

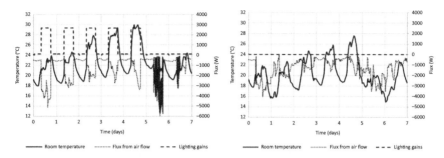

Figure 11.10   Results before (left) and after obFMU activation

conditions. Exploring this variability would support flexibility in environmental controls that might lead to higher-value uses of marginal spaces.

11.3   Testing life cycle resilience

The destiny of building performance simulation is as realistic emulator, not a predictor, of future reality. This requires a change in BPS+ application intent from performance prediction to operational resilience testing. Instead of requiring tool users to define the assessment conditions and subsequently analyse assessment outcomes, they would submit their model for perpetual annual simulation, with randomised perturbations imposed to represent events such as severe weather, equipment breakdown, and tariff price change. Time series outcomes over appropriate intervals (short, medium, and long term) would be probed and evaluated against regulated expectations relating to performance aspects such as indoor environmental conditions, running cost, and emissions. The simulation would conclude only when all test conditions had been experienced and performance requirements met or failed.

Whilst the capabilities and features of such an approach should be the same for all users, it could be powered by any BPS+ 'engine' that is able to represent the causal relationships underlying the particular resilience test to be applied. For example, when testing the resilience of a net-zero energy building, brightening sky conditions would cause greater daylight penetration to indoor spaces resulting in artificial light dimming. This, in turn, would reduce the electrical load, perhaps requiring the power flow from a local PV array to be exported, resulting in a power quality impact on the low voltage network and an export refusal or tariff adjustment – and so on. Such a simulation would result in an entirely different outcome than one that merely determined the output power from the PV array for a given irradiance and would lead naturally to a robust practical scheme.

Although the functionality of BPS+ tools covers an extensive range of performance aspects, the technology is not particularly easy to apply in practice because of the burden placed on users in relation to model creation, simulation coordination, and results analysis. A way to alleviate this problem is to adapt the manner in which simulation is accessed by providing cloud-based services offering industry-standard performance assessments. Imagine a future in which simulation services exist that:

- operate on the basis of design proposals delivered as high-integrity building information models;
- require no user involvement in the simulation process;
- automatically initiate standard performance assessments that cover lifetime operation;
- evaluate all relevant performance aspects against standard criteria;
- judge overall acceptability via models of building users/operators (as opposed to relying on simulation tool users as now); and
- facilitate the unambiguous comparison of alternative proposals.

This is the approach demonstrated in a prototype Resilience Testing Environment (RTE) named Marathon[8] (Clarke 2018, Clarke and Cowie 2020) as depicted in Figure 11.11.

The prototype RTE is powered by ESP-r when operated in scripted mode (refer to Section 4.5) to automate the tasks of the required resilience tests. As summarised in Figure 11.12, the aim is to render outcomes unequivocal by standardising performance assessments corresponding to different targets (e.g., low energy dwellings, estate facility management, community energy schemes, and critical environments) and levels of test stringency (e.g., current regulations, PassivHaus standard, and net-zero CO_2).

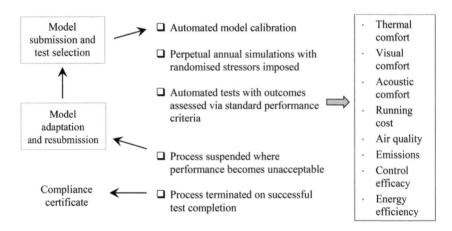

Figure 11.11   Components of the prototype RTE

---

[8]https://www.esru.strath.ac.uk/applicatios

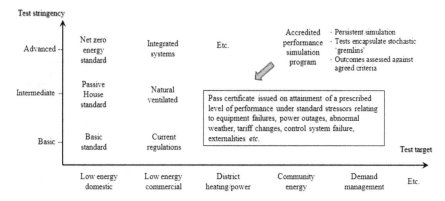

Figure 11.12 Progressive resilience test levels

RTE external view

The process commences when a user uploads a model via a Web page as depicted in the Marathon screen image of Figure 11.13, which shows four uploaded projects and one selected comprising three available models. One of these models has then been selected for a resilience test. The number of projects and sub-models that can be uploaded is limited only by the server capacity.

It is envisaged that users will generate multiple models within a given project, corresponding to different building configurations, portions of an estate, design options, or legislative standards. It is the user's responsibility to remove or replace such models as required in response to resilience test outcomes. As shown in Figure 11.13, summary information is associated with each model at this stage, including the name of the model file for upload. In the RTE prototype, this model corresponds to the format required by ESP-r although this could be replaced by the model format required by any approved BPS+ tool. In the longer term, the model will be BIM compliant once the standard has evolved to support the data model requirement for high-resolution BPS+ programs (Clarke *et al.* 2012).

As depicted in Figure 11.14, a selected model is associated with a resilience test chosen from those on offer for the model type (domestic, non-domestic, community energy, etc.). A model can be associated with multiple resilience tests (here two), or different models to the same test. Within Marathon, example resilience tests have been implemented for domestic buildings, non-domestic buildings, and district heating schemes and correspond to increasing levels of performance stringency. The RTE supports the addition of tests for other cases in future (e.g., health care, smart grids, and community energy schemes). At this stage, the model/test combination is placed in a queue behind prior submissions to await simulation processing when the server resource becomes available.

The model/test pairings are subjected to perpetual annual simulations under weather conditions synthesised using a stochastic weather generator (see below). Note that simulations have no predetermined end time and will continue under the

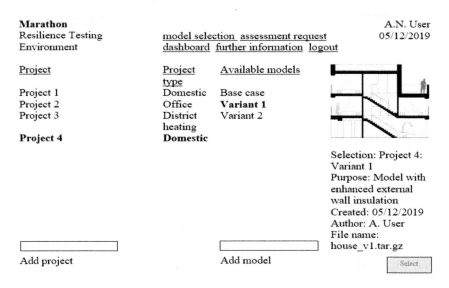

Marathon   A.N. User
Resilience Testing   model selection assessment request   05/12/2019
Environment   dashboard further information logout

Project	Project type	Available models
Project 1	Domestic	Base case
Project 2	Office	**Variant 1**
Project 3	District heating	Variant 2
Project 4	**Domestic**	

Selection: Project 4: Variant 1
Purpose: Model with enhanced external wall insulation
Created: 05/12/2019
Author: A. User
File name: house_v1.tar.gz

Add project   Add model   Select

Figure 11.13   Model upload and selection in Marathon

rules defined for the enacted test until the test-specific operational contingencies relating to weather and operational stresses have been encountered.

As a simulation proceeds, the spatially and temporally varying system states are probed to ensure that the performance is compliant with the test criteria. When, at some point in time, all performance criteria are deemed acceptable, the simulation is terminated and a compliance certificate is issued. The model is archived along with the certificate for possible later use in compliance checking.

Where, at some stage, the criteria indicate performance failure, the test is suspended (but not terminated) and the user is given summary feedback to enable appropriate remedial actions to be taken. As and when appropriate, an adjusted model that encapsulates such actions can be uploaded to allow the resilience test to recommence.

Figure 11.15 shows the Marathon dashboard indicating test progress and various possible outcomes.

RTE internal view

On receipt of a model, the intention is that the RTE automatically appends the required simulation constructs (e.g., gridding, heat transfer, occupant behaviour models, etc.) and commissions an automated calibration exercise to ensure that the model is fit for purpose. The user is given the opportunity to accept or decline the suggested model parameter adjustments. At the time of writing, these adjustments correspond to the requirements of ESP-r and remain a work in progress.

When a computational resource becomes available, the model is subjected to perpetual annual simulations, each one utilising a different hourly weather file generated by the Indra synthetic weather generator (Rastogi 2016). Based on one or more years of historical weather data, Indra creates plausible and physically valid

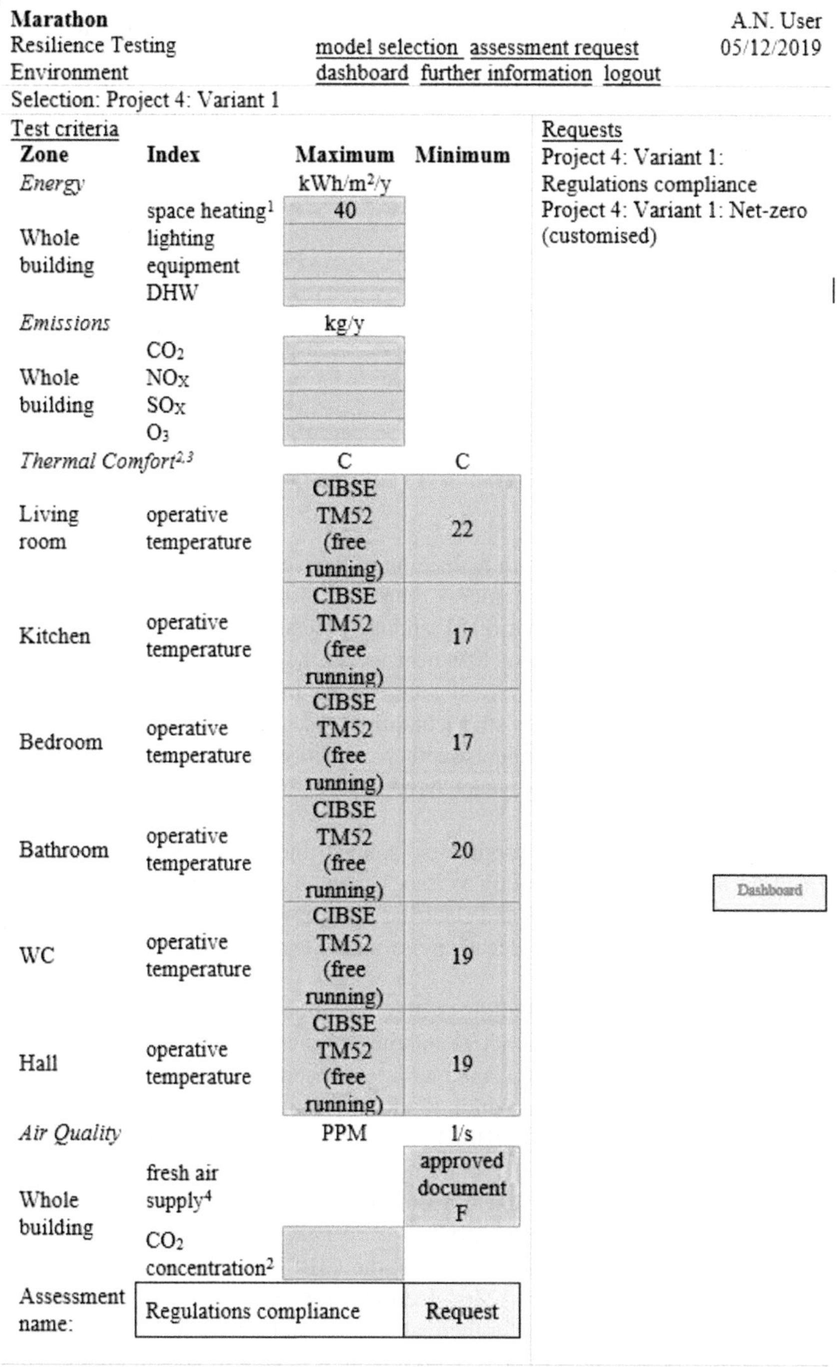

Marathon A.N. User
Resilience Testing model selection assessment request 05/12/2019
Environment dashboard further information logout
Selection: Project 4: Variant 1

Test criteria				Requests
Zone	**Index**	**Maximum**	**Minimum**	Project 4: Variant 1:
Energy		kWh/m²/y		Regulations compliance
	space heating[1]	40		Project 4: Variant 1: Net-zero
Whole	lighting			(customised)
building	equipment			
	DHW			
Emissions		kg/y		
	CO_2			
Whole	NO_x			
building	SO_x			
	O_3			
Thermal Comfort[2,3]		C	C	
Living room	operative temperature	CIBSE TM52 (free running)	22	
Kitchen	operative temperature	CIBSE TM52 (free running)	17	
Bedroom	operative temperature	CIBSE TM52 (free running)	17	
Bathroom	operative temperature	CIBSE TM52 (free running)	20	
WC	operative temperature	CIBSE TM52 (free running)	19	
Hall	operative temperature	CIBSE TM52 (free running)	19	
Air Quality		PPM	l/s	
Whole building	fresh air supply[4]		approved document F	
	CO_2 concentration[2]			
Assessment name:	Regulations compliance		Request	

Dashboard

Figure 11.14   Associating resilience tests with an uploaded model

			A.N. User
Marathon			
Resilience Testing Environment	model selection assessment request		05/12/2019
	dashboard further information		
	logout		

ID	Request	Status	Actions
74	Project 1: Base case: Regulations compliance	passed	certificate
92	Project 2: Base case: Regulations compliance	cancelled	
93	Project 2: Variant 1: Regulations compliance	abandoned	feedback detailed results
95	Project 2: Variant 2: Regulations compliance	passed	certificate feedback detailed results
96	Project 2: Variant 2: BREEAM recommended	suspended	revise model abandon
110	Project 3: Base case: Regulations compliance	running	cancel
123	Project 4: Base case: Regulations compliance	passed	certificate
124	Project 4: Variant 1: Regulations compliance	passed	certificate
125	Project 4: Variant 1: Net-zero (customised)	running	cancel

Figure 11.15   Resilience test status and outcome feedback

variations of weather: that is, each realisation of the weather generator is a plausible future year in that the weather patterns represented therein are valid for the location. Through the process of controlled random generation, Indra is able to produce periods of weather phenomena such as heat waves, hot and dry summers, and mild winters as required for life cycle resilience testing. In this way, the model is subjected to typical and extreme weather conditions occurring at different times each year whilst ensuring that the annual weather collection overall is representative of the model location (i.e., some years have seasons/months/days that are colder/ warmer/windier and so on).

Each annual simulation progresses against the operational contingencies defined for a particular test relating to issues such as utility supply failure, context changes, control system failure, and electricity export refusal. Such events might be randomly imposed on several occasions to stress test a design in a manner that reflects reality.

For the case of domestic buildings, the evolution of the spatially and temporally varying performance parameters relating to energy and environment is continuously probed to ensure compliance with test-specific performance criteria. The following criteria are employed within the RTE prototype: Chartered Institution of Building Services Engineers (CIBSE) overheating criteria relating to adaptive comfort (CIBSE 2013), the UK building regulations ventilation criteria (Gov 2019), and the Scottish Government's Building Standards technical handbook of energy and emissions criteria (SG 2017).

For the case of non-domestic buildings, the CIBSE criteria for CO_2 concentration (CIBSE 2018) and BS EN 12464-1 criteria for visual comfort (BS 2011) are added to the above. Alternatives for thermal comfort and air quality – ISO 7730 (ISO 2005) and BS EN 15251 (BS 2007), respectively – have also been implemented to facilitate research into the impact of standards on resilience test outcomes.

The concept of a simulation-based environment that applies standardised performance assessments to uploaded proposals, and provides a means to issue compliance certificates corresponding to increasing levels of resilience stringency, has the potential to benefit design practitioners, policymakers, estate managers, and researchers. It harmonises the use of performance simulation in practice as a means to explore dynamic behaviour over a scheme's lifetime and thereby ensure a high-performance standard.

The use of automatic performance assessment provides access to sophisticated simulation scenarios with minimum user effort because the approach incorporates model creation and assessment knowledge alongside automated simulation tool control. Users can concentrate on improving building performance rather than expending effort and resources on complex model calibration and simulation process control.

The approach also provides a mechanism for building regulation compliance that overcomes the constraints of the present rudimentary methods that fail to respect the thermodynamic integrity of buildings. That said it is emphasised that simulation models may still be configured against standard prescriptions to obtain outputs from simulation tools for legislative compliance purposes, to size system components for peak demand, or to obtain deeper performance insight – that is, the traditional application approach.

11.4   Thermal bridges

Thermal bridges have often been considered as 'noise-in-the-system' even though programs exist to undertake realistic assessments of the heat and mass transfers within building constructions when exposed to time-varying weather conditions. Two notable examples include WUFI[9] and AnTherm[10]. With the advent of disruptive standards that rely heavily on the identification of facade faults such as PassivHaus[11], thermal bridges have become a noise that cannot be ignored. Indeed, thermal bridge assessment is now considered best practice.

One approach is to match building sections with published data. The UK Building Research Establishment, for example, maintains a database[12] and some new building codes suggest values for common details. Unfortunately, coverage is not extensive. Where data are missing, numerical[13] and analytical (Tang and Saluja 1998, Hassid 1989) applications exist to calculate 2D or 3D conduction and report this in terms of Ψ-values (pronounced psi) that can be included in a BPS+ model. This is a measure of the additional heat transmittance per metre length of the thermal bridge (W/m.K).

Some BPS+ tools provide facilities to define thermal bridge types and locate these within a building model as required. Figure 11.16 shows such a session

[9]https://wufi.de/en/
[10]https://www.antherm.at/antherm/EN/
[11]https://passivehouse.com/
[12]https://tools.bregroup.com/certifiedthermalproducts/
[13]Therm: https://windows.lbl.gov/therm-software-downloads

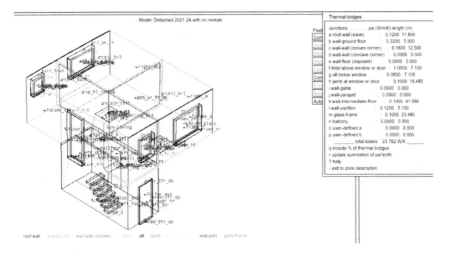

Figure 11.16   Thermal bridges located in a portion of a residence

Table 11.5   Energy flow (kWh) without/with thermal bridges

Flowpath	Without		With	
	Gain	**Loss**	**Gain**	**Loss**
Infiltration	0	−3784	.013	−3719
Casual gains	1109	0	1109	0
Thermal bridges	0	0	0	−2550
Heat storage (air point)	23	−23	23	−23
Convection external opaque	79	−5199	188	−5197
Convection other opaque	302	−3583	413	−3613
Convection transparent	33	−389	36	−385
Heating system	11,433	0	13,717	0
Total	12,979	−12,979	15,487	−15,487

underway in ESP-r. Here, junctions are identified automatically from the model geometry and the Ψ-values selected from a database as a function of the defined constructions.

To demonstrate the importance of thermal bridges, consider Table 11.5, which presents an energy breakdown for a portion of an ESP-r model with and without thermal bridges. In this case, the magnitude of the heat loss associated with thermal bridges over the year is significant.

Treating a facade as a collection of 2D geometric entities with 1D conduction heat transfer ignores their interconnections and the thermal bridges that can occur at junctions. Higher dimensional information can be added to a model to address this issue as well as enable an assessment of intra-construction heat and vapour flow. Whilst ESP-r's default treatment of construction conduction is uni-directional, an adaptive multi-dimensional gridding treatment is also possible to represent intersectional materials (Nakhi 1995). This allows localised multi-dimensional modelling. For example, building constructions can be modelled in 1D except in one room where the constructions are upgraded to 2D or 3D. Furthermore, within a zone, the constructions may be defined as a mix of 1D, 2D, and 3D representations. As shown in Figure 11.17, this is achieved via the use of massless connector surfaces (at node 4 to join nodes 1 and 4 in the example shown).

Whilst defining geometrical information can be complex, it can, fortunately, be generated automatically based on user directives defining the location and level of enhancement required. Table 11.6 lists the data required to enhance the con-structional resolution of a model.

This information, along with data defining the vapour diffusivity of the con-struction materials, supports the study of thermal bridges, surface condensation, and mould growth.

11.5 Design detail

Real buildings have copious exceptions to standard sections. For example, a timber frame system may have an average timber fraction of 10% whilst some areas of the facade can have double this fraction. In commercial buildings, embedded structural elements may be abstracted to the point where the local condensation risk cannot be assessed. Indeed, the full complexity of facade detailing and the specific points of failure intrinsic to those details, as illustrated in the thermographic image of Figure 11.18, are too often abstracted into an aggregate composition with equiva-lent performance. Creating model facades based on standard sections may suffice for long-term estimates but fail to deliver information on the variability and dis-tribution of surface temperatures needed for condensation analysis.

Abstract representations of interior and exterior reveals at windows and doors are a modelling legacy with consequences. In deep facades, reveals represent substantial surface areas and paths for heat transfer as well as interrupting the distribution of light. A thermographic survey easily demonstrates the divergent temperatures, and yet reveals are absent from many models as if their ubiquity rendered them unre-markable to the point of invisibility. Thermal bridge (TB) representation helps to capture the additional heat flow in a zone energy balance but the absence of divergent surface temperatures does not allow an assessment of local discomfort and dampness.

The ability to include a representation of design detail has been explored in several ESP-r projects. Consider the model of Figure 11.19.

In the upper images, the facade is represented as a collection of flat entities with the bounds of the room at the inside face. The window frame is therefore flush with the inside wall face whilst the outside reveals are included to support shading calculations. An insightful analysis requires that the frame be properly located in relation to the wall elements as shown in the lower images.

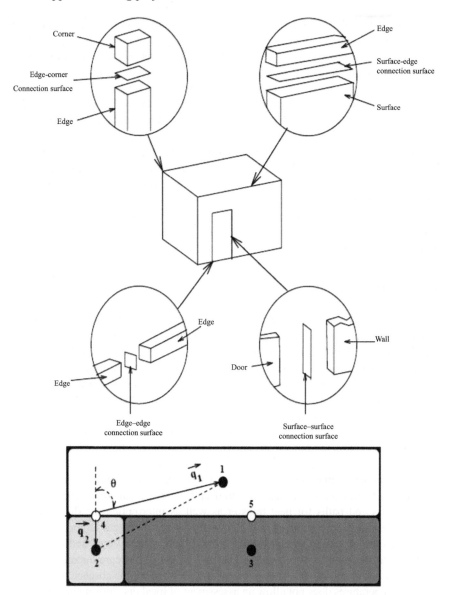

Figure 11.17   Construction issues requiring multi-dimensional treatment (credit: Nakhi 1995)

Of course, the inclusion of thermal bridges in the manner described in the previous section does not fully compensate for the missing surface area. To illustrate the impact of including reveals, results for three cases – flush frames and with 100 mm and 200 mm reveals – are listed in Tables 11.7 and 11.8.

Table 11.6   Data corresponding to resolution enhancement

For each surface participating in the 3D grid:

- whether lumped parameter or discretised
- edge definitions for the bounding box of the 3D grid
- adjacency with other surfaces
- grid distribution and dimensions
- material properties for each cell
- directives for conduction only or conduction and radiation

Figure 11.18   Facade temperature distribution

Another issue is the treatment of intermediate floors. In models where the floor structure thickness is ignored, it is likely that the exposed facade at the intermediate floor is also missing. Ceilings and floors take on aspects of reveals as they approach the outer facade in that they include heat flow paths that are not fully described. Thermal bridges help with the bookkeeping but do not help with the diverging surface temperatures adjacent to the facade.

Figure 11.20 shows a terrace model created within the INDU-Zero off-site manufacturing project[14]. Here, the horizontal bands between the ground and first floor, and between the first floor and attic, are treated as explicit thermal zones.

[14]https://northsearegion.eu/indu-zero

*Figure 11.19   Conventional facade definition (upper) and with window frame
detailed*

The advent of high-performance standards such as PassivHaus is indicative of
the changing nature of what can be considered 'noise' in the assessment of performance. When faults in facades, either in the form of poor detailing or faults
accrued during the construction process, which was previously ignored, nullify
other high-performance investments, traditional approaches to model abstraction
are ripe for questioning. Where compliance-oriented approaches provide limited
performance insight, BPS+ tools need to evolve to include detail.

Table 11.7   Volume/surface area with/without reveals

Room	No reveal	100 mm reveal	200 mm reveal
Living	81.5 m^3	82.1 m^3	82.7 m^3
	48.5 m^2	50.2 m^2	51.8 m^2
Non-living	131.4 m^3	132.3 m^3	133.1 m^2
	83.4 m^2	86.1 m^2	88.7 m^2

Table 11.8   Annual heating demand (kWh/m^2.y)

Case	Living	Non-living
No reveals or TB	198.3	231.7
No reveals + TB	252.3	278.0
100 mm reveals + TB	258.3	284.0
200 mm reveals + TB	264.9	291.2

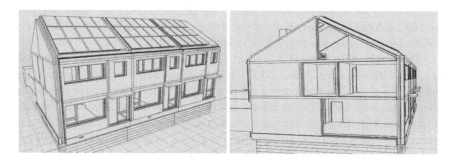

Figure 11.20   Views of facade features and intermediate floors

11.6   Ensuring occupant wellbeing

Multiple factors influence wellbeing in the indoor environment, and there are many ways to evaluate these factors (Fantozzi and Rocca 2020). A critical issue is dampness and mould growth, which are recognised as problems affecting a significant proportion of buildings worldwide. This is a cause for concern given that individuals can spend up to 90% of their day indoors. Approximately 2.5 million UK residences are affected, with well-documented cases in Europe and North America (Martin *et al.* 1987, Morris *et al.* 1989, Lewis *et al.* 1989, Smith *et al.* 1992, Hendry and Cole 1993) and epidemiological evidence that mouldy housing has a detrimental effect on the physical and mental health of children and adults (Dales *et al.* 1991). Singh (1995) described how building design, occupancy behaviour, and management can affect the incidence of allergic factors and estimated the cost of repairing the damage caused by timber decay in the UK housing stock to be approximately £400M per annum.

A notable feature of BPS+ is its ability to bring focus to the spatial distributions of indoor environmental conditions. This is in contrast to the usual focus on regulatory compliance or CO_2 emissions reduction. Clarke *et al.* (1996) undertook a multidisciplinary project to enable ESP-r to assess the likelihood of surface mould growth due to poor construction and inadequate control of the indoor environment.

High levels of airborne spores may occur due to the growth of fungi on walls. Data from the 1991 Scottish Housing Condition Survey (Scottish Homes 1993), for example, indicated that around 12.3% of Scottish houses were affected, with inadequate heating, insulation, and ventilation cited as the principal causal factors. Respiratory and/or allergenic symptoms, principally in children, have been diagnosed, particularly in atopic individuals (Burr *et al.* 1988, Hunter *et al.* 1988, Strachan 1988). Whilst the precise mechanisms for these symptoms are not clear, toxic fungal metabolites, particularly mycotoxins and possibly volatile organic compounds produced by moulds have been implicated.

The inhalation of airborne microorganisms and their metabolites may cause a range of respiratory symptoms depending on the immunological status of the host and the type of organism present. Some are specific building-related diseases, which can be identified by immunological or microbiological tests; others are recognised as syndromes with no readily identifiable cause. Yet others are non-specific reactions to components of the airborne dust or are poorly defined (e.g., chronic fatigue syndrome or increase in coughing).

The traditional approach to mould alleviation is to treat the building construction with water repellents and fungicidal agents. This simply treats the effect and not the cause. An engineering approach requires that the interior finishes be considered along with their associated multi-layered constructions, with the heat, air, and moisture transport processes manipulated to establish an appropriate surface relative humidity (RH) and temperature regime. Tackling mould growth problems in the domestic and work environments requires manipulation of the indoor climate through good building design, construction, and management (particularly in relation to ventilation and thermal bridge elimination).

Fungi can exist over a wide range of temperatures, below $0\,°C$ fungal cells survive but rarely grow, whilst above $40\,°C$ most cells cease growing and die. Between $0\,°C$ and $40\,°C$, fungal activity depends on the effect of temperature on enzyme activity. Not all moulds grow over the same temperature range, and the ability of a particular mould to develop at low or high temperatures depends on its classification. Psychrophilic fungi have their maximal growth rates below $20\,°C$, mesophilic fungi have their maximal growth rates at temperatures in the range of $20\,°C–40\,°C$, thermo-tolerant fungi can grow in temperatures above $40\,°C$, whilst thermophilic fungi have their maximal growth rate above $50\,°C$.

Grant *et al.* (1989) showed that within the temperature range of $5\,°C–25\,°C$, increasing temperature permits growth at lower water activity levels. Other researchers have described this phenomenon, where the maintenance of a surface temperature removed from the optimum results in a reduction in the range of water levels permitting germination and subsequent growth (Smith and Hills 1982). Whilst intermittent condensation at the wall surface does increase the probability of mould proliferation, the traditional view that condensation has to occur is unfounded. Indeed, condensation (100% RH) by itself is unsuitable for sustainable mould growth as pure water is theoretically devoid of essential nutrients. Most filamentous fungi are known to have an optimum moisture requirement that is often significantly below saturation (Hunter *et al*. 1988, Grant *et al*. 1989).

With the knowledge that the provision of sufficient water in the form of surface RH is the key to ultimately controlling indoor mould development, any parameter that influences surface RH has the potential to affect fungal growth. Likewise, any parameter that affects the geometry, moisture, and/or nutritional composition of a material, particularly at its surface (e.g., pore size distribution, capillary condensation, etc.), can influence the development of a growing fungus. Interior finishes are influenced by their adjacent multi-layered constructions, with the heat, air, and moisture transported through the building fabric affecting the surface RH.

Factors known to govern RH at a wall surface include the hygrothermal properties of constructional materials, the presence of thermal bridges, zone ventilation rate, HVAC system capabilities, weather variations, and moisture sources. Materials and decorative finishes become moist when they have a water or vapour open porosity (i.e., they are hygroscopic) – materials such as glass, which have zero porosity, exhibit surface wetting but no mass wetting.

In damp constructional materials, once colonisation has begun, and if the temperature is sufficiently high, progressive deterioration of the substrate will proceed with increasing rapidity as moisture is released via respiration. Colonisation follows a succession in which mould species occupy distinct positions that are strictly determined by moisture content. Marked population changes occur as the moisture content increases and species that are more competitive at higher moisture content develop.

The key question then is how best to alleviate the problem. Although the use of biocidal materials on walls and carpets may be appropriate in some circumstances, it is generally agreed that removal of the conditions that promote mould growth, principally dampness, is the preferred strategy. Whilst resources are being directed to improve the housing stock, and will undoubtedly limit condensation and mould growth, there remains an inadequate understanding of how the interaction of building materials with their environments can create microclimates that promote the initiation and spread of moulds. The failings associated with a building's construction or environmental control systems often do not appear until post-occupancy. One approach to the problem is to undertake rigorous performance appraisals before construction. The International Energy Agency, for example, has undertaken two major research programmes – *Condensation and Energy* (Hens and Sneave 1991), and *Heat, Air, and Moisture Transport* (Hens and Kumaran 1994)– both of which have resulted in algorithms for the prediction of localised environments within buildings. BPS+, when endowed with such algorithms, can be used to:

- determine the likelihood of mould occurrence for a given case;
- rank order measures for non-biological alleviation of mould problems; and
- evaluate the applicability of a given solution more generally.

Moulds that appear on a wall surface can be classified as either xerophilic (fungi capable of growth under dry conditions with RH <85%) or hydrophilic (those that require greater amounts of free water to sustain growth). This variation in water requirement frequently results in successive colonisation of a surface by a variety of different moulds (Grant *et al.* 1989). Certain moulds of the genus *Eurotium* or *Aspergillus* are xerophiles and can grow at RH values less than 75%, whilst members of genera *Penicillium*, *Cladosporium*, *Ulocladium*, and *Stachybotrys* require higher moisture levels. From previous studies on the mould flora of UK houses (Hunter *et al.* 1988, Flannigan and Hunter 1988, Lewis *et al.* 1989), it is known that the predominant mould genera are *Penicillium*, *Cladosporium*, *Aspergillus*, and *Sistotrema*, with the first three genera particularly important because of their toxigenic potential. The ability of these moulds to utilise nitrogen-poor substrates and survive at relatively low moisture levels is the feature that contributes to their success in colonising the internal surfaces of domestic dwellings. Moreover, their ability to produce aerial spores allows their rampant spread.

Based on a review of the literature, six main mould species were identified and their limiting conditions for growth were established. The species are *Aspergillus repens*, *Aspergillus versicolor*, *Penicillium chrysogenum*, *Cladosporium sphaerospermum*, *Ulocladium consortiale*, and *Stachybotrys atra*. These moulds were selected to represent species with differing requirements for moisture. For example, *A. repens* was chosen as an example of a xerophilic mould requiring around 75% RH to grow, whilst, at the other extreme, *S. atra* (a wet-loving fungus) requires an RH in excess of 97%. The other species chosen for incorporation into the model fall into RH growth zones between these two extremes. Consequently, whilst the model is based on the growth characteristics of six moulds, it can be considered to represent the behaviour of different moisture-requiring physiological groups of moulds and, as such, can be used for predicting the growth of many other species. The moulds selected for modelling, with the exception of *A. repens*, are known producers of mycotoxins. Figure 11.21 summarises the growth limits for each mould group.

The isopleth labels in this figure correspond to the following moulds.

(A)   Highly xerophilic (dry loving).
(B)   Xerophilic.
(C)   Moderately xerophilic.
(D)   Moderately hydrophilic.
(E)   Hydrophilic.
(F)   Highly hydrophilic (wet loving).

Isopleths define the minimum combination of RH and temperatures for which mould growth will occur. Below these limits, growth is not sustainable. Note that moulds corresponding to higher-level curves may initiate growth because of the respiratory-related moisture release of already-established moulds corresponding to lower-growth curves. This phenomenon – local surface RH elevation – should ideally be included within a mould prediction algorithm.

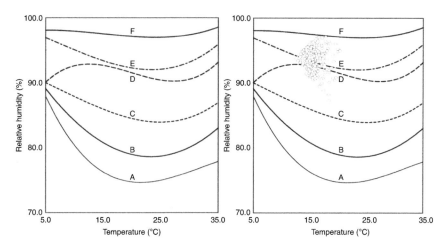

Figure 11.21   Predominant mould growth isopleths and (right) an example of superimposed simulation predictions over time

The mould prediction facility as encapsulated within ESP-r comprises three components: a database of mould isopleths as above, the tracking of the time evolution of local surface temperature and RH, and a means to relate surface condition to mould initiation. The growth curves are held within an ESP-r database in the form of coefficients defining the temperature/RH isopleths. To assess the risk of mould infestation, uncertainty bands are applied to these curves as shown in Figure 11.22 for the case of *P. chrysogenum*. Here, the C line corresponds to the case of standard building materials, C'' to a highly nutritious substrate such as foodstuffs or laboratory culture media (the optimum growth condition), and C' a building material where the nutritive status has been enhanced by adulteration with a carbon source. Based on these data, it is possible to determine the growth probability zones as indicated.

It should be noted that these risk categories are not based on experimentally derived data but on expert opinion interpreting the likely effect on the growth of the six modelled species as conditions move in incremental steps above and below the isopleth curves of Figure 11.22, which have been constructed from best available experimental data.

To test the ESP-r feature, the isopleth model was compared with real data obtained from a mould-contaminated house (MacGregor and Taylor 1995) as shown in Figure 11.23.

Samples were taken at the wall surface where mould growth had occurred. It is generally the case that mould-infected areas contain various species of moulds that develop during the prolonged colonisation period. Different species can develop since each type can take advantage of suitable temperature and water activity levels that occur transiently at susceptible wall surfaces. Thus more xerophilic types gain

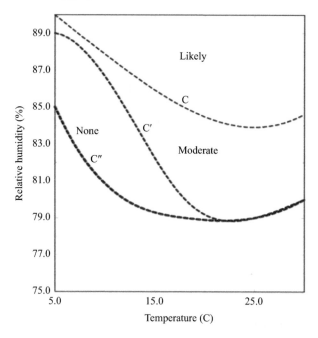

Figure 11.22   Risk categories for Penicillium chrysogenum

Figure 11.23   Mould growth in a monitored house

an advantage when moisture levels decrease and conversely more hydrophilic types can reactivate and regrow when moisture levels rise during periods of condensation.

The mould species that develop at the different RH levels were identified by conventional microbiological techniques. At RH values of 85.5% or less only the xerophilic species *A. versicolor* and *Eurotium herbariorum* developed. At RH values of 88.5% and above, additional moulds that grew were *Penicillium* spp., *Aureobasidium pullulans*, *C. sphaerospermum*, and *Alternaria alternata*. At RH values of 94.5% and above, the previously mentioned species grew together with yeasts that appeared at these high RH values.

In addition to identifying the mould isolates, the speed of mould development on the range of RH-adjusted malt extract agar (MEA) media was monitored as summarised in Table 11.9.

It took 2 days for mould to appear on the 98.9% RH plates, 7 days at 94.5%, 16 days at 88.5%, 20 days at 85.5%, 58 days at 81%, and 97 days at 78.5% (the mean surface RH as measured on site). No mould development occurred at an RH value lower than 78.5% over a 120-day incubation period. Based on the evolving growth risk data, the length of time that conditions remain within each of the five growth categories can be tracked at simulation time. This, in turn, supports an assessment of the overall risk of mould growth under changing conditions.

As summarised in Figure 11.24, an ESP-r model of the test case shown in Figure 11.23 was established (Kelly 1996) and predictions were compared with measurements.

The house is a steel-framed, three-bedroom, semi-detached residence built to a 1940s pre-fabricated style. In places the steel frame results in a major thermal bridge, leading to condensation and mould growth. Insulation levels are

Table 11.9   Period required for mould appearance on 2% MEA plates

Days	RH (%)								
at 25 °C	98.9	94.5	88.5	85.5	81.0	78.5	74.5	71.2	67.8
0	−	−	−	−	−	−	−	−	−
2	+	−	−	−	−	−	−	−	−
7	+	+	−	−	−	−	−	−	−
16	+	+	+	−	−	−	−	−	−
20	+	+	+	+	−	−	−	−	−
58	+	+	+	+	+	−	−	−	−
97	+	+	+	+	+	+	−	−	−
120	+	+	+	+	+	+	−	−	−

− no growth; + appearance of mould

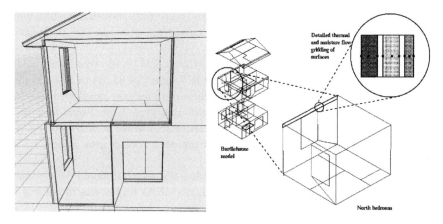

Figure 11.24   ESP-r model of the test house

generally low with excessive infiltration rates resulting from warped window frames. The building is located on an exposed housing estate. The lower floor comprises a living room, hall, kitchen, bathroom, and store, whilst the upper floor comprises three bedrooms and an upper hall. The loft space was modelled as a separate zone. In the north bedroom, the model was constructed to a resolution which enables the explicit tracking of air and vapour flow. Of particular interest in the simulation was the junction of the north wall and ceiling where mould growth had occurred.

Simulations were performed[15] based on weather data collected at the site. The results for the problematic north bedroom are shown in Figure 11.25. The predictions showed reasonable agreement given that several parameters were highly uncertain (e.g., the contribution of occupants to moisture generation and the hygroscopic properties of the building materials).

A BPS+ mould alleviation study would typically involve the ranking of measures on a least cost basis: wall insulation upgrade; elimination of thermal bridges; moisture removal at source; improved ventilation; modifications to heating systems; modified control strategy; alternative construction, materials, and surface finishes; and user behaviour changes. Consider the case when ESP-r was applied to contrast two options – increased heating against improved insulation. Figure 11.26 shows results for the base case model and for cases corresponding to 200W continuous heating, 500W continuous heating, and 500W continuous heating with improved insulation.

For the base case, the environmental conditions are within the mould growth zone for a considerable period so mould growth was deemed likely (as observed in

---

[15]ESP-r's treatment of construction moisture flow is described in detail elsewhere (Nakhi 1995, Clarke 2001).

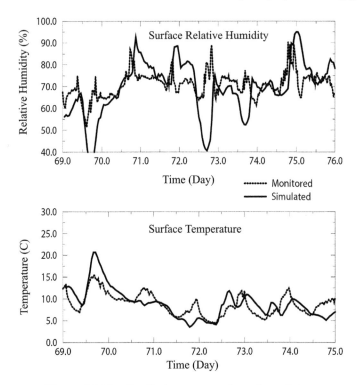

Figure 11.25 Predictions vs. monitored conditions

the house). The results for the other cases show the mould problem can be alleviated by increasing the heat input and improving insulation.

11.7 Chapter summary

High-resolution modelling and simulation are able to tackle the performance gap between design intent and what is realised in practice. This is achieved by including consideration at the design or renovation stage of parameter sensitivity, occupant behaviour, and resilience to system disturbances. Where the aim is to deliver solutions that engender wellbeing it is necessary to go further by including consideration of design detail and mould species growth characteristics.

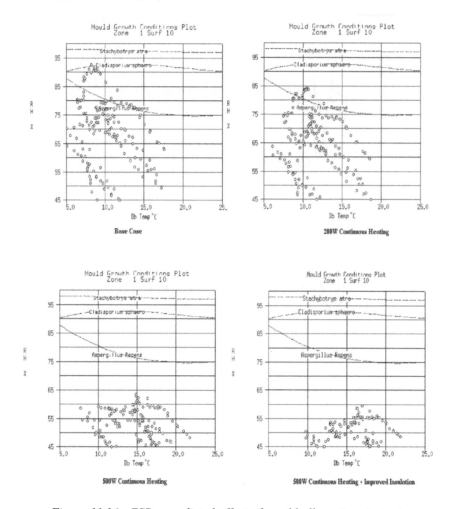

Figure 11.26   ESP-r predicted effect of mould alleviation strategies

References and further reading

Belafi Z, Hong T and Reith A (2016) 'A library of building occupant behaviour models represented in a standardised schema', *Proc. Fourth European Conf. Behaviour and Energy Efficiency*, Coimbra, Portugal.

Bourgeois D (2005) 'Detailed occupancy prediction, occupancy-sensing control and advanced behavioural modelling within whole-building energy simulation', *PhD Thesis*, Université Laval, Québec, Canada.

Bourgeois D, Reinhart C and Macdonald I (2006) 'Adding advanced behavioural models in whole building energy simulation: a study on the total energy impact of manual and automated lighting control', *Energy and Buildings*, 38, pp. 814–823.

BS (2007) *Standard 15251, Indoor Environmental Input Parameters for Design and Assessment of Energy Performance of Buildings Addressing Indoor Air Quality, Thermal Environment, Lighting and Acoustics*, British Standard Institute.

BS (2011) *Standard EN 12464-1, Light and Lighting: Lighting of Work Places*, British Standard Institute, 2011.

Burr M I, Mullins J, Merrett T G and Stott N C H (1988) 'Indoor moulds and asthma', *Journal of the Royal Society of Health*, 3, pp. 99–101.

CIBSE (2013) *Technical Manual 52: The Limits of Thermal Comfort: Avoiding Overheating in European Buildings*, Chartered Institute of Building Services Engineers, London.

CIBSE (2018) *Guide A: Environmental Design*, Chartered Institute of Building Services Engineers, London.

Clarke JA (2001) *Energy Simulation in Building Design* (2nd ed.), Butterworth-Heinemann, Oxford, ISBN 0 7506 5082 6.

Clarke J A (2018) 'The role of building operational emulation in realizing a resilient built environment', *Architectural Science Review*, 61(5), pp. 358–361.

Clarke J A and Cowie A (2019) 'Development and application implications of enabling occupant behaviour modelling within building performance simulation', *Proc. Building Simulation'19*, Rome, Italy.

Clarke J A and Cowie A (2020) 'A simulation-based procedure for building operational testing', *Proc. CIBSE/ASHRAE Technical Symposium*, University of Strathclyde, Glasgow, 16–17 April 2020.

Clarke J A and Cowie A (2020) 'A simulation-based procedure for building operational resilience testing', *Proc. CIBSE/ASHRAE Technical Symposium*, 16–17 April, Glasgow, United Kingdom.

Clarke J A, Hand J W, Kelly N, *et al.* (2012) 'A data model for integrated building performance simulation', *Proc. Building Simulation and Optimisation*, Loughborough, England.

Clarke J A, Johnstone C M, Kelly N J, *et al.* (1996) 'Development of a technique for the prediction/alleviation of conditions leading to mould growth in houses', *Final Report for Project* 68017, Scottish Homes.

Clarke J A, Johnstone C M, Macdonald I A, *et al.* (2000) 'The deployment of photovoltaic components within the lighthouse building in Glasgow', *Proc. 16th European Photovoltaic Solar Energy Conf.*, Glasgow.

Cowie A, Hong T, Feng X and Darakdjian Q (2017) 'Usefulness of the obFMU module examined through a review of occupant modelling functionality in building performance simulation programs', *Proc. Building Simulation '17*, San Francisco, USA.

Dales R E, Burnett R and Zwanenburg H (1991) 'Adverse health effects among adults exposed to house dampness and molds', *American Reviews of Respiratory Disease*, 143, pp. 505–509.

de Wilde P (2014) 'The gap between predicted and measured energy performance of buildings: a framework for investigation', *Automation in Design*, 41, pp. 40–49.

Fantozzi F and Rocca M (2020) 'An extensive collection of evaluation indicators to assess occupants' health and comfort in indoor environment', *Atmosphere*, 11(1).

Flannigan B and Hunter C A (1988) 'Factors affecting airborne moulds in domestic dwellings', *Indoor and Ambient Air Quality*, R Perry and P W Kirk (Eds.), pp. 461–468, London.

Flett G (2017) 'Modelling and analysis of energy demand variation and uncertainty in small-scale domestic energy systems', *PhD Thesis*, ESRU, University of Strathclyde.

Fritsch R, Kohler A, Nygård-Ferguson M and Scartezzini J-L (1990) 'A stochastic model of user behaviour regarding ventilation', *Building and Environment*, 25 (2), pp. 173–181.

Gov (2019) *Approved Document F1 Means of ventilation*, gov.uk/government/publications/ventilation-approved-document-f, HM Government.

Grant C, Hunter C A, Flannigan B and Bravery A F (1989) 'The moisture requirements of moulds isolated from domestic dwellings', *International Biodeterioration*, 25, pp. 259–284.

Gunay H B, O'Brien W and Beausoleil-Morrison I (2015) 'Implementation and comparison of existing occupant behaviour models in EnergyPlus', *Building Performance Simulation*, 9(6), pp. 567–588.

Haldi F and Robinson D (2008) 'On the behaviour and adaptation of office occupants', *Building and Environment*, 43(12), pp. 2163–2177.

Haldi F and Robinson D (2009) 'Interactions with window openings by office occupants', *Building and Environment*, 44(12), pp. 2378–2395.

Hassid S (1989) 'Thermal bridges in homogeneous walls: a simplified approach', *Building and Environment*, 24(3), pp. 259–264.

Hendry K M and Cole E C (1993) 'A view of mycotoxins in indoor air', *Toxicology and Environment Health*, 36, pp. 183–198.

Hens H and Kumaran M K (1994) 'Computer modelling of heat, air and moisture transport through building components: state-of-the-art', *Newsletter Issue 19*, International Energy Agency, Energy Conservation in Buildings and Community Systems Programme.

Hens H and Sneave E (Eds.) (1991) 'Annex 14; condensation and energy', *Final Report Vol. 1* Sourcebook International Energy Agency, Conservation in Buildings and Community Systems Programme.

Hong T, Chou S K and Bong T Y, (2000) 'Building simulation: an overview of developments and information sources', *Building and Environment*, 35(1), pp. 347–361.

Hong T, Sun H, Chen Y, Taylor-Lange S C and Yan D, (2016) 'An occupant behavior modeling tool for co-simulation', *Energy and Buildings*, 117, pp. 272–281.

Hong T, D'Oca S, Turner W J N and Taylor-Lange S C (2015) 'An ontology to represent energy-related occupant behavior in buildings. Part I: introduction to the DNAs Framework', *Building and Environment*, 92, pp. 764–777.

Hong T, D'Oca S, Turner W J N, Taylor-Lange S C, Chen Y and Corgnati S P (2015a) 'An ontology to represent energy-related occupant behavior in buildings. Part II: implementation of the DNAS framework using an XML schema', *Building and Environment*, 94, pp. 196–205.

Hunt D R G (1979) 'The use of artificial lighting in relation to daylight levels and occupancy', *Building and Environment*, 14, pp. 21–33.

Hunter C A, Grant C, Flannigan B and Bravery A F (1988) 'Mould in buildings, the air spora of domestic dwellings', *International Biodeterioration*, 24, pp. 81–101.

ISO (2005) *Standard 7730, Ergonomics of the Thermal Environment – Analytical Determination and Interpretation of Thermal Comfort Using Calculation of the PMV and PPD Indices and Local Thermal Comfort Criteria*, International Standardization Organization.

Judkoff R and Neymark J (1995) 'International energy agency building energy simulation test (BESTEST) and diagnostic method', *Report TP-472-6231*, National Renewable Energy Laboratory, USA, 1995.

Kelly N (1996) 'ESP-r analysis of a mould infested house', *Project Report*, ESRU, University of Strathclyde.

Lewis C W, Anderson J G, Smith J E, Morris G P and Hunt S M (1989) 'The incidence of moulds within 525 dwellings in the United Kingdom', *Environmental Studies*, 25, pp. 105–12.

Love J A (1998) 'Manual switching patterns in private offices', *Lighting Research and Technology*, 30, pp. 45–50.

Macdonald I A (2002) 'Quantifying the effects of uncertainty in building simulation', *PhD Thesis*, ESRU, University of Strathclyde.

MacGregor K and Taylor A (1995) 'Breathing Sunshine into Scottish Housing', *Report to Scottish Homes*, Dept. of Mechanical Engineering, Napier University

Martin C J, Platt S D and Hunt S M (1987) 'Housing conditions and ill health', *British Medical Journal*, 294, pp. 1125–7.

Monari F, (2016) 'Sensitivity analysis and bayesian calibration of building energy models', *PhD Thesis*, University of Strathclyde, Glasgow.

Morris G P, Murray D, Gilzean I M, *et al.* (1989) 'A study of dampness and mould in Scottish public sector housing', *Airborne Deteriogens and Pathogens*, pp. 163–73.

Nakhi A E (1995) 'Adaptive construction modelling within whole building dynamic simulation', *PhD Thesis*, ESRU, University of Strathclyde, Glasgow.

Newsham G R (1994) 'Manual control of window blinds and electric lighting: implications for comfort and energy consumption', *Indoor Environment*, 3 (3), pp. 135–144.

Nicol J F (2001) 'Characterising occupant behavior in buildings: Towards a stochastic model of occupant use of windows, lights, blinds, heaters and fans', *Proc. Building Simulation '01*, Rio de Janeiro, Brazil, pp. 1073–1078.

O'Brien W (2020) 'An international review of occupant-related aspects of building energy codes and standards', *Building and Environment*, 179.

ONS (2023) Families and households in the UK: 2022, Office for National Statistics, https://ons.gov.uk/peoplepopulationandcommunity/birthsdeathsandmarriages/families/bulletins/familiesandhouseholds/2022/.

Page J, Robinson D, Morel N and Scartezzini J-L (2008) 'A generalised stochastic model for the simulation of occupant presence', *Energy and Buildings*, 40, pp. 83–98.

Rastogi P, (2016) 'On the sensitivity of buildings to climate: the interaction of weather and building envelopes in determining future building energy consumption', *PhD Thesis*, Ecole Polytechnique Fédérale de Lausanne, Switzerland.

Reinhart C F (2004) 'Lightswitch 2002: a model for manual control of electric lighting and blinds', *Solar Energy*, 77(1), pp. 15–28.

Reinhart C F and Voss K (2003) 'Monitoring manual control of electric lighting and blinds', *Lighting Research and Technology*, 35, pp. 243–258.

Ren X, Yan D and Wang C (2014) 'Air-conditioning usage conditional probability model for residential buildings', *Building and Environment*, 81, pp. 172–182.

Richardson I, Thomson M and Infield D (2008) 'A high-resolution domestic building occupancy model for energy demand simulations', *Energy and Buildings*, 40(8), pp. 1560–1566.

Rida M, Kelly N and Cowie A (2018) 'Integrating a human thermo-physiology model with a building simulation tool for better occupant representation', *Proc. BauSIM '18*, Karlsruhe, Germany.

Rijal H B, Tuohy P G, Nicol J F, *et al.* (2008) 'Development of adaptive algorithms for the operation of windows, fans and doors to predict thermal comfort and energy use in Pakistani buildings', *ASHRAE Transactions*, 114(2), pp. 555–573.

Scottish Homes (1993) *Scottish Housing Condition Survey 1991*, A Report to the Scottish Office and Scottish Homes Board, Edinburgh.

SG (2017) *Building Standards technical handbook 2017: domestic buildings (section 7 sustainability)*, Scottish Government, https://gov.scot/publications/building-standards-2017-domestic.

Singh J (1995) 'The built environment and the developing fungi', *Building Mycology*, Singh J (Ed.), E and F Spon, London, pp. 1–21.

Smith JE, Anderson JGA, Lewis CR and Murad YM (1992) 'Cytotoxic fungal spores in the indoor atmosphere of the damp domestic environment', *FEMS Microbiology Letters*, 100, pp. 337–44.

Smith S L and Hill S T (1982) 'Influence of temperature and water activity on germination and growth of *Aspergillus restrictus* and *A. versicolor*', *Transactions of the British Mycological Society*, 79, pp. 558–559.

Strachan D P (1988) 'Damp housing and childhood asthma: validation of reporting of symptoms', *British Medical Journal*, 297, pp. 1223–1226.

Tang D and Saluja G (1998) 'Analytic analysis of heat loss from corners of buildings', *Heat and Mass Transfer*, 41(4–5), pp. 681–689.

Tuohy P G, Humphreys M A, Nicol J F, Rijal H B and Clarke J A (2009) 'Occupant behaviour in naturally ventilated and hybrid buildings', *ASHRAE Trans.*, 115(1).

Wang C, Yan D and Jiang Y (2011) 'A novel approach for building occupancy simulation', *Building Simulation*, 4(2), pp. 149–167.

Yan D, O'Brien W, Hong T, *et al.* (2015) 'Occupant behavior modeling for building performance simulation: current state and future challenges', *Energy and Buildings*, 107, pp. 264–278.

Yan D, Hong T, Dong C, *et al.* (2017) 'IEA EBC Annex 66: definition and simulation of occupant behavior in buildings', *Energy and Buildings*, 156, pp. 258–270.

Yun G Y and Steemers K (2008) 'Time-dependent occupant behaviour models of window control in summer', *Building and Environment*, 43(9), pp. 1471–1482.

Zhang Z and Barrett P (2012) 'Factors influencing the occupants' window opening behaviour in a naturally ventilated office building', *Building and Environment*, 50, pp. 125–134.

Chapter 12

Virtual world to reality

This chapter considers the use of BPS+ within building upgrade schemes and the subsequent extraction of data aspect models in formats that support the off-site manufacture of building components. It also addresses the use of BPS+ in support of real-time control and facilities management, and as a low-cost replacement for physical prototype testing. Finally, it speculates on the future role of artificial intelligence and the possibility of design appraisal in near real time. The ultimate aim is to support an extension of the application capability of BPS+ in both scale and depth (Clarke 2015).

12.1 Upgrade plan quality assurance

Societies are confronted by myriad challenges in relation to the transition to cleaner, greener energy solutions. A core challenge is to reduce the energy use of existing buildings many of which are sub-standard. Taking Glasgow as an example of the challenge, the city has approximately 314,500 dwellings of which around 107,000 (34%) are socially owned and, of these, 44,000 (42%) have no wall insulation and are in the hard-to-heat category. In a typical year, Glasgow City Council in partnership with local Housing Associations will renovate around 800 ± 200 dwellings (PC 2021). This implies that it would take around 30 years to renovate the dwellings in the hard-to-heat category alone. The question is how the upgrade rate might be improved.

The European Commission funded INDU-ZERO[1] project described here was focused on the North Sea Region, which is home to approximately 60 million inhabitants spread across several European territories. A large number of outdated dwellings, some 22 million built between 1950 and 1985, possess poor insulation and high energy demands. The project set out to develop a blueprint for a production factory capable of producing 15,000 dwelling renovation packages per year at a reasonable cost. The project consortium comprised 14 partners from The Netherlands, Belgium, Germany, the United Kingdom, Norway, and Sweden.

In practice, the retrofit rollout is complicated by the need to accommodate building- and site-specific needs. To address this issue, the Energy System Research Unit's (ESRU) role in the project was to develop an upgrade assessment procedure based on high-resolution BPS+ modelling and deliver digital part models as required by factory

[1]https://northsearegion.eu/indu-zero/

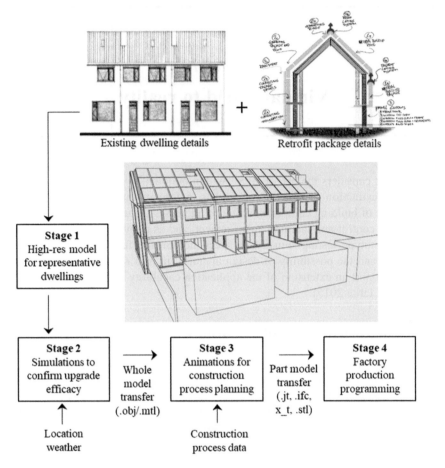

Existing dwelling details   Retrofit package details

Stage 1
High-res model
for representative
dwellings

Stage 2
Simulations to
confirm upgrade
efficacy

Whole
model
transfer
(.obj/.mtl)

Stage 3
Animations for
construction
process planning

Part model
transfer
(.jt, .ifc,
x_t, .stl)

Stage 4
Factory
production
programming

Location
weather

Construction
process data

Figure 12.1   The INDU-ZERO renovation-package data chain

production equipment. Figure 12.1 summarises the four-stage data chain underpinning the off-site renovation-package selection, appraisal, and production procedure.

Stage 1

Two high-resolution digital models are established for representative dwellings within an estate due for an upgrade: one corresponds to the existing dwellings, and the other with the INDU-ZERO renovation package applied[2]. The information required to establish a high-resolution dwelling model is extensive:

- 3D geometry;
- hygro-thermal properties of construction materials;

---

[2]In a competitive market alternative renovation packages would exist for all building archetypes.

- details on thermal bridges, air leakage distribution, and internal thermal mass (e.g., furnishings and fittings);
- operational data relating to occupancy, lighting, and small power usage;
- definitions of heating, ventilation, domestic hot water, and embedded new and renewable energy components; and
- weather time series.

Whilst such data are often readily available, they usually are collated from disparate sources such as construction drawings, site surveys, manufacturers' data, building standards, city cadastres, and existing computer models in a variety of formats (progressively a building information model (BIM) in future). To reiterate a point made previously, this information collation exercise is required only for the small number of dwellings deemed representative of the targeted estate. The BPS+ model definition effort is therefore modest compared with the resources being directed to the upgrade overall. In addition, it may be expected that pre-constructed models will be more widely available in future, as the BIM standard evolves and new buildings replace old.

Stage 2

The Stage 1 models are subjected to performance simulations in a manner that stress tests the upgrade proposal, with outcomes used to compare the efficacy of alternative upgrade options and guide the selection of the features that offer an acceptable cost-performance solution. Within the project, ESP-r was used to assess upgrade combinations as proposed by partners.

Stage 3

Once the efficacy of a proposed upgrade has been confirmed, the ESP-r high-resolution digital model is transformed to <.obj/.mtl> format and passed to the Blender application[3] where the planned retrofit can be animated and site activities motion-tracked to confirm constructability.

Stage 4

Assuming the outcomes from Stages 2 and 3 are satisfactory, discrete model parts may be extracted from Blender and various file formats passed to the manufacturing facility to support production equipment programming.

To confirm robustness, the procedure was applied to three typical but hypothetical dwelling types built in the period 1965–67 (Figure 12.2, upper) and, separately, to three refurbished, monitored dwellings located at Enschede in The Netherlands (Figure 12.2, lower).

The hypothetical dwelling models related to before and after the application of the INDU-ZERO upgrade package as summarised in Figure 12.3 in relation to fabric components.

Figure 12.4 summarises the three technical systems available for selection. These included various combinations of solar thermal (ST) panels, photovoltaic

---

[3]https://blender.org/

Terraced   Semi-detached   Apartments

Before   After

Figure 12.2   Hypothetical (upper) and existing (lower) dwellings

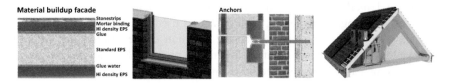

Figure 12.3   Fabric components of the INDU-ZERO upgrade package

(PV) panels, mechanical ventilation with heat recovery (MVHR), air-source heat pumps (ASHP), and electric boilers (EB). The objective was to assess the contribution of an upgrade to the attainment of net-zero carbon performance.

With the real dwellings, where the upgrade corresponded to an early version of the INDU-ZERO fabric upgrades[4], only post-upgrade simulations were undertaken to facilitate a comparison between ESP-r predictions and field measurements to establish the veracity of the former.

The following assessment results correspond to the application of different upgrades to one of the demonstration dwellings located at Enschede. Annual simulations were undertaken for Amsterdam, Glasgow, and Oslo covering five weather years per location: one similar to regulatory standards plus four representing ± variations; the following results correspond to the average weather condition. To ensure realism in the representation of occupant behaviour, stochastic annual profiles for lighting, appliance, cooking, hot water, and heating system use were generated via the OccDem application[5] and typical profiles selected matched to regulatory standards (e.g., the Standard Assessment Procedure (SAP) for the UK

[4]https://rcpanels.nl/
[5]https://www.esru.strath.ac/applications/

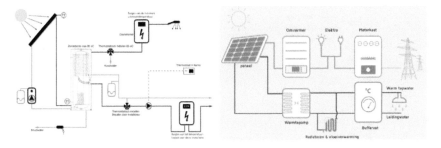

3 ST, 27 PV, EB, and MVHR applied to all   8 PVT, 16 PV, ASHP, and MVHR
hypotheticals and Enschede No. 28   applied to Enschede No. 27

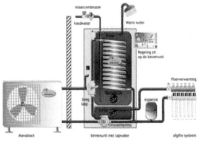

8 PVT, 16 PV, ASHP, and MVHR applied to Enschede No. 27

Figure 12.4   INDU-ZERO technical systems and dwelling assignments

climate). Figure 12.5 shows an outcome for lighting usage equating to 4.9 kWh/m^2.y (IQR = interquartile range).

For each dwelling type (terraced, semi-detached, and apartment), four variants were analysed: (1) pre-renovation; (2) with the renovation package applied; (3) as (2) with additional air tightness and MVHR; and (4) as (3) with additional under-floor insulation[6]. The simulation procedure was automated to enable application by individuals with no specialist BPS+ knowledge, thus helping to propagate the approach in future.

Table 12.1 summarises the annual space heating energy requirements for direct electric (DE) and ASHP, and for each dwelling type, upgrade variant, and climate location.

Table 12.2 gives the corresponding net-at-the-meter annual energy data, which is defined as the summation of the energy required for space heating, lighting, appliances, cooking, and hot water minus the photovoltaic supply. The corresponding annual energy supply predictions for the 27 PV panels (with an installed capacity of 9 kWp), when normalised to the dwelling floor area, are 59.6 kWh/m^2.y (Amsterdam), 45.3 kWh/m^2.y (Glasgow), and 49.4 kWh/m^2. y (Oslo).

---

[6]Possible only with suspended floors with access.

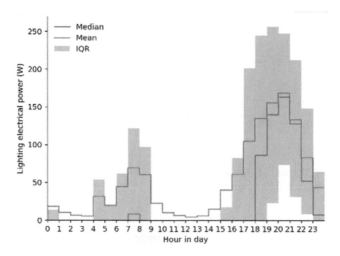

Figure 12.5   Identification of a typical lighting profile

Table 12.1   ESP-r annual heating energy demand ($kWh/m^2.y$)

Archetype/upgrade	Amsterdam		Glasgow		Oslo	
	DE	**ASHP**	**DE**	**ASHP**	**DE**	**ASHP**
1. Terrace						
Pre-upgrade	120.0	–	126.0	–	171.0	–
INDU-ZERO panels upgrade	52.4	30.4	54.6	32.6	74.0	49.2
+ airtightness & MVHR	36.1	19.5	33.5	19.7	43.9	26.5
+ underfloor insulation	4.4	3.5	4.1	3.3	11.2	7.8
2. Semi-detached						
Pre-upgrade	159.2	–	174.3	–	208.5	–
INDU-ZERO panels upgrade	65.1	38.7	73.9	45.5	81.4	53.5
+ airtightness & MVHR	35.6	20.2	38.5	19.9	47.7	28.4
+ underfloor insulation	6.2	4.0	6.0	3.8	13.9	8.4
3. Apartment						
Pre-upgrade	82.5	–	90.2	–	119.1	–
INDU-ZERO panels upgrade	23.9	12.9	27.3	14.6	35.1	19.9
+ airtightness & MVHR	7.5	4.5	8.3	4.7	13.2	7.9

In summary, the upgrade makes a significant contribution towards the attainment of net-zero energy performance although further progress towards that goal will require attention to the non-heating loads (including electric vehicle charging in future). For example, removing the PV contribution of 45.3 $kWh/m^2.y$ in the case of a Glasgow terrace dwelling with INDU-ZERO panels, air tightness, and MVHR applied would increase the results in Table 12.2 to 92.5 $kWh/m^2.y$ and 74.4 $kWh/m^2.y$, respectively, which contrasts favourably with the current UK average of around 165 $kWh/m^2.y$.

Table 12.2   ESP-r net-at-the-meter energy (kWh/m².y)

Archetype/upgrade	Amsterdam		Glasgow		Oslo	
	DE	**ASHP**	**DE**	**ASHP**	**DE**	**ASHP**
1. Terrace						
Pre-upgrade	120.0	–	126.0	–	171.0	–
INDU-ZERO panels upgrade	47.3	25.5	63.8	41.8	79.1	54.5
+ airtightness & MHVR	31.0	14.3	**42.7**	**29.1**	49.0	31.6
+ underfloor insulation	−0.7	−1.5	13.3	12.5	16.3	13.0
2. Semi-detached						
Pre-upgrade	159.2	–	174.3	–	208.5	–
INDU-ZERO panels upgrade	60.0	33.8	83.1	54.7	85.5	58.6
+ airtightness & MHVR	30.5	15.1	47.7	29.1	52.8	33.7
+ underfloor insulation	1.1	−1.1	15.2	12.9	19.0	13.7
3. Apartment						
Pre-upgrade	82.5	–	90.2	–	119.1	–
INDU-ZERO panels upgrade	78.4	67.4	81.8	69.1	89.6	74.4
+ airtightness and MHVR	62.0	54.5	62.8	59.2	67.7	62.4

Overall, the following observations were made.

- The INDU-ZERO upgrade package would result in a reduction in DE space heating demand of approximately 57%, 60%, and 70% for the terrace, semi-detached, and apartment cases, respectively (under the three climate conditions studied). Where an ASHP is deployed, the corresponding reductions in heating demand rise to approximately 74%, 75%, and 84%. That is, the upgrade makes a significant contribution to space heating energy reduction.
- With additional airtightness and MVHR added to the upgrade, the corresponding figures equate to a 73%, 78%, and 90% reduction (for DE) and 84% and 87% reduction (for ASHP; there is no ASHP in apartments). That is, the upgrade facilitated good progress towards net-zero energy.
- With underfloor insulation applied where applicable, the corresponding figures are 95% and 96% (DE), and 96% and 97% (ASHP), that is, the upgrade results in near net-zero energy for the terrace and semi-detached cases, and especially so in the Amsterdam climate.

The following section continues with Stages 3 and 4 of the INDU-ZERO data chain shown in Figure 12.1.

12.2   Supporting off-site manufacture

A high-resolution BPS+ model can support the product design and production processes. The following example is taken from the INDU-ZERO project (as described in the previous section) to illuminate the downstream use of extracted data. The process commences with the exporting of relevant portions of the ESP-r high-resolution model to Blender – wall and roof panels, windows, and connecting mechanisms. This export adheres to the Wavefront Object (.obj) and Material

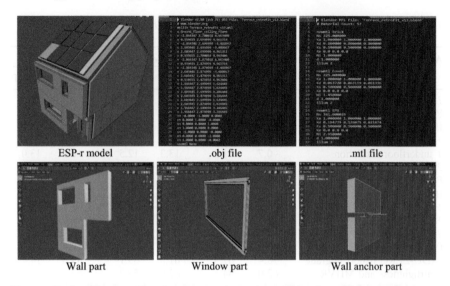

ESP-r model	.obj file	.mtl file
Wall part	Window part	Wall anchor part

Figure 12.6   ESP-r-to-Blender exports (upper) and Blender-to-factory part exports

Template Library (.mtl) file formats[7] as depicted in Figure 12.6 (upper). Blender is then used to map these digital models to a variety of file formats (.jt, .ifc, x_t, .stl, etc.)[8] as required by the robotic entities and manufacturing procedures deployed in the INDU-ZERO off-site production factory (Figure 12.6 lower).

The process enables objects of any size to be morphed and scaled to the required dimension via the .obj file structure. The thermal properties of each object are contained within the .mtl file which, when coupled with digital models of a target dwelling, can be extracted to simulate building energy performance using preferred applications. If the upgraded building's energy performance is considered adequate, collision detection procedures can be initiated ensuring the new objects can be physically installed onto the existing building without issue.

The files exported from Blender are utilised within the factory to establish instruction-sets to control the automated manufacturing process – as depicted in Figure 12.7 for the case of a robotic arm where a delivered industry foundation classes (IFC) file is further processed as indicated.

12.3   Smart facilities management

The diffusion of smart devices into the built environment is creating new opportunities for the monitoring and control of energy. Examples of the types of data that

---

[7]https://paulbourke.net/dataformats/
[8]https://okino.com/solutions/parasolid.htm/

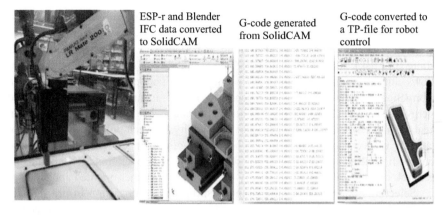

Figure 12.7 Robot programming (credit: Struck 2017)

may be made widely available to underpin new commercial services include energy usage by appliance; equipment condition monitoring; indoor air quality, space temperatures, humidity, and comfort conditions; and the presence of pollutants such as smoke and CO. The widespread adoption of such 'e-services' has the potential to substantially improve the cost, performance, and reliability of estate management systems. Further, it may be anticipated that the direct linking of professionals and citizens, through e-services, will assist society to find a better balance between energy efficiency, behavioural change, new energy systems deployment, and regional/community economic benefit. BPS+ has an important role to play in generating e-service content.

At the regional level, aggregate energy consumption data may be used to formulate energy action plans or to support national policy (e.g., the targeting of energy efficiency action plans). These data may also be used to support high-level, time-critical control actions to facilitate the best use of available new and renewable energy resources. The availability of multi-frequency energy usage information will allow utilities to make informed business decisions on the development of new renewable generation capacity. Detailed demand information can also assist utilities to plan for future changes to electricity supply infrastructure, such as the evolution of microgrids.

At the community level, e-services allow bi-directional interactions, facilitating both the collection of data and the implementation of control actions based on these data. Options that may be enabled include the ability to manipulate large numbers of local small power, heating, hot water, and cooling loads to facilitate the deployment of renewable energy sources throughout the built environment. Community-scale e-services also provide a means to drive down energy demands through the implementation of predictive, weather-responsive control of heating and cooling systems. E-services at this level can also ensure acceptable environmental conditions for vulnerable citizens by alerting local care bodies to potential health problems relating to domestic energy usage: low internal temperatures (hypothermia) or high humidity levels (condensation and mould growth). In addition, embedded sensors can perform a valuable safety role in detecting and warning building occupants and relevant authorities of abnormal circumstances: the presence of smoke, or the malfunctioning of boilers (CO risk), or electrical goods failure (fire risk). At the domestic level, the availability of energy consumption data allows citizens to take control of their energy usage. Examples of the types of useful information that may be made generally available include community-averaged energy consumption, time histories of household energy use, and energy consumption breakdowns. Evidence exists that the availability of such information can substantially reduce energy consumption by encouraging competition and behavioural change. It is also possible to provide data on appliance operation, giving valuable feedback to the development of energy-efficient and customer-oriented appliances.

Selecting appropriate solutions and ensuring robust operation in practice are tasks well-served by BPS+. Further, by allowing disparate organisations to access the same information resource, the e-service approach to building control and performance monitoring will nurture inter-disciplinarity. This will serve the needs of sustainable development because it will bring together the different stakeholders' viewpoints and so encourage the innovative developments that exist at the interface between the disciplines.

Figure 12.8 depicts an e-service platform as established within an Engineering and Physical Sciences Research Council (EPSRC) funded project (EPSRC 2010) based on the EnTrak and ESP-r programs, and BuildAx hardware[9].

The BuildAX platform comprises monitoring devices corresponding to environmental conditions, occupancy states, and power usage, all linked wirelessly to a logger/router from where data are fetched at the frequency required to enact information services tailored to the needs of specific users. Also shown are the sensor specifications.

For each e-service, BuildAX field sensors comprising the following components deliver relevant data. Figure 12.9 shows the LRS and ENV sensors.

LRS – a logger/router to store locally and transmit monitored data to a remote location where it may be utilised by EnTrak. The device also acts as a Web server.
ENV – for the monitoring of indoor environmental conditions, including temperature, relative humidity, illuminance, movement, and surface contact (e.g., door opening).
CO₂ – for the monitoring of CO_2 concentration.
GAS – for the monitoring of gas consumption.
PWR – for the monitoring of electricity consumption.
PSW – a remotely controllable electrical switch.

An example EnTrak e-service definition utilising two ENV sensors is shown in Figure 12.10; the lower image is the operational outcome. Here, time-series data relating to light level are scrutinised and an alert is issued when a low level is detected (e.g., due to lamp failure).

As summarised in Figure 12.11, the LRS receives data from paired sensors and fetches requests from EnTrak via a shell script (Enget), which leverages the Linux

---

[9]https://www.esru.strath.ac.uk/applications/

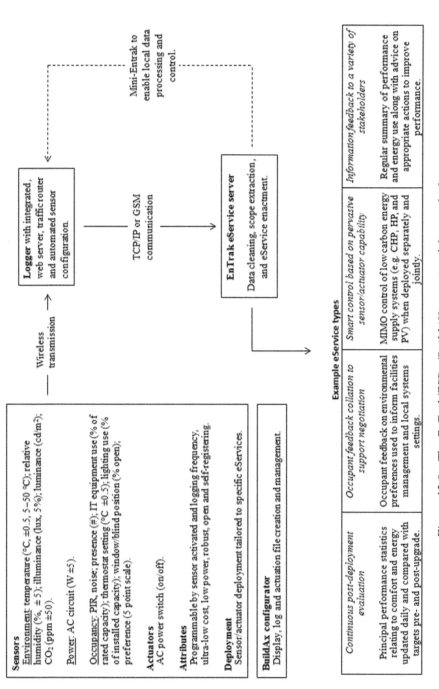

Sensors
Environment: temperature (°C, ±0.5, 5–50 °C); relative humidity (%, ±5); illuminance (lux, 5%); luminance (cd/m²); CO_2 (ppm ±50).

Power: AC circuit (W ±5).

Occupancy: PIR, noise; presence (#); IT equipment use (% of rated capacity); thermostat setting (°C ±0.5); lighting use (% of installed capacity); window/blind position (% open); preference (5 point scale).

Actuators
AC power switch (on/off).

Attributes
Programmable by sensor activated and logging frequency, ultra-low cost, low power, robust, open and self-registering.

Deployment
Sensor/actuator deployment tailored to specific eServices.

BuildAx configurator
Display, log and actuation file creation and management.

Logger with integrated, web server, traffic router and automated sensor configuration.

— Wireless transmission →

Mini-Entrak to enable local data processing and control.

TCP/IP or GSM communication

EnTrak eService server
Data cleaning, scope extraction, and eService enactment.

Example eService types

Continuous post-deployment evaluation	Occupant feedback collation to support negotiation	Smart control based on pervasive sensor/actuator capability	Information/feedback to a variety of stakeholders
Principal performance statistics relating to comfort and energy updated daily and compared with targets pre- and post-upgrade.	Occupant feedback on environmental preferences used to inform facilities management and local systems settings.	MIMO control of low carbon energy supply systems (e.g. CHP, HP, and PV) when deployed separately and jointly.	Regular summary of performance and energy use along with advice on appropriate actions to improve performance.

Figure 12.8   The EnTrak/ESP-r/BuildAX e-service delivery platform

Figure 12.9   BuildAX LRS (left) and ENV sensor

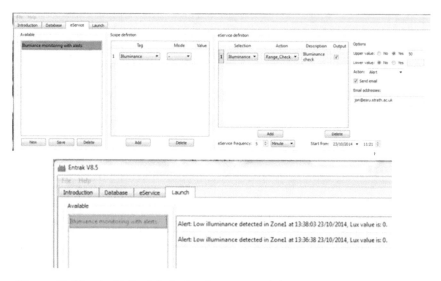

Figure 12.10   An EnTrak light-level e-service definition (upper) and outcome

Wget protocol, at a time frequency associated with the e-service. EnTrak may be located within the same or different network alongside an SQL database associated with the e-service.

The scripting facilities, command shell, and applications found on Linux automate the data transfers. These tools are also available on Windows computers via a lightweight toolset named MinGW[10] (Minimalist GNU for Windows). This

[10]https://Mingw-w64.org/

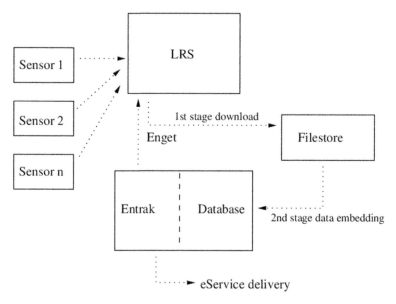

Figure 12.11   The e-service communication paths

allows for the deployment of common scripts for data-gathering tasks across a variety of computer platforms and operating systems. Depending on the requirements of the e-service, sensors and LRS devices are deployed by pairing groups of the former with one of the latter.

The definition of an e-service follows a 3-stage procedure. First, a database is populated as shown in Figure 12.12 for the case of a simple 2-sensor deployment.

Here, the sensors each have one static attribute and five dynamic attributes, with the latter based on online data capture from a matched BuildAX deployment. In this way, deployments of arbitrary complexity are defined, including data capture from different locations at variable frequencies.

Second, the e-service content is defined by scoping on the entity attributes as required and defining the actions to be applied to the data returns when the e-service is running (Figure 12.10, upper). In this way, separate e-services may relate to the same database.

Last, the required e-service is launched and the output is directed to the display type required, for example, as shown in Figure 12.13 for the case of a real-time monitoring e-service.

The power of the e-service approach can be enhanced in several ways by deploying BPS+. For example, the field data captured by BuildAX sensors can be replaced by simulated data corresponding to alternative approaches to facilities management. In addition, an e-service can be deepened by utilising monitored data to calibrate a BPS+ input model before use to generate higher-order performance data such as the distribution of glare and the mean age of air.

Figure 12.12   Defining the database underpinning an e-service

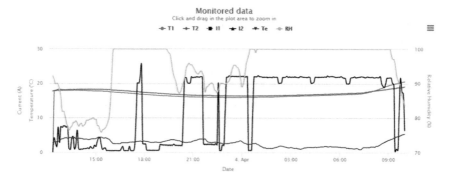

Figure 12.13   Output delivered by an online monitoring e-service

12.4   Creation of physical prototypes from BPS+ model

BPS+ tools require a superset data model to assess built environment performance in a realistic manner. This enables them to export part models to other tools and thereby support non-energy assessments (e.g., as illustrated in Section 12.2). There is a rich history of physical modelling in building design with 3D mock-ups used to communicate complex building forms. Whilst users of building simulation tools are usually presented with wireframe images, it is possible to leverage composition and contiguity information to create 3D solid representations of projects. What remains is to bundle this enhanced information into a form digestible by 3D printers, which are driven by low-level instructions.

The project described here (Hand 2023) explored this possibility in relation to structural work applied to a portion of a historic building constructed circa 1855. As shown in Figure 12.14, the stone facade and internal walls (brick plus plaster) are characteristically massive. The floor structure had sagged and the sound deafening had developed faults. Ensuring that remedial work is properly carried out would be greatly helped by physical mock-ups.

An ESP-r model already existed for energy and acoustic analysis so the task was to enhance this to represent the 3D nature of the facade and floor. Figure 12.9 (left) shows the usual wireframe presentation of the model where polygons follow the inside face of the external facade and internal structures. A simple extrusion of model polygons gets only part of the way to the 3D characteristics of the facade, which also needs fittings to communicate scale and placement. Figure 12.15 (right) has these 3D entities in place. From this, it is possible to produce useful renderings and feed the production of a physical prototype.

To support the remedial work, it was necessary to delineate the internal structures and acoustic materials as shown in wireframe and solid representations in Figure 12.16. These additions do not affect the thermal characteristics of the model. The quality assurance (QA) implications are obvious, especially where the composition highlights inconsistencies.

Creating physical mock-ups requires taking into account the constraints of 3D printing, including avoiding printing a solid over the air and the minimal resolution of the printing engine. For the overall model, the first constraint can be dealt with if the print is sliced at the lower face of the ceiling and the apartment is printed in two parts as indicated in Figure 12.17.

Figure 12.14   Historic building facade and floor structure

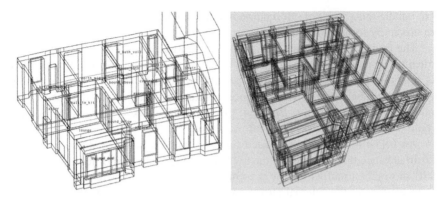

Figure 12.15   Existing thermal simulation model (left) and with 3D facets expressed

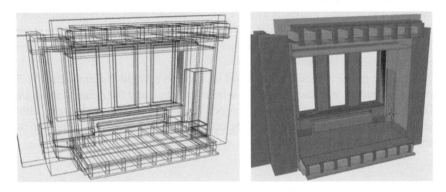

Figure 12.16   Focused model with structure and acoustic treatment added

Figure 12.17   Printing the overall model (left) and the finished prototype

Printing the whole apartment loses fine detail (the plaster and flooring were below the resolution of the print), so a model variant can be created with the thickness of plaster and the flooring slightly thickened. For a focused facade model, the prototype resolution is sufficient to identify construction details associated with the bay window as shown in Figure 12.18.

The image on the right is a close-up of the ceiling structure with a traditional acoustic treatment applied (ceiling plaster/lath, air space, thin boards with ash above, another air space, and the flooring of the apartment above).

The teaching and QA implications of physical mock-ups are significant. Imagine focusing on the junctions of a building as in Figure 12.19. The chaos that

Figure 12.18   Physical prototype with structural and acoustic details added

Figure 12.19   Details clarified by physical prototypes

would later be evident on site but not noticed in sectional representations might be avoided.

The complexities of this traditional facade are no less than the complexities of modern facades and point to what could be a new normal for BPS+: true 3D representations of buildings further reducing the abstraction of numerical models. The descriptive rigour needed is balanced by the clarity delivered. The default treatment of heat conduction in 1D might then evolve to something more fit for tracking the complex heat flows within high-performance facades.

12.5 Energy-efficient shipping

Given the first principles nature of BPS+, it can be applied to artefacts other than buildings. From a thermodynamic viewpoint, there is little difference between a building and a ship. Application of BPS+ will be straightforward as long as the tool supports the temporal variation of principle parameters. For example, it will be necessary to change the ship's position and orientation over time as it traverses waypoints. Likewise, weather files will need to be substituted to represent different climate zones and sea temperature states as the journey progresses.

One project (Dodworth *et al.* 2009) utilised ESP-r to investigate novel heating, ventilation, and air conditioning (HVAC) configurations and controls that would serve the requirements of cruise liners when following specific routes. The aim, in response to energy-related regulation, was to optimise ship design to achieve CO_2 emission reduction whilst maintaining passenger comfort. The challenge was to model complex geometries, vessel navigation, and variable occupant requirements. A significant consideration was maximising solar access for recreational areas. Multi-zone models were established as shown in Figure 12.20, and ESP-r's temporal definition feature was used to adjust model parameters (weather, sea reflectivity, space usage schedules) as required.

Scenario simulations were conducted to determine the impact of design change scenarios on energy use and comfort levels as follows.

- Reference case – minimum volume flow rate 40% of maximum fan speed, temperature set point of 22 °C.
- S1 – minimum fan speed reduced to 30% of maximum.

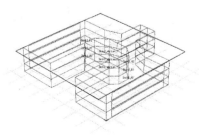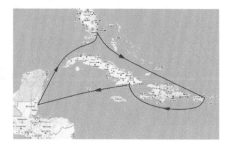

Figure 12.20   A 3-deck ship portion attributed to reflect cruise parameters

- S2 – energy-saving period between 12:00 and 18:00 when temperatures 'float'.
- S3 – thickness of the insulating material of the outside walls was reduced by 50%.

Figure 12.21 summarises the predicted energy savings and impact on comfort for the best scenario.

The result in this particular case indicated that reducing the fan speed had the largest impact on energy use and an acceptable impact on comfort.

12.6   Virtual car test facility

The project summarised here (Kelly *et al.* 2003) utilised ESP-r to appraise design options for a planned transport test facility at the University of Liege within which the internal climate could be closely controlled, including the creation of fog of various densities for the testing of automotive collision avoidance equipment under conditions of degrading visibility. Options to be considered included construction materials, HVAC system regulation, and injection nozzle placement.

At project commencement, ESP-r's computational fluid dynamics (CFD) module was modified to include consideration of the distribution and concentration of small water droplets suspended in the air as occurs within fog (Kelly and Macdonald 2004). This entailed adding a droplet concentration equation to the CFD domain against the assumption that the droplets move with the surrounding air (a zero-slip condition). Figure 12.22 shows a prediction of droplet concentration where fog is produced by two sprinkler nozzles.

To build confidence in the approach, a model of an existing test facility at the Laboratoire Regional des Ponts et Chausées (LRPC) was developed and calibrated against measurements. Figure 12.23 shows the model and an example of the level of agreement eventually obtained.

A second development involved the refinement of the existing ESP-r driver for Radiance to include the properties of different types of fog (liquid water content and droplet size) to enable the visualisation of non-homogeneous fog conditions. The required new data included the fog albedo, extinction coefficient, and scattering eccentricity (defining the backward light scattering). Figure 12.24 shows an example of a Radiance output with fog scattering (Rushmeir 1994) based on data passed from ESP-r.

An ESP-r model of the planned test facility was then created as shown in Figure 12.25. This model comprised a central test chamber surrounded by a conditioned buffer space and utility zones (office, roof space, plant room). An HVAC network was included to deliver heating, cooling, humidification, and de-humidification. An airflow network was added to facilitate the modelling of wind-induced airflows between zones and through interaction with the HVAC network. The test chamber comprised a CFD domain with an array of fog production nozzles.

The model was used to explore approaches to the maintenance of stable temperatures and high humidity (the conditions amenable to the production of fog) as shown in Figure 12.26:

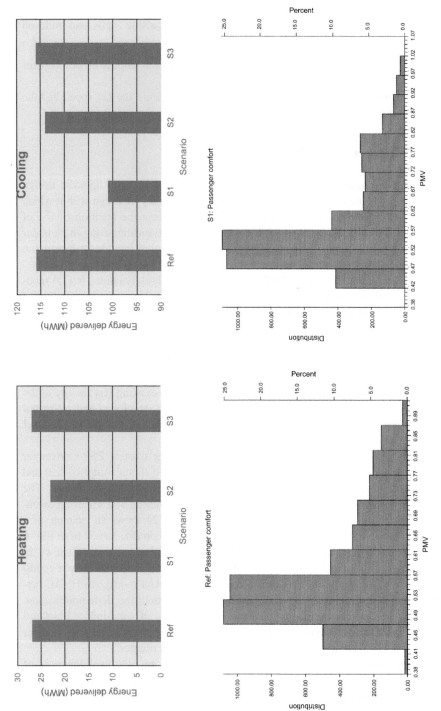

Figure 12.21   Impact of design changes

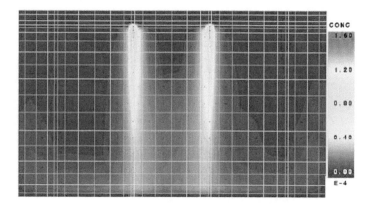

Figure 12.22   ESP-r prediction of droplet concentration

- direct control of temperature and humidity inside the chamber;
- direct cooling of the test room air to the dew point temperature;
- use of underfloor heating/cooling to maintain stable temperatures; and
- indirect control of test room temperatures through conditioning of the buffer space.

It was concluded that direct control is the most effective means of achieving stable conditions with low-energy expenditure. Overall, it was demonstrated that creating stable fog conditions would require a 4-hour preheating/cooling period within the chamber and a 2-hour period within the buffer zone. In addition, air would require to be supplied to the chamber from the HVAC system at a low flow rate of 350 m^3/h to maintain fog homogeneity. After some iteration, it was possible to devise a chamber control strategy that would maintain stable fog conditions.

As it existed at the end of the project, the ESP-r/Radiance pairing was capable of implementing different automotive test scenarios within a virtual fog chamber. A significant application would be the assessment of headlamp performance and laser-based safety equipment as supported by simulation results corresponding to cases such as illustrated in Figure 12.27.

In this way, BPS+ can be utilised as a virtual test bed to allow car safety testing under different weather conditions.

12.7   Big data

New technologies and systems are routinely mooted as potential solutions for low-energy/carbon cities. Examples include innovative insulation products, advanced glazing, context-aware control, combined heat and power plant, heat pumps, solar thermal/electric systems, fuel cells, urban wind power, smart lighting, smart grids, and biomass/district heating. Given the complexity of the problem domain, it is

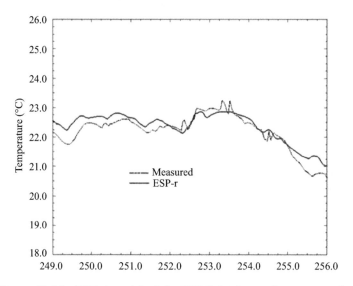

Figure 12.23   ESP-r model of the LRPC facility and agreement level

unlikely that fiscal measures alone will bring about solutions comprising effective blends. This notion gives rise to two aphorisms.

- If a proposal is not simulated at the design stage, then it is unlikely to deliver the required performance when built.
- If post-occupancy performance is not routinely monitored, then the gap between operational performance and design intent will grow.

These issues could be tackled by a data-centric approach that blends virtual and real data, making available information that may be acted upon by interested stake-holders, including designers, planners, property managers, and citizens. This approach has been explored in several research projects, including:

Figure 12.24   A car entering a fog bank

- Hit2Gap[11] funded under the EC's Horizon 2020 R&D program;
- SmartHomes projects funded by the EC and the Technology Strategy Board (Clarke 2001, Doran 2008); and
- EnTrak project funded by EPSRC (Clarke and Hand 2015).

Figure 12.28 summarises the approach when applied at the city scale.

Data are collected from estate monitoring devices – smart meters, weather stations, and pervasive environmental sensors – and used to quantify multi-variate estate performance. Relevant performance information is then delivered at a suitable frequency to stakeholders in appropriate formats, for example, spatial maps depicting clean technology deployment opportunities at the city level, and timely advisories to building operators. To support action planning, BPS+ scenario simulations are undertaken to determine the impact of interventions such as property upgrades, demand management/response initiatives, or the introduction of a disruptive technology such as electric vehicles.

As depicted in Figure 12.29, a feature of BPS+ is its ability to generate disaggregated demand profiles at high resolution (per property, technical system, or connected estate, etc.) for building stock models generated automatically on the basis of rules derived from surveys (Clarke *et al.* 2004) (see Section 7.7).

The approach respects the underlying thermodynamic complexity, links energy use to wider issues such as comfort and air quality, enables life cycle assessment, and accommodates uncertainty – all whilst providing an insight into multi-variate performance. Whilst traditional building energy management systems (BEMS) are able to provide a portion of the required estate information, it is unlikely that this will be complete in several important respects. Because the focus will be on HVAC state measurement and control, issues such as occupancy presence and behaviour,

[11]https://cordis.europa.eu/project/id/680708

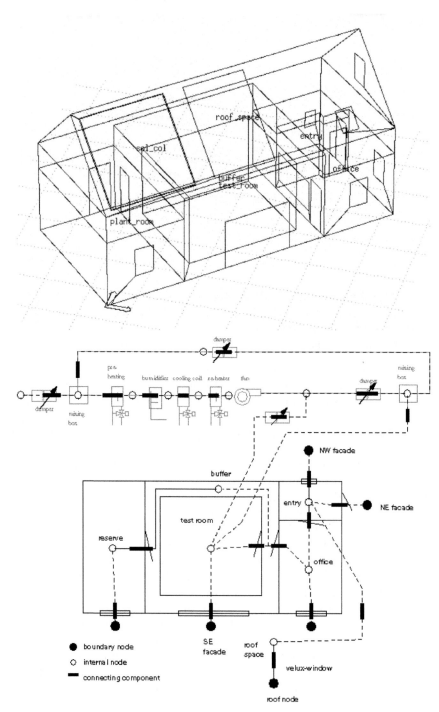

Figure 12.25   ESP-r building, HVAC, and airflow model of the proposed test facility

Figure 12.26 Test chamber before/after non-homogeneous fog production

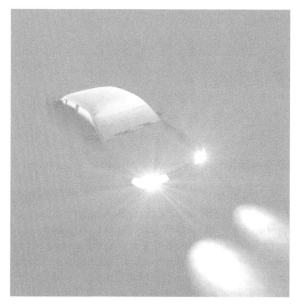

Figure 12.27 Radiance depiction of a car entering a fog bank

the spatial distribution of indoor conditions, disaggregation of load profiles, and local weather will typically be absent. It is for these reasons that the low-cost BuildAX[12] monitoring system (Figure 12.9) was developed within a project funded

---

[12]https://www.esru.strath.ac.uk/applications and https://digitalinteraction.github.io/openmovement/buildax/site/

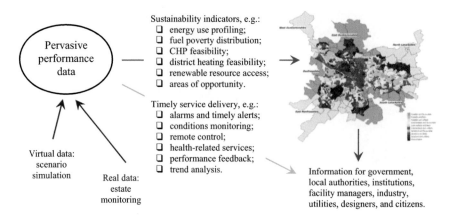

Figure 12.28   A data-centred approach to city energy management

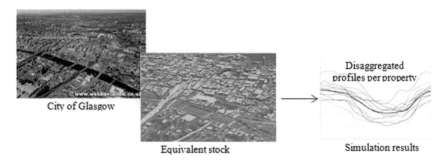

Figure 12.29   Disaggregated load profiles generated from a building stock model

by the UK Engineering and Physical Sciences Research Council (Clarke and Hand 2015). This comprises a logger/router/server (left) fed by distributed multi-sensors.

A multi-sensor integrates sensors for temperature, relative humidity, movement, illuminance, contact (e.g. door/window opening), and battery state. These data are broadcast to the logger wirelessly at 2.4 GHz from whence they may be collected by remote agents as described below. The logger encapsulates a Web server that enables immediate display of the monitored data as shown in Figure 12.30 for the case of a deployment of six sensors in an office.

Whether real or virtual, the collected data are transformed to useful information. This requires the imposition of data processing rules that depend on the service being enacted. This transformation is performed by the EnTrak system (Clarke *et al.* 2014) via a three-stage process as follows.

As shown in Figure 12.31, Stage 1 involves the formal definition of the entities to be monitored – here buildings on the Strathclyde University campus. In another application, an entity might be a utility meter, a vehicle, a plant component, and so on.

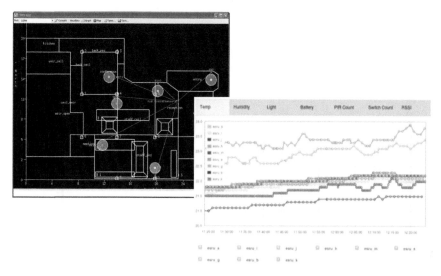

Figure 12.30   BuildAX data superimposed on a plan view and a graph of environmental data

Entities are defined in terms of descriptive and time-series attributes, where each attribute is a tuple comprising a tag/value pair. This attribution is restricted to only those data required to enact the targeted service, i.e., EnTrak is not a general-purpose database management system. Each time-series attribute has an associated data collection definition, such as by file exchange with a remote server or by the direct querying of monitoring devices deployed in the field such as a BuildAX logger. The required fetch frequency is specified per attribute and a test connection is made; later, usually after completion of Stage 2, the overall monitoring scheme is commenced with all data stored in a MySQL[13] database.

In Stage 2, services are established by associating actions with all or part of the entity attribute schema as required. For example, in the upper part of Figure 12.32, an operational Energy Performance Certificate (EPC) has been defined by applying a set of actions to electricity and gas meter readings, whilst in the lower portion, a high-temperature alert is defined by range-checking all dynamic attributes with tag 'Temperature' and value 'Lecture Hall'.

In Stage 3, individual services are started and run at the required frequency (e.g., monthly for the EPC service, 5 minutely for the temperature alert service). This results in the repetitive application of the stage rules to the incoming monitored/simulated data until the service is stopped. As shown in Figure 12.33, the outcome is delivered as an XML file to support alternative delivery formats, styles,

---

[13]https://mysql.com

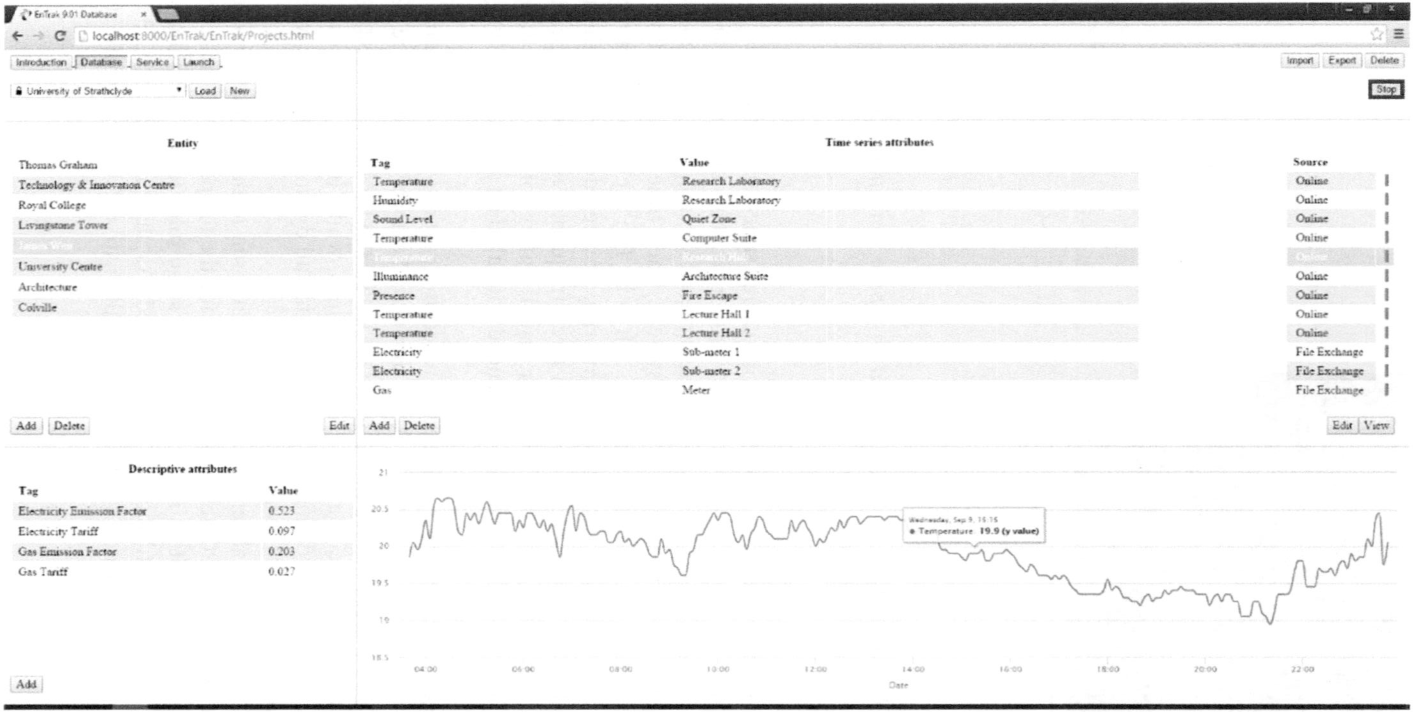

Figure 12.31   Entity definition in EnTrak

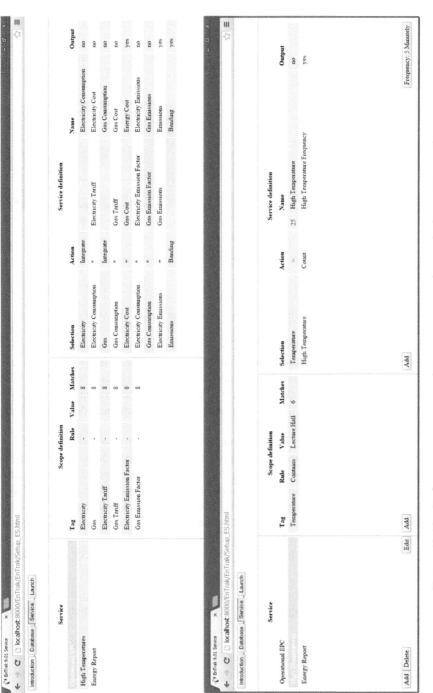

Figure 12.32   Defining e-services via data processing rules applied to entity attributes

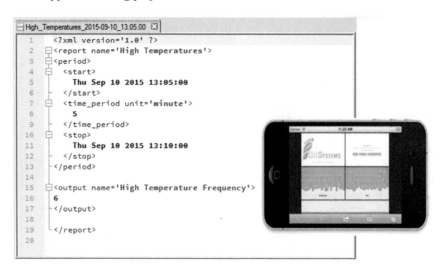

Figure 12.33   A service outcome example

and devices. In the example shown here, the final delivery platform is a smartphone app.

To support 'what-if' studies, it is possible to replace the incoming data from field monitoring with prediction time series emanating from BPS+ or to mix real and virtual data. For example, one service might deliver an operational EPC, whilst another might deliver a virtual EPC corresponding to some post-upgrade scenario. The difference quantifies the potential and informs the upgrade decision-making process.

The EnTrak system, including its BuildAX and ESP-r components, was applied to 75 homes as part of the Innovate UK Future Cities Demonstrator project (Allison *et al.* 2015). Based on the monitoring of energy use, indoor conditions and weather parameters, and stock simulation to generate benchmarks, a service was established to assure the quality of insulation upgrades applied to hard-to-heat homes throughout Glasgow. Figure 12.34 depicts the service outcome as delivered to the housing department of Glasgow City Council.

Other deployments have been demonstrated in a variety of contexts including:

- commercial buildings undertaken as part of the EPSRC's digital transformation program targeting digitally mediated occupant negotiation in facilities management;
- large building stock performance monitoring in support of energy management in Scottish Local Authorities; and
- online assessment of novel building designs and systems as deployed within the BRE Innovation Park network.

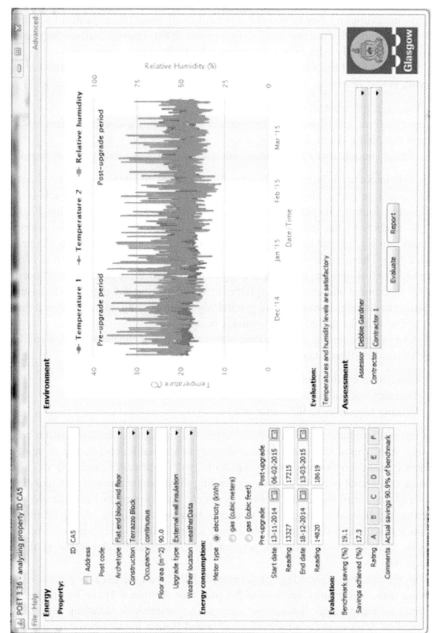

Figure 12.34   The upgrade quality-assurance tool delivered to Glasgow City Council

12.8   Artificial intelligence

Before the current millennium, the uptake of AI systems was gradual due to limitations in computing power and the availability of programs capable of solving real-world problems (Roman *et al.* 2020). Since 2000, interest and funding have grown exponentially with major investments by technology companies.

AI has reached the point where it is sometimes difficult to discern whether an offering is real or virtual. Some notable examples of AI applications include:

- language models that can imitate the linguistic skills of experts;
- imaging models that can create artwork or manipulate existing images in photorealistic detail; and
- clinical models that can undertake imaging analysis accurately and at speed.

AI is able to collate information from multiple sources and deliver outputs to users in any language and to any level of detail. Its key feature is the ability to self-learn and thereby improve its performance over time (Xiang *et al.* 2022, de Wilde 2023).

In relation to the built environment, buildings and their associated energy systems can be endowed with an AI capability to achieve various goals (Debrah *et al.* 2022). These might include the monitoring of local weather data as part of an adaptive control system; reducing energy consumption through the regulation of HVAC equipment and electrical appliances; fault detection and diagnosis to ensure systems are running correctly; and general signal processing and pattern recognition to improve the underlying knowledge base.

A key component of an AI system is its self-learning capability, which can be characterised by three stages:

- stimulation by the local environment;
- detecting variations in system parameters; and
- forming responses to the variations.

The internal structure of an AI system consists of interconnected information processing units, called neurons, corresponding to functional levels relating to inputs, inference based on synaptic weightings to impart significance, and outputs. The training of a neural network involves generating input/output datasets and adjusting the weightings until the simulated outputs match. The learning rate is controlled by determining how the weightings fluctuate between iterations. A performance parameter terminates the learning phase once a threshold has been reached. Further tests are then performed using independent datasets to evaluate the precision of the trained AI.

In the context of building performance, BPS+ can be used to generate AI training data. Uncertainty analysis techniques (see Section 4.4) can be applied to generate an input/output dataset that covers a representative sampling range. By this method, long-term training data are obtained from a physics-based model that emulates reality. The final stage is to test the AI by measuring its performance when confronted by a new problem for which the results are known. An interesting prospect is to continue AI training by coupling it to big data sources as described in the previous section. Once established, AI agents could provide useful inputs at the design, construction, and operation phases of buildings as typified by the following brief examples.

During the design phase, AI can be used for building performance assessment. Todeschi *et al.* (2021), for example, used a machine-learning model trained using performance predictions to estimate the hourly energy consumption of buildings with a claimed annual mean error of 12.8%. AI could be used to evolve a high-resolution BPS+ model by conflating concept information with site images. Rahimian *et al.* (2020) explored this issue by capturing daily site images and mapping these to a BIM model. As these images would normally contain objects such as materials, equipment, and people, the AI was taught to remove these from the scene.

During the construction phase, safety is paramount and machine-learning algorithms can be employed to predict high-risk issues allowing prevention measures to be implemented (Pan and Zhang 2023). AI techniques can also be used to manage construction logistics, avoid conflicts, and ensure the smooth running of supply chains. Any detected departures from the design intent can be highlighted and the design phase BIM model updated accordingly. The time evolution of the 3D model can be incorporated within an augmented reality environment to allow stakeholders to view the construction process virtually.

During the operation phase, a large resource is directed towards problem resolution (Zhang and Ashuri 2018). AI can help facility managers to identify cost-effective solutions in a timely manner and even control the post-occupancy application of BPS+ in the process.

An example of AI-driven design software is Dreamcatcher[14], which allows users to input materials, constraints, and preferences, and generates solutions based on cost, structure, aesthetics, and other requirements. Another example is Building Commander[15], which regulates HVAC control parameters by utilising building modelling and weather/energy price prediction to ensure buildings are kept ventilated and comfortable. A third example is Spacio[16], which represents a building as connected elements. This allows each element to react in concert with the other elements. The system can analyse building information in the form of floor plans or digital models. This combination of relational data and building information enables the system to exploit topological relationships to provide solutions whenever a design is changed.

12.9   Real-time design appraisal

To be even more speculative, it is possible to envisage a future modelling environment where the time delay between problem description and performance

---

[14]https://research.autodesk.com/projects/project-dreamcatcher/
[15]https://mymesh.co.uk/aibuildingcommander/
[16]https://spacio.ai/

feedback is greatly reduced. By endowing the objects that are used to describe a problem with behaviour, it is possible to generate performance information as soon as relevant object groupings are created. For example, a sun, wall, and obstruction object would give shading information as each is moved relative to the other. The prerequisite of such an environment is the existence of a set of objects that encapsulate both description and behaviour.

Such a future is possible because the idea has already been explored. A virtual drag-and-drop building physics environment was operational three decades ago. Function did literally follow the description. The computational limits of the day precluded its general adoption, and the community went in other directions. However, the limits on computation no longer apply and computer science has evolved so it is worth exploring how it evolved and the lessons learned.

The creation of an object set, and the means to manipulate it, was explored in the Energy Kernel System (EKS) project funded by the UK Engineering and Physical Sciences Research Council (Charlesworth *et al.* 1991, Hammond and Irving 1992, Clarke *et al.* 1992). The project identified the real and abstract (mathematical) entities underlying building energy modelling, explored the feasibility of representing these entities using the object-oriented (OO) programming paradigm, and established a system demonstrator[17]. The essence of the OO approach is that a program can be composed of independent objects communicating via messages. The main challenge was to decompose building modelling into object classes and subclasses and define the properties of these classes in terms of their data members, behaviour, and inter-relationship.

The core concept of the OO approach is data abstraction (Kim and Lochovsky 1990). Since the EKS was intended to be a construction environment for BPS+ programs, objects correspond to building parts whilst object data relate to object behaviour. This required a functional decomposition of the building (Clarke *et al.* 1989). The important point about the EKS approach is that it builds programs that are exactly matched to the problem in hand whilst providing a mechanism to deal with the physical/abstract mix so dominant in performance simulation.

The building performance domain was decomposed into primitive functional elements, such as sun position tracking, conduction, convection, radiation, equation solving, and polygon operations. These are the functions, albeit at different levels of abstraction, to be found within all BPS+ programs. An initial research task was to review a representative range of existing programs: the functions so identified are described elsewhere (Tang 1990, Wright *et al.* 1990, Clarke *et al.* 1990). A minimum level of data requirement was then associated with each function. For example, a material conduction function will require a set of thermophysical properties irrespective of the underlying mathematical model and so its minimum data requirement is a 'Material'[18] object (or strictly speaking a pointer to such an object) and a 'Dimension' object. Likewise, a matrix inversion function requires the matrix topology and coefficient values irrespective of the inversion technique to

[17]https://www.esru.strath.ac.uk/applications/
[18]Henceforth single-quoted and capitalised names signify classes or their instances, termed objects.

be used. Each function is now associated with a physical class which, logically, knows the context of the function. For example, 'Construction' logically knows about thermal resistance, whilst 'Room' logically knows about shortwave response. More contentious perhaps, 'Construction' may know about reference U-values (which have prescribed surface resistances), whilst only 'Building' may know about actual U-values because to evaluate this parameter requires knowledge of the properties and thermodynamic state of several objects (air volumes, surfaces, constructions, adjacent spaces, and so on). It follows that class functions must relate only to the intrinsic data and properties of a class and not require the existence of, or assume data or properties of, another class. This ensures that a class will encapsulate only data that are pertinent to that class and that the data members of each class, as required to support its functions, can be guaranteed to be available at run time to the object made from the class.

Abstract classes are now identified by gathering related functionality as implied by the functions of the physical classes. For example, the convection function of 'Room' requires surface area, hydraulic diameter, and heat flow direction, all of which are geometrical entities and so are gathered into an abstract class 'Polygon'. This ensures that classes will not possess functionality where that functionality could be made more generally available by encapsulation within another class. Where a class has several domain theory functions (e.g., convection, shortwave response, and occupant behaviour in the case of the 'Room' class, or different matrix solution techniques in the case of the 'Solver' class), these functions are implemented as links to abstract classes containing the required functionality of the domain theory. The different formulations of any given domain theory can then be handled by classes derived from these abstract classes. This approach facilitates the handling of the multiplicity of domain theories, without incurring combinatorial explosion in the parent classes.

The C++ language and an OO database (ONTOS 1989) were utilised to define and contain a class taxonomy from which models of different functionality could be built. The classes were organised into 'used by' and 'derived from' hierarchies and placed under the control of an instantiation mechanism. This means that programs offering different modelling capabilities can be automatically constructed by merely selecting the required class variants.

Figure 12.35 shows the EKS prototype's 'used by' hierarchy. This specifies how the classes interrelate and define the information flows. Orthogonal to this plane is the EKS inheritance hierarchy used to represent the alternative domain theories, for example, the alternative conduction theories used by the 'Layer' class. The dilemma between extensibility (i.e., the ability of an existing class to use, or be used by a new class) and security (i.e., the guarantee that classes are compatible) is solved by the use of meta-classes. These define the behaviour of each EKS class and have knowledge of other dependent classes. By insisting that programs can only be built from meta-classes, it is possible to have an extensible and secure system without paying the performance penalty of run-time type checking.

EKS classes are of three types: base, principal, and intrinsic. Base classes provide the interface to the principal classes and are generic in that they cannot be

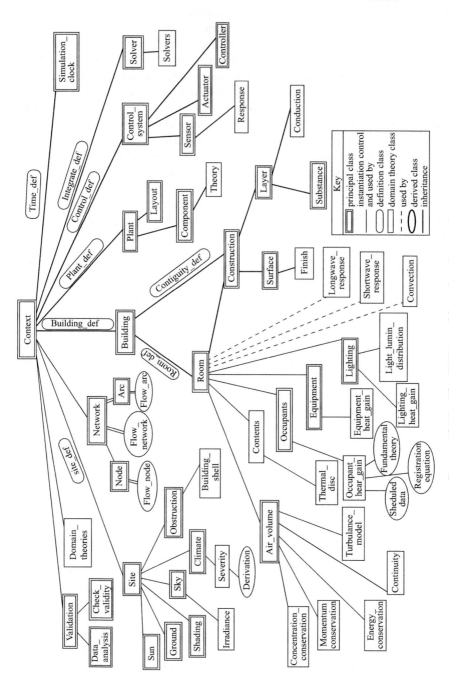

Figure 12.35   The EKS 'used by' class hierarchy

instantiated because their functions are virtual, being implemented only in the principal classes derived from them. They exist to allow the future extension of the class taxonomy into other domains without the need to carry the thermodynamic functionality of the existing implementation. In addition, they guarantee that future derived classes will interoperate with the other EKS classes.

Principal classes are selected by a program developer to define the capabilities of some programs. Principal classes are derived from a corresponding base class to offer a particular implementation of the virtual functions of its parent. That is, these classes represent the physical or thermodynamic states of the entities they represent. In the case of the domain theory classes, alternative principal classes will exist to represent the different possible theoretical approaches.

Intrinsic classes are the internal workhorses of the EKS, serving to transport data between the principal classes, control the class selection process at template specification time, contain support data such as weather time series, and dimension all properties of state. Because they are not selected by the program builder, intrinsic classes are not shown in the Figure 12.36 taxonomy. The separation of the underlying functions in this way gives maximum flexibility and code reuse when creating new programs. The base classes provide a common semantic for the principal classes and permit classes derived from them to be reliably used in different contexts.

Derived classes are subtypes of their parent. This means that such a class may be transparently replaced (even at run time) by any class derived from it. Thus, as shown in Figure 12.36, if the functionality of 'DynamicConduction' is required, either finite difference or response function approaches can be selected whereas, if only the lesser functionality of 'Conduction' is called for, both steady state or dynamic conduction classes can be selected.

It is this facility that gives the EKS its flexibility: in model building (support of theoretical variants); in run-time features such as dynamic model substitution; in support for program validation; and in program maintenance. Where different program architectures handle the same task in different ways, derived classes provide a means to incorporate that functionality whilst ensuring minimal impact on the rest of the system. Behavioural inheritance also reduces coding and improves

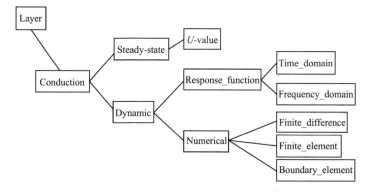

Figure 12.36   Use of derived classes

reliability since new classes gain access to their parent's code and so need only add the code for the extra functionality they provide.

Extensibility is also assisted because new classes and variants of existing classes can be added without requiring any changes to existing classes. Actual program building is carried out by a special template class. Once the required program capability is defined, a 'Template' is constructed as a collection of meta-class instances. Since the 'Template' is the program specification, there is no need to generate an executable program at this stage. Instead, once the problem data is available, the 'Template' uses its meta-class objects to create a program tailored exactly to the problem being addressed.

As a program development platform, it is expected that the EKS will need to support a wide range of systems represented by different mathematical categories, for example, hyperbolic partial differential models for aerodynamic systems, parabolic partial differential models for diffusion systems, elliptic partial differential models for wave and vibration systems, ordinary differential models for lumped parameter systems and so on. One way to achieve this would be to provide representation schemes for all the different possible approaches – a finite difference 'Layer' and a finite difference 'Conduction'; a steady state 'Layer' and a steady state 'Conduction' and so on. Apart from being inelegant, this would give rise to the problems of interface complexity and combinatorial explosion, resulting in duplication of code/functionality and creating intractable maintenance difficulties. Instead, the EKS employs a theory representation based on a vectorised state-space equation method, implemented using sparse matrix techniques. This makes it possible to represent most of the equation-based theories in the domain of building energy modelling in a consistent manner as vectorised state-equations. The details of this method are given elsewhere (Tang 1990, Clarke *et al.* 1992). Theory, in vectorised form, is encapsulated within four special transport classes: 'Equation_set', 'Equation', 'Coefficient' and 'State_variable'.

As shown in Figure 12.37, an OO database lies at the heart of the EKS and has three roles.

First, it holds persistent objects relating to entities such as weather and material properties. Second, it holds the problem description and results as objects, enabling the interfaces to be separated from the body of the performance prediction engine. Last, it holds the 'Template' and the meta-class objects. The programs surrounding the database are used to specify a particular program architecture. The indicated modules have the following roles.

- EKS_cb builds the problem context in terms of a definition of the features required of a program.
- EKS_tb builds a given 'Template' by allowing class variants to be selected as required. As each class is selected, the corresponding meta-class defines the dependent classes so that the process is automated.
- EKS_dd takes a 'Template' as input and outputs the corresponding OO product model.
- EKS_dm supports the user definition of the product model and holds the data in the form of 'X_def' ('something_definition') objects. Typically, a program

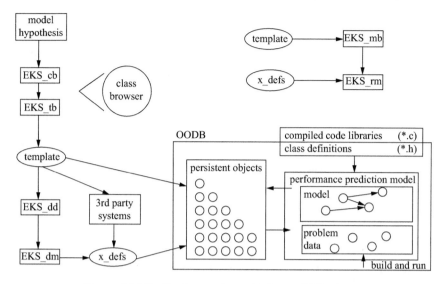

Figure 12.37   Storing and manipulating EKS objects

will require several 'X_def' objects to define building geometry, construction, operation, etc.

● EKS_mb builds a program based on a given 'Template'.

● EKS_rm controls a simulation by associating the program with its corresponding 'X_def' objects.

An environment such as the EKS opens up the prospect of a radical change to the design support process. Imagine, for example, a computer system that allowed the definition of a design hypothesis by the graphical selection of component parts representing a site, walls, windows, radiators, shading devices, sensors, the sun, and so on. If these components were related to EKS classes then the user would effectively be constructing, in real time, a model that is matched to the envisaged design. Given the functionality of the EKS, it would then be possible to arrange that the object instances externalise their behaviour immediately on selection.

By bringing together hypothesis manipulation and performance appraisal, a real-time computer-supported design environment is enabled. This, in turn, would enable the application of simulation at the conceptualisation stage of the design process where the potential benefits are greatest. Realising such an environment is a challenge issued to the next generation of BPS+ researchers.

12.10   Chapter summary

It has been demonstrated that the richness of a high-resolution model enables support for upgrade quality assurance, smart facilities management, off-site component manufacture, and the 3D printing of physical prototypes. BPS+ can complement large-scale performance monitoring by providing information on future prospects in support of facilities management and policymaking. Likewise, BPS+ and AI techniques can collaborate, the latter assisting with input model creation, and the former providing training data relating to future states. Finally, by recasting the methods of BPS+ to a form that is amenable to manipulation, it should be possible to produce applications that deliver near-real-time computational support, a possibility that has been demonstrated in the past but not taken up.

References and further reading

Allison J, Cameron G, Clarke J A, Cockroft J, Markopoulos A and Samuel A (2015) 'Confirming the effectiveness of insulation upgrades applied to Glasgow housing', *Final Report to Glasgow City Council for Innovate UK's Future City Demonstrator Project*.

Clarke J A (2001) 'On-line energy services for smart homes', *Final Report for Project NNE5-2000-00258*, European Community EESD-Energy Programme (1998–2002), available from ESRU, University of Strathclyde.

Clarke J A (2015) 'A vision for building performance simulation; a position paper prepared on behalf of the IBPSA board', *Building Performance Simulation*, 8 (2), pp. 39–43.

Clarke J A and Hand J W (2015) 'An overview of the EnTrak/BuildAX eService delivery platform', *Report for EPSRC Project EP/I000739/1O1*-2015, ESRU, University of Strathclyde.

Clarke J A, Tang D, James K and Mac Randal D F (1992) 'The Energy Kernel System', *Final Report for Grant GR/F/07880*, Science and Engineering Research Council, Swindon.

Clarke J A, Hand J W, Kim J M, *et al.* (2014) 'Pervasive sensing as a mechanism for improving energy performance within commercial buildings', *Proc. Building Simulation and Optimization*, London.

Clarke J A, James K and Tang D (1989) 'Simulation methods concerning building performance prediction: a general review, *EKS Project Report*, ESRU, University of Strathclyde.

Clarke J A, James K and Tang D (1990) 'EKS prototype status review and issues arising', *EKS Project Report*, Energy Systems Research Unit, University of Strathclyde.

Clarke J A, Johnstone C M, Kondratenko I, *et al.* (2004) 'Using simulation to formulate domestic sector upgrading strategies for Scotland', *Energy and Buildings*, 36, pp. 759–70.

Charlesworth P, Clarke J A, Hammond G, *et al.* (1991), 'The energy kernel system: the way ahead?', *Proc. of BEP'91*, Canterbury, April 10–11, pp. 223–236.

de Wilde P (2023) 'Building performance simulation in the brave new world of artificial intelligence and digital twins: a systematic review', *Energy and Buildings*, 292, p. 113171.

Debrah C, Chan A P C and Darko A (2022) 'Artificial intelligence in green building', *Automation in Construction*, 137, p. 104192.

Dodworth K P, Pennycott A R W, Doherty N, Kwon T G and Hwang I J (2009) 'The application of dynamic energy modelling to the design of cruise vessel HVAC systems', *Proc. Int. Symposium On Marine Engineering*, Busan.

Doran S (2008) 'Internet-enabled monitoring and control of the built environment', *Completion Report for TSB Project TP/3/PIT/6/I/*16934, Building Research Establishment.

EPSRC (2010) 'Part 2: building management linking energy demand, distributed conversion and storage using dynamic modelling and a pervasive sensor Infrastructure', https://gow.epsrc.ukri.org/NGBOViewGrant.aspx?GrantRef= EP/I000739/1.

Hammond G and Irving I (1992) 'Validation studies appropriate to the development of the energy kernel system for building environmental analysis', *Final Report for Grant GR/F/*07880, Science and Engineering Research Council, Swindon.

Hand J (2023) 'Coming full circle from virtual simulation models to 3-D printed architectural models', *Proc. Building Simulation and Optimisation'22*, Newcastle.

Kelly N J and Macdonald I (2004) 'Coupling CFD and visualisation to model the behaviour and effect on visibility of small particles in air', *Proc. eSim'04*, pp. 153–160, Vancouver.

Kelly N, Clarke J A and Strachan P A (2003) 'Improvement of transport safety by simulation-based control of fog production in a chamber', *Final Report for Project G6RD-CT-2000-00211*, European Commission.

Kim W and Lochovsky F H (Eds.) (1990) *Object-Oriented Concepts, Databases, and Applications* ACM Press, New York.

ONTOS (1989) *Object Database Documentation*, Ontologic Inc., Burlington, MA.

Pan Y and Zhang L (2023) 'Integrating BIM and AI for smart construction management: current status and future directions', *Archives of Computational Methods in Engineering*, 30, pp. 1081–1110.

PC (2021) *Private communication*, 'Neighbourhoods, Regeneration and Sustainability Services,' Glasgow City Council.

Rahimian F P, Seyedzadeh S, Oliver S, Rodriguez S, and Dawood N (2020) 'On-demand monitoring of construction projects through a game-like hybrid application of BIM and machine learning', *Automation in Construction*, 110, p. 103012.

Roman N D, Bre F, Fachinotti V D and Lamberts R (2020) 'Application and characterization of metamodels based on artificial neural networks for building performance simulation: a systematic review', *Energy and Buildings*, 217, p. 109972.

Rushmeir H E (1994) 'Rendering participating media: problems and solutions from application areas', *5th Eurographics Workshop on Rendering*, pp. 35–36.

Struck C (2017) *Private Communication: Factory Programming Example*, Research Centre for Urban & Environmental Development, Saxion University of Applied Science and Technology, The Netherland.

Tang D (1990) 'The EKS theory representation', *EKS Project Report*, Energy Systems Research Unit, University of Strathclyde.

Todeschi V, Boghetti R, Kämpf J H and Mutani G (2021) 'Evaluation of urban-scale building energy-use models and tools – application for the city of Fribourg, Switzerland', *Sustainability*, 13.

Wright A J, Charlesworth P, Clarke J A, *et al.* (1990) 'The use of object-oriented programming techniques in the UK energy kernel system for building simulation', *Proc. European Simulation Multiconference*, Nuremberg, pp. 548–552.

Xiang Y, Chen Y, Xu J and Chen Z (2022) 'Research on sustainability evaluation of green building engineering based on artificial intelligence and energy consumption', *Energy Reports*, 8, pp. 11378–11391.

Zhang L and Ashuri B (2018) 'BIM log mining: discovering social networks,' *Automation in Construction*, 91, pp. 31–43.

Chapter 13

Strategic renewables

The world is not running out of energy resources although there are diminishing options that are deemed acceptable in many jurisdictions. From an energy planning perspective, the future need not be viewed in a negative manner as something that is too uncertain to enable informed action: it should be viewed as an opportunity. Simulation offers a practical way to improve upon what has gone before by planning for a more integrated energy system at all scales.

This chapter considers the possibility of establishing large-scale renewable energy assets in a manner that complements matched loads located nearby or remotely distributed but synchronised. To this end, BPS+ can be used to determine the load profiles of target communities (local, remote, existing, or planned) and size corresponding renewable energy technology (RET) installations. In the latter case, this might entail the use of a commercial package such as PVsyst[1] or a free package such as RETScreen[2] in cases where a BPS+ tool does not offer such a modelling feature. This chapter summarises the nature of supply-side models and indicates how they can be utilised to complement BPS+. The presented models, although rudimentary, are considered suitable for inclusion in BPS+ to provide a first-order estimate of the available renewable resources. The models correspond to assessment methods as taught by Energy Systems Research Unit (ESRU) staff to students in the Renewable Energy Systems and the Environment MSc course at the University of Strathclyde. Three technologies are briefly covered corresponding to tidal stream, wind, and solar power capture.

National policy increasingly favours the development of RET (Gov 2003, SG 2003). The trend to date has been to develop technologies that are stochastic in nature, principally hydro and land-based wind/solar power initially, then offshore wind power in the medium to longer term. This implies increasing levels of vulnerability within the electricity supply network that, in turn, necessitates increased levels of energy storage, standby plant, and network expansion/control to prevent supply disruption and maintain power quality. Other RET options include biomass, tidal stream, wave, geothermal, and energy from waste.

RET technologies are usually deployed strategically, with connection to the electricity transmission/distribution network as opposed to local load matching. In

---

[1]https://pvsyst.com/help/license_buy.htm
[2]https://natural-resources.canada.ca/maps-tools-and-publications/tools/modelling-tools/retscreen/7465

the United Kingdom, for example, wind farms were initially established at sites that could be connected to the distribution network at minimal costs. Later deployments, and in particular offshore wind, had to be undertaken in areas remote to centres of population and with a less robust or absent network. This required investment capital to develop/upgrade distribution and transmission networks. Network development costs, based on a conservative estimate[3] of £1.5M/km, equate to a required investment of billions of pounds to open up remote geographical areas. Assuming a 70% network utilisation factor, and a capital repayment time of 7 years at 5% net present value, this would equate to a transmission charge in excess of 12p/MWh.km. This additional cost will affect the financial viability of future developments where the resource is located hundreds of kilometres from demand centres. One mitigation approach is to deploy other energy sources alongside wind. This raises additional questions as to who should pay these upfront costs, who would become the owners, and how this investment should be recouped.

A value around 5 W/m^2 (peak) is typically assumed for the power density of land-based wind farms although estimates vary. At this level, some 7,000 km^2 of land would have been required to attain a previous UK target of generating one-third of all electrical energy from wind by 2020 (equivalent to an installed capacity of 35 GW allowing for a 33% capacity factor). Is it any wonder that this target was missed?

Whilst the use of approximately 3% of the UK land area to site around 15,000 large turbines is technically feasible, policy constraints may be expected to prevail due to public unwillingness to countenance the industrialisation of the landscape on such a vast scale. The problem is mitigated by a shift to offshore wind because of the increased power density and more favourable public opinion. However, new barriers then arise, not least the economic challenge of shallow/deep sea installation, the need for new and more expensive transmission lines, the impact on the maritime environment, and myriad maintenance-related issues.

It is for such reasons that many observers have called for a balanced energy portfolio encompassing fossil, nuclear, and renewable, with wind energy leading the renewable rollout at a less hurried pace. Although subject to reform at present, two mechanisms have been used in recent years to incentivise the industry to deploy low-carbon generation. Contracts for Difference[4] agree to a 'strike price' for electricity generated from renewable sources. When prices are below this level, generators receive a top-up and when higher they pay back the difference. The intention is to provide revenue certainty for generators. A Capacity Market[5] mechanism ensures that the system overall has sufficient reliable generating capacity by auctioning generator and demand side response agreements in return for payments for capacity to meet security of supply.

---

[3]https://acer.europa.eu/electricity/infrastructure/network-development/transmission-infrastructure-reference-costs
[4]https://gov.uk/government/collections/contracts-for-difference
[5]https://gov.uk/government/collections/electricity-market-reform-capacity-market

13.1   Tidal stream

It has been estimated that the UK exploitable tidal stream resource is around 18 TWh/y (Black and Veatch 2005, Johnstone *et al.* 2013). Two principal technologies have been researched: the oscillating aerofoil driving hydraulic accumulators, and horizontal-axis turbines evolved from wind power technology, with prototypes installed off the coasts of Norway and the UK. Tidal current turbines operate in a more determinate environment than wind turbines, allowing the maximum flow velocities to be predicted with reasonable accuracy. Likewise, the dynamic loading caused by velocity shear and misalignment is largely predictable although stochastic inputs will arise when storm surges increase current velocities and introduce dynamic loading due to surface wave action. In addition, sites will generally be close to land so that the fetch for surface wave development will be limited. Incoming turbulence will also generate fluctuating loads, although the range of excursions, particularly in the direction of flow, will be relatively small. For these and other reasons, research is continuing to determine the operational performance envelope of tidal stream rotors under real operating conditions (Myers and Bahaj 2004, Calautit and Johnstone 2022). Such research enables design solutions to be matched to specific site conditions.

Because a tidal stream offers firm power, whereby the quantity and timing of outputs can be precisely known in advance, it is possible to match the output from a tidal stream farm to the energy demand of nearby communities and thereby avoid the need for new transmission infrastructure and associated losses. To this end, BPS+ can be used to quantify the temporal energy demand of such communities and explore measures to adjust the profile to align better with the tidal power supply. Alternatively, by establishing tidal stream power generation at locations with complementary output phasing, a more uniform supply should be achievable that will reduce the grid impact.

Consider the rudimentary model applied by Clarke *et al.* (2005), which is suitable for the present purpose: more sophisticated models are available elsewhere (e.g., the hybrid dynamic approach developed by McCombes 2014). The instantaneous power, P, available to a single tidal stream turbine is given by $P = \frac{1}{2} \rho A V^3$ where ρ is the fluid density, A the rotor swept area, and V the velocity of the fluid stream. If the variation of V with time is assumed to be sinusoidal, P will vary as shown in Figure 13.1 (upper profile), which covers a typical tidal half-cycle of about 6 hours 12 minutes. The lower profile corresponds to the turbine power output, $C_p P$, for arbitrary cut-in and rated stream velocities and assumes that the coefficient of performance, C_p, is maintained at a high value between these conditions.

By phase coordinating suitably located tidal energy power stations, the aggregate power output, although not constant, could match a substantial portion of the base load. This idea is not new and has interesting resource planning implications: e.g., with growing electricity demands, phasing could contribute to the maintenance of network stability.

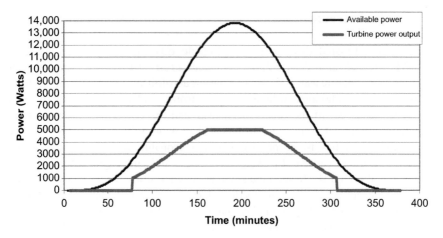

Figure 13.1   Tidal stream power (upper) and turbine output

Figure 13.2   Sites for phase analysis

Consider three geographically dispersed but complementary Scottish coastal sites as shown in Figure 13.2: Cape Wrath off the northwest coast, Crinan at the Sound of Jura off the west coast, and Sanda off the Mull of Kintyre, a peninsula to the southwest. The distance separating Cape Wrath from Sanda is 237 miles.

The hourly stream velocities at each site can be determined from admiralty charts during both spring and neap tide conditions, which arise due to the gravitational interactions between the Sun, Earth, and Moon (D'Oliveira *et al.* 2002). Figure 13.3 depicts site velocities over a 24-hour period: note the departures from a sinusoidal curve.

To illustrate the power extraction potentials, a 10m diameter turbine is located at each site. For simplicity, a constant power coefficient of 0.4 is applied throughout the speed range and no rated power limit is imposed. Figure 13.4 shows the resulting power outputs and a summation for all turbines under spring and neap conditions.

For spring tides, a significant base load is evident, amounting to about one-third of peak power output. A similar trend is apparent for neap tides but the outputs are lower. Some changes between successive cycles are evident. Sanda, for example, is

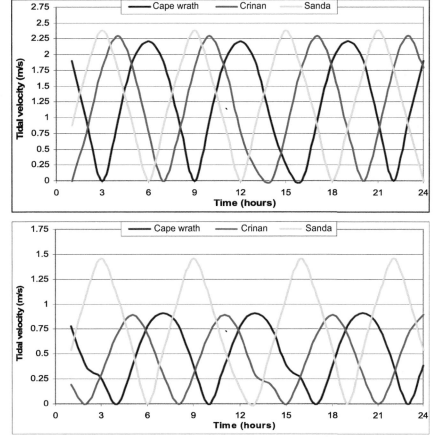

Figure 13.3   Spring (upper) and neap tide velocities

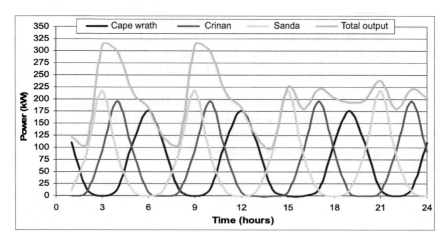

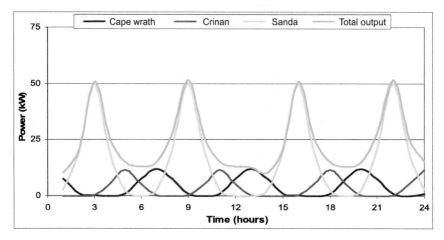

Figure 13.4   Site and aggregate power output for a 10m diameter tidal turbine in spring (upper) and neap tides

cycling at a higher frequency than the other two, a phenomenon that will reverse at some point in the lunar cycle. Three main conclusions emerged from the study.

- The aggregate power output is sensitive to the characteristics of its component parts.
- Given the natural variations that will occur between successive tidal cycles (under the influence of weather or the longer cycle of spring and neap tides), some irregularity in the aggregate power output is likely.
- Accurate prediction of performance for aggregated systems of this kind may be problematic. This is not to infer that the output is unpredictable but that accurate data are needed to make the prediction.

It is evident that the maximum power delivered under maximum spring tide velocities is around 220 kW, and such a turbine would operate at this capacity for approximately 10% of the time over the lunar cycle. This results in a large variation in peak power output between spring and neap tides, 220 kW and 50 kW, respectively, for Sanda. To reduce this extreme range, the turbine's delivered power may be limited to a rated capacity, achieved by designing the turbine so that the hydrofoil sections are either pitch or stall regulated above a specific tidal current velocity. In such cases, the power delivered by the turbine is proportional to the speed of the tidal current until the limiting capacity is reached; beyond this, power output is held constant even though the available power within the tidal stream continues to increase. Regulating/limiting the power delivered from a tidal turbine enables a lower capacity generator to be used thereby incurring a cost saving and achieving a higher capacity factor because the turbine will operate at its rated capacity for longer periods of time, especially during spring tides as shown in Figure 13.5. This will improve the economics of power delivery from such systems.

The disparity between the total output from 80 kW limited capacity tidal turbines and tidal turbines without limiting capacity located at the three sites during spring tides is depicted in Figure 13.6. Limiting the capacity produces a more uniform power delivery with substantially reduced variation, thus facilitating the use of tidal energy as a contribution to base load power supply. In the case of turbines without limiting capacity, the larger variation in power output is due to the cubic relationship between power and stream velocity.

Since the components connecting the tidal turbine to an electrical network are sized and costed in accordance with the maximum capacity delivered, limiting the capacity reduces connection costs, gives a more uniform power delivery and reduces the payback period associated with system connection costs. This is due to the reduced capacity rating of the required hardware, and the higher capacity factors achieved because peak power is delivered for longer periods of time, especially during spring tides.

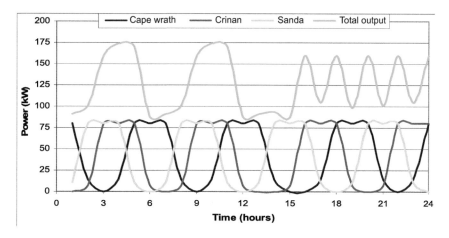

Figure 13.5   Site and aggregate power output for a 10m diameter tidal turbine with limiting capacity in spring tides

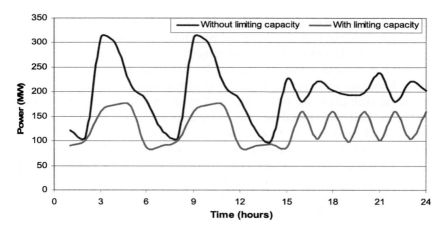

Figure 13.6   Site and aggregate power output for a 10m diameter tidal turbine with and without limiting capacity in spring tides

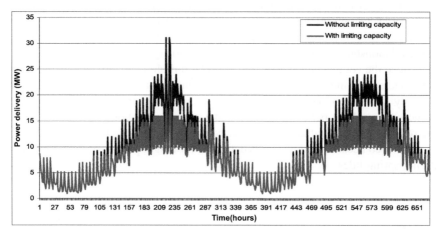

Figure 13.7   Fluctuations in power output between spring and neap tides for a tidal farm with and without limiting capacity

To demonstrate firm power of sufficient capacity, 100 tidal turbines were deployed virtually at each of the three tidal sites. For the case where no limiting capacity is implemented, each turbine was rated at 220 kW giving an installed capacity of 66 MW: the energy yield from this installation is 6.5 GWh and the average capacity factor is 14.7% over a lunar cycle. Limiting the capacity of the turbines to 80 kW gives an installed capacity of 24 MW, reduces the energy yield to 5.25 GWh, and increases the average capacity factor to 32.5%. This demonstrates that an installation with 36% of the capacity rating of a non-limited installation will deliver 80% of the energy, give a better payback, and result in a more uniform power. Figure 13.7 compares the power delivered from a 300 unit tidal farm for both 80 kW and 220 kW turbines over the lunar cycle.

It is apparent that limiting the capacity of the turbines reduces the magnitude of the variation between spring and neap tidal cycles as well as producing a more uniform power delivery. When combining the power output from tidal farms located at three complementary sites, and where a continuous base load is required, the limiting capacity of the turbines may be determined by the tidal conditions experienced during neap tides. This however is unlikely to be the most cost-effective way of exploiting the resource. Setting the limiting capacity at a higher value implies the need for an additional backup or storage facility, options that could be explored using BPS+ tools. From Figures 13.4 (upper), 13.5, and 13.7, it is clear that some level of base load provision can be achieved. The twice-daily cycle could be further smoothed by the use of hydraulic pumped storage, or the phased operation of a conventional hydropower plant, which could be especially useful in a local context.

The fluctuations in output due to the lunar cycle are more intractable, as they affect all sites simultaneously. Here, the power output could be regulated so that the maximum rated output is that experienced during neap tides so that only a small proportion of a site's resource is utilised. This is a doubtful economic proposition. Alternatively, long-term (say weekly) energy storage could be introduced so that the excess capacity captured during spring tides can be stored for use during neap tides to maintain the capacity of delivered power. A third and more likely option during neap tides would be the introduction of other sources of energy to meet the shortfall. In any event, the predictability of tidal stream power output may be regarded as an asset in energy supply management.

13.1.1 Turbine design

Modelling and simulation are apt methods for use in turbine design. Most turbine prototypes follow conventional wind turbine practice, with 2- or 3-bladed horizontal-axis rotors (Stobart 1983, Fraenkel 2002). Although the fundamental fluid dynamic interactions between rotor and stream are the same as in wind energy conversion, there are differences that are likely to cause divergences in technological development. Some of these are obvious and influence materials selection and structural design, for example, the higher density of the fluid medium and the increased surface fouling and corrosion. Some are less obvious and will have a profound impact on the take-up of marine power: the predictable range of the current velocities at a given site, and the relatively low levels of turbulence in the stream. It may be beneficial to adopt ideas that were first postulated for wind energy but not fully developed.

This section draws on an Engineering and Physical Sciences Research Council (EPSRC) funded project that developed a contra-rotating marine turbine (CoRMat) based on co-axial rotors in close proximity with a dissimilar number of blades on the upstream and downstream rotors (Clarke *et al.* 2007, 2009). It demonstrates the cooperative use of modelling and experiment in support of the translation of a research idea to full-scale deployment.

The contra-rotating action substantially reduces the reactive torque that is an inherent feature of such devices. The dissimilar blade numbers eliminate power

Figure 13.8   A 1/30th scale prototype undergoing tow-tank testing

lulls, which are produced when the downstream blades are in the shadow of the upstream blades. The CoRMat configuration comprises three blades (120° apart) on the upstream rotor and four blades (90° apart) on the downstream rotor. Each rotor drives a shaft, the upstream rotor in an opposite direction to the downstream rotor. The relative rotational speeds between the two shafts are considerably faster than a conventional turbine, eliminating the need for a gearbox and enabling direct drive of a contra-rotating generator design developed later in the project. This arrangement extracts more energy from the flowing stream than a single-rotor device because it captures the energy in the swirl within the downstream wake.

The specific objectives of the project were to prove the technical feasibility of contra-rotating turbine technology and identify a route to full-scale commercial deployment. The first task was to employ the BEM[6] program that utilises the widely applied blade element, momentum method (Fu *et al.* 2021). This was used to design a 1/30th scale, 0.8m diameter prototype turbine for tow-tank testing as shown in Figure 13.8 to verify predictions. Program inputs comprise blade profile, water density, stream speed, number of blades in upstream and downstream rotors, blade tip radius, blade root radius, tip speed ratio, chord length, and pitch angle. Outputs include C_p value, power, shaft torque, rotational speed, thrust loading per rotor, and blade root bending moment.

The modelling approach was to specify the upstream rotor geometry and then use BEM to predict the geometry of the downstream rotor and the performance of the complete turbine over a range of tip speed ratios. Under all circumstances, a condition of zero net torque was imposed: the torques exerted on the two rotors by the flowing stream were equal and opposite. Figure 13.9 shows the alignment between predicted and measured values of the power coefficient as a function of tip speed ratio.

---

[6]https://www.esru.strath.ac.uk/applications/

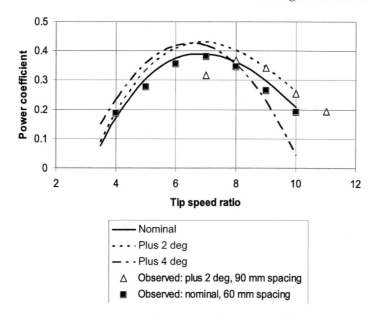

Figure 13.9   Predicted vs. measured performance of the 1/10th scale prototype

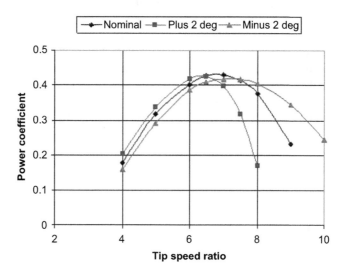

Figure 13.10   Predictions for a range of blade pitch angles

The next task was to design a 1/10th scale prototype for in-sea testing. To facilitate deployment from an anchored vessel (as opposed to a fixed mooring), the maximum rotor size was limited by the handling capacity of the contracted vessel, and a diameter of 3m was chosen. The BEM performance predictions for the turbine are given in Figure 13.10. As with the smaller prototype, provision is made for adjusting the blade pitch when fitting into the hubs, and predictions for three different settings are shown. It can be seen that refinements to the design led to improved peak power coefficients despite a higher rotor solidity.

Large blade bending loads are a feature of horizontal-axis turbines. For the present design, they are particularly severe for the downstream, 4-bladed rotor with its relatively slender blades. This rotor would also be subject to substantial intermittent loads from blade-to-blade and blade-to-strut interactions.

A finite element analysis of alternative blade designs for this rotor was carried out using ANSYS Fluent[7], with glass reinforced plastic and steel spar, and aluminium alloy constructions considered. A single blade was modelled using 20-node isoparametric solid elements. The finite element mesh applied is shown in Figure 13.11 (left). The model was loaded using mean bending moment distributions as predicted by BEM. The analysis (right) indicated that the GRP/stainless steel design experienced unacceptably high stresses in the spar and at the interface between the spar and GRP.

Consequently, an alternative rotor geometry was investigated using the BEM program. Chord and blade thickness were increased by 25% over the entire span of the blade. Hydrodynamic performance and blade loadings were examined for a range of operating conditions as shown in Figure 13.12 – here 'std' refers to the initial rotor configuration and 'fat' to the high-solidity blades of the downstream rotor.

As expected, the peak power coefficient occurs at a lower tip speed ratio, but its value is almost identical. Blade thrust coefficients follow the same pattern. The outcome should be a reduction in mean out-of-plane bending stresses to about 50% of previous values.

This turbine configuration is intended for use with a contra-rotating generator, but that was still to be developed and an alternative system had to be found for the

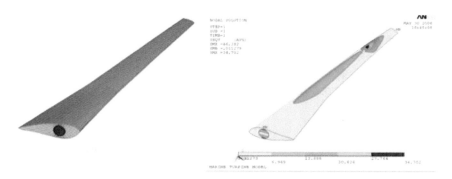

Figure 13.11   Finite element mesh (left) and stress plot for a downstream blade

---

[7]https://ansys.com/products/fluids/ansys-fluent/

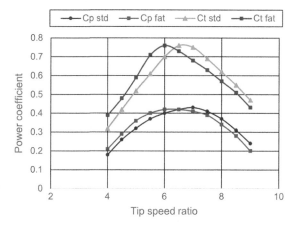

Figure 13.12   Power and axial thrust coefficients for the standard and high-solidity blade cases

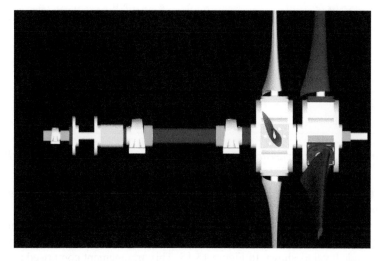

Figure 13.13   Layout of the 3m test turbine – blades, hubs, shafts, and torque measuring brakes

3m prototype. The choice was constrained by a number of factors, principally the limited space available in the turbine nacelle, the lifting capacity of the tender vessel, and cost. The final configuration is shown in Figure 13.13.

Dedicated loggers attached to each rotor stored data from strain gauges to measure in-plane and normal blade root bending moments. Real-time measurements of braking torques and mooring line forces were transferred by umbilical cable to the tender vessel. Operation and stability were monitored by underwater cameras, and visualisation of the wake flow was captured. Figure 13.14 shows an

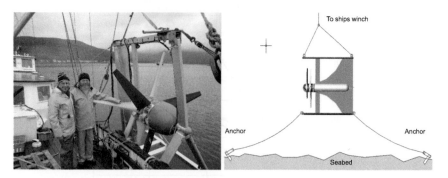

Figure 13.14   Schematic of the 3m turbine in-sea test configuration

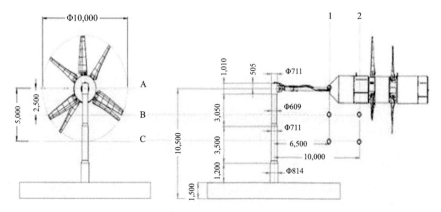

Figure 13.15   Anchor arrangement and CFD output points (1A, 1B, 1C, 2B, 2C)

in-sea test underway and the arrangement for station keeping: mooring lines taken to seabed anchors and to the surface[8].

In a related project (Cockroft and Samuel 2014), ANSYS Fluent was used to investigate the impact of an alternative anchoring arrangement on water velocity at the turbine blade level as shown in Figure 13.15. This arrangement comprised a support mast upstream of the turbine. Velocity fluctuations in the form of vortices shed from the mast caused a change in velocity (linear motion) or vorticity (local rotation).

Fluent simulations were undertaken using a velocity-type inlet boundary condition of magnitude 2, 3, and 4 m/s and a pressure-type outlet boundary. Simulations over 60s at 0.2s time steps were conducted for each inlet case. Results were analysed in terms of the drag coefficient, Cd, and vortex generation as typified by the outputs of Figure 13.16. Here, Cd corresponds to point B for the 2 m/s case. A steady cycling is observed (left) after an initial unsteady start indicating regular vortex shedding; the

---

[8]In 2009, the project attracted the Energy Institute's Technology Award for 'The Development and In-sea Testing of a Contra-rotating Marine Turbine'.

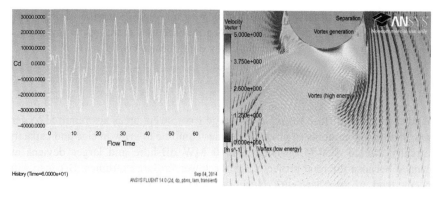

Figure 13.16   Coefficient of drag (left) and vortex generation

Figure 13.17   CoRMaT being prepared for sea trails at EMEC (credit: Johnstone 2018)

right image corresponds to the region immediately downstream of the mast at 41s, showing flow separation, vortex generation, and the rotation of shed vortices.

It was concluded that velocity decreases on average by 16%–27% of the free stream velocity, with some residual disturbance reaching the turbine blades.

Following on from the in-sea testing phase, and as illustrated in the collage of Figure 13.17, Johnstone (2010) established a spinout company, Nautricity, to develop a full-scale prototype of the CoRMat turbine and undertake trials within a grid-connected test bed at the European Marine Energy Centre in Orkney[9].

The research and development journey continues at ESRU.

13.2 Wind farms

Wind energy is a diffuse resource and large rotor swept areas are required to generate significant amounts of power. Turbines continue to grow in size and capacity, currently around 4 MW onshore and 10 MW offshore and larger devices are planned because the total project costs are lower for large turbines. Because wind speeds tend to increase with elevation, taller machines experience relatively favourable conditions and can produce maximum power for a greater proportion of the time (i.e., they have a higher capacity factor).

Most large machines have variable-pitch blades, which maximise energy capture in light winds and can be used to provide aerodynamic braking when required. Variety is exhibited in drive train design, with competition between geared-up and direct drive generators, and the increasing use of power electronics to enable variable-speed operation. The technology is still maturing and reliability remains a concern, although availability figures for onshore wind farms are often in excess of 97%.

For most sites, wind is an unpredictable resource and the potential must be assessed using statistical methods. Wind speed measurements gathered over extended periods are used to determine their frequency of occurrence over a usable range of velocities as expressed by a Weibull distribution: Figure 13.18 shows the probability of occurrence, p(u), of different wind velocities. The median wind speed is indicated by the vertical line, which divides equal areas.

A rudimentary estimate of the energy yield for a single wind turbine would proceed in four stages as follows. First, the probability density distribution for the

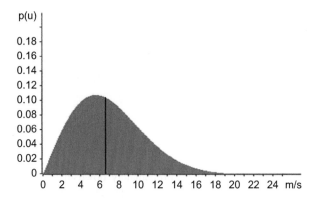

Figure 13.18 Probability density of wind velocities

[9]https://emec.org.uk/

site is used to construct a velocity exceedance curve as illustrated by the upper exceedance curve in Figure 13.19.

The exceedance curve indicates the number of days in a typical year on which the wind speed exceeds a specified value. Superimposed on the curve are three wind speeds that relate to the performance of the wind turbine to be deployed:

- a cut-in speed below which it would be difficult to run the turbine smoothly and the power output would be small;
- a cut-out speed above which the turbine is shut down to avoid structural damage; and
- a rated speed at which the turbine produces its rated power output – at higher wind speeds the power output is limited to the rated value.

Second, the exceedance curve is converted to an equivalent power curve. For a wind turbine rotor of swept area A (m^2), operating in a wind of velocity V (m/s), the power, P_t (W), theoretically available is

$$P_t = 0.5\rho A V^3$$

where ρ is air density (kg/m^3).

In practice, the actual power output, P_a, will be significantly less as characterised by a power coefficient, C_p, defined as the ratio of actual to theoretical power. Betz (Wilberforce *et al.* 2023) demonstrated that C_p is unlikely to exceed 0.593 and no turbine yet produced has approached this figure: modern turbines might reach a C_p of about 0.45.

The tip speed ratio of a turbine is defined as $\lambda = \Omega R/V$, where Ω is the rotor angular velocity (rad/s) and R is the radius (m) at the blade tip. A typical variation of C_p with tip speed ratio is shown in Figure 13.20.

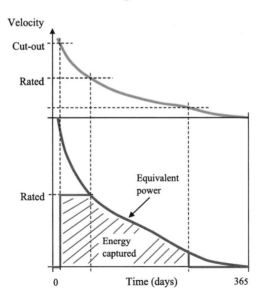

Figure 13.19 Site velocity exceedance curve (upper) and equivalent power curve

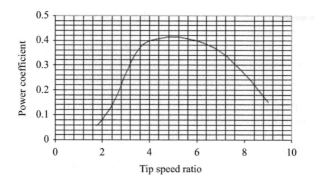

Figure 13.20   Power coefficient vs. tip speed ratio

A wind turbine connected to the grid is usually constrained to turn at a fixed angular velocity, so if the wind speed changes, so will the tip speed ratio. For each wind speed in the velocity exceedance curve, the power, P, that the turbine would produce is calculated and an equivalent power curve is drawn to the same time base – as depicted by the lower curve in Figure 13.19. When the wind speed exceeds its rated value, the power output is limited to the rated power. Below cut-in and above cut-out speeds, the power output is zero.

Third, the annual energy output is calculated as $\int P.dt$, with the integration limits set to correspond to the day limits of the shaded area under the power curve. In this way, an annual energy production estimate can be made for a single turbine and scaled up for a wind farm. Of course, the blade element and computational fluid dynamics (CFD) methods as applied in the previous section are also applicable here. In addition, there are commercial software tools for the appraisal of turbine layout within a wind farm.

Fourth, the effectiveness of a turbine is indicated by its capacity coefficient defined as the average power delivered in a typical year of operation divided by the rated power of an installation.

Last, potential deployment sites can be determined from spatial planning tools such as GOMap (see Section 10.3) that take into consideration technical and policy constraints to the use of land for this purpose.

One GOMap-type application (Clarke *et al.* 1996) examined the feasibility of wind farm proposals in the Caithness region of Scotland, with planning applications assessed under technical and environmental categories. The former, as listed in Table 13.1, relate to whether a proposal is technically 'Unlikely', 'Likely', or 'Favourable' according to various physical factors. These were scored 3, 2, and 1, respectively, for each of the subjects loaded into the GOMap program. This constituted a stage 1 analysis.

The Caithness region will then be covered by areas that score between a minimum of 3 (most favourable) and a maximum of 9 (most unlikely) on technical grounds. In this way, the GIS tool can produce a stage 1 opportunity map as shown in Figure 13.21.

The environmental factors of Table 13.2 are policy matters for which land is defined as 'Sensitive', 'Intermediate', or 'Possible', and scored as indicated.

Table 13.1 Wind farms, technical issues

Unlikely (3)	Likely (2)	Favourable (1)
>15 km from a grid line	Between 5 and 15 km from a grid line	<5 km from a grid line
Within MLURI 'Woodland', 'Wetland', 'Montane', and 'Build-up'	Within MLURI 'Peatland'	Within MLURI 'Rough Grazings' and 'Agricultural'
> 6 km from any road or vehicle track	Between 3 and 6 km from any road or vehicle track	<3 km from any road or vehicle track

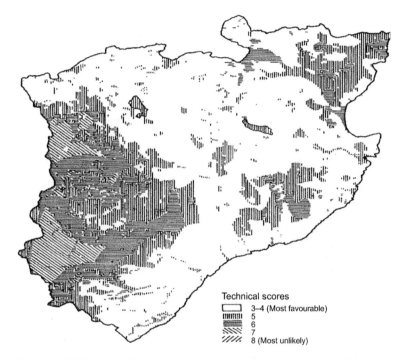

Technical scores

- 3–4 (Most favourable)
- 5
- 6
- 7
- 8 (Most unlikely)

Figure 13.21 Technical opportunity map for wind farms in Caithness

All areas in Caithness will then have a score ranging from a possible maximum of 25 to a possible minimum of 9. Figure 13.22 shows the result where a score of 11 or less is classed as being 'Possible' and 17 or more as 'Sensitive'.

The weight to be attached to the various environmental factors differs markedly with respect to single/small cluster wind turbines for community use. The general effect is a major liberalisation of policy concerning their location. Furthermore, whereas wind farms should be kept some distance away from dwellings, by their nature, community turbines will need to be close to settlements to be technically favourable.

Table 13.2   Wind farms, policy issues

Sensitive (3)	Intermediate (2)	Possible (1)
All bird areas	Not assigned	All non-bird areas
Landscape types 3, 5, and 7	Landscape types 2, 4, and 8	Landscape types 1, 1b, and 6
AGLVs and designated landscapes	Not assigned	The rest
SSSIs	Not assigned	The rest
Areas of exceptional archaeological interest	Areas of above-average archaeological interest	The rest
Not assigned	Caution areas for airports and masts	The rest
Within 200m of BT links	Not assigned	The rest
Within 1 km of a dwelling (as defined in relation to public roads)	Between 1 km and 2 km from a dwelling	Greater than 2 km from a dwelling
Not assigned	Within 2 km of a railway line	The rest

Policy scores
8 - 11
12
13
14
15-16
17-20 (Sensitive)

Figure 13.22   Policy opportunity map for wind farms in Caithness

The two opportunity maps can be combined to give an overall opportunity rating as shown in Figure 13.23. The dots on this map show the sites being targeted by developers at the time. It is clear that the higher energy yield sites are sensitive from a policy viewpoint leading to likely application refusal. This situation gives

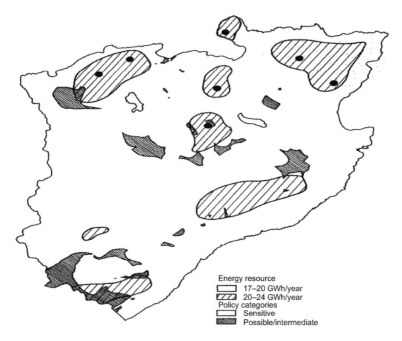

Figure 13.23 A combined opportunity map for Caithness

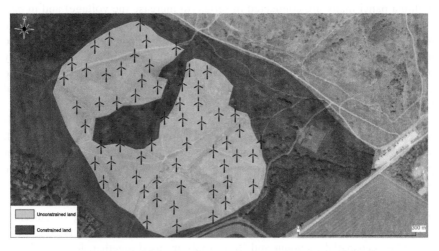

Figure 13.24 Wind turbine siting on unconstrained land

rise to the radical idea that GOMap-like tools be made available to developers so that areas of opportunity can be identified by them in advance of a planning application to improve the chance of success.

Figure 13.24 shows a GOMap unconstrained, semi-urban site with a wind farm superimposed taking account of turbine size and interspacing requirements.

13.3   Photovoltaic farms

Electricity is produced directly from sunlight by a photovoltaic cell by which solar radiation falling on two semiconductors in close contact generates an electrical voltage (Kazmerski 2006). An incident photon with sufficient energy will knock an electron out of its normal condition of being bound in the crystal structure of the semiconductor. These free electrons move across the junction between the two materials more easily in one direction than the other, giving one side a negative charge and therefore a potential difference relative to the other (Li *et al.* 2015). The amount of energy generated by a photovoltaic (PV) cell depends on the amount of sunlight that falls on it, and this in turn depends on location and exposure (Rhodes *et al.* 2014). As only some photons carry the required amount of energy, only a portion of the radiation can be converted to useful electricity, typically 4%–25%, although this is an evolving area and cells with over 40% efficiency are under development (NREL 2020). Commercially available solar cells can be made of monocrystalline silicon grown from a seed crystal (efficient but expensive), poly-crystalline silicon made from grains of the monocrystalline version (less efficient but cheaper), or a thin film of amorphous silicon (least efficient and cheapest) (Dhass *et al.* 2020). PV technology is widely available with increasing efficiency providing reasonable energy generation at decreasing production cost (Ahmed *et al.* 2009, Devabhaktuni *et al.* 2013).

The voltage and current generated by an individual cell are small so that within a panel many are connected in series (to increase the voltage) and in par-allel (to increase the current). Multiple panels are then connected in series and in parallel to form an array. An array needs to be held in place by a frame and fixed in place by either piled foundations or concrete weights on the legs. The PV array produces direct current, which is passed through an inverter to convert it to alternating current with a frequency of 50 Hz before being fed into the National Grid (Drax Group 2017).

When considering a possible layout, an important consideration is shading. If one panel is shaded, the output of the array feeding an inverter will drop. Overshading can arise from surrounding buildings and trees, and from other parts of the PV installation if arrays are not correctly spaced. Determining the optimal inter-row spacing is therefore necessary to maximise power output.

As described in some of the examples in Chapter 6, BPS+ can be used to predict the power output from any proposed scheme, freestanding or building-integrated. For this reason, no further consideration of PV modelling is given here.

When geospatial mapping and BPS+ tools are used together, it is possible to identify unconstrained land areas for PV farm deployment and available rooftops for building-integrated schemes. Figure 13.25 gives an example of sites identified in a project focused on Glasgow as described in Section 10.3.

Figure 13.25   Land and rooftop PV deployment

13.4   Chapter summary

BPS+ can be applied to support decisions on strategic renewable energy systems deployment by locating energy demand centres in close proximity to minimise transmission losses and/or operate synchronistically to improve network reliability. The design of such systems can be facilitated by exploiting GIS functionality to identify unconstrained areas of opportunity after screening for technical and policy constraints. BPS+ can be also applied to develop communities whose demand is matched to the available resources, either through new build, refurbishment, or a mix at each site.

References and further reading

Ahmed N A, Miyatake M and Al-Othman A K (2009) 'Hybrid solar photovoltaic/ wind turbine energy generation system with voltage-based maximum power point tracking', *Electric Power Components and Systems*, 37, pp. 43–60.

Black and Veatch (2005) *UK Tidal Stream Energy Resource Assessment*, The Carbon Trust, London.

Calautit K and Johnstone C (2022) 'A numerical CFD study on the integration and performance enhancement of oscillating aerofoil wind energy converters into buildings', *Proc. IBPSA Scotland uSim Conf.*, Glasgow, 3, pp.1–9.

Clarke J A, Connor G, Grant A D and Johnstone C M (2005) 'The impact of limiting tidal turbine capacity on the output characteristics from three complementary tidal sites in Scotland over the lunar cycle', *Proc. World Renewable Energy Congress*, Aberdeen.

Clarke J A, Connor G, Grant A D and Johnstone C M (2007) 'Design and testing of a contra-rotating tidal current turbine', *Power and Energy*, 221(A2), pp. 171–179, May.

Clarke J A, Connor G, Grant A D and Johnstone C M (2009) 'Development and in-sea performance testing of a single point mooring supported contra-rotating tidal turbine', *Proc. Ocean, Offshore and Arctic Engineering Conf.*, Hawaii, USA.

Clarke J A, Evans M, Grant A D, *et al.* (1996) 'Integration of Renewable Energies in European Regions', *Final Report for Project RENA-CT94-0064*, ESRU, University of Strathclyde.

Cockroft J and Samuel A (2014) 'Computational fluid dynamics study of a subsea turbine installation', *Consultancy Report E472*, ESRU, University of Strathclyde.

D'Oliveira B, Goulder B and Lee-Elliott E (2002) *The Macmillan Reeds Nautical Almanac*, ISBN 0 333 781821.

Devabhaktuni V, Alam M, Depuru S S, Green R C, Nims D and Near C (2013) 'Solar energy: trends and enabling technologies', *Renewable and Sustainable Energy Reviews*, 19, pp. 555–564.

Dhass A D, Kumar R S, Lakshmi P, Natarajan E and Arivarasan A (2020) 'An investigation on performance analysis of different PV materials', *Materials Today*, 22, pp. 330–334.

Drax Group (2017) *Why We Need the Whole Country on the Same Frequency*, https://drax.com/energy-policy/need-whole-country-frequency/.

Fraenkel P L (2002) 'Power from marine turbines', Power and Energy, 216(A1), pp. 1–14.

Fu S, Ordonez-Sanchez O, Martinez R, Johnstone C, Allmark M and O'Doherty T (2021) 'Using blade element momentum theory to predict the effect of wave-current interactions on the performance of tidal stream turbines', *Marine Energy*, 4(1), pp. 25–35.

Gov (2003), *Our Energy Future – Creating a Low Carbon Economy*, Department of Trade and Industry, URN 03/658.

Johnstone C M (2010) *https://emec.org.uk/about-us/our-tidal-clients/nautricity/*.

Johnstone C M (2018) Private Communication.

Johnstone C M, Pratt D, Clarke J A and Grant A (2013) 'A techno-economic analysis of tidal energy technology', *Renewable Energy*, 49, pp. 101–106.

Kazmerski L L (2006) 'Solar photovoltaics R&D at the tipping point: a 2005 technology overview', *Electron Spectroscopy and Related Phenomena*, 150(2–3), pp. 105–135.

Li Y, Grabham N J, Beeby S P and Tudor M J (2015) 'The effect of the type of illumination on the energy harvesting performance of solar cells', *Solar Energy*, 111, pp. 21–29.

McCombes T (2014) 'An unsteady hydrodynamic model for tidal current turbines', *PhD Thesis*, ESRU, University of Strathclyde.

Myers L and Bahaj AS (2004) 'Basic operational parameters of a horizontal axis marine current turbine', *Proc. 8th World Renewable Energy Congress*, Denver, USA.

NREL (2020) '*Best research-cell efficiency chart*', NREL National Center for Photovoltaics, https://nrel.gov/pv/assets/pdfs/best-research-cell-efficiencies.20200406.pdf.

Rhodes J D, Upshaw C R, Cole W J, Holcomb C L and Webber M E (2014) 'A multi-objective assessment of the effect of solar PV array orientation and tilt on energy production and system economics', *Solar Energy*, 108, pp. 28–40.

SG (2003) 'Securing a renewable future: Scotland's renewable energy', Scottish Executive, ISBN 0-7559-0766-3.

Stobart A F (1983) 'Wind energy: some notes on its collection, storage and application', *Energy World*, pp. 4–6.

Wilberforce T, Olabi A, Sayed E, Alalmi A and Abdelkareem M (2023) 'Wind turbine concepts for domestic wind power generation at low wind quality sites', *Cleaner Production*, 394.

Chapter 14

Conclusions and future perspectives

Some three decades ago, the Latham report (Latham 1994) recognised the chronic shortage of skilled workers within the construction industry, identified a lack of training and education, and called for team working within a process of continuous improvement. Such shortages are still apparent, and upskilling of the construction workforce remains a major challenge to improve the quality of future buildings and ensure the proper installation of new and renewable technologies. BPS+ provides a cost-effective means to improve productivity. It does this by amplifying user intellect to allow complex problems to be analysed in a timely and rigorous manner. In addition, it supports distributed team working on projects of any scale.

Figure 14.1(a) summarises the present situation with integrated, building performance simulation (IBPS) tool-sets used by in-house or external specialists to appraise design solutions proposed by others (in this diagram IBPS is a synonym for BPS+). Such a situation is ultimately unsatisfactory because the inherent delay in performance feedback cannot readily accommodate the decision-making requirements of practitioners.

A more desirable situation is summarised in Figure 14.1(b) and (c). In the former case, a tool-independent, intelligent interface is deployed to cater for the needs of diverse user types whilst ensuring adherence to common data and process models; in the latter case, a common approach to software engineering is adopted to ensure that BPS+ is underpinned by harmonised modelling methods and common standards for tool development, validation, and use.

There are compelling reasons why BPS+ developers can no longer afford to work as independent groups creating non-interchangeable software. A principal reason is the need to ensure that different tools produce compatible outputs when applied to the same problem. Although this can be achieved by subjecting each tool to formal validation procedures, the effort required to implement procedures and ensure ongoing conformance is formidable. A more elegant approach is to devise a repository of validated methods and a procedure for method combination, which may be used by developers to create customised simulation systems that embody an appropriate level of detail and utilise the most up-to-date techniques. This approach would allow the creation of tools on the fly as design solutions are described and evolved; the key to providing CAD-based performance appraisals in realtime. The benefits of such an approach are many and varied.

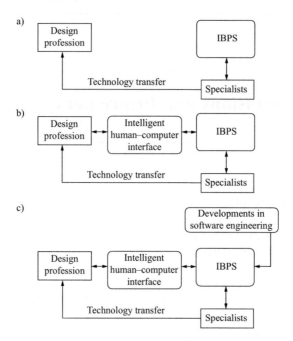

Figure 14.1   Progressive options for BPS+ deployment

- Developers need not be experts in all areas to formulate a BPS+ tool. Conversely, limited scope, special purpose programs can be created as required.
- New methods can be established and added once proven to enhance the scope or depth of future tools.
- Alternative abstractions of each method, detailed and simplified, micro and macro, and so on, can be made available, with blends selected at tool construction time.
- Domain-integrated tools can be readily constructed, or, alternatively, domain decoupling can be introduced to facilitate model reduction.
- Tool construction can be dependent on problem composition with methods added as the problem state evolves to support new appraisals.
- Different methods for the same modelling function can be attached to different problem parts. For example, the majority of constructions might be represented by a one-dimensional transient conduction scheme, and the remainder by a three-dimensional scheme to facilitate a local condensation study.
- Tools can have variable scope, from a single zone through a whole building or community.

To these ends, Mazzarella and Pasini (2009) reflected on the achievements of, and future prospects for using an object-oriented approach to BPS+ tool development.

The international BPS+ development community comprises diverse technical and business interests and has not yet evolved mechanisms by which long-term development goals can be agreed and collaboratively pursued. This situation is exacerbated by the 'Seven Deadly Sins' of software tool development as described by Maver (1995) in the context of Computer-Aided Architectural Design and recast in Table 14.1 for building performance simulation.

Table 14.1   The Seven Deadly Sins of BPS+

Macro-myopia:	The claim that a tool is widely applicable and easy to use but with no acknowledgement of its limitations. This situation derives from commercial and academic pressures on tool creators that make it difficult to admit that their product is other than state-of-the-art, to acknowledge that other tools may be more advanced in some respects, and to engage with the contributions from the wider community. Addressing this issue would, at the very least, act as a catalyst for developer collaboration and better tool interoperability.
Déjà vu:	The re-emergence of ideas that have striking similarities to earlier work but with no attempt to openly acknowledge or build upon what went before. Inappropriately mentored newcomers to the field often proffer solutions that have been previously tried and rejected or, more problematic, expend considerable effort implementing methods that do not contribute new functionality. This situation is compounded by 'hiding' such solutions behind user-friendly but deceptive user interfaces.
Xenophilia:	The importing of concepts from other disciplines (most typically computer science) that divert intellectual effort from researching what lies at the heart of the BPS+ challenge. A common example is a tool with an elegant optimisation algorithm that acts on results from a simplified core that gives misleading outputs by design. The absence of performance simulation as a core discipline makes it difficult to justify R&D funding resulting in slow progress and low impact.
Non-sustainability:	Where the R&D effort is devoted to over-indulgent tool development, such as the reimplementation of existing methods corresponding to a new software engineering paradigm, with little attention given to researching design solutions that yield improved quality of performance to building clients and users. This results in 'new' tools of diminished capability when compared to what went before. Indeed, tool vendors, commercial or academic, are more likely to announce a 'stunning new feature' – BIM model import, legislation compliance support, user plug-in capability, etc. – than invest effort in understanding how their tool can improve building performance.
Failure to validate:	Where a plethora of exotic claims relating to predictive preciseness is not subjected to any independent verification. In most other disciplines, this situation is considered unacceptable. The existence of an independent tool accreditation agency or, at the very least, the requirement that tools encapsulate standard validation tests that can be activated by users, would do much to eliminate spurious claims and improve tool fidelity vis-à-vis the real world.

(Continues)

Table 14.1   (Continued)

Failure to evaluate:	Where there is weak investigation of tool ease of use and applicability to real problems. The absence of credible user feedback means that future R&D is undirected and vulnerable to academic drift. The professional bodies could usefully take the lead in activities focused on application requirements capture to identify necessary new functionality, bring forward application standards, and inform the content of training provisions for practitioners.
Failure to criticise:	Where a community conspires to condone and even encourage self-indulgent speculation and solipsism: a bad example to set for the next generation of researchers and developers. A useful role for construction sector bodies would be to initiate activities that bring constructive criticism to bear on the capabilities and application deficiencies of all tools as a means to influence the funding bodies and thereby ensure a better future.

BPS+ tool creators are forced to address disparate requirements relating to user interfaces, data model manipulation, mathematical models, numerical methods, database management, software engineering, validation, documentation, and so on. Because there is limited development task sharing, and since no single organisation will possess the necessary expertise in all areas, contemporary tools can have substantial deficiencies vis-à-vis reality. To compound the problem, tools are promoted in a manner that hides these deficiencies and, thereby, implicitly undermines the development effort expended by competitors. This is clearly an intolerable situation and one that serves only to fragment a relatively small development community. The consequence of such behaviours is a slow pace of change, lack of standards, unnecessary duplication of effort, tension between developers, and a plethora of applications all with substantial shortcomings.

One professional body – the International Buildings Performance Simulation Association (IBPSA[1]) – has taken action to address the above issues through the publication of a future vision for the discipline (Clarke 2015) and through the fostering of activities to direct the called-for developments. The need now is for those bodies who represent the construction industry to take a proactive role in directing tool evolution and application by addressing pertinent questions such as the following.

- What are the costs and benefits of the high-resolution simulation approach?
- How can a business identify the correct software tools for its needs?
- Who should provide independent tool validation and accreditation?
- How can modelling tools best be embedded within a business?
- What are the different roles required from members of a simulation team?
- What training will staff require and who can provide this?
- In what ways will business work practices need to be adapted?

[1]https://IBPSA.org

- How are high-resolution models constructed and quality-assured?
- Where will I find approved databases for use in model definition?
- How are models calibrated before use and documented/archived thereafter?
- What are the requirements for standard performance assessments?
- What performance criteria should be used to appraise overall performance?
- What are the business risks and rewards associated with investing in the technology?

Some progress in these regards has already been made with professional bodies such as the Chartered Institution of Building Services Engineers (CIBSE) and the American Society of Heating, Refrigerating, and Air-Conditioning Engineers (ASHRAE) establishing mechanisms to support tool use in practice – such as the work of the Building Simulation and Energy Modelling group and the Building Energy Modelling Professional Certification programme, respectively. In a recent project, a CIBSE-led initiative involving industry and academic partners set out to explore an approach to the automated assessment of the operational resilience of submitted proposals (Clarke and Cowie 2020) based on long-term simulation. A prototype environment exists and awaits further development and exploitation.

Only by guarding against the Seven Deadly Sins and finding answers to such pertinent questions, will performance simulation tools become demonstrably quicker, cheaper, and better than the traditional approaches they seek to replace. That is an exciting prospect for the construction industry: improved productivity through an easily accessed, low-cost, computational approach to build environment design and refurbishment.

Before BPS+ programs can be routinely applied in practice, there are four issues to be addressed. First, since all design assumptions are subject to uncertainty, programs must be able to apply uncertainty bands automatically to their input data (Macdonald 1996). Performance risk can then be assessed based on predicted ranges. Second, validation procedures must be agreed and routinely applied as the modelling systems evolve in response to user requirements. Third, program inter-operability must be enabled so that design support environments evolve in response to interdisciplinary design needs. Finally, a mechanism is required to place program development on a task-sharing basis to ensure the integrity and extensibility of future systems. To this end, next generation researchers are invited to improve on specialised development platforms such as COMBINE (Augenbroe 1992), Modelica (Fritzson 2014), EKS (Clarke *et al.* 1992), and the like.

The evolution of BPS+ will benefit from a proactive approach by its user base. What is needed is an industry-led mechanism to define development priorities and application standards whilst demanding closer collaboration between vendor organisations and academia, and the production of students with core modelling skills.

The potential for the further evolution of BPS+, in terms of both capability and applicability, is immense. Realising such potentials will require a more proactive industry that acts collaboratively to specify the required functionality and approve application standards. At the same time, the BPS+ development community, commercial and academic, must find a way to replace unhelpful competition with task-sharing collaboration. Indeed this is the implicit message of Proposition 1 (of 17) in the above-referenced future vision paper:

IBPSA will evolve, through researcher and practitioner consultation, a requirements specification for future BPS tool application functionality, use this to periodically agree where the state-of-the-art falls short, and foster specific developments that address deficiencies and may be shared by all.

The modelling capabilities of BPS+ are expected to evolve apace as practitioners seek to address the complexities underpinning the transition to more sustainable energy systems. The hope is that the construction industry will come to recognise the built environment as a complex energy system and embrace the computational approach to design. Were the industry to resile reliance on simplified design tools and checkbox approaches and move to simulation they would acquire the ability to test virtual prototypes rapidly and at low cost. This powerful design paradigm would impart a new competitive edge in a world that is likely to be dominated by smart software agents.

The authors would like to conclude this book by quoting Tom Maver, Emeritus Professor of Computer Aided Design at the University of Strathclyde.

"Today's justifiable concern with the energy conscious design of buildings is a recurring echo of a fundamental theme dating back to the first human settlements: the need for shelter from an inclement environment. Of the factors that determine the degree of effective shelter provided by a building, none are more important, or more worthy of the designer's consideration, than its form, fabric and systems. These are also the determinants, for better or worse, of the range of attributes that determine the quality of the built environment – from the lifecycle cost of the building to its visual impact on the site."

"The exciting prospect that emerges, then, is of an integrated computer-based system in which the exploration of form, fabric and operation leads to a design solution embodying a balance between cost and performance, between investment and return, between need and aspiration."

"The new generation of integrated energy models exemplify the benefits that a sustained intellectual commitment can bring to the quality of the built environment. It is to be hoped that this endeavour will itself be a model, and an inspiration, for future developments in computer-aided design."

References and further reading

Augenbroe G (1992) 'Integrated building performance evaluation in the early design stages', *Building and Environment*, 27(2), pp. 149–61.

Clarke J A (2015) 'A vision for building performance simulation: a position paper prepared on behalf of the IBPSA Board', *Building Performance Simulation*, 8 (2), pp. 39–43.

Clarke J A and Cowie A (2020) 'A simulation-based procedure for building operational resilience testing', *Proc. CIBSE/ ASHRAE Technical Symposium*, 16–17 April, Glasgow, United Kingdom.

Clarke J A, Tang D, James K and Mac Randal D F (1992) 'Energy Kernel System', *Final Report for Grant GR07880*, Engineering and Physical Science Research Council, Swindon.

Fritzson P (2014) *Principles of Object-Oriented Modeling and Simulation with Modelica 3.3: A Cyber-Physical Approach*, IEEE, ISBN 9781-118-859124.

Latham M (1994) 'Constructing the team', *Joint Review of Procurement and Contractual Arrangements in the United Kingdom Construction Industry*, HMSO, ISBN 011752994X.

Macdonald I (1996) 'Development of sensitivity analysis capability within the ESP-r program', *PhD Progress Report*, ESRU, University of Strathclyde.

Maver T W (1995) 'CAAD's Seven Deadly Sins', *Proc. Computer Aided Architectural Design Futures*, National University of Singapore, https://papers.cumincad.org/cgi-bin/works/Show?35ac/.

Mazzarella L and Pasini M (2009) 'Building energy simulation and object-oriented modelling: review and reflections upon achieved results and further developments', *Proc. Building Simulation'09*, Glasgow.

Appendix A

Theoretical basis of ESP-r

ESP-r is a multi-physics modelling tool, which is primarily focused on modelling the performance of the built environment. The generic nature of the tool allows considerable latitude in application scale and detail – for example, to represent explicitly a thermostat, a computer monitor, a vehicle, or a portion of a ship as well as buildings at variable levels of detail. It holds a super-set data model supporting multiple domain solvers as well as acting as a repository for the descriptors needed by third-party tools.

The objective of the ESP-r project (as undertaken over five decades to date) has been to demonstrate the benefits of a computational approach to design whereby building performance can be emulated to test resilience in use. System evolution has been enabled through funded research and PhD projects, with the evolving source code made available under a no-cost, open-source licence. It has been extensively applied as an exploration tool within research projects, and commercially within consultancy projects undertaken by Energy Systems Research Unit (ESRU) staff and many others.

The system has been the subject of sustained development since 1974 and its performance has been scrutinised in numerous international projects (e.g., Strachan and Baker 2008, and many more since). The aim has always been to explore the possibility of emulating performance in a manner that corresponds to reality, supports early-through-detailed design stage application, and enables integrated performance assessments in which no single issue is unduly prominent. The approach taken respects the nature of the design process wherein the focus of interest evolves over time and each additional investment in descriptive detail may yield new insights. Given sufficient motivation, expert users can implement models, which can be viewed as a digital twin. As ESP-r is made available under a no-cost, open-source licence[1], it is possible to modify the numerical solution being undertaken and to add additional methods to support bespoke design questions.

The system attracted the Royal Society ESSO Energy Award in 1989[2] for its 'outstanding contributions to the advancement of science or engineering or technology leading to the more efficient mobilisation, use or conservation of energy resources'.

---

[1]https://www.esru.strath.ac.uk/Applications/esp-r/
[2]https://royalsociety.org/grants-schemes-awards/awards/past-awards/

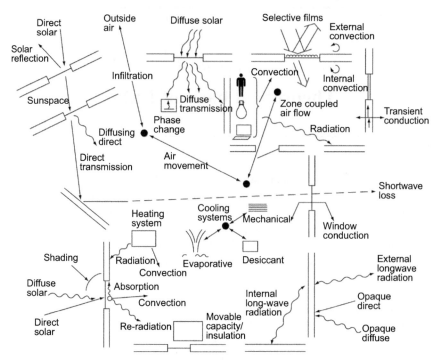

Figure A.1   Building energy flow paths (a diagram constructed in 1984 using a Rotring pen ;-).

Consider Figure A.1, which illustrates the interacting energy flow paths occurring within and between building spaces.

Within ESP-r, such a system is discretised and the small material and fluid volumes to result assigned a 'node' and characterised by state variables (temperature, pressure, moisture content, concentration, voltage, etc.), and thermo-fluid properties (conductivity, density, specific heat, viscosity, permeability, etc.). Nodes are then linked by time-dependent resistances that govern heat and mass exchanges. Mass, momentum, and energy balance equations are then established as appropriate and solved simultaneously at successive time steps as a simulation proceeds. The essential attribute of this approach is that the problem parameters can vary over time to reflect the impact of temperature/moisture changes, occupant behaviours, and control system action. In this way, the entire system is contained within a single numerical framework enabling an emulation of reality. The theoretical basis and software architecture of ESP-r is reported elsewhere (Clarke 2001[3]).

An ESP-r input model comprises coupled polyhedral zones that describe the geometry and related construction materials. Augmenting these zones is a series of networks, each of which describes an energy subsystem such as heating,

---

[3]and within related PhD theses available at www.esru.strath.ac.uk/publications.htm

ventilation, and air-conditioning plant, and fluid/electricity flow. The combination of the thermal zones and associated networks, together with occupancy characteristics, user-defined control strategies, and weather time series, forms a complete model of the building and its operational context. Models may start partially and simplified where information is not yet available and then be progressively evolved thereafter.

The problem, as described, is broken down into small control volumes and each is represented by a set of conservation equations for mass, energy, momentum, and/or species balance as appropriate. A model will comprise many control volumes, each assigned one or more conservation equations depending on the physical processes present in the associated region. A feature of the control volume approach is that the same physical process can be modelled at different levels of detail. For example, the air within a zone may initially be modelled as a single control volume against the assumption that the air is well mixed. At a later stage, this single control volume can be replaced by an airflow network (see below), a zonal airflow representation (Inard *et al.* 1996), or enhanced resolution within a computational fluid dynamics domain (Negraõ 1995) by which issues such as thermal stratification and contaminant distribution are addressed. Similarly, a 1D representation of the building fabric can be upgraded to a 3D representation to include the effects of the thermal bridges associated with construction details. The addition of a moisture balance equation to the control volumes would then facilitate the modelling of coupled heat and moisture flow as required in a condensation and mould growth study. The point is that it is possible to enhance resolution by adding control volumes and/or deepen the analysis scope by adding additional conservation equations to control volumes.

Control volume equations related to specific physical domains (building zones, environmental control systems, energy supply systems, fluid/power flow, etc.) are grouped, and each group is processed by equation solvers optimised for the equation types. The coordinated solution of all domain equation sets, with realistic weather data and control actions imposed, gives rise to a detailed and integrated view of performance evolution over time. In this way, the domain equation groups are solved independently but under supervisory control that respects the physical couplings between domains (e.g., conjugate heat and mass transfer). Examples of important couplings include: building thermal processes and daylight illuminance distribution; building/plant thermal processes and distributed fluid flow; building thermal processes and intra-room air movement; building distributed airflow and intra-room air movement; electrical demand and embedded power systems (renewable energy based or otherwise); and construction heat and moisture flow. The integrated solution of all domains therefore requires the coordinated application of the domain solvers applicable to the particular model being used. By allowing different domains to be represented at different levels of detail and solved at different frequencies, the approach ensures that the user has access to decision support that is in phase with the information to hand at each design stage.

A.1   Zone representation

The solution procedure within a simulation proceeds as follows. The conductive, convective, and radiation exchanges associated with a building's constructions are established as a set of energy balance equations and a direct solution method applied. The approach is based on a semi-implicit scheme, which is second-order time accurate, unconditionally stable for all space and time steps and allows time-dependent and/or state variable dependent boundary conditions and problem parameters. Iteration is employed for the case of non-linearity where system parameters (e.g., heat transfer coefficients) depend on state variables (e.g., air and adjacent surface temperatures). An optimised numerical technique is employed to solve the system equations simultaneously, whilst keeping the required computation to a minimum. Consider the energy balance equation set for the simple room of Figure A.2 when expressed in matrix notation: $A_{n+1}\theta_{n+1} = B_n\theta_n + C$, where A and B are coefficients matrices corresponding to the future $(n + 1)$ and present (n) time rows, respectively, θ a vector of node temperatures and flux injections, and C a known boundary conditions vector. (In this representation the 'x' symbols are place holders for the equation coefficient values establishes for a given simulation time step.) Since all parameters on the right-hand side are known, this matrix equation simplifies to $A_{n+1}\,\theta_{n+1} = D$, where D is a vector $(= B_n\,\theta_n + C)$.

Here, the form of A is for the case of unidirectional conduction and a single air node. The top left corner sub-matrix corresponds to the nodes (finite volumes) inside construction 1, and the sub-matrix (single equation) immediately below the diagonal corresponds to the construction 1 surface node. Similarly, there are sub-matrices corresponding to constructions 2 through 6. The last coefficient on the diagonal of A corresponds to the air node within the room. The coefficients on the upper and lower off-diagonals are radiation exchange coefficients connecting the inner surfaces.

Such a system of equations is solved efficiently by partitioning and reordering operations. Figure A.3 shows the outcome when applied to the coefficient matrix of Figure A.2. Note that null matrices are not shown; sub-matrices in the even rows and the last row are single equations; and θ_i, θ_{is}, and θ_a correspond to the temperatures of the intra-construction, surface, and air nodes, respectively.

The sub-matrices may be rearranged by changing rows such that the vectors corresponding to the intra-construction nodes are moved to the upper part of the system matrix, and the vectors (single equations) corresponding to the surface nodes are moved to the lower part. This gives rise to the matrix of Figure A.4, from which it may be observed that

- the block matrix at the top left corner consists of sub-matrices of internal construction nodes, and is block tri-diagonal;
- the block at the lower right corner is a full block matrix, comprising surface nodes and the air node(s); and
- the block matrices at the top right and lower left corners represent the connections between the innermost construction nodes and the corresponding surface node.

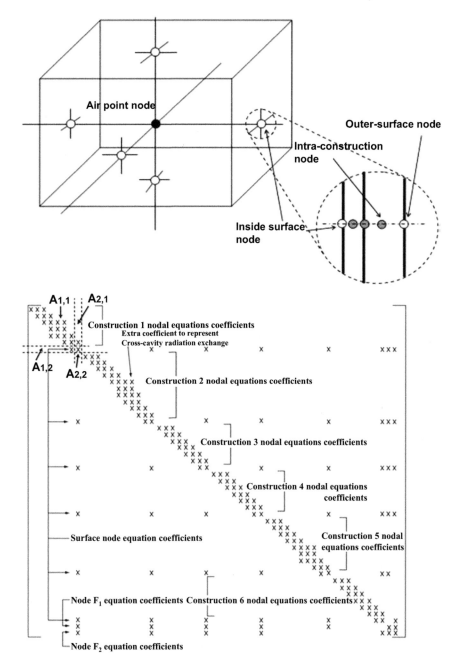

Figure A.2   The future time row matrix, A, for a simple room model

$$
\begin{bmatrix}
A_{1,1} & A_{1,2} & & & & & & & & & & & \\
A_{2,1} & A_{2,2} & & A_{2,4} & & A_{2,6} & & A_{2,8} & & A_{2,10} & & A_{2,12} & A_{2,13} \\
 & & A_{3,3} & A_{3,4} & & & & & & & & & \\
 & A_{4,2} & A_{4,3} & A_{4,4} & & A_{4,6} & & A_{4,8} & & A_{4,10} & & A_{4,12} & A_{4,13} \\
 & & & & A_{5,5} & A_{5,6} & & & & & & & \\
 & A_{6,2} & & A_{6,4} & A_{6,5} & A_{6,6} & & A_{6,8} & & A_{6,10} & & A_{6,12} & A_{6,13} \\
 & & & & & & A_{7,7} & A_{7,8} & & & & & \\
 & A_{8,2} & & A_{8,4} & & A_{8,6} & A_{8,7} & A_{8,8} & & A_{8,10} & & A_{8,12} & A_{8,13} \\
 & & & & & & & & A_{9,9} & A_{9,10} & & & \\
 & A_{10,2} & & A_{10,4} & & A_{10,6} & & A_{10,8} & A_{10,8} & A_{10,10} & & A_{10,12} & A_{10,13} \\
 & & & & & & & & & & A_{11,11} & A_{11,12} & \\
 & A_{12,2} & & A_{12,4} & & A_{12,6} & & A_{12,8} & & A_{12,10} & A_{12,11} & A_{12,12} & A_{12,13} \\
 & A_{13,2} & & A_{13,4} & & A_{13,6} & & A_{13,8} & & A_{13,10} & & A_{13,12} & A_{13,13}
\end{bmatrix}
\times
\begin{bmatrix}
\theta_1 \\ \theta_{1s} \\ \theta_2 \\ \theta_{2s} \\ \theta_3 \\ \theta_{3s} \\ \theta_4 \\ \theta_{4s} \\ \theta_5 \\ \theta_{5s} \\ \theta_6 \\ \theta_{6s} \\ \theta_a
\end{bmatrix}
$$

Figure A.3 Partitioning of the A coefficients matrix

$$
\left[\begin{array}{cccccc|ccccccc}
A_{1,1} & & & & & & A_{1,2} & & & & & & \\
 & A_{3,3} & & & & & & A_{3,4} & & & & & \\
 & & A_{5,5} & & & & & & A_{5,6} & & & & \\
 & & & A_{7,7} & & & & & & A_{7,8} & & & \\
 & & & & A_{9,9} & & & & & & A_{9,10} & & \\
 & & & & & A_{11,11} & & & & & & A_{11,12} & \\
\hline
A_{2,1} & & & & & & A_{2,2} & A_{2,4} & A_{2,6} & A_{2,8} & A_{2,10} & A_{2,12} & A_{2,13} \\
 & A_{4,3} & & & & & A_{4,2} & A_{4,4} & A_{4,6} & A_{4,8} & A_{4,10} & A_{4,12} & A_{4,13} \\
 & & A_{6,5} & & & & A_{6,2} & A_{6,4} & A_{6,6} & A_{6,8} & A_{6,10} & A_{6,12} & A_{6,13} \\
 & & & A_{8,7} & & & A_{8,2} & A_{8,4} & A_{8,6} & A_{8,8} & A_{8,10} & A_{8,12} & A_{8,13} \\
 & & & & A_{10,9} & & A_{10,2} & A_{10,4} & A_{10,6} & A_{10,8} & A_{10,10} & A_{10,12} & A_{10,13} \\
 & & & & & A_{12,11} & A_{12,2} & A_{12,4} & A_{12,6} & A_{12,8} & A_{12,10} & A_{12,12} & A_{12,13} \\
 & & & & & & A_{13,2} & A_{13,4} & A_{13,6} & A_{13,8} & A_{13,10} & A_{13,12} & A_{13,13}
\end{array}\right]
\times
\begin{bmatrix}
\theta_1 \\ \theta_2 \\ \theta_3 \\ \theta_4 \\ \theta_5 \\ \theta_6 \\ \theta_{1s} \\ \theta_{2s} \\ \theta_{3s} \\ \theta_{4s} \\ \theta_{5s} \\ \theta_{6s} \\ \theta_a
\end{bmatrix}
$$

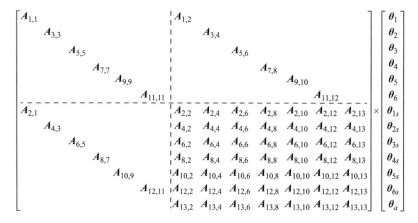

Figure A.4 Block partitioning of A

The lower left block matrix can be eliminated (as shown below) and thus the internal construction nodes and the surface nodes are decoupled. Therefore, only *m+1* equations need to be solved to obtain the nodal temperatures of the wall surfaces and the air within the room (where *m* denotes the number of constructions bounding the room). Once the temperatures of the wall surfaces and the air are obtained, the temperatures of the internal construction nodes may be obtained using backward substitution. This is done by taking into account the particular features of the equation set, that is the sub-matrices on the diagonal with odd numbers ($A_{1,1}$, $A_{3,3}$, …) are tri-diagonal, whilst even numbers are single equations; most of the off-diagonal sub-matrices have only a single coefficient.

Without losing generality, the notion of matrix inversion is used. However, here the inversion requires only the elimination of the lower diagonal coefficients in the first pass. Consider the following system of equations.

$$A_{1,1}\theta_1 + A_{1,2}\theta_{1s} = D_1 \tag{A.1a}$$

$$A_{2,1}\theta_1 + A_{2,2}\theta_{1s} + A_{2,4}\theta_{2s} + A_{2,6}\theta_{3s} + A_{2,8}\theta_{4s} + A_{2,10}\theta_{5s} + A_{2,12}\theta_{6s}$$
$$+ A_{2,13}\theta_a = D_{1s} \tag{A.1b}$$

$$A_{3,3}\theta_2 + A_{3,4}\theta_{2s} = D_2 \tag{A.1c}$$

$$A_{4,2}\theta_{1s} + A_{4,3}\theta_2 + A_{4,4}\theta_{2s} + A_{4,6}\theta_{3s} + A_{4,8}\theta_{4s} + A_{4,10}\theta_{5s} + A_{4,12}\theta_{6s}$$
$$+ A_{4,13}\theta_a = D_{2s} \tag{A.1d}$$

$$\vdots$$

To eliminate the θ_1 term from (A.1b), the θ_2 term from (A.1d), and so on, it is assumed that inverse matrices A^{-1} exist such that

$$\theta_1 = A_{1,1}^{-1}\left(D_1 - A_{1,2}\theta_{1s}\right) \tag{A.2a}$$

$$A_{2,1}\theta_1 + A_{2,2}\theta_{1s} + A_{2,4}\theta_s + A_{2,6}\theta_{3s} + A_{2,8}\theta_{4s} + A_{2,10}\theta_{5s} + A_{2,12}\theta_{6s}$$
$$+ A_{2,13}\theta_a = D_{1s} \tag{A.2b}$$

$$\theta_2 = A_{3,3}^{-1}\left(D_2 - A_{3,4}\theta_{2s}\right) \tag{A.2c}$$

$$A_{4,2}\theta_{1s} + A_{4,3}\theta_2 + A_{4,4}\theta_{2s} + A_{4,6}\theta_{3s} + A_{4,8}\theta_{4s} + A_{4,10}\theta_{5s} + A_{4,12}\theta_{6s}$$
$$+ A_{4,13}\theta_a = D_{2s} \tag{A.2d}$$

$$\vdots$$

Substituting (A.2a) into (A.2b), (A.2c) into (A.2d), and so on, gives

$$Sch\left(A_{2,2}\right)\theta_{1s} + A_{2,4}\theta_{2s} + A_{2,6}\theta_{3s} + A_{2,8}\theta_{4s} + A_{2,10}\theta_{5s} + A_{2,12}\theta_{6s} + A_{2,13}\theta_a$$

$$= D_{1s} - A_{2,1}A_{1,1}^{-1}D_1$$

$$A_{4,2}\theta_{1s} + Sch\left(A_{4,4}\right)\theta_{2s} + A_{4,6}\theta_{3s} + A_{4,8}\theta_{4s} + A_{4,10}\theta_{5s} + A_{4,12}\theta_{6s} + A_{4,13}\theta_a$$

$$= D_{2s} - A_{4,3}A_{3,3}^{-1}D_2$$

$$\vdots$$

$$A_{12,2}\theta_{1s} + A_{12,4}\theta_{2s} + A_{12,6}\theta_{3s} + A_{12,8}\theta_{4s} + A_{12,10}\theta_{5s} + Sch\left(A_{12,12}\right)\theta_{6s}$$

$$+ A_{12,13}\theta_a = D_{6s} - A_{6,5}A_{6,6}^{-1}D_6$$

where $Sch(A_{2,2})$ is the Schur complement (Zhang 2005) for $A_{2,2}$:

$$Sch\left(A_{2,2}\right) = A_{2,2} - A_{2,1}A_{1,1}^{-1}A_{1,2}$$

This results in seven equations with a full coefficient complement to be solved simultaneously. The solution gives the temperatures of the air and surface nodes. Back substituting the temperatures of the surface nodes gives the temperatures of the internal construction nodes. Taken together, this procedure gives the simultaneous solution of the complete matrix equation for the room. Considering the dimensions of $A_{1,1}$, $A_{3,3}$, and so on and assuming each is approximately 10×10, the complete matrix A will be 67×67.

Given that the method outlined above solves only seven simultaneous equations per zone in this case, the computational saving is substantial. The implementation of this procedure is complicated by the presence of intra-construction phenomena such as moisture transfer or the imposition of control-regulated heat injections/extractions corresponding to solar penetration and novel devices such as hybrid photovoltaic components and phase change materials. These phenomena require that certain terms of the corresponding conservation equations be not eliminated at matrix reduction time. The solution process can be extended to any number of thermal zones, with selected parts of the model treated in more or less detail.

A.2   Construction heat and moisture flow

To facilitate the modelling of inter-constructional moisture flow, the governing thermal equation associated with construction control volumes can be augmented with a vapour transport equation. Temperature and water vapour pressure are then the transport potentials and the coupling variables. As implemented in ESP-r (Nakhi 1995), the vapour flow within a homogeneous, isotropic control volume is given by

$$\rho_o \varphi \frac{\partial (P/P_s)}{\partial t} + \frac{\partial \rho_l}{\partial t} = \frac{\partial}{\partial x} \left(\delta_P^\theta \frac{\partial P}{\partial x} + D_P^\theta \frac{\partial \theta}{\partial x} \right) + S$$

where ρ is the density, o and l denote porous medium and liquid respectively, φ the moisture storage capacity, P the partial water vapour pressure, P_s the saturated vapour pressure, δ the water vapour permeability, D the thermal diffusion coefficient, and S a moisture source term. θ and P denote temperature and pressure driving potentials respectively, with the principal potential given as the subscript.

When converted to its finite volume equivalent, the above equation is non-linear and so the equations for this domain are solved by an iterative method, with linear under-relaxation employed to prevent convergence instabilities where discontinuities occur in the moisture transfer rate at the maximum relative humidity due to condensation.

A.3   Airflow

ESP-r is typically used for multi-zone simulations. In this case, a nodal network may be established to model inter-zone airflow, including infiltration and mechanical ventilation. The approach is based on the solution of the 1D,

Navier–Stokes equation assuming mass conservation and incompressible flow. The result is a set of non-linear equations representing the conservation of mass as a function of pressure difference across flow restrictions. To solve these equations, each non-boundary node is assigned an arbitrary pressure and the connecting components' flow rates are determined from a corresponding mass flow model. The nodal mass flow rate residual (error), R_i, for the current iteration is then determined from

$$R_i = \sum_{k=1}^{N} \dot{m}_k$$

where $\dot{m}_k$ is the mass flow rate along the k_{th} connection to node i, and N the total number of connections linked to node i. These residuals are used to determine nodal pressure corrections, P^*, for application to the current pressure field, P:

$$P^* = P - C$$

where C is a pressure correction vector. The process, which is equivalent to a Newton–Raphson technique, iterates until convergence is achieved. C is determined from

$$C = J^{-1}R$$

where R is the vector of nodal mass flow residuals, and J^{-1} is the inverse of the square Jacobian matrix whose diagonal elements are given by

$$J_{n,n} = \sum_{i=1}^{L} \left(\frac{\partial \dot{m}}{\partial \Delta P} \right)_i$$

where L is the total number of connections linked to node n. This summation is equivalent to the rate of change of the node n residual with respect to the node pressure change between iterations. The off-diagonal elements of J are the rate of change of the individual component flows with respect to the change in the pressure difference across the component (at successive iterations):

$$J_{n,m} = \sum_{i=1}^{M} -\left(\frac{\partial \dot{m}}{\partial \Delta P} \right)_i ; n \neq m$$

where M is the number of connections between node n and node m. To address the sparsity of J, its solution is achieved by lower-upper (LU) decomposition with implicit pivoting, known as Crout's method with partial pivoting (Press *et al.* 1986). Conservation considerations applied to each node then provide the convergence criterion: $\sum \dot{m} \rightarrow 0$ at all internal nodes. As noted by Walton (1982), there may be occasional instances of low convergence with oscillating pressure corrections required at successive iterations. A relaxation factor is therefore applied using a process similar to Steffensen Iteration (Conte and de Boor, 1972).

A.4   Intra-zone air movement

The above network approach may also be used to model airflow within zones, where the zone air is represented by multiple control volumes (Inard *et al.* 1996). However, there are limitations to this approach. First, there is no way to account for the conservation of momentum in the flow so the method is not applicable for driving flows. Second, the inter-volume couplings are characterised as a function of pressure difference and this requires the specification of a discharge coefficient, C_d, and little research has been done to determine applicable values.

To address these limitations, ESP-r has an in-built computational fluid dynamics (CFD) model for intra-zone airflow simulation. Flow inside a room is characterised by a set of time-averaged conservation equations for the three spatial velocities (U, V, W), temperature, θ, and concentration, C. For turbulent flows two additional equations are added, for turbulence intensity (k) and its rate of dissipation (ε). This is the well-known $k - \varepsilon$ model. As with the building thermal domain, these conservation equations are discretised by the finite volume method (Negraõ 1995, Versteeg and Malalasekera 1995) to obtain a set of linear equations of the form

$$a_p \emptyset_p = \sum_i a_i \emptyset_i + b$$

where $\emptyset$ is the relevant variable of state, p designates a cell of interest, i designates the neighbouring cells, b relates to the source terms applied at p, and a_p, a_i are the self- and cross-coupling coefficients, respectively. Because these equations are strongly coupled and non-linear, they are solved iteratively for a given set of boundary conditions. The SIMPLEC method is employed (Patankar 1980, Van Doormal and Raithby 1984) in which the pressure of each cell is linked to the velocities connecting with surrounding cells in a manner that conserves continuity. The method accounts for the absence of an equation for pressure by establishing a modified form of the continuity equation to represent the pressure correction that would be required to ensure that the velocity components determined from the momentum equations move the solution towards continuity. This is done by using a guessed pressure field to solve the momentum equations for intermediate velocity components U, V, and W and then using these velocities to estimate the required pressure field correction from the modified continuity equation. The energy equation and any other scalar equations (e.g., for species concentration) are then solved and the process iterates until convergence is attained. To avoid numerical divergence, under-relaxation is applied to the pressure correction terms. The solution of the discretised flow equations is achieved using the tri-diagonal matrix algorithm, favoured because of its modest storage requirements and computational speed.

A.5   HVAC systems

Within ESP-r, a heating, ventilation, and air conditioning (HVAC) system comprises coupled plant component models, each described by one or more control volumes. McLean (1982), Tang (1984), Hensen (1991), and Aasem (1993) have applied the control volume modelling technique to a variety of plant components. The resulting library of models allows the simulation of air-conditioning systems, hot water heating systems, and mixed air/hot water systems. Chow (1995) extended the capability of the control volume approach to plant modelling by identifying 27 'primitive parts' describing basic thermodynamic processes underpinning air-conditioning systems. This allows new plant component models to be synthesised by combining these parts as required.

The equations derived for plant systems are of a form similar to equations (A.1a) through (A.1d), which allows their solution using an efficient, direct method. The solution of the plant matrix is dictated by control interaction, where the state of system components is adjusted to bring about a desired control objective. The sensed node for plant control may be another plant component or a node in the building model, for example, the flow rate of cold water into a chiller coil may be controlled based on the relative humidity in a thermal zone. The plant-side matrix equation is substantially smaller than its counterpart on the building side. For example, the total number of equations for a domestic central heating system is approximately 150, whilst a building-side model for an average-sized house will require approximately 1000 equations. It is therefore possible to process the plant model as two equation sets for energy and mass balance without the application of partitioning to accommodate sparsity. These equation sets appear as additional sub-matrices in the A matrix of Figure A.2.

A.6   Electrical power flow

The approach to network airflow, as summarised above, can be applied to resolve the electrical power flows associated with the building (Kelly 1998). The electrical circuit is conceived as a network of nodes representing the junctions between conducting elements and locations where power is extracted to feed loads, or added from the supply network or embedded renewable energy components. Application of Kirchhoff's current law to some arbitrary node, i, with N connected nodes, forms the basis for the network power flow solution:

$$\sum_{j=1}^{n} \bar{I}_{i,j} = 0$$

where $\bar{I}$ is a complex number describing the magnitude of the current and its phase angle. The following two equations (adapted from Gross 1986) can be derived for any node i in an AC power systems model if the impedance characteristics of the connecting electrical equipment are known:

$$\sum_{p=1}^{n} V_i V_p Y_{i,p} \cos\left(\theta_i - \theta_p - \alpha_{i,p}\right) = -\sum_{q=1}^{y} P_{G_i^q} + \sum_{r=1}^{z} P_{L_i^r}$$

This equation represents a real power balance associated with node i. V is voltage, Y admittance (the inverse of impedance Z); θ and α are the phase angles associated with V and Y, respectively, P is real power (W), subscript p refers to some node connected to I, G is associated with power injected into node i, and L represents power drawn from the node. A similar expression can be derived for the reactive power, Q (VA_r), associated with node i:

$$\sum_{p=1}^{n} V_i V_p Y_{i,p} \sin\left(\theta_i - \theta_p - \alpha_{i,p}\right) = -\sum_{q=1}^{y} P_{G_i^q} + \sum_{r=1}^{z} P_{L_i^r}$$

The solution procedure is identical to that employed for inter-zone airflow, except that here the state variable is voltage, not pressure, and two equation sets must be solved corresponding to real and reactive power flows.

A.7   Solution procedure

If the whole system (building + plant + controls) equation set were set down in matrix equation form, these matrices would be topologically sparse and diagonally dominated. In ESP-r, the problem of processing a matrix of this type is overcome by subdividing the matrix into a number of connected matrices, each one representing a portion of the overall system. The solution of the entire system – now held in partitioned form – need only address fully populated matrices as opposed to sparse ones. Throughout a simulation, all finite volume time constants are evaluated at each time step. The form of the related equation type (mixed implicit/ explicit or pure implicit), and/or the processing frequency of the corresponding matrix partition, can then be adjusted to ensure a stable solution of high accuracy. As an added advantage, the method can accommodate any time-varying excitation and represent accurately control systems in terms of the spatial position of control sensors and actuators as well as the temporal factors inherent within controllers.

An ESP-r model will comprise constituent parts corresponding to an apt level of detail (constructions, moisture flow, electrical networks, airflow networks, and so on). However, as each part is based on the same finite volume considerations, when connected together they form a consistent mathematical description of the building and its systems, no matter the level of detail adopted.

Whole system representation is attained as the coordinated solution of the domain equations under control action that links model parameters as required (e.g., room air temperature to the mass flow rate induced by a fan). Essentially, the method links the different technical domain solution methods (direct or iterative) through a process of iterative 'handshaking' at key linkage points. This handshaking involves solving the subsystem equation sets separately, based on previous time step values of the coupled variables and then iterating to a solution. Figure A.5 summarises the approach.

At each building-side time step and for a given weather boundary condition, the air/liquid flow networks corresponding to the building and plant are established, control considerations imposed, and the equations solved. The solution of these

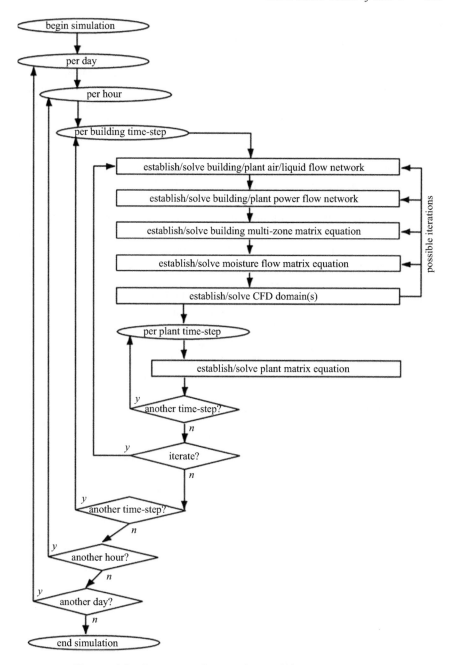

Figure A.5   Iterative solution of nested domains in ESP-r

networks gives the airflow and working fluid flow rates throughout the building and within the plant system respectively.

The electrical power flow network representing building-side entities (lighting, small power, PV facades, and so on) and plant components (fans, pumps, combined heat and power (CHP) plant, and so on) is established, constrained by control action, and the equations solved. The facility may be used to impose demand-side actions on load-consuming systems. (This network model and the preceding one for air/liquid flow may also be invoked at a higher frequency from within the HVAC solution loop.)

The building-side, multi-zone matrix equation is then established using the latest estimates of the fluid/power flows and plant-induced flux injections/extractions. Equation solution is achieved as indicated previously to obtain the building's temperatures and heat flows throughout the discretised model. Using the newly computed intra-construction temperatures, the moisture matrix equation is established and solved. This gives the moisture distribution within the building fabric. Using the building temperatures and airflow rates as boundary conditions, the CFD model is established and solved. This gives the intra-zone distribution of temperature, velocity, pressure, and contaminants. The building temperatures and air/liquid flow rates are then used, along with relevant control loops, to establish and solve the plant heat and mass flow matrix equations. The solution of these equations gives the plant temperatures and flow rates.

The various equation sets for the building model represent systems with very different time constants: in the fabric of the building, temperatures change slowly over a period of hours, whilst in the plant system, temperatures and flow change more rapidly and typically within minutes. A feature of the modular solution process is the ability to vary the solution frequency of each equation set, enabling the solution to capture the dynamics of a particular system. The zonal energy equations can be processed at hourly or sub-hourly intervals, while the plant and flow equations sets can be processed at a higher frequency that suits their smaller time constants. The advantage of this process is that plant and flow can be rigorously simulated without adding to the computational overhead of solving the much larger number of building equations at an unnecessarily high frequency.

To orchestrate the solution process, domain-aware conflation controllers are imposed on the different iterations. Consider, for example, the linking of the building thermal, network airflow, and CFD models. This employs a controller that ensures that the CFD model is appropriately configured at each time step (Beausoleil-Morrison 2000). At the start of a time step, the zero-equation turbulence model of Chen and Xu (1998) is employed in investigative mode to determine the likely flow regimes at each surface (forced, buoyant, fixed, fully turbulent, or weakly turbulent). This information is then used to select appropriate surface boundary conditions, whilst the estimated eddy viscosity distribution is used to initialise the k and ε fields. A second CFD simulation is then initiated for the same time step.

On the basis of the investigative simulation, the nature of the flow at each surface is evaluated from the local Grashof (Gr) and Reynolds (Re) Numbers, which indicate how buoyant and how forced the flow is respectively:

- $Gr/\text{Re}^2 \ll 1$ – forced convection effects overwhelm free convection;
- $Gr/\text{Re}^2 \gg 1$ – free convection effects dominate; and
- $Gr/\text{Re}^2 \approx 1$ – both forced and free convection effects are significant.

Where buoyancy forces are insignificant, the buoyancy term in the z-momentum equation is discarded to improve solution convergence. Where free convection predominates, the log-law wall functions are replaced by the Yuan *et al.* (1993) wall functions and constant boundary conditions imposed where the surface is vertical; otherwise, a convection coefficient correlation is prescribed (this means that the thermal domain will influence the flow domain but not the reverse). Where convection is mixed, the log-law wall functions are replaced by a prescribed convection coefficient boundary condition. Where forced convection predominates, the ratio of the eddy viscosity to the molecular viscosity (μ_t/μ), as determined from the investigative simulation, is examined to determine how turbulent the flow is locally:

- $\mu_t/\mu \leq 30$ – the flow is weakly turbulent and the log-law wall functions are replaced by a prescribed convection coefficient; and
- $\mu_t/\mu > 30$ – the log-law wall functions are retained.

The iterative solution of the flow equations is re-initiated for the current time step. For surfaces where heat transfer, *hc*, correlations are active, these are shared with the building thermal model to impose the surface heat flux on the CFD solution. Where such correlations are not active, the CFD-derived convection coefficients are inserted into the building thermal model's surface energy balance. Where an airflow network is active, the node representing the room is removed and new connection(s) are added to effect a coupling with the appropriate domain cell(s) (Clarke *et al.* 1995, Negraõ 1995). A technique by Denev (1995) is employed to ensure the accurate representation of both mass and momentum exchange in a situation where CFD cells and network flow components are of dissimilar size. Similar conflation mechanisms exist to coordinate the handshaking between other domain pairings.

A.8   Future developments

ESP-r is made available under a no-cost, open-source license[4] enabling source code modification as required. The modelling of new phenomena often only requires minor changes to the source code. For example, modelling of new thermal processes occurring in the building fabric requires only the implementation of a new source term, the adjustment of existing equation coefficients, or incorporating both together through a control function. This was the approach taken when implementing phase change materials and the nuances of double-skin facade modelling. In such cases, no solver adaptations were required. The system has also proved suited to the modelling of innovative components such as light-sensitive shading

[4]https://opensource.org/

devices and hybrid PV facades, both of which required enhancements to the resolution of existing models to allow slat angle adjustment in the former case and heat transfer surface geometry modelling in the latter. Future developments may require the creation of a new domain model. The issue then becomes the ease with which the new domain can be connected to those that exist.

Referring to the requirements for health-promoting buildings, specific solver developments could include an extension of construction moisture flow modelling from 1D to 3D, to facilitate a more accurate representation of moisture transport. This would be an important addition to ESP-r's capabilities in the modelling of surface phenomena such as condensation and growth of biological contaminants. It would not be appropriate to couple a detailed surface model to a lumped air volume, as the detail of the surface model predictions would be lost. Instead, adequate representation of surface and air vapour distribution dictates that a CFD domain couples to the detailed surface model. Such an extension affects the treatment of surface (de)absorption in the presence of an active CFD domain. This will require extensions to the linkages between the building thermal/moisture and CFD domains to handle the possible gridding cases: 1D/3D, 2D/3D and 3D/3D.

Concerning CFD, this is already widely applied to problems in building air movement, and the integrated CFD domain solver within ESP-r can be readily applied to many air quality issues with minor adaptation. Other developments will require additional equations and this will affect the solution procedure. For example, Kelly and Macdonald (2004) implemented a continuum-based model to enable the tracking of very small particles in flows with Stoke's number significantly less than 1. This new feature then enabled the modelling of fog as reported in the example application of Section 12.6.

The modelling of fire/smoke requires that extra equations be added to handle combustion reactions and the transport of combustion products (e.g., via the implementation of mixture fraction and gas radiation models). Such adjustments can be readily implemented within the code since the governing equations are of a diffusive type and so can be treated in the same way as the energy and species diffusion equations. In conjunction with the network flow model, the transient distribution of fire and smoke may then be applied as a boundary condition for the prediction of the movement of occupants during a fire.

In relation to airflow modelling using the network approach, possible improvements include the introduction of a pressure capacity term into the mass balance equations to facilitate the modelling of compressible flows and the introduction of transport delay terms to the network connections. Indeed, if the fluid velocity and geometrical characteristics of a particular connection are known, then a transport delay can be calculated automatically.

Future developments may also relate to the electrical domain, especially in relation to the modelling of micro-generation systems. These will often be subjected to control actions based on electrical as opposed to thermal criteria. Examples include voltage regulation, network stability, and phase balancing. Control of this type requires the creation of high-level controllers to couple the electrical and other associated domains. Such controllers will need to be intelligent enough to balance the conflicting demands of local comfort and community benefit within a microgrid. This type of control will likely include some form of finance-induced decision-making.

The detailed, dynamic modelling of microgrids will require the simultaneous modelling of buildings and micro-generation systems along with the electrical system to which they are connected. Modelling at this scale will benefit from parallel processing, where different aspects of the model can be allocated to different processor threads.

In the long term, additional solver developments may be implemented to bring about computational efficiencies and thereby assist with the translation of simulation to the early design stage. For example:

- context-aware solution accelerators may be embedded within the solvers to control their appropriate invocation;
- parallelism may be introduced to allow the different domains to be established and solved in tandem to reduce simulation times; and
- network computing might be exploited to allow different aspects of the same problem to be pursued at different locations as an aid to team working.

Such developments might be built upon entirely new methods such as 'intelligent matrix patching', whereby the coupling information between domain models is stored in a 'patch matrix' allowing the numerical model of the coupling components to be activated only when the actual coupling takes place. Furthermore, a greater level of coefficients management may be introduced to ensure that the matrix coefficients are only updated when required and otherwise never reprocessed. Such devices would lead to significant reductions in computing times.

Finally, it is worth mentioning that no matter how well the technical domains of the building are modelled, the effects of occupancy can vastly alter the physical behaviour and ultimately the predicted energy performance. These effects arise from two avenues: behaviour (e.g., the occupants' response to window opening) and attitude (e.g., the rejection of facilities on grounds other than performance). To facilitate the modelling of occupant behaviour in ESP-r two approaches are possible: 'typical' interactions may be included within a controller that has the authority to adapt the parameters of the affected domain models before solution; or, more realistically, an occupancy response model may be introduced by which the response to stimulus is explicitly represented.

References and further reading

Aasem E O (1993) 'Practical simulation of buildings and air conditioning systems in the transient domain', *PhD Thesis*, ESRU, University of Strathclyde.

Beausoleil-Morrison I (2000) 'The adaptive coupling of heat and air flow modelling within dynamic whole-building simulation', *PhD Thesis*, ESRU, University of Strathclyde.

Chen Q and Xu W (1998) 'A zero-equation turbulence model for indoor airflow simulation', *Energy and Buildings*, 28(2), pp. 137–144.

Chow T T (1995) 'Air-conditioning plant component taxonomy by primitive parts', *PhD Thesis*, ESRU, University of Strathclyde.

Clarke J A (2001) *Energy Simulation in Building Design* (2nd ed.), Butterworth-Heinnemann, Oxford.

Clarke J A, Dempster W M and Negraõ C (1995) 'The implementation of a computational fluid dynamics algorithm within the ESP-r system', *Proc. Building Simulation '95*, pp. 166–175, Madison.

Conte S D and de Boor C (1972) *Elementary Numerical Analysis: An Algorithmic Approach*, McGraw-Hill, New York.

Denev J A (1995) 'Boundary conditions related to near-inlet regions and furniture in ventilated rooms', *Proc. Application of Mathematics in Engineering and Business*, pp. 243–248, Institute of Applied Mathematics and Informatics, Technical University of Sophia.

Gross C A (1986) *Power Systems Analysis* (2nd ed.), Wiley, New York.

Hensen J L M (1991) 'On the thermal interaction of building structure and heating and ventilating system', *PhD Thesis*, Eindhoven University of Technology.

Inard C, Bouia H and Dalicieux P (1996) 'Prediction of air temperature distribution in buildings with a zonal model', *Energy and Buildings*, 24, pp. 125–132.

Kelly N J (1998) 'Towards a design environment for building-integrated energy systems: the integration of electrical power flow modelling with building simulation', *PhD Thesis*, ESRU, University of Strathclyde.

Kelly N J and Macdonald I (2004) 'Coupling CFD and visualisation to model the behaviour and effect on visibility of small particles in air', *Proc. eSim '04*, Vancouver, pp. 153–160.

McLean D J (1982) 'The simulation of solar energy systems', *PhD Thesis*, ESRU, University of Strathclyde.

Nakhi A E (1995) 'Adaptive construction modelling within whole building dynamic simulation', *PhD Thesis*, ESRU, University of Strathclyde.

Negraõ C O R (1995) 'Conflation of computational fluid dynamics and building thermal simulation', *PhD Thesis*, ESRU, University of Strathclyde.

Patankar S V (1980) *Numerical Heat Transfer and Fluid Flow*, Hemisphere, New York.

Press W H, Flannery B P, Teukolsky S A and Vettlering W T (1986) *Numerical Recipes: The Art of Scientific Computing*, Cambridge University Press.

Strachan P and Baker P (2008) 'Outdoor testing, analysis and modelling of building components', *Building and Environment*, 43(2), pp. 127–128.

Tang D C (1984) 'Modelling of heating and air conditioning systems', *PhD Thesis*, ESRU, University of Strathclyde.

Van Doormal J P and Raithby G D (1984) 'Enhancements of the SIMPLE method for predicting incompressible fluid flows', *Numerical Heat Transfer*, 7, pp. 147–163.

Versteeg H K and Malalasekera W (1995) *An Introduction to Computational Fluid Dynamics: The Finite Volume Method*, Longman, Harlow.

Walton G N (1982) 'Airflow and multiroom thermal analysis', *ASHRAE Transactions*, 88(2).

Yuan X, Moser A and Suter P (1993) 'Wall functions for numerical simulation of turbulent natural convection along vertical plates', *Heat and Mass Transfer*, 36(18), pp. 4477–4485.

Zhang F (Ed.) (2005) *The Schur Complement and Its Applications*. New York, NY: Springer.

Appendix B

Model file organisation and data quality assurance

B.1 Model folders and files

BPS+ tool vendors have adopted different approaches to model content organisation. EnergyPlus, for example, contains model data in a single Input Definition File (IDF) and, along with weather data files and a separate folder for simulation results, presents this to the user via a conceptually simple layout as shown in Figure B.1.

Here, there are IDF files that support different simulation periods and a bash script created by the user to run assessments. In the Output folder, the file names are tool generated and thus care is needed not to overwrite these files when a different IDF file is used. By convention, the weather file is in the same folder as the IDF files.

ESP-r employs a scheme in which the model files are grouped by type as shown in Figure B.2 for a model representing a Trombe–Michel wall included in an office facade. On the right are some of the images held with the model.

The model folders are as follows.

- *cfg* holds model configuration files that define the content of a model and simulation directives such as pre-set simulation periods and suggested weather files. This folder also holds the topology of the 3D geometry.
- *ctl* holds control definitions. Here focused on controlling the facade vents.
- *dbs* holds model-specific databases. Here, a clone of the standard construction composites with added Trombe wall compositions.
- *doc* holds model documentation. Here in html and markdown format as well as the log file for the project.
- *images* holds project images. Here the flow network, expected flow rates, and a Blender file as well as the Wavefront files from which the Blender model was generated.
- *nets* holds plant and airflow network descriptions. Here an airflow network implementing the Trombe wall flow regime.
- *rad* holds Radiance files as generated by ESP-r in support of visualisations and luminance distribution tracking within a simulation.
- *tmp* holds simulation results. Since these files are usually large, they are normally deleted after processing since they can be readily recreated.
- *zones* holds the geometry, construction, and operational definitions for each zone.

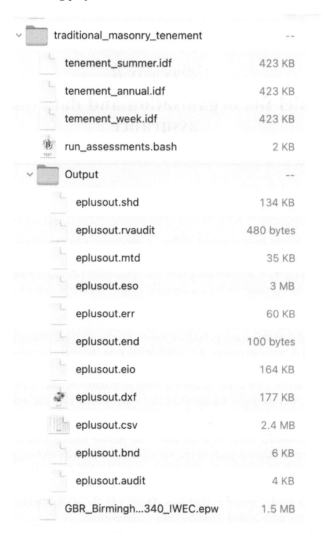

Figure B.1   EnergyPlus model folders and files

Consider a parametric study that examined different combinations of flows, controls, and problem composition. Assigning a short notation for each change variant might result in files and model entities such as Tr_FlA_CtD_CnB, which convey intent but are somewhat cryptic.

Another issue is file archiving as a project proceeds. Figure B.3 shows some possibilities.

Some are compressed single file archives of the whole model where the naming scheme identifies the date of creation. The 'office_ctl' folder is the current working version of the model, and 'office_ctl_last' is the last working version. During the rapid evolution of a model, access to a recent set of files supports visual comparisons (see below) and acts as a repository to draw from if a file is corrupted.

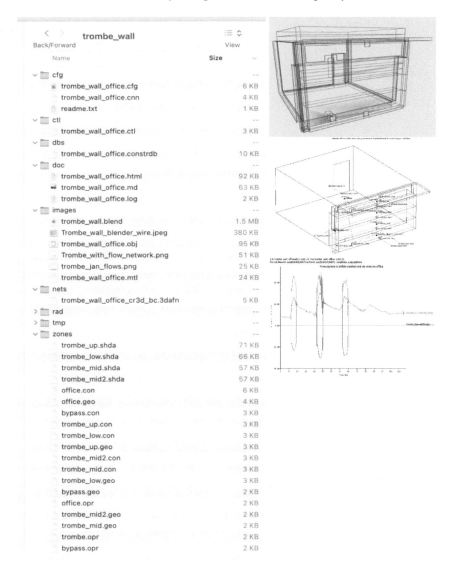

Figure B.2   ESP-r model folders and files

Such models can be viewed and manipulated within a tool's interface (Project Manager in the case of ESP-r), and it is possible to use external applications to directly view and compare problem files. In Figure B.4, for example, the files of two ESP-r models are compared using the Meld[1] application to determine files that have changed.

---

[1]https://meldmerge.org/

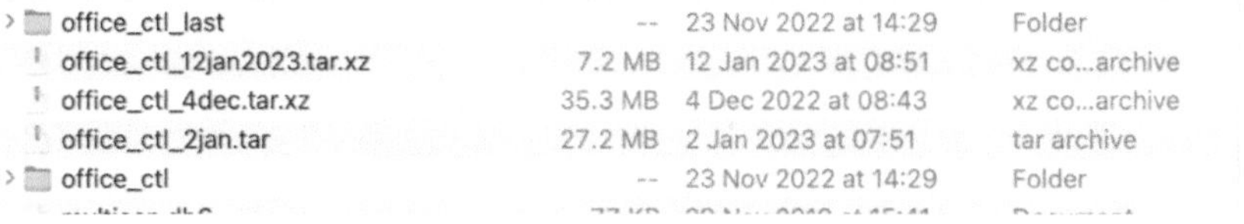

> 🗀 office_ctl_last	--	23 Nov 2022 at 14:29	Folder
⌐ office_ctl_12jan2023.tar.xz	7.2 MB	12 Jan 2023 at 08:51	xz co...archive
⌐ office_ctl_4dec.tar.xz	35.3 MB	4 Dec 2022 at 08:43	xz co...archive
⌐ office_ctl_2jan.tar	27.2 MB	2 Jan 2023 at 07:51	tar archive
> 🗀 office_ctl	--	23 Nov 2022 at 14:29	Folder

Figure B.3   Examples of model archiving

It is then possible to drill down to discover the specific changes. Figure B.5, for example, identifies the addition of thermal mass to a zone.

B.2   Model quality assurance

A BPS+ input model is typically extensive and embodies many complex dependencies. Consequently, there are reasons why a model might fail to represent adequately the design intent. Problems can stem from semantic misunderstandings derived from the different conceptual outlooks of interface and building designers, or from uncertainties associated with design parameter values and their evolution throughout the design process. Several actions can assist with the quality assurance (QA) of models before use.

The act of opening a model initiates myriad checks, for example, ensuring that zones are bounded, that attribution is complete, that domain couplings are sensible, and that data are within sensible ranges. In the last case, this might involve data processing to derive quantities that are directly relevant to the user – such as reference *U*-values, zone volumes, and surface areas. Any further checks will require user participation. Where problems are detected, it is likely that the program will give an indication of the remedial actions required.

The ability to display and manipulate 3D model images is a helpful feature. Less obvious is the need to adopt a clear and consistent naming convention that helps other members of the design team and client to comprehend the virtual embodiment of the project. It is a truism that to name something is to own it. That said naming areas in a building model 'room 1', 'room 2', etc. convey no information on use and design intent.

Most models are built on assumptions that should be documented to aid model adaptation and results interpretation. For example, where a building is large, only a representative portion might be modelled under the assumption of an adiabatic condition at carefully chosen cut planes. This assumption may need to be removed when the model scope is extended. The issue of model opacity has broad implications. Practitioners are often required to shift attention between projects and there are limits to their ability to maintain a coherent mental model of multiple projects, especially if a project has been idle for several months and there is an urgent need to reactivate it.

Some faults only become evident at simulation time. To counter this, it is possible to subject the model to standard tests and scrutinise the results. For

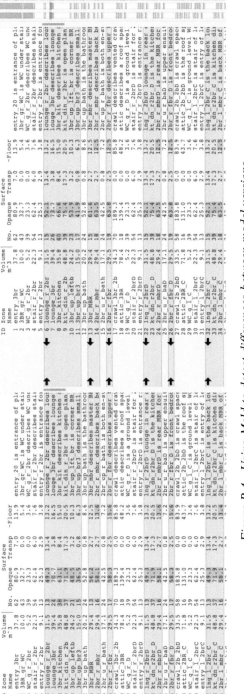

Figure B.4   Using Meld to detect differences between model versions

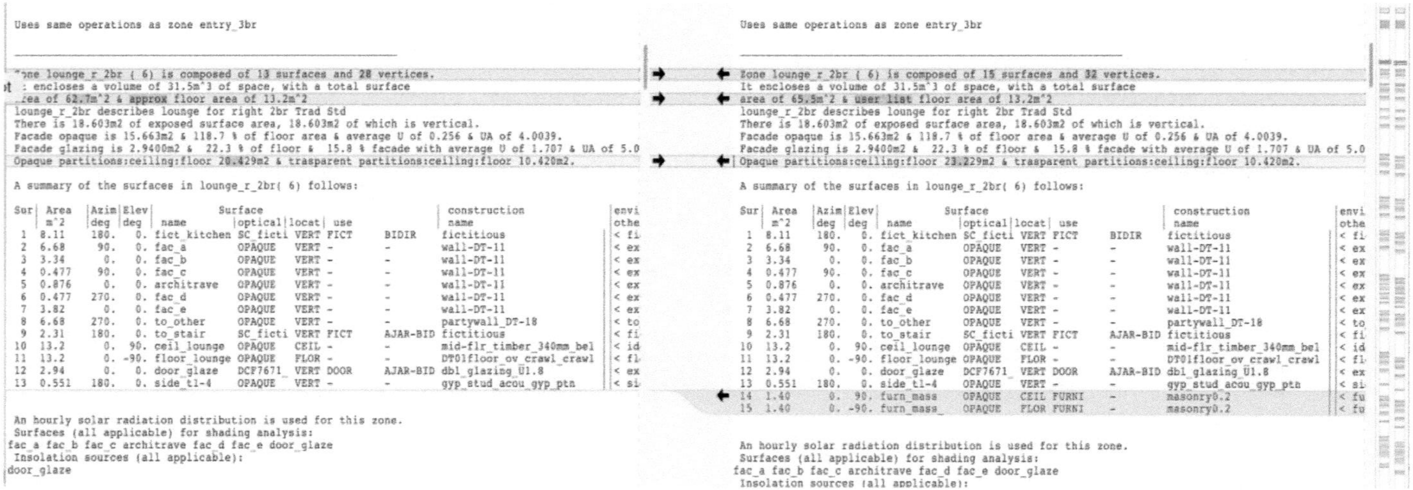

Figure B.5 Using Meld to identify a specific model change

example, an inappropriate control system definition will be obvious in a graph of the controlled parameter variation over time. Likewise, inappropriate data in a weather file will have a noticeable impact.

It is important to retain the association between a model and the simulations that can be sensibly applied to it. ESP-r, for example, includes the concept of 'simulation parameter sets' that define the pre-conditioning period, the simulation period and time step, the weather file to use, and the performance data to be recorded.

Various levels of QA support are normally available within a BPS+ program. For example, in Figure B.6, a wireframe view of a thermal zone is accompanied by a list of surface attributes. When the focus moves to a specific surface, the displayed attribution changes appropriately.

Because a model is typically created using accelerators – for example, by selecting a pre-constructed casual gain profile or control system definition from a menu – ESP-r provides facilities to scan the model contents, generate reports that can be used to expose the underlying data, and check model consistency. This includes options to adapt the level of detail depending on the topic. The ESP-r project manager exports model aspect descriptions as a Markdown document that can be converted to a Word document. In Figure B.7, a user has made selections that guide report scope and detail.

Best practice dictates that models are extensively documented. Users might be required to rapidly reacquaint themselves with a model's assumptions, goals, and current state. In addition to information that might be gleaned from the tool interface, there are parallel streams of information that help maintain a model's integrity.

One technique is to embed within the model an 'elevator pitch' that summarises the project aims. This should then be augmented as the project progresses based on analysis outcomes and in response to design evolution requiring model adaptation. The following is an example of a model overview indicating intent and content.

This model represents one level of an office block with reception, meeting room, an open plan space, and manager's office. It is intended to test the operation of a mixed mode ventilation system (natural + mechanical). It has a flow network to represent infiltration, inter-zone air mixing, and VAV system terminal boxes located in the ceiling plenum. The occupied spaces include typical furnishings and fittings. Alternative control system definitions are included relating to natural ventilation only, mechanical ventilation only, and mixed mode operation.

The building is located in Ottawa in a commercial center location. The associated weather file corresponds to Ottawa International Airport. The range of annual air temperature at this site is extreme ($-25\,°C$ to $35\,°C$). Some of the construction details are specific adaptations for these extremes. The transition seasons often include significant swings in temperatures so that the building might need to switch from heating to cooling operation several times a week. This eventuality is included in the model's simulation parameter sets.

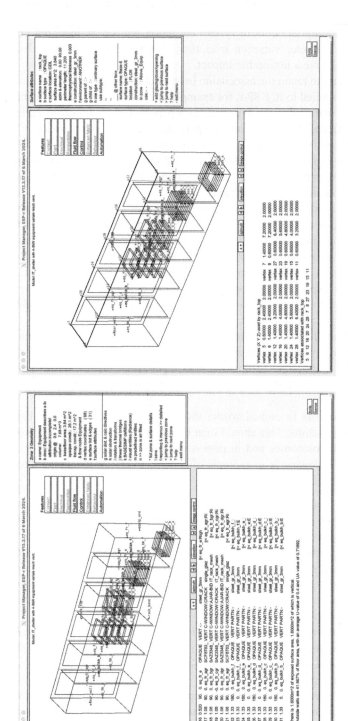

Figure B.6 Model attribution linked to geometric focus

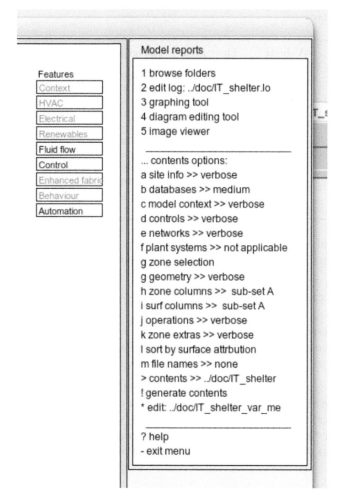

Figure B.7   Defining QA report content

Summaries related to model integrity are essential. In the following example the zone named 'recep_tb' is highlighted as having incomplete attribution and will cause the model to be rejected by the simulator. Where this ten-entity model was intended to be later replicated to represent other floors (for example), it would be better to adopt another naming convention: for example, 'manager' to 'manager_accounting', 'conference' to 'conference_floor2'.

Zone summary

Zone	Volume	Floor area	Description
manager	40.5	13.5	Manager's office
general	175.5	58.5	Open plan space with 4 desks
conference	141.8	47.2	Conference room for 10–12 people
reception	182.3	60.8	Reception adjacent to conference room
ceil_void	89.5	180.0	Ceiling void
vav_box_m	0.2	0.6	VAV terminal box for 'manager'
floor_void	36.0	180.0	Floor void
ceiling_belo	36.0	180.0	ceiling below floor as boundary condition
recep_tb	0.2	0.6	**Attribution incomplete!**
conf_tb	0.2	0.6	VAV terminal box for 'general'
all	702.0	721.8	–

It should also be possible to expand zone summaries and include familiar derived values whether used or not in a simulation:

Floor area 721.8 m^2
Wall area 82.9 m^2
Window area 69.2 m^2
External surface area 152.1 m^2
External walls are 11.5% of floor area with a standard U-value of 0.267
Glazing is 45.5% of external wall area with a standard U-value of 2.81

Models typically comprise a hierarchy of entities and QA reports allow anomalous patterns to be detected. In the following list of the surfaces in 'manager', for example, there are surfaces named 'spandral' and 'vent' that are associated with the construction 'insul_frame'. Is this the correct assignment? A critical review of facade details might suggest a different treatment of the ventilation device.

Surfaces in 'manager'

Surface	Construction	Other side
pt_general	gyp_gyp_ptn	pt_mgrs:general
part_frame	insul_frame	part_frame:general
door	Door	door:general
pt_other	gyp_gyp_ptn	identical environment
ceiling	Ceiling	ceiling:ceil_void
floor	flr_ov_pln	man_floor:floor_void
south_spandral	extern_wall	external
vent	insul_frame	external
south_frame	insul_frame	external
south_glazing	dbl_glz	external

(Continues)

(*Continued*)

Surface	Construction	Other side
part_glaz	dbl_glz	part_glaz:general
adesk_t	Door	adesk_t_:manager
adesk_t_	Door	adesk_t:manager
...	...	...
...	...	...
ccab_fr	steel_gr_3 mm	ccab_fr_:manager
ccab_fr_	steel_gr_3 mm	ccab_fr:manager
cpaper_fr	file_papers	cpaper_fr_:manager
cpaper_fr_	file_papers	cpaper_fr:manager
office_inlet	steel_gr_3 mm	office_inlet:ceil_void
office_ret	steel_gr_3 mm	office_ret:ceil_void

A terse level of reporting, as above, can be counterproductive but so too can long names that are difficult to parse visually. Further, literal naming such as 'south_ spandrel' and 'south_frame' will only confuse if the building is subsequently rotated. In such cases 'front_*' or 'right_*' might be less ambiguous. It is sometimes helpful to request further details such as in the following example.

Construction aspect areas (m^2)

Construction	Total	To ambient	To other zone	Back-to-back	To ground	To similar
extern_wall	2.3	2.3	0.0	0.0	0.0	0.0
insul_frame	2.3	1.4	1.0	0.0	0.0	0.0
door	5.3	0.0	3.0	2.3	0.0	0.0
dbl_glz	10.4	5.3	5.0	0.0	0.0	0.0
ceiling	13.2	0.0	13.2	0.0	0.0	0.0
gyp_gyp_ptn	27.0	0.0	13.5	0.0	0.0	13.5
flr_ov_pln	13.5	0.0	13.5	0.0	0.0	0.0
...	...	...	...	...	...	...
...	...	...	...	...	...	...

The feedback of calculation directives supports the interpretation of results at a later stage. For example:

Zone solar radiation distribution is explicitly calculated including shading from 6 external obstructions.

Zone view factors are explicitly calculated including an allowance for contents.

A model can include entities for use with other tools. For example, objects representing zone content that will participate thermodynamically can also have associated visual entities for use in Radiance and Blender:

Visual entities

name	composition
desk_top	door
desk_leg_	steel_pl_3
desk_leg_	steel_pl_3
desk_leg_	steel_pl_3
...	...
...	...

Models also include schedules of entities that contribute heat and/or moisture to zones. In the case of prescribed casual gains (for use where stochastic prediction is not employed), this will include attributes such as the sensible and latent component of each gain and the radiant/convective split of the former. Summary tables allow inconsistencies to be spotted. For example, in the following example, the radiant/convective spit in the first entry is suspicious because it violates the pattern for the 'people' type. The report might also raise the question, 'is lighting set at 10 W/m^2 consistent with best practice'? Furthermore, if position-specific comfort were of interest, this report would signal that explicit occupant modelling should be activated in place of this prescription. Where remote working is possible, there should be alternative weekday schedules.

Casual gains

Day type	Gain type	Unit	Hours	Sensible gain	Latent gain	Radiant component	Convective component
Weekday	people	W	0–7	10.0	10.0	0.5	0.5
Weekday	people	W	7–8	20.0	10.0	0.6	0.4
Weekday	people	W	8–9	60.0	30.0	0.6	0.4
Weekday	people	W	9–12	100.0	50.0	0.6	0.4
Weekday	people	W	12–14	65.0	32.5	0.6	0.4
Weekday	people	W	14–17	100.0	50.0	0.6	0.4
Weekday	people	W	17–24	0.0	0.0	0.5	0.5
Weekday	lighting	W/m^2	0–8	0.0	0.0	0.5	0.5
Weekday	lighting	Wm^2	8–18	10.0	0.0	0.3	0.7
Weekday	lighting	Wm^2	18–24	0.0	0.0	0.5	0.5
Weekday	equipment	Wm^2	0–8	0.0	0.0	0.5	0.5
Weekday	equipment	Wm^2	8–18	5.0	0.0	0.4	0.6
Weekday	equipment	Wm^2	18–24	0.0	0.0	0.5	0.5
Saturday	people	W	0–7	0.0	0.0	0.5	0.5
Saturday	people	W	7–8	20.0	10.0	0.5	0.5
Saturday	people	W	8–9	60.0	30.0	0.5	0.5
...	...	...	...	...	...	...	...
...	...	...	...	...	...	...	...

Where ancillary spaces in a building are being explicitly modelled, their attributes will also be of interest. In the following example, a ceiling void is detailed and this includes the surface 'Use' attribute set to 'grill' indicating that the surface is involved in an airflow network. The report verbosity level is increased in this example to include geometry – important where a warning has been issued indicating an inconsistent edge definition or duplicate entity names.

Summary of zone ceil_void

Surface	Use	Construction	Environment other side
structure	–	conc_250	identical
south_edge	–	extern_wall	external
east_edge	–	extern_wall	external
north_edge	–	extern_wall	external
core_b	–	ceiling_rev	identical
core_a	–	gyp_blk_ptn	identical
core_c	–	gyp_blk_ptn	identical
ceiling	–	ceiling_rev	ceiling:manager
office_inlet	grill	steel_gr_3 mm	office_inlet:manager
office_return	grill	steel_gr_3 mm	office_ret:manager
ceiling_gen	–	ceiling_rev	ceiling:general
op_a_inlet	grill	steel_gr_3 mm	op_a_inlet:general
op_b_inlet	grill	steel_gr_3 mm	op_b_inlet:general
mb_o_front	–	hvac_case	mb_o_front:mix_op_man
mb_o_right	–	hvac_case	mb_o_right:mix_op_man
mb_o_left	grill	hvac_case	mb_o_left:mix_op_man
...	...	...	...
...	...	...	...

A summary of the edges

Vertices	List	Name	Perimeter
10	19,20,21,22,23,24,25,26,27,28	structure	60.0
8	1,2,4,5,6,21,20,19	south_edge	25.0
11	6,7,8,18,11,12,13,24,23,22,21	east_edge	37.0
6	13,14,15,16,25,24	north_edge	19.0
6	16,17,9,27,26,25	core_b	25.0
17	2,1,3,34,35,38,37,36,35,34,33 ...	ceiling	22.74
...	...	...	...
...	...	...	...

Summary of vertices

Vertex	X	Y	Z	Vertex	X	Y	Z
1	0.0	0.0	3.0	53	10.4	4.9	3.0
2	3.0	0.0	3.0	54	10.4	4.5	3.0
3	0.0	4.5	3.0	55	10.0	4.5	3.0
4	6.0	0.0	3.0	56	6.0	13.5	3.0
5	9.0	0.0	3.0	57	5.4	14.5	3.0
6	12.0	0.0	3.0	58	5.0	14.5	3.0
...	...	...	...	...	...	...	...
...	...	...	...	...	...	...	...

A model will also optionally include definitions corresponding to airflow, plant, control, and electricity networks. In the case of an airflow network, the entities represent air nodes within thermal zones, plant components, and at external boundaries, linked by flow components corresponding to flow paths and sources/ sinks of pressure. Confirming that the network attributes are a reasonable abstraction of the building is assisted by outputs such as the following. In this example, some of the node names are obscure and should be explained in the project documentation.

> *The model includes a mass flow network representing inter-zone airflow, infiltration via facade leakage, flows from ceiling mounted VAV boxes, and plenum air extract. The network comprises 37 nodes, 46 components and 45 connections.*

Summary of network nodes

Node	Fluid	Node Type	X	Y	Z
manager	Air	Internal, unknown	1.5	2.25	1.5
general	Air	Internal, unknown	6.0	3.0	1.5
conference	Air	Internal, unknown	7.5	15.0	1.5
reception	Air	Internal, unknown	7.5	9.75	1.5
mb_recep	Air	Internal, unknown	4.5	9.3	3.25
mb_conf	Air	Internal, unknown	4.5	15.3	3.25
BW-Op01:008	Air	Boundary wind	1.5	−0.9	0.7
BW-Op02:010	Air	Boundary wind	4.5	−0.9	0.7
BW-Cr02:012	Air	Boundary wind	4.5	−0.9	1.95
...	...	...	...	...	...
...	...	...	...	...	...

Summary of network components

Component	Fluid	DB ref.	Associated with	X	Y	Z	Description
DABiz01:003	Air	130	manager: door	0.5	4.5	1.5	door width 0.1m; height 3.0m; reference height 1.5m; discharge factor 0.5
BiCrz01:003	Air	120	manager: door	0.5	4.5	0.0	crack width 0.002m; crack length 8.0m
GrOpz01:008	Air	40	manager: vent	1.5	0.0	0.7	orifice opening area 0.08 m^2; discharge factor 0.65
GrMEI01:044	Air	30	manager:office_inlet	1.7	1.2	3.0	constant volume fan 0.056 m^3/s
GrOpz01:045	Air	4	manager:office_return	1.7	3.7	3.0	orifice opening area 0.16 m^2; discharge factor 0.65
...	...	...	...	...	...	...	...
...	...	...	...	...	...	...	...

When airflow predictions do not match expectations, it is often related to the network being ill-formed. For example, a window-framing element associated with office modules on the same facade may have a leakage characteristic that can be represented as a simple orifice but in one office this definition has been omitted. The connection between 'mix_op_manager' and 'manager' below is acting as a supply rather than an extract.

Connection summary

Node @ +	ΔHght	Node @ −	ΔHght	Component	Z @ +	Z @ −
Manager	0.0	general	0.0	DABiz 01:003	1.5	1.5
BW-Op01:008	0.0	manager	−0.8	GrOp z01:008	0.7	0.7
mix_op_man	−0.25	manager	1.5	GrMEI01:044	3.0	3.0
Manager	1.5	ceil_void	−0.25	GrOpz01:045	3.0	3.0
General	0.0	reception	0.0	FABiz02:004	1.5	1.5
BW-Op02:010	0.0	general	−0.8	GrOpz02:010	0.7	0.7
BW-Cr02:012	0.0	general	0.45	WiCrz02:012	1.95	1.9
...	...	...	...	...	...	...
...	...	...	...	...	...	...

Details of the control system can be output in summary or detailed format. In ESP-r, the algorithm that defines the control action at each time step is called a control function. The following extracts correspond to control functions operating to regulate space temperature and airflow network components.

Control function 1: 'mixer_manager' senses dry bulb temperature in 'mix_op_man' and actuates the air point in 'mix_op_man'. During weekdays, there are 3 control periods.

Period	Start	Sensing	Actuating	Control law
1	0.00	db T	flux	Free-floating
2	7.00	db T	flux	Remote sensor: heating 3,000W, cooling 3,000W, maintain 20 °C–25 °C in 'manager' and 'general'.
3	18.00	db T	flux	Free-floating

Control function 4: 'ceiling_temp_match' senses dry bulb temperature in 'ceil_void' and forces 'ceiling_belo' to this value (i.e., a symmetry boundary condition).

Period	Start	Sensing	Actuating	Control law
1	0.00	db T	flux	Match temperature: heating 5,000W, cooling 5,000W. Remote sensor dry bulb temperature in 'ceil_void'; scale 1.00, offset 1.00

The airflow network has control applied to 19 components with facade vents regulated via control functions.

Control Function 1: senses node (1) manager and actuates the flow path: 2 BW-Op01:008 – manager via GrOpz01:008 to force it closed at all hours.

Period	Start	Sensing	Actuating	Control law
1	0.00	db T	Flow	On/off setpoint 0.00, inverse action ON, fraction 1.0.

Control Function 14: senses node (1) manager and actuates the flow path: 3 mix_op_man – manager via GrMEI01:044.

Period	Start	Sensing	Flow	Control law
1	0.00	db T	Flow	Fan normally off with 2 sensors: 'manager' setpoint 20.0 °C inverse action or 'manager' setpoint 25.0 °C direct action.

Control Function 15: senses node (2) general and actuates flow path: 18 general – ceil_void via GrOpz02:093.

Period	Start	Sensing	Actuating	Control law
1	0.00	db T	Flow	Damper normally closed with 2 sensors: general setpoint 20.0 °C inverse action or general setpoint 25.0 °C direct action.

During a project, it is important to note the provenance of a model, the extent to which it draws from standard data sources, and where there are placeholders that need to be updated as and when new information becomes available:

As-distributed databases used with the exception of constructions and contents, which use bespoke databases that include innovative facade configurations.
Local databases associated with the model:
constructions../dbs/office_vent_constr.db4
predefined objects../dbs/office_vent_bc.predef.c

The following table provides details of the thermophysical properties that have been instantiated on selection of a construction. In addition to re-instantiating a construction, it is possible to change selected entries in this table.

Composition of 'extern_wall'

Material	Thickness (mm)	Conductivity (W/m.K)	Density (kg/m^3)	Specific heat (J/kg.K)	Emissivity	Absorptivity	Diffusivity (m^2/s)
Brick	100	0.960	2000.	650.	0.90	0.70	25.
Glasswool	150	0.040	250.	840.	0.90	0.30	4.
Air gap	50	–	–	–	–	–	–
Concrete block	100	0.240	750.	1000.	0.90	0.65	10.
Gypboard	12	0.190	950.	840.	0.91	0.22	11.

Such data may be selected to impose a required *U*-value or to match a specific product.

Summary of 'drei3holz', a composite wood and cork Passive House grade window frame set to give a U-value of 0.73

Material	Thickness (mm)	Conductivity (W/(mK))	Density (kg/m^3)	Specific heat (J/kg.K)	Emissivity	Absorptivity
Wester Larch	25.0	0.14	590.	1800.	0.9	0.65
Cork insulation	24.0	0.04	105.	1800.	0.9	0.60
Air gap	5.0	–	–	–	–	–
Fir (20% mc)	10.0	0.14	419.	2720.	0.9	0.65

In similar manner, the properties of glazing can be scrutinised.

Summary of 'tripglz_089' with 'tr_Kgl_arg' optics

Material	Thickness (mm)	Conductivity (W/m.K)	Density (kg/m³)	Specific heat (J/kg.K)	Emissivity	Absorptivity
Pilkington K glass	6.0	1.050	2500.	750.	0.83	0.05
Air gap	12.0	–	–	–	–	–
Clear float glass	6.0	1.050	2500.	750.	0.83	0.05
Air gap	12.0	–	–	–	–	–
Clear float glass	6.0	1.050	2500.	750.	0.83	0.05

Overall solar transmission and layer absorption

	Incidence angle (° from surface normal)				
	0	40	55	70	80
Layer	absorption				
1	0.260	0.264	0.264	0.255	0.195
2	0.001	0.001	0.001	0.001	0.001
3	0.103	0.111	0.115	0.107	0.083
4	0.001	0.001	0.001	0.001	0.001
5	0.098	0.095	0.086	0.061	0.028
Transmission	0.382	0.361	0.313	0.193	0.077

BPS+ tools have syntactic rules for geometric entities. For example, ESP-r requires that all thermal zones be fully bounded. During the creation of the zone, a missing surface may not be visually apparent, however, an incomplete zone may have any number of pernicious impacts. Figure B.8 shows a ceiling surface marked as missing. In the normal wireframe view, this could easily be missed.

Switching the display to include 'outward' facing surface normal arrows provides visual confirmation of correct geometry. There is also a topology-checking option in the menu that provides warnings of unbounded edges. ESP-r uses the ordering of surface vertices to determine the outward-facing normal vector. A reversed surface will not necessarily cause an assessment to fail but will inject unintended consequences such as a volume subtraction and incorrect solar relationship that will be difficult to detect. Consider the highlighted surface in Figure B.9. Because it is reversed (normal pointing inward), the ordering of the constructional layers is incorrect with the brick facing the room and plasterboard facing the elements! It would require advanced pattern-matching skills to detect where predictions are tainted.

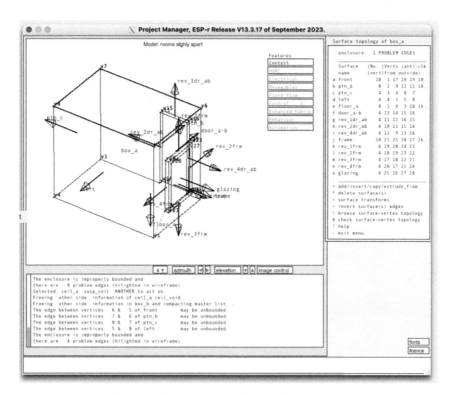

Figure B.8   Missing surface in thermal zone

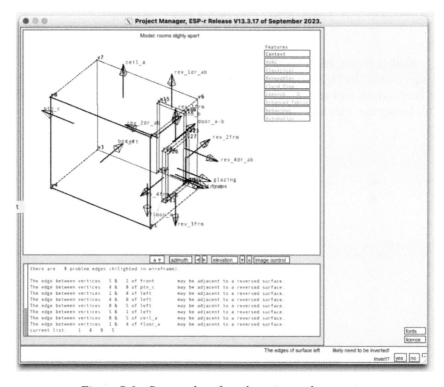

Figure B.9   Reversed surface detection and correction

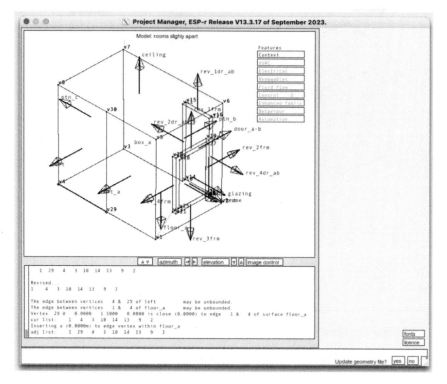

Figure B.10   Unbounded edges detection

Models evolve and a surface might need to be subdivided to accommodate a design change. In Figure B.10, the leftmost surface has been split via the addition of new vertices and re-definition of edges but the floor edge no longer matches. This breaks the rules and the checker offers to correct this.

Example ESP-r automation script

The following scripts should be considered in conjunction with the description given in Section 4.5.

Script 1

```
if test "X$Edir" = "X" then          # Variable Edir enables script to
   echo "Please set up 'Edir' variable" # be run from any directory.
   exit
fi

clear                                # Clear the window and
echo "ESP-r overheating assessment." # display header.

comfort_index="res"                  # Comfort index is resultant temperature, with
criteria=27                          # values above 27C deemed unacceptable.

building=$Edir/tmp/system_cfg        # Default input file.
weather=$Edir/tmp/weather            # Default weather file.

comfort_lock=$Edir/tmp/comfort_lock      # Lock file controls script sequencing.
comfort_results=$Edir/tmp/comfort_results # Outputs placed here.

start="17 7"                         # 17 July is the warmest day in the default
finish="17 7"                        # ESP-r weather file.

timesteps=1                          # Default time step is 1 hour.
information="no"                     # Help off by default.
demo="no"                            # Script demo mode off.

if test $# -ne 0 then                # Process command line options.
   for i do
      case "$i" in
      -h)    information="yes"; shift;;
      -help) information="yes"; shift;;
      -i)    comfort_index=$2;shift;shift;;
      -o)    criteria=$2;shift;shift;;
      -f)    comfort_results=$2; shift; shift;;
      -c)    weather=$2; shift; shift;;
      -b)    building=$2; shift; shift;;
      -p)    start="$2 $3"; shift; shift;
             finish="$2 $3"; shift; shift; shift;;
      -t)    timesteps=$2; shift; shift;;
      -d)    demo_mode=$2; shift; shift;;
      --)    shift;;
      -*)    echo "Unknown option: $i. Type comfort -h"
             exit 2;;
      esac
   done
fi
```

```
if test $information = yes then        # Respond to help request.
        echo
        echo " Command line options:
        echo "   information (help)      -h or -help"
        echo "   comfort index          -i index"
        echo "   comfort criteria       -o criteria"
        echo "   results file           -f filename"
        echo "   climate file           -c filename"
        echo "   building               -b filename"
        echo "   period                 -p sd sm fd fm"
        echo "   time steps             -t timestep"
        echo "   demo mode              -d yes/no"
        echo
        echo " Default: comfort -d no -i res -o 28"
        exit
fi


echo " Files used:      ${building}"      # echo files used
echo "                  ${weather}"
echo "                  from ${start} to ${finish}"
echo " Files created:   ${comfort_results}"
echo
echo " Comfort index: ${comfort_index} at ${criteria}"
echo
echo " please wait ..."
echo

rm -f $Edir/tmp/comfort_*
rm -f ${comfort_results}
                                # Run ESP-r's 'clm' module to determine
                                # weather file as a function of location.
if ${demo_mode} = no
then
        clm >>$Edir/tmp/comfort_trace 2>$ESPdir/tmp/weather <<~
        -6
        selection_request       # Commands to 'drive' the
        ${building}             # climate module.
        f

        ~
        ${start}='cat $Edir/tmp/comfort_ps'      # The analysis period is
        ${finish}='cat $Edir/tmp/comfort_pf'     # set from the results of
        ${timesteps}='cat $Edir/tmp/comfort_pts' # the 'clm' run.
fi

bps >>$ESPdir/tmp/comfort_trace <<END     # Run ESP-r's 'bps' module,
-6                                        # taking inputs from this script
${weather}                                # and save output for later analysis.
1
${building}
```

```
y
1
3
${comfort_results}
${start}
${finish}
${timesteps}
s
n
y
>
-
f
END
```

```
echo " Simulation complete; analysis commencing"
echo " with output directed to separate windows."
```

```
if test $comfort_index = res then
        comfort_index="d"
        graphpic="f"
fi
```

```
if test $comfort_index = air then
        comfort_index="c"
        graphpic="a"
fi
```

```
if test $comfort_index = set then
        comfort_index="e"
        graphpic="g"
fi
```

```
export comfort_results comfort_lock criteria comfort_index        # Export data for use by
other scripts.
```

```
newwin -t "worst zone" -f title -x 0 -y 8 -w 653 -h 600 -ix 944 -iy 656 -iw 56 -ih 56 -u --
$Edir/scripts/script2                                 # Run script 2 in a new window.
```

```
while test ! -f ${comfort_lock}                       # Script 1 now sleeps until the
do                                                    # script 2 lock file appears.
sleep 5
done
if test -s ${comfort_lock} then                       # It appears! Check for discomfort;
                                                      # lock file identifies worst zone.
newwin -t "cause" -f title -x 3 -y 352 -w 550 -h 500 -ix 944 -iy 592 -iw 56 -ih 56 -u --
$Edir/scripts/script3                                 # Determine cause of overheating.
```

```
WZ=`cat ${comfort_lock}`                              # Use 'res' module to get frequency
                                                      # distribution from results file.
```

```
res >>$ESPdir/tmp/comfort_trace 2>$ESPdir/tmp/comfort_graph <<END
-6
${comfort_results}
y
4
y
${WZ}
c
&
${graphpic}
10
y
1
4
-
-
f
END
                                                    # Draw frequency distribution plot.
graph -t "frequency distribution" <$Edir/tmp/comfort_graph
fi
```

Script 2

```
                                                    # Use 'res' to get comfort statistics.
res >>$Edir/tmp/comfort_trace  2>$Edir/tmp/comfort_pt <<END
-6
${comfort_results}
y
b
${comfort_index}
n
-
f
END
                                        # Use awk to get overheating zones
                                        # ranked in PZ. File comfort_awk_pt has
                                        # awk driver commands.
PZ=`(echo OHT $criteria ; cat $Edir/tmp/comfort_pt) | awk -f comfort_awk_pt `

if test "X$PZ" = "X" then
    echo "Congratulations!"                         # Whole building is comfortable.
    echo "No overheating occurs."
    echo
    echo "Zone summary table follows ...."
    echo
    cat $Edir/tmp/comfort_pt > ${comfort_lock}
    echo
    read xa                                         # Hang around until told to go away.
    exit 0                                          # Normal exit indicates no discomfort.
fi
```

```
echo "The following zones overheat, $PZ."        # Output problem zones.
echo

WZ=                                              # Locate worst occupied zone.
for i in $PZ
do
   echo "Now checking zone $i for occupants."
   if test "X$WZ" = "X"
   then
        WZ=$i                                    # Worst zone.
   fi
   impb  <<END                                   # ESP-r module 'impb' is used to
     $i                                          # test a zone for occupants.
   END
   if test $? -eq 1 then                         # Impb error exits if zone is occupied.
                                                 # $? is last exit status.
        echo "Occupied."
        echo "Worst discomfort occurs in occupied zone $i."

        OZF=1                                    # Worst zone is occupied.
        WZ=$i
        break
   else
        echo "Zone not occupied."
   fi
done
if test "X$OZF" = "X"
then
                                                 # No impb error exit.
   echo "Worst discomfort occurs in unoccupied zone $WZ."
fi
echo
echo "Zone summary table follows ...."
cat $Edir/tmp/comfort_pt
echo $WZ >$comfort_lock                          # Put WZ in lock file for access by
read xa                                          # script 3 and hang around until
                                                 # told to continue.
```

Script 3

```
WZ=`cat $comfort_lock`                           # Get worst zone from lock file.
                                                 # and create graph for worst zone.
res >>$Edir/tmp/comfort_trace 2>$Edir/tmp/comfort_wzt <<END
-6
${comfort_results}
y
c
4
y
${WZ}
```

```
a
f
g
b
!
-
f
~

graph -t "worst zone profiles" <$Edir/tmp/comfort_wzt  # Draw graph in new window and
                                                       # get energy balance for worst zone.
res >>$Edir/tmp/comfort_trace 2>$Edir/tmp/comfort_eb <<END
-6
${comfort_results}
y
b
p
${WZ}
2
-
f
END
cat $Edir/tmp/comfort_eb                              # Display energy balance and use awk
                                                      # to get offending energy flowpaths
OHC=`awk -f $Edir/scripts/comfort_awk_eb $Edir/tmp/comfort_eb`
echo
echo "The cause of the zone ${WZ} overheating is"
echo "$OHC."

for i in $OHC                        # Insert plot commands into variable GP
do                                   # for later use to select graph.
   case $i in
   "Infilt")    GP=$GP'l
';;                                  # 's needed to put \n into GP.
   "Vent")      GP=$GP'm
';;
   "WcondE")    GP=$GP'n
';;
   "WcondI")    GP=$GP'n
';;
   "DcondE")    GP=$GP'o
';;
   "DcondI")    GP=$GP'o
';;
   "Solair")    GP=$GP'p
';;
   "CasConv")   GP=$GP'q
';;
   "Surfconv")  GP=$GP'r
';;
```

```
   "Plant")    GP=$GP'k
';;
   esac
done
GP=$GP'!'                               # Add draw command then get
                                        # graph of causal flowpaths ...
res >>$Edir/tmp/comfort_trace 2>$Edir/tmp/comfort_cause <<END
-6
${comfort_results}
y
c
$GP
-
f
END
                                        # ... and display it.
graph -t "causal flowpaths" <$Edir/tmp/comfort_cause

export $OHC                             # Export cause and initiate sensitivity
                                        # analysis in new window.
newwin -t "sensitivity" -f title -x 3 -y 300 -w 550 -h 500 -ix 944 -iy 500 -iw 56 -ih 56 -u --
$Edir/scripts/script4
```

Script 4

```
   # For each of the causal flowpaths identified in script 3 (and exported to this
   # script in variable OHC), scripts 1 and 2 are re-run against appropriate changes
   # to the building description to establish the sensitivity of the flowpaths to design
   # intervention. The possible changes are passed to module 'impc' in table form.
   # This table contains the percentage changes that may be applied to the relevant
   # parameters contained in the building model

   for i in $OHC
   do
      case $i in
      "Infilt")    parameter=1;;
      "Vent")      parameter=2;;
      "WcondE")   parameter=3;;
      "WcondI")   parameter=4;;
      "DcondE")   parameter=5;;
      "DcondI")   parameter=6;;
      "Solair")    parameter=7;;
      "CasConv")  parameter=8;;
      "Surfconv")  parameter=9;;
      "Plant")     parameter=10;;
      esac

      echo "Comfort analysis; modifying description for $parameter ..."
   >>$Edir/tmp/comfort_trace
```

```
impc >>$Edir/tmp/comfort_trace <<END    # The 'impc' module is used to modify
-6                                       # the building description for the
$Edir/tmp/sensitivity_table              # sensitivity analysis.
$Edir/tmp/building
END

echo "Comfort analysis; re-simulate for $parameter ..." >>$Edir/tmp/comfort_trace
bps >>$Edir/tmp/comfort_trace  <<END        # Re-simulations are initiated.
-6
${weather}
1
${building}
y
1
3
${comfort_results}
${start}
${finish}
${timesteps}
s
n
y
>
-
f
END
                                        # New energy balance output is fetched.
res >>$Edir/tmp/comfort_trace 2>>$Edir/tmp/comfort_sensitivity <<END
-6
${comfort_results}
y
b
p
${WZ}
2
-
f
END
done

cat $Edir/tmp/comfort_sensitivity            # Display results of sensitivity analysis.
```

Appendix D
Material properties

This appendix presents the outcome of a review of existing datasets of thermo-physical properties of building materials and an assessment of data reliability in terms of the underlying test procedures (Clarke and Yaneske 2009). Scrutiny of the data gave rise to the following points.

- The range of properties for which values are quoted is frequently limited to thermal conductivity, density, and vapour resistivity as required for simple steady-state heat loss and condensation calculations.
- The sources of much of the data are not identified, and little information is given on the underlying experimental conditions or procedures. Consequently, it is often impossible to check compatibility between different values.
- Many of the data values are derived from work carried out with non-standard apparatus or date from a time that precedes modern standards of equipment and operation.
- Much of the agreement that does exist between different data sets may be attributable to historical 'borrowing'. This may lead, erroneously, to an optimistic assessment of the inherent uncertainty.
- No guidance is given on the variation in properties such as density and internal structure inherent in the production of many building materials, and there is no agreement on the procedure for determining the thermal conductivity of materials as the moisture content varies. Such variations can lead to large differences in reported material properties.
- There is no consensus on the manner in which materials are grouped for the presentation of the data. What is needed is a common system such as the CIB[1] Master list of materials (Eldridge 1974), which integrates thermal properties within a broad material classification system.
- Many values are quoted without any statement as to whether they correspond to single or multiple measurements. A random inspection of several referenced works would suggest that values are usually derived from the work of a single researcher based on a small sample size.
- There is no agreement on the procedure for the determination of the thermal conductivity of materials in the moist state or guidance given on the variation of density inherent in the production process.

---

[1]https://cibworld.org

- There is tacit agreement that the uncertainty within the data is use-context dependent. The traditional calculation methods are clearly expected to yield no more than ballpark estimates of real conditions.

The work distinguished between two contexts in which the data might be used. The first is in studies where the aim is to compare different buildings made of ostensibly the same materials. In such cases, the absolute accuracy of data is not paramount and 'reference' data may be used. The second is in the calculation of real building performance where the variations of properties with moisture content, and the inherent uncertainties in the manufacture and use of building materials, are of key importance.

The following material properties were examined: conductivity (W/m.K), density (kg/m^3), specific heat (J/kg.K), surface emissivity ($-$), surface shortwave absorptivity ($-$), and vapour resistivity or resistance (MNs/kg.m or MNs/kg). From consideration of the use context of the materials, and the reliability/scope of their underlying test procedures, the data were classified as follows.

Category 1 – Impermeables

Materials that act as a barrier to water in the vapour and/or liquid states and do not alter their hygro-thermal properties by absorbing or being wetted by water.

Category 2 – Non-hygroscopic

Lightweight insulations, such as mineral wools and foamed plastics, which display water vapour permeability, zero hygroscopic water content, an apparent thermal conductivity, and operate under conditions of air-dry equilibrium normally protected from wetting by rain.

Category 3 – Inorganic-porous

Masonry and related materials that are inorganic, porous, and may contain significant amounts of water due to hygroscopic absorption from the air or wetting by rain, which affects their hygro-thermal properties and their thermal conductivity in particular.

Category 4 – Organic-hygroscopic

Organic materials such as wood and wood-based products that are porous and strongly hygroscopic and display a non-linear water vapour permeability characteristic.

Further information is available in the literature:

- on the underlying test methods and their accuracy (Shirtcliffe and Tye 1985, Bomberg and Solvason 1985, Hager 1985);
- on the correlation between conductivity and density (Jakob 1949, Ball 1968, Arnold 1970, Van Geem and Fiorato 1983);
- on the correlation between conductivity and moisture content/temperature (Jespersen 1953, Billington 1952, Valore 1980, Stuckes and Simpson 1986);
- on the correlation between conductivity and thickness (ASHRAE 1985, Siviour 1985);
- on the determination of specific heat capacity (Hens 1984);
- on the determination of vapour permeability (MacLean and Galbraith 1988); and
- on the determination of surface properties (Holden and Greenland 1951).

Category 1 – Impermeables

Material	Conductivity (W/m.K)	Density (kg/m^3)	Specific heat (J/kg.K)
Asphalt			
Poured	1.20	2100.	920.
Reflective coat	1.20	2300.	1700.
Roofing	1.15	2330.	840.
Bitumen			
Flooring	0.85	2400.	1000.
Insulation	0.20	1000.	1700.
Ceramics			
Glazed	1.40	2500.	840.
Glass			
Cellular sheet	0.048	140.	840.
Foamed	0.052	140.	840.
4-mm clear float	1.05	2500.	750.
6-mm antisun	1.05	2500.	750.
Block	0.70	3500.	840.
Ceramic	1.40	2500.	840.
Plate	0.76	2710.	840.
Brick	1.40	2500.	840.
Mirror	2.80	2500.	840.
Linoleum			
Regular	0.19	1200.	1470.
Metal			
Aluminium	203.	2700.	880.
Brass	110.	8500.	390.
Bronze	64.	8150.	–
Copper	384.	8600.	390.
Duraluminium	160.	2800.	580.
Iron	72.	7900.	530.
Cast iron	56.	7500.	530.
Lead	35.	11340.	130.
Steel	45.	7800.	480.
Stainless steel	29.	7850.	480.

(Continues)

(*Continued*)

Material	Conductivity (W/m.K)	Density (kg/m³)	Specific heat (J/kg.K)
Tin	65.	7300.	240.
Zinc	113.	7000.	390.
PVC			
Regular	0.16	1380.	1000.
Tiles	0.19	1200.	1470.
Roof covering			
Felt	0.19	960.	840.
Rubber			
Regular	0.17	1500.	1470.
Hard	0.15	1200.	1000.
Expanded board	0.032	70.	1680.
Tiles	0.30	1600.	2000.

Category 2 – Non-hygroscopic

Material	Conductivity (W/m.K)	Density (kg/m³)	Specific heat (J/kg.K)
Carpet			
Cellular rubber underlay	0.10	400.	1360.
Synthetic	0.06	160.	2500.
Foam			
Phenol	0.04	30.	1400.
Phenol, rigid	0.035	110.	1470.
Polyisocyanate	0.03	45.	1470.
Polyurethane	0.028	30.	1470.
Polyvinyl chloride	0.035	37.	1470.
Urea formaldehyde foam	0.054	14.	1470.
Glass fibre			
Quilt	0.04	12.	840.
Slab	0.035	25.	1000.
Strawboard	0.085	300.	2100.
Wool	0.04	12.	840.
Insulating board			
Mineral fibre, wet felted	0.051	290.	800.
Mineral fibre, wet moulded	0.061	370.	590.
Mineral fibre, resin binder	0.042	240.	710.
Loose fill			
Cellulosic insulation	0.042	43.	1380.
Exfoliated vermiculite	0.069	260.	880.
Glass, granular	0.07	180.	840.
Gravel	0.36	1840.	840.
Perlite, expanded	0.051	100.	1090.

(*Continues*)

(*Continued*)

Material	Conductivity (W/m.K)	Density (kg/m³)	Specific heat (J/kg.K)
Roof gravel or slag	1.44	880.	1680.
Sand	1.74	2240.	840.
Stone chippings	0.96	1800.	1000.
Dry render	0.50	1300.	1000.
Mineral wool			
Regular	0.038	140.	840.
Fibrous	0.043	96.	840.
Resin bonded	0.036	99.	1000.
Rock wool	0.033	100.	710.
Miscellaneous			
Acoustic tile	0.057	290.	1340.
Cratherm board	0.05	176.	840.
Perlite	0.046	65.	840.
Perlite panel	0.055	170.	840.
Vermiculite	0.058	350.	840.
Vermiculite panel	0.082	350.	840.
Felt sheathing	0.19	960.	950.
Mineral board, preformed	0.042	240.	760.
Mineral fibre	0.04	100.	1800.
Mineral fibre slab	0.035	30.	1000.
Polystyrene, rigid	0.036	16.	1210.
Silicon	0.18	700.	1000.
Plastic tiles	0.50	1050.	840.
Polyisocyanurate board	0.02	32.	920.
Polystyrene			
Extruded	0.035	25.	1470.
Expanded	0.035	23.	1470.
Expanded PVC	0.04	100.	750.
Polyurethane			
Cellular board	0.023	24.	1590.
Expanded	0.023	24.	1590.
Unfaced	0.023	32.	1590.

Category 3 – Inorganic-porous

Material	Conductivity (W/m.K)	Density (kg/m³)	Specific heat (J/kg.K)
Asbestos			
Cement	1.02	1750.	840.
Cement board	0.58	1920.	1010.
Cement decking	0.36	1500.	1050.

(*Continues*)

(*Continued*)

Material	Conductivity (W/m.K)	Density (kg/m³)	Specific heat (J/kg.K)
Brick			
Aerated	0.3	1000.	840.
Breeze block	0.44	1500.	650.
Burnt	0.75	1300.	840.
Inner leaf	0.62	1800.	840.
Outer leaf	0.96.	2000.	650.
Paviour	0.96	2000.	840.
Reinforced	1.1	1920.	840.
Cement and Plaster			
Cement, regular	0.72	1860.	840.
Cement blocks	0.33	520.	2040.
Cement fibreboard	0.082	350.	1300.
Cement mortar	0.93	1900.	840.
Limestone mortar	0.7	1600.	840.
Plaster	1.5	1900.	840.
Plaster/sand aggregate	0.72	1860.	840.
Cement screed	1.4	2100.	650.
Plaster	0.51	1120.	960.
Plasterboard	0.17	800.	1090.
Plaster, dense	0.5	1300.	1000.
Plaster, lightweight	0.16	600.	1000.
Rendering	0.79	1330.	1000.
Plaster, vermiculite	0.2	720.	840.
Ceramics			
Tiles	1.2	2000.	850.
Clay			
Tile	0.85	1900.	840.
Tile, burnt	1.3	2000.	840.
Tile, hollow	0.623	1120.	840.
Tile, paver	1.803	1920.	840.
Concrete			
Heavyweight	1.3	2000.	840.
Lightweight	0.2	620.	840.
Medium lightweight	0.32	1050.	840.
Very lightweight	0.14	370.	840.
No fines	0.96.	1800.	840.
Aerated, cellular	0.7	1000.	840.
Aerated roofing slab	0.16	500.	840.
Block, lightweight	0.64	1660.	840.
Block, mediumweight	0.86	1970.	840.
Block, heavyweight	1.31	2240.	840.
Block, aerated	0.24	750.	1000.
Block, hollow, lightweight	0.58	720.	840.
Block, hollow, mediumweight	0.86	930.	840.
Block, hollow, heavyweight	1.35.	1220.	840.
Block, partially filled, lightweight	0.67	1090.	840.
Block, part filled, mediumweight	0.85	1260.	840.

(Continues)

(*Continued*)

Material	Conductivity (W/m.K)	Density (kg/m³)	Specific heat (J/kg.K)
Block, part filled, heavyweight	1.35	1570.	840.
Block, perlite filled, lightweight	0.17	770.	840.
Block, perlite filled, medium.	0.2	900.	840.
Cast	1.28	2100.	1010.
Cast, dense	1.4	2100.	840.
Cast, dense, not reinforced	1.7	2400.	840.
Cast, dense, reinforced	1.9	2500.	840.
Cast, lightweight	0.38	1200.	1000.
Cement or lime based, aerated	0.21	580.	840.
Cinder	0.69	1410.	840.
Foamed	0.08	400.	920.
Glass reinforced	0.9	1950.	840.
Refractory insulating	0.25	10.	840.
Tiles	1.1	2100.	840.
Vermiculite aggregate	0.17	450.	840.
Masonry			
Heavyweight	0.9	1850.	840.
Lightweight	0.22	750.	840.
Mediumweight	0.32	1050.	840.
Quarry stones, calcareous	1.4	2200.	840.
Semi-heavy blocks	0.60.	1350.	840.
Very light blocks	0.19	470.	840.
Miscellaneous			
Calcium silicate brick	1.5	2000.	840.
Dried aggregate	1.31	2240.	840.
Granolithic	0.87	2085.	840.
Lime stone	1.8	2420.	840.
Siporex	0.12	550.	1000.
Thermalite	0.19	750.	840.
Roofing			
Tile	0.84	1900.	800.
Terracotta	81	1700.	840.
Soil			
Earth, common	1.28	1460.	880.
Earth, gravel	0.52	2050.	180.
Alluvial clay, 40% sands	1.21	1960.	840.
Stone			
Basalt	3.5	3000.	840.
Gneiss	3.49	2880.	840.
Granite	3.49	2880.	840.
Limestone	2.9	2750.	840.
Marble	2.9	2750.	840.
Porphyry	3.49	2880.	840.
Red granite	2.9	2650.	900.
Sandstone	1.3	2150.	840.
Sandstone tiles	1.2	2000.	840.
Slate	1.72	2750.	840.

(Continues)

(*Continued*)

Material	Conductivity (W/m.K)	Density (kg/m³)	Specific heat (J/kg.K)
Slate shale	2.1	2700.	840.
White calcareous stone	2.09	2350.	840.
White marble	2	2500.	880.

Category 4: Organic-hygroscopic

Material	Conductivity (W/m.K)	Density (kg/m³)	Specific heat (J/kg.K)
Cardboard/paper			
Bitumen impregnated	0.06	1090.	1000.
Laminated	0.072	480.	1380.
Cloth, carpet, gFelt			
Bitumen/felt layers	0.50	1700.	1000.
Carpet, simulated sheep wool	0.06	200.	1360.
Carpet, wilton	0.06	190.	1360.
Felt, semi-rigid, organic bonded	0.035	48.	710.
Jute fibre	0.067	330.	1090.
Wool felt underlay	0.04	160.	1360.
Cork			
Board	0.04	160.	1890.
Expanded	0.044	150.	1760.
Expanded and impregnated	0.043	150.	1760.
Tiles	0.08	530.	1800.
Straw			
Board	0.057	310.	1300.
Fibre board or slab	0.10	300.	2100.
Thatch	0.07	240.	180.
Miscellaneous			
Afzelia, minunga, meranti	0.29	850.	2070.
Expanded ebonite	0.035	100.	1470.
Expanded perlite board, organic bonded	0.052	16.	1260.
Glass fibre board, organic bonded	0.036	100.	960.
Thermalite turbo block	0.11	480.	1050.
Weatherboard	0.14	650.	2000.
Wood			
Fir, pine	0.12	510.	1380.
Maple, oak	0.16	720.	1260.
Beech, ash, walnut, meranti	0.23	650.	3050.
Spruce, Sylvester	0.12	530.	1880.
Willow, birch	0.14	520.	2280.
Softwood	0.17	550.	1880.
Hardwood	0.05	90.	2810.

Absorptivity and emissivity

These data are grouped into four categories (one null) as follows.

Category/material	Absorptivity (–)	Emissivity (–)
Impermeables		
Aluminium (polished)	0.10–0.40	0.03–0.06
Aluminium (dull, rough polish)	0.40–0.65	0.18–0.30
Aluminium (anodised)	–	0.72
Aluminium surfaced roofing	–	0.216
Asphalt (new)	0.91–0.93	–
Asphalt (block)	0.85–0.98	0.90–0.98
Asphalt (weathered)	0.82–0.89	–
Asphalt (pavement)	0.852–0.928	–
Bitumen felt/roofing sheets	0.86–0.89	0.91
Bitumen (parking lot)	0.86–0.89	0.90–0.98
Brass (polished)	0.30–0.50	0.03–0.05
Brass (dull)	0.40–0.065	0.20–0.30
Brass (anodised)	–	0.59–0.61
Bronze	0.34	–
Copper (polished)	0.18–0.50	0.02–0.05
Copper (dull)	0.40–0.065	0.20–0.30
Copper (anodised)	0.64 0.60	–
Glass	–	0.88–0.937
Iron (unoxidised)	–	0.05
Iron (polished/bright)	0.40–0.65	0.20–0.377
Iron (oxidised)	–	0.736–0.74
Iron (red rusted)	–	0.61–0.65
Iron (heavily rusted)	0.737	0.85–0.94
Iron, cast (unoxidised/polished)	–	0.21–0.24
Iron, cast (oxidised)	–	0.64–0.78
Iron, cast (strongly oxidised)	–	0.95
Iron, galvanised (new)	0.64–0.66	0.22–0.28
Iron, galvanised (aged/very dirty)	0.89–0.92	0.89
Lead (unoxidised)	–	0.05–0.075
Lead (old/oxidised)	0.77–0.79	0.28–0.281
Rubber (hard/glossy)	–	0.945
Rubber (grey/rough)	–	0.859
Steel (unoxidised/polished/stainless)	0.2	0.074–0.097
Steel (oxidised)	0.2	0.79–0.82
Tin (highly polished/unoxidised)	0.10–0.40	0.043–0.084
Paint, aluminium	0.30–0.55	0.27–0.67
Paint, zinc	0.3	0.95
PVC	–	0.90–0.92
Tile (light)	0.3–0.5	0.85–0.95
Varnishes	–	0.80–0.98

(Continues)

(Continued)

Category/material	Absorptivity (–)	Emissivity (–)
Zinc (polished)	0.55	0.045–0.053
Zinc (oxidised)	0.05	0.11–0.25
Non-hygroscopic		
Because insulants and the like are never used as surface finishes, there are no entries under this heading.		
Inorganic-porous		
Asbestos board	–	0.96
Asbestos paper	–	0.93–0.94
Asbestos cloth	–	0.9
Asbestos cement (new)	0.61	0.95–0.96
Asbestos cement (very dirty)	0.83	0.95–0.96
Brick (glazed/light)	0.25–0.36	0.85–0.95
Brick (light)	0.36–0.62	0.85–0.95
Brick (dark)	0.63–0.89	0.85–0.95
Cement mortar, screed	0.73	0.93
Clay tiles (red/brown)	0.60–0.69	0.85–0.95
Clay tiles (purple, dark)	0.81–0.82	0.85–0.95
Concrete and plain concrete tile	0.65–0.80	0.85–0.95
Concrete block	0.56–0.69	0.94
Plaster	0.30–0.50	0.91
Stone, granite (red)	0.55	0.90–0.93
Stone, limestone	0.33–0.53	0.90–0.93
Stone, marble	0.44–0.592	0.90–0.93
Stone, sandstone	0.54–0.76	0.90–0.93
Stone, slate	0.79–0.93	0.85–0.98
Stone, quartz	–	0.9
Organic-hygroscopic		
Paper	–	0.091–0.94
Paper (white, bond)	0.25–0.28	–
Cloth, cotton, black	0.67–0.98	–
Cloth, cotton, deep blue	0.82–0.83	–
Cloth, cotton, red	0.562	–
Cloth, wool, black	0.87–0.88	–
Cloth, wool, black	0.749	–
Cloth, felt, black	0.775–0.861	–
Cloth, all fabrics	–	0.89–0.92
Wood, beach	–	0.94
Wood, oak	–	0.89–0.90
Wood, spruce	–	0.82
Wood, walnut	–	0.83

Vapour resistivity

These data are grouped into the same four categories as follows.

Category/material	Vapour resistivity (MNs/g.m)
Impermeables	
Asphalt (laid)	∞
Bitumen roofing sheets	2000–60,000
Bituminous felt	15,000
Glass, cellular	∞
Glass, sheet/mirror/window	∞
Glass, expanded/foamed	∞
Glass brick	∞
Linoleum (1200 kg/m^3)	9000
Metals and metal cladding	∞
Paint, Gloss (vapour resistant)	40–200
Plastics, PVC sheets on tile	800–1300
Plastics, hard	45,000
Rubber (1200–1500 kg/m^3)	4500
Rubber tiles (1200–1500 kg/m^3)	∞
Tiles, ceramic	500–5000
Tiles, glazed ceramic ¥	∞
Plastics, PVC sheets on tile	800–1300
Plastics, hard	45,000
Non-hygroscopic	
Mineral fibre, glass fibre/wool	5–7
Mineral fibre/wool	5–9
Mineral fibre, rock wool	6.5–7.5
Phenolic (closed cell)	150–750
Phenol formaldehyde	19–20
Polystyrene, expanded	100–750
Polystyrene, extruded	600–1500
Polystyrene, extruded without skin	350–400
Polyethylene foam	20,000
Polyurethylene foam	115–1000
PVC foam (rigid)	40–1300
Urea formaldehyde foam	5–20
Inorganic-porous	
Asbestos cement (800 kg/m^3)	70
Asbestos cement, sheeting, substitutes (1600–1900 kg/m^3)	185–1000
Brick, blast furnace slag (1000–2000 kg/m^3)	350–500
Brick, calcium silicate (<1400 kg/m^3)	25–50
Brick, calcium silicate (>1400 kg/m^3)	75–125
Brick, dense (>2000 kg/m^3)	100–250
Brick, heavyweight (>1700 kg/m^3)	45–70
Brick, lightweight (<1000 kg/m^3)	25–50
Brick, mediumweight (>1300 kg/m^3)	23–45
Brick, sand lime (<1400 kg/m^3)	25–50

(Continues)

(Continued)

Category/material	Vapour resistivity (MNs/g.m)
Brick, sand lime (>1500 kg/m³)	75–200
Concrete, cellular (450–1300 kg/m³)	9–50
Concrete, cast (<1000 kg/m³)	14–33
Concrete, cast (>1000 kg/m³)	30–80
Concrete, cast (>1900 kg/m³)	115–1000
Concrete, expanded clay (500–1000 kg/m³)	25–33
Concrete, expanded clay (1000–1800 kg/m³)	33–75
Concrete, foamed steam hardened (400–800 kg/m³)	25–50
Concrete, natural pumice (500–1400 kg/m³)	25–75
Concrete, no fines (1800 kg/m³)	20
Concrete, polystyrene foamed (400 kg/m³)	80–100
Concrete, porous aggregate (1000–2000 kg/m³)	15–50
Concrete, porous aggregate without quartz sand	25–75
Concrete, close textured	350–750
Concrete, slag, and Rhine sand (1500–1700 kg/m³)	50–200
Concrete, insulating	23–26
Concrete blocks (very light)	15–1507
Plaster/mortar, cement based (1900–2000 kg/m³)	5–20
Plaster/mortar, lime based (1600–1800 kg/m³)	545–205
Plaster/mortar, gypsum, gypsum plasterboard	30–60
Stone, basalt, porphyry, bluestone	∞
Stone, granite, marble	150–∞
Stone, slate	150–450
Stone, slate shale	>3000
Stone, limestone, firm	350–450
Stone, limestone, soft	130–160
Stone, limestone, soft tufa	25–50
Stone, sandstone	75–450
Stone, clay	75
Tiles, clay tile, ceramic	750–1500
Tiles, floor tile, ceramic	115
Tiles, terracotta roof tile	180–220
Organic-hygroscopic	
Carpet, normal backing	7–20
Carpet, foam backed, or foam underlay	100–3002
Chipboard	30–500
Chipboard, bonded with cement	19–50
Chipboard, bonded with U.F.	200–700
Chipboard, bonded with melanine	300–500
Chipboard, bonded with P.F.	250–750
Corkboard	50–200
Cork insulation	25–50
Cork, expanded	23–50
Cork, expanded and impregnated	45–230
Cork, expanded with bitumous binding	45–230
Hardboard	230–1000
Fibreboard	150–375
Fibreboard, hardwood fibres	350

(Continues)

(*Continued*)

Category/material	Vapour resistivity (MNs/g.m)
Fibreboard, porous wood fibres	25
Fibreboard, bitumous	25
Fibreboard, cement based	19–50
Mineral and vegetable fibre insulation	5
Multiplex (800 kg/m^3)	200–2000
Multiplex, light pine	80
Multiplex, North Canadian Gaboon	80
Multiplex, red pine	875–250
Paper	500
Particle board, softwood	25
Plywood	150–2000
Plywood, decking	1000–6000
Plywood, marine	230–375
Plywood, sheathing	144–1000
Strawboard	45–70
Triplex–multiplex (700 kg/m^3)	200–500
Wood, ash	200–1850
Wood, balsa	45–265
Wood, beech	200–1850
Wood, beech, soft	90–700
Wood, birch	90–700
Wood, fir	45–1850
Wood, gaboon, North Canadian	45–1850
Wood, oak	200–1850
Wood, pine	45–1850
Wood, pine, Northern red; Oregon	90–200
Wood, pitch pine	200–1850
Wood, spruce	45–1850
Wood, teak	185–1850
Wood, walnut	200–1850
Wood, willow	45–1850
Wood wool slabs	15–40
Wood wool/cement slabs	15–50
Wood wool/magnesia slabs	19–50
Wood lath	4

Simulations focused on specialist topics will need further data that are normally available in external databases and/or from suppliers. Examples include the embodied energy database produced by Hammond *et al.* (2011), the acoustic absorption coefficients charts available online[2,3], and the sorbate diffusion coefficients available from KoBra[4].

[2]https://acoustic-supplies.com/absorption-coefficient-chart/
[3]https://acoustic.ua/st/web_absorption_data_eng.pdf
[4]https://sourceforge.net/projects/kobra/

References and further reading

Arnold PJ (1970) 'Thermal conductivity of masonry materials', *BRS Current Paper CP 1/70*.

ASHRAE (1985) 'Thermal insulation and water vapor retarders', *Handbook of Fundamentals*, Chapter 20, American Society of Heating, Refrigerating, and Air Conditioning Engineers, Georgia, USA.

Ball EF (1968) 'Measurements of thermal conductivity of building materials', *JIHVE*, 36, pp. 51–56.

Billington NS (1952) *Thermal Properties of Buildings*, Cleaver-Hume Press Ltd., London.

Bomberg M and Solvason KR (1985) 'Discussion of heat flow meter apparatus and transfer standards used for error analysis', *ASTM Publication 879*, pp. 140–153.

Clarke JA and Yaneske PP (2009) 'A rational approach to harmonisation of the thermal properties of building materials', *Building and Environment*, 44(10), pp. 2046–2055.

Eldridge HJ (1974) Properties of building materials, *Medical and Technical Publishing Co.*, Lancaster.

Hager NE (1985) 'Recent developments with the thin-heater thermal conductivity apparatus', *ASTM Publication 879*, pp. 180–190.

Hammond G, Jones C, Lowrie F and Tse P (2011) *Embodied Carbon: The Inventory of Carbon and Energy*, BSRIA, ISBN 978 0 86022 703 8.

Hens H (1984) 'Kataloog van Hygrothermische Eigenschappen van Bouwen Isolatiematerialen', *Technical Report* 1, Catholic University of Louvain: Laboratory for Building Physics.

Holden TS and Greenland JJ (1951) 'Coefficients of solar absorptivity and low temperature emissivity of various materials – a review of the literature', *Report R.6*, CSIRO: Division of Building Research.

Jakob M (1949) *Heat Transfer, Part 1*, Chapman and Hall, London.

Jespersen HB (1953) 'Thermal conductivity of moist materials and its measurement', *JIHVE*, 21, pp. 157–174.

MacLean RC and Galbraith GH (1988) 'Interstitial condensation: applicability of conventional vapour permeability values', *Building Services Engineering Research and Technology*, 9(1), pp. 29–34.

Shirtcliffe CJ and Tye RP (Eds) (1985) 'Guarded hot plate and heat flow meter methodology', *ASTM Publication 879*, pp. 140–153, American Society for Testing and Materials, Philadelphia.

Siviour JB (1985) 'Thermal performance of mineral fibre insulation', *Building Services Engineering Research and Technology*, 6(2), pp. 91–92.

Stuckes AD and Simpson A (1986) 'Moisture factors and the thermal conductivity of aerated concrete', *Building Services Engineering Research and Technology*, 7(2), pp. 73–77.

Valore RC (1980) 'Calculation of U-values of hollow concrete masonry', *Concrete International*, 2(2), pp. 40–63.

Van Geem MG and Fiorato AE (1983) 'Thermal properties of masonry materials for passive solar design - a state-of-the-art review', *Report DOE/CE/30739*, Construction Technology Laboratories, Skokie, IL.

Appendix E
Knowledge-based user interfaces

As the construction industry becomes more assertive in its demands of BPS+, it is likely that high on the agenda will be the harmonisation of user interfaces. This will help regularise training at all levels and ensure that users can switch between applications as required. This, in turn, will facilitate the prospect of an intelligent interface (Clarke 1990, 1991) that can:

- handle the complex transactions that occur within the design process;
- support design concurrency;
- preserve audit trail (who did what, when, and why?);
- offer a more constructive user dialogue in terms of interaction style, feedback, and tutoring;
- incrementally evolve the product model and offer intelligent defaults; and
- accommodate application-to-user and application-to-application semantics.

To explore such issues the COMBINE project (Augenbroe 1995) established an intelligent, integrated building design system (IIBDS; Clarke *et al.* 1998) as summarised in Figure E.1.

The IIBDS[1], which was constructed using an intelligent front-end authoring system developed in a previous, EPSRC-funded project (Clarke and Mac Randal 1989, 1991), comprises the following elements.

- A 'Blackboard' to serve as a communication centre for its various clients. This supports concurrency and traceability through the collection, organisation, and storage of the session chronicle.
- An 'Application Handler' to control the various design tools, pass them their data, and receive their returns.
- A 'Knowledge Handler' to control design tool access to the product model and the communication with the designer (verification of entries, supplemental inferencing, and feedback/guidance).
- A 'Dialogue Handler' to converse with the user by means of appropriate interface tools employing acceptable concepts that relate to the different user types and levels of expertise.

---

[1]https://www.esru.strath.ac.uk/Applications

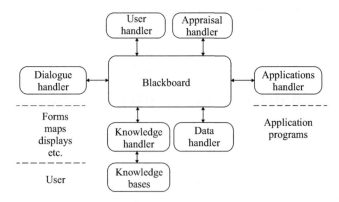

Figure E.1   Architecture of the IIBDS

- A 'User Handler' to track the user's progress and ensure that the system responds in an appropriate manner when difficulties are encountered.
- An 'Appraisal Handler' to hold the BPS+ control syntax against standard performance assessment control scripts.
- A 'Data Handler' to extract an application's data from the Blackboard and map these data to the required format.

In dealing with the design process, the IIBDS must support the flow of data/information between work steps (WS) and handle events in terms of starting and stopping the design tools (DT). The mechanism adopted to handle these issues within the IIBDS is as follows.

The required sequencing of DTs (i.e., the process model) is captured as a Petri-Net[2] (D'Angelo 1983, Javor 1993). This Petri-Net File (PNF) is read into a knowledge base associated with the Knowledge Handler – the Application Knowledge Handler (AKH) – where it is transformed into Prologue facts and used by an inference engine to animate the session. By modifying this process knowledge base, it is possible to control the rigidity of the system and its handling of parallelism. Consider three examples.

1. DT invocation is not sequenced nor functionally constrained so that the designer is able to invoke the DTs in any order and activate their internal functions as required. That is, the PNF is used only for DT access control.
2. The DTs are sequenced but not functionally constrained so that DT selection is prescribed whilst function invocation is not. That is, the PNF controls DT ordering but DT use is opportunistic; concurrency is allowed.
3. The DTs are both sequenced and functionally constrained so that the system, not the designer, controls the order of DT selection and the invocation of the WS. The user remains in control of the process and decides if the outcome of a WS is acceptable: the PNF enforces rigid DT use but no concurrency is allowed.

---

[2]Petri-Nets provide a formal technique for the representation and analysis of concurrent, discrete events such as found within work groups engaged in building design activities.

The process model corresponding to the first case relates to 'shallow' control whereby DT transactions are managed, whilst the model corresponding to the last case relates to 'deep' control whereby knowledge is introduced in relation to design purpose. The use of an external Petri-Net description and the dynamic loading feature makes it possible to change the process being enacted at any time should it become necessary to adapt the rigidity of the process in response to events.

Each node in the Petri-Net corresponds to a WS and triggers a knowledge predicate that 'knows' what should be done at this point in the process. This is where the problematic issue of concurrency is handled. The knowledge base has access to the Blackboard (i.e., the design process state) and to the DES (i.e., the state of the problem description). This knowledge base will either be established to react only to the Petri-Net (when in the prescriptive mode of operation) or to react to the design process state (when in the reactive mode of operation). When a WS is commenced, the knowledge base ensures that (a) the data required for the task is available, (b) starts the appropriate DT and then (c) hands control to the user. It then monitors what the DT is doing and finally ensures that the results of the WS are captured and propagated via the Blackboard. Whilst the DES is responsible for the handling of the data, the AKH is responsible for driving the process (i.e., triggering state changes in the Petri-Net), propagating information to other knowledge bases, and keeping track of the design status and history.

The process and design tool knowledge bases are event driven and operate asynchronously. This enables concurrency. Event-driven controllers can handle any amount of concurrency, subject only to their ability to 'understand' what the other controllers are 'saying'. In practice, unconstrained concurrency is of little value as it is inherently unstable and unpredictable. The DT knowledge base is therefore made subordinate to the process knowledge base, which activates/de-activates the former as appropriate. By activating more than one process knowledge base at a time, Petri-Net handling can effectively move from single to coloured token passing[3]. Furthermore, DT knowledge bases can be forced to listen only to the Petri-Net, giving a slavish compliance to the specified process, or encouraged to react to other DTs giving a more dynamic, context-sensitive system. Figure E.2 shows a Petri-Net comprising an arrangement of DTs corresponding to the intermediate case process model (DT invocation sequenced but use unconstrained).

On entering the design session, the user is required to use AutoCAD or MicroStation to create a new problem geometry. On exiting the CAD DT, the ATTRIBUTE DT is used to complete the site, composition, and operational characteristics of the problem. When attribution is complete, the user is presented with a choice of compliance checking or thermal/lighting performance appraisal. In the case of compliance checking, the conclusions provided will influence a user's choice to modify the problem geometry (via the CAD DT), its composition (via the ATTRIBUTE DT), or invoke either an energy/comfort assessment (ESP-r) or a lighting/visual evaluation (Radiance). Finally, the user can either revisit the CAD, ATTRIBUTE DTs or exit the session. Although this process model supports a

---

[3]https://en.wikipedia.org/wiki/Coloured_Petri_net

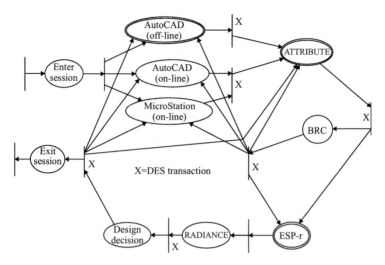

Figure E.2   Petri-Net for an intermediate case process model

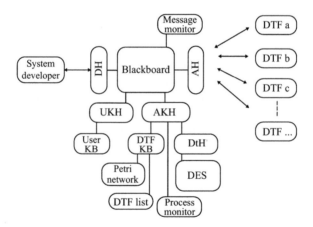

Figure E.3   Internal structure of the IIBDS

cooperative dialogue between the user and the active DTs, the session possesses no knowledge about the design purpose.

Figure E.3 shows the corresponding internal structure of the IIBDS. There are two knowledge handlers, relating to user and application control, communicating through the Blackboard as indicated.

The message passing between the user and application domains is isolated within a 'transaction' area of the Blackboard. The aim of introducing knowledge about design purpose is supported by the addition of a 'journal' area on the Blackboard. This is a repository for the aggregate log of transactions within the system and is used to feed the Prologue predicates of the design session knowledge base. In particular, the nature of the DTs presented to the user, and how these DTs are sequenced and constrained, are supported by the addition of design process knowledge to the AKH. This has been achieved by arranging for the AKH to load the Petri-Net representations as implied by the user's choice of design session.

Between the AKH and the Data Handler/DES resides the Process Monitor, which presents the current position of the token in the Petri-Net and the passing of STEP-compliant files[4] to and from the DES. Also shown is the Transaction Monitor (TM), which observes the transactions between the knowledge handlers, the DTs and DES. The TM is used to observe and analyse an active design session.

To explain the working of the IIBDS, a series of snap-shots follow that record a user's progress when the active design session corresponds to the intermediate-level process model previously outlined. Within these snap-shots, the arrows show the potential flows of information: a single arrow indicates a notification, a double arrow indicates sending and listening. The 'user_dialog' area of the Blackboard is reserved for user interaction transactions, the 'application_dialog' area is reserved for transactions related to the DTs. The 'journal' area receives messages from the various knowledge handlers and organises these for subsequent analysis and process control.

Figure E.4 defines the state of the system after the user has selected a design tool function (DTF) and the corresponding actions have been triggered: a message passes to the Dialogue Handler (DH) indicating the requested interaction, the DH passes the message to the 'user_dialog' area, and the User Knowledge Handler (UKH) tells the Blackboard to start DTF_x.

Figure E.5 is the state of the Blackboard after the UKH has issued a message 'application_ dialog start DTF_x' to the transaction area. The AKH finds the actual

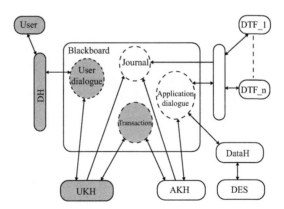

Figure E.4   State of the IIBDS after user action

[4]ISO 10303 file format.

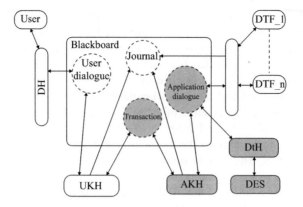

Figure E.5   User knowledge handler requesting a DTF

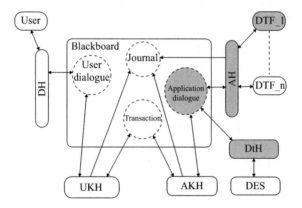

Figure E.6   DES returns STEP file and DTF_x starts

application and posts the message 'new_application DTF_x' to the 'application_ dialog' area. The Data Handler (DtH) then queries the DES, 'get_data_for DTF_x', and the DES returns the appropriate data for design tool 'x'.

In Figure E.6, after the DES issues 'data_for DTF_x file':

- the DtH posts 'new_application application_parameters' to the 'application_ dialog' area;
- the Application Handler (AH) starts the DT and establishes a pipe to receive the performance return(s);
- when the application is complete, the AH records this and sends 'closed DTF_x revised_data_file' to the 'application_dialog' area; and
- the AKH posts 'closed DTF_x' to the transaction area, which is received by the UKH for transmission (not shown) to the user.

Consider a session that incorporates knowledge in relation to design purpose. With reference to the Petri-Net shown in Figure E.7, the initial portion of the design

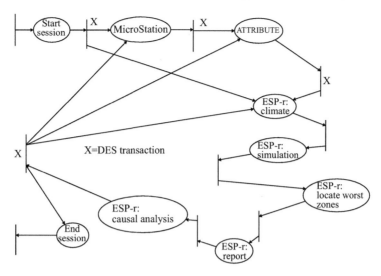

Figure E.7 Summer overheating Petri-Net

session might proceed as in the previous unconstrained case (in terms of geometry specifications and attribution). For those users who enter the summer overheating design session with an existing model, there is a direct path to the overheating assessment.

Simulation environments, such as ESP-r, provide facilities to enable direct access to their internal functions. In the current example, some of these functions are accessed, with the user involved as the arbitrator of acceptability or otherwise of the performance returns. Here, the process model involves the determination of the weather patterns that would constitute an acceptable test of summer overheating risk. Whilst such a decision is implicit in most simulation-based studies, here it has been made explicit. Next, the focus is shifted to a simulation of the current model and then the determination of what constitutes the worst spaces in terms of overheating. The search rules operate based on the highest operative temperature in an occupied space as determined from a series of inquiries of the database of simulation results. It is possible to have alternative rules governing this DT function.

Assuming that overheating has been detected, two presentations are made to the user. First, a frequency binning of temperatures and a graph of temperatures in the worst zone is displayed. This sets the context that violated the rules of the assessment. Second, the process model calls for the presentation of information on the likely causal factors. For example, if high internal gains were the cause of the overheating then only this information would be presented. The user can then either exit the design session, select a different weather condition, or return to the CAD or ATTRIBUTE DTs. A typical session is shown in Figure E.8.

Whilst systems such as the IIBDS are entirely possible, their emergence as robust, useful tools will only occur if the construction industry adopts a simulation

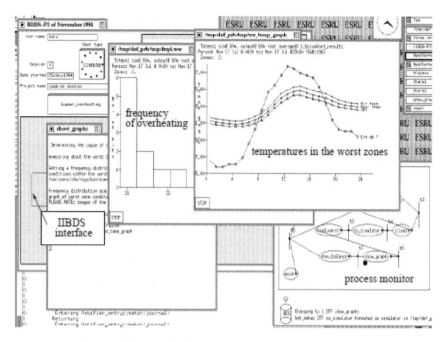

Figure E.8   A summer overheating evaluation session

approach to the design/management of the built environment and demands a uniform approach to user interfaces. It is likely that such a viewpoint will be fostered more by the emergence of a ubiquitous Internet than by altruistic considerations of best practice.

References and further reading

Augenbroe G L M (Ed.) (1995) 'Computer models for the building industry in Europe, Second Phase (COMBINE 2)', *Final Report for EC Contract JOU2-CT92-0196*, Delft University of Technology, Faculty of Civil Engineering.

Clarke J A (1990) 'Advanced design tools for energy conscious building design', *Proc. 1st World Renewable Energy Congress, V4*, pp. 2265–2277, Reading, MA.

Clarke J A (ed.), (1991) 'Intelligent front ends for engineering applications', *Artificial Intelligence in Engineering*, (Special Issue), Computational Mechanics Publications, 6(1).

Clarke J A and Mac Randal D (1989) 'An intelligent front-end for computer-aided building design', *Proc. 3rd European Simulation Congress*, Edinburgh.

Clarke J A and Mac Randal D (1991) 'An intelligent front-end for computer-aided building design', *Artificial Intelligence in Engineering, Computational Mechanics Publications*, 6(1), pp. 36–45.

Clarke J A, Hand J, Mac Randal D F and Strachan P A (1998) 'Design tool integration within the COMBINE Project', *Proc. 2nd European Conf. on Product and Process Modelling in the Building Industry*, BRE, Garston, 19–21 October.

D'Angelo G J (1983) 'Tutorial on Petri Nets', Bell Laboratories, New Jersey.

Javor A (1993) 'AI controlled high level Petri nets in simulating FMS', *Proc. AI, Simulation and Planning in High Autonomy Systems*, 20–22 September, Tucson, USA.

Index